Wildlife Biodiversity Conservation

Susan C. Underkoffler • Hayley R. Adams
Editors

Wildlife Biodiversity Conservation

Multidisciplinary and Forensic Approaches

Springer

Editors
Susan C. Underkoffler
Wildlife Forensic Sciences
and Conservation, College of Medicine
University of Florida
Gainesville, FL, USA

Hayley R. Adams
Wildlife Forensic Sciences
and Conservation, College of Medicine
University of Florida
Gainesville, FL, USA

ISBN 978-3-030-64681-3 ISBN 978-3-030-64682-0 (eBook)
https://doi.org/10.1007/978-3-030-64682-0

Cover illustration: Top: A mother ring-tailed lemur carries her infant on her back in Nahampoana Reserve, Madagascar. Bottom: Primate skull on the ground at an illegal hunter camp in Equatorial Guinea's Pico Basile National Park. © Jessica Weinberg McClosky.

This Springer imprint is published by the registered company Springer Nature Switzerland AG.
The registered company address is: Gewerbestrasse 11, 6330 Cham, Switzerland

Preface

For most of history, man has had to fight nature to survive; in this century he is beginning to realize that, in order to survive, he must protect it.
— *Jacques-Yves Cousteau*

We are writing this introduction in the midst of a global pandemic. Over the past several months, humans have experienced history in the making. We have each come to more intimately understand and appreciate the power of nature, through the replication of a microscopic organism, to redefine global standards virtually overnight, bringing life as we knew it to a halt and connecting us as a global species more than ever before. The virus causing this pandemic, a coronavirus, originated in free-ranging wildlife, and jumped species to humans as a result of close and unnatural contact, with a dense interface created by humans and their cultural preference for the consumption of bushmeat.

There is perhaps no better time to implement strategies that will prevent such catastrophic events from reoccurring. It is time to learn the lessons hidden within the challenge, lest we doom ourselves to repeat such catastrophes through future outbreaks of zoonotic disease, resulting in countless deaths, global crisis, economic downfall, and immeasurable emotional and psychological damage.

Nature is serving a grand wake-up call. Conservationists are on the front lines to answer the call by digging deeper into global ecological challenges, and to resolve them with solutions that are innovative, sustainable, and above all, culturally relevant. This is extremely important, as there is no one-size-fits-all approach to conservation issues.

This book represents the evolution of conservation to a new paradigm, one that combines highly technical skill sets with the most fundamental and arguably the most critical aspect of conservation—the recognition of and inclusion of humans as a species in the natural equation. Conservation is as much about the human animal as it is about the infinite planetary life forms we strive to protect and preserve. The application of sophisticated technologies to biodiversity conservation represents a new era, one in which we will see somewhat divergent disciplines merging and producing beneficial results that are not mutually exclusive, but rather all-encompassing. Forensic science is just one of these subjects, which for quite a long time was only applied to human crime. Now, we are realizing its utility in fighting the destruction of the planet and the loss of other species. Wildlife and environmental crime are being met with an arsenal of scientific advancements, ushering in the new field of conservation forensics. And we are only at the beginning.

At the heart of conservation are human beings—our needs, preferences, cultural beliefs and practices, struggles, desires, joys, and sorrows. Too often humans have been placed on the "negative" side of the equation, by blaming and shaming them for destroying Mother Nature. But the paradigm is shifting, with a more conscious approach emerging, whereby in order to see positive change, conservationists recognize that people at the interface with nature must be included as part of the solution. Humans are beginning to realize that our existence and long-term survival is tied inextricably with the survival of biodiversity as a whole. Indeed, we are seeing people reevaluate their place on Earth, appreciating the life all around them, spending more time in nature, and embracing and valuing not only the large, charismatic species but even the tiniest of creatures normally overlooked. We hope this continues, for as Henry Beston wrote in his 1928 memoir, *The Outermost House*: "the world to-day is sick to its thin blood for lack of elemental things, for fire before the hands, for water welling from the earth, for air, for the dear earth itself underfoot." Those words were never more true, almost one hundred years later.

We are starved for connection to nature, for inspiration from the wild, for the awe that comes from realizing just how small we each are. But despite our diminutive status on Earth, we as humans have important roles to play in nature stewardship. As this pandemic has demonstrated, our actions have consequences, not just for our own species—we need to accept that the decisions we make and the actions we take have a ripple effect through an entire ecosystem. The resources on this planet are finite; we cannot continue to exhaust them without ramifications. We must be cognizant that conservation begins with us and that it is not a solitary endeavor. Conservation is about listening to the needs of the people at the interface, understanding the drivers behind conservation challenges, and including locals—both in the decision-making process as well as in the implementation of mitigations and strategies that benefit both people and nature. Conservation successes occur when the approach is not only transdisciplinary but multidisciplinary and community-based as well. We are truly—all species—in this together.

Our hope is that, in addition to serving as an educational resource, you will find inspiration in this publication, and perhaps even enlightenment regarding what it truly means to be a conservationist. Perhaps, it will serve as a catalyst to transform your mindset and approach to conservation, through recognizing that humans are not only the source of the challenges we face, but are also the key to radical transformations and sustainable solutions. As a conservationist, you hold a great and honorable responsibility to be the change you wish to see in the world. The contributors to this textbook are indeed silent heroes in a world desperate for fearless advocates of conscious change. It is our honor to share the inspiring work of the authors featured in this textbook, who have tirelessly devoted their lives to protecting Mother Nature.

Gainesville, FL

Hayley R. Adams
Susan C. Underkoffler

Contents

About the Editors

Susan Underkoffler, MFS, holds an Associates of Fine Arts degree, two Bachelor of Arts degrees in Conservation Biology and Scientific Illustration from Arcadia University, and a Master of Forensic Science degree from Drexel University College of Medicine. She has spent over ten years applying forensic and criminalistic techniques to the investigation of animal crimes, first as Forensics Manager for the Pennsylvania SPCA, where she developed the first Forensics Unit and handled all forensic responsibilities associated with humane law enforcement animal cruelty cases, and then as a consultant to wildlife, livestock, and companion animal organizations. For many years, she was an environmental scientist, performing wetland delineations, Phase I site assessments, and wildlife and habitat evaluations and monitoring in both Pennsylvania and Delaware, and prior to that she researched vaccine development using molecular biotechnology approaches with plants as vector systems. She is currently Director of the Wildlife Forensic Sciences and Conservation graduate program at the University of Florida.

Susan has traveled nationally and internationally as a professional consultant and educator for wildlife and animal cruelty investigations and has conducted published international research in several African countries. Her artwork and illustrations have received numerous awards and have been featured in many scientific publications and private collections. A passionate science communicator, she enjoys bringing to life her background, experiences, and commitment to conservation through her art.

Hayley R. Adams has over 20 years of experience in conservation medicine and One Health. She has a particular interest in the multidisciplinary and community-based approach to conservation challenges and enjoys exploring the human dimensions of conservation. She is the Founder and Director of Silent Heroes Foundation, a private foundation devoted to supporting conscious conservation initiatives around the globe.

She received her Doctor of Veterinary Medicine from the University of Tennessee College of Veterinary Medicine in 2001. In 2007, she received her PhD from the University of Tennessee College of Veterinary Medicine's Comparative Medicine Program, with a concentration in epidemiology and virology. Her PhD research focused on the molecular epidemiology and diagnosis of lentiviruses of free-ranging lions in southern Africa.

She is a board-certified Diplomate in the American College of Veterinary Preventive Medicine as well as the American College of Veterinary Microbiology. She teaches conservation medicine and related courses at the University of Florida, Department of Wildlife Forensic Sciences and Conservation. She serves as adjunct professor at the University of Tennessee College of Veterinary Medicine. She is certified as an Advanced Meditation Instructor and a Compassion Fatigue Therapist. She is an author with her first nonfiction book, *Conscious Conservation: Less Doing, More Being*, and is currently writing her first novel.

Part I

The Multidisciplinary Approach to Conservation

Building Peace to Save Nature: Multi-disciplinary Approaches to Managing Conflicts in Conservation

Isla D. Hodgson, Jeremy Cusack, Isabel Jones, Jeroen Minderman, Lovisa Nilsson, Rocío A. Pozo, O. Sarobidy Rakotonarivo, and Nils Bunnefeld

Abstract

Conservation conflicts are highly complex and challenging issues to understand and address. Such problems involve ecological and economic arguments that are centred around conservation, such as disputes over land rights and access, or the management of wildlife. However, conflicts often have roots that go beyond conservation, concerning social, cultural, historical and political matters. These dimensions are typically latent, requiring significant expertise and resources to disentangle and overcome. The management of conservation conflicts should therefore be carefully considered and enacted in a way that reflects the multi-dimensional nature of such phenomena.

In this chapter, we provide an overview of several disciplines and perspectives that view conflict through different lenses and offer a multitude of methodologies and tools that can be used to advance our knowledge and improve our approaches to researching and tackling them. Using illustrative examples, we include insights from the social and political sciences, the natural sciences, armed conflicts and peace studies. Current research and management efforts tend to focus on superficial aspects of conflicts, such as tangible or measurable wildlife impacts or clashes of interest, and overlook deeper-rooted issues. In this chapter, we hope to demonstrate that taking a multi-disciplinary perspective can contribute to a more holistic approach, beneficial to both conflict management and broader challenges in conservation.

Keywords

Conservation conflicts · Conflict management · Multi-disciplinary · Decision-making · Human dimensions

An Introduction to Conflicts in Conservation

The conservation of nature is a subject that is inherently susceptible to conflict. The Concise

I. D. Hodgson (✉) · I. Jones · J. Minderman · N. Bunnefeld
Biological and Environmental Sciences, University of Stirling, Stirling, UK
e-mail: isla.hodgson@stir.ac.uk

J. Cusack
Biological and Environmental Sciences, University of Stirling, Stirling, UK

Centre for Ecosystem Modelling and Monitoring, Universidad Mayor, Santiago, Chile

L. Nilsson
Grimsö Research Station, Department of Ecology, Swedish University of Agricultural Sciences, Uppsala, Sweden

R. A. Pozo
Center for Climate and Resilience Research, CR2, Santiago, Chile

O. S. Rakotonarivo
École Supérieure des Sciences Agronomiques, Antananarivo, Madagascar

S. C. Underkoffler, H. R. Adams (eds.), *Wildlife Biodiversity Conservation*,
https://doi.org/10.1007/978-3-030-64682-0_1

Oxford English Dictionary defines conflict as 'a state of opposition or hostilities', 'a fight or a struggle' or a 'clashing of opposed principles' (COED 2011). People hold strong and often disparate views of the natural world and how it should (or should not) be managed. Such views are a reflection of different cultures, beliefs and values (Hill et al. 2017; Redpath et al. 2013). Conflicts therefore frequently emerge between individuals—or groups—with different interests and opinions relating to conservation and can arise in a wide variety of situations (Baynham-Herd et al. 2018). For example, rural farmers may kill protected carnivores in retaliation after losing livestock to predation, which clashes with the priorities of conservation bodies. Or, the establishment of a new reserve that places restrictions on an existing land-use—such as fishing or logging—may cause hostilities between those implementing the restrictions and the resource managers affected by them.

Conflicts in conservation are widespread problems of global concern, with often severe and negative consequences for both people and nature (d'Harcourt et al. 2017; Redpath et al. 2015a). They can significantly limit progress towards worldwide goals of biodiversity preservation and sustainable development (Young et al. 2016a; UN 2015). Evidence suggests that additional environmental pressures such as climate change, increasing habitat degradation and human population expansion may serve to increase the frequency and severity of conflicts (Vargas et al. 2019; Mason et al. 2018). Therefore, in recent years, conflicts have received growing attention from the scientific community, as well as international governments and conservation bodies. For example, the International Union for the Conservation of Nature (IUCN) formally designated a task force of experts to research and implement more preventative and mitigative strategies for conflicts involving wildlife (IUCN 2020b). However, current research indicates that there are still substantial gaps in the knowledge and practice of conflict management, suggesting a necessary transformation in how we understand and approach such challenges (Hodgson et al. 2020).

Understanding Conservation Conflicts

Because of the myriad places and scenarios in which conflicts can occur, we often see multiple terms used to describe them. These include human-wildlife conflicts (Dickman and Hazzah 2016), biodiversity conflicts (Young et al. 2010), land-use conflicts (Bax et al. 2019) and people-park conflicts (De Pourcq et al. 2019), to name but a few. For the purpose of this chapter, we refer to the broader term 'conservation conflicts'—defined by Redpath et al. (2013)—as it encompasses a variety of situations where conservation and other competing interests clash, often with one party asserting its interests at the expense of another.

Conflict occurs fundamentally between humans, rather than between humans and wildlife or humans and nature. The term 'human-wildlife conflict' (HWC) is used widely by researchers, organisations, policy-makers and the media and has been criticised for the implication that wildlife are aware of their role as antagonists of humans (Peterson et al. 2013). Further, it is argued that terms like HWC suggest that once the adverse effects of wildlife on people—or vice versa—are eliminated, then conflicts will be resolved (Young et al. 2010; Peterson et al. 2013; Redpath et al. 2015b). But the reality is often much more complex.

Human-nature interactions can indeed have negative consequences, such as the predation of livestock, damage to human property, injury or loss of life, the hunting or exploitation of wild species and habitat destruction (Dickman and Hazzah 2016). It is thus necessary to implement measures to lessen these impacts. However, it is now widely recognised among the scientific and conservation community that more is required to address the full complexity of what is now referred to as a conservation conflict (Madden and McQuinn 2014; Redpath et al. 2015b). At the heart of such conflicts lies an amalgamation of social, political, cultural and economic dimensions which continually evolve over time.

Figure 1 provides a visual model for different levels of conflict and how they may be managed.

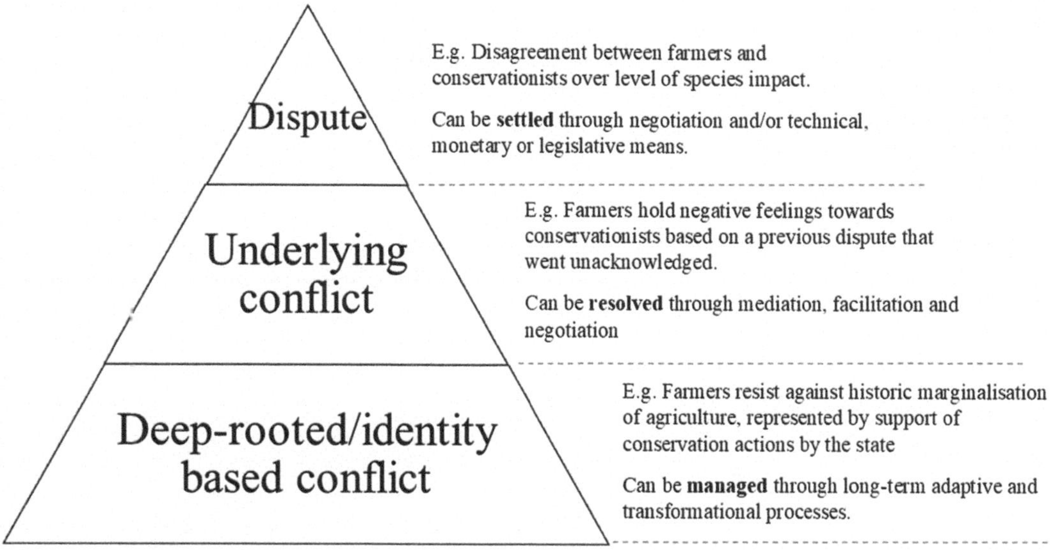

Fig. 1 The 'levels of conflict' as identified by the CICR (2000), with corresponding examples and suggested forms of management. Adapted from Madden and McQuinn (2014)

First described by the Canadian Institute for Conflict Resolution (CICR 2000), it was later adapted for conservation by Madden and McQuinn (2014) and is considered to be a pivotal piece of work that has been instrumental in developing our understanding of conservation conflicts. This model splits conflict into three 'levels': dispute, underlying conflict and identity or deep-rooted conflict. The most superficial level, known as the 'dispute', is the surface disagreement, which in the case of conservation may centre around environmental management, or the impact that wildlife has on human lives and livelihoods caused by species. Disputes can be settled, often through technical or monetary means—such as fencing, control of 'so-called problem' animals or compensation for livestock loss or crop damage. If disputes are unresolved or ignored, then an underlying conflict may develop. This lends a history to the conflict. Over time, tension grows, and emotions play a role in deepening the context of the issue. With each additional dispute, anxieties increase, and there is often an imbalance of power between parties, whether real or perceived, as one interest attempts to assert dominance over the other. At this stage, we begin to see coalitions form between individuals with similar interests and values.

Beneath the surface dispute and underlying conflict, however, there often exists another dimension, known as deep-rooted or identity-based conflict. This level stems from issues that at first seem distantly related to conservation, such as social inequalities, injustices, cultural barriers, political histories, fractured relationships, value systems and more systemic asymmetries in power (Bhatia et al. 2019; Hodgson et al. 2020). These aspects collectively pertain to individual or group identity and thus can be significantly entrenched within society. They contribute to decision-making, prejudices, assumptions and therefore how an individual reacts towards conservation or conflict management efforts. This is further complicated when conflicts involve multiple stakeholders[1] at different levels of society (Fig. 2). For instance, in Malta, conflicts relating to the spring hunting of

[1] The term 'stakeholder' refers to individuals or groups who influence, and are affected by, decisions relating to the management of wildlife and natural resources (Sterling et al. 2017).

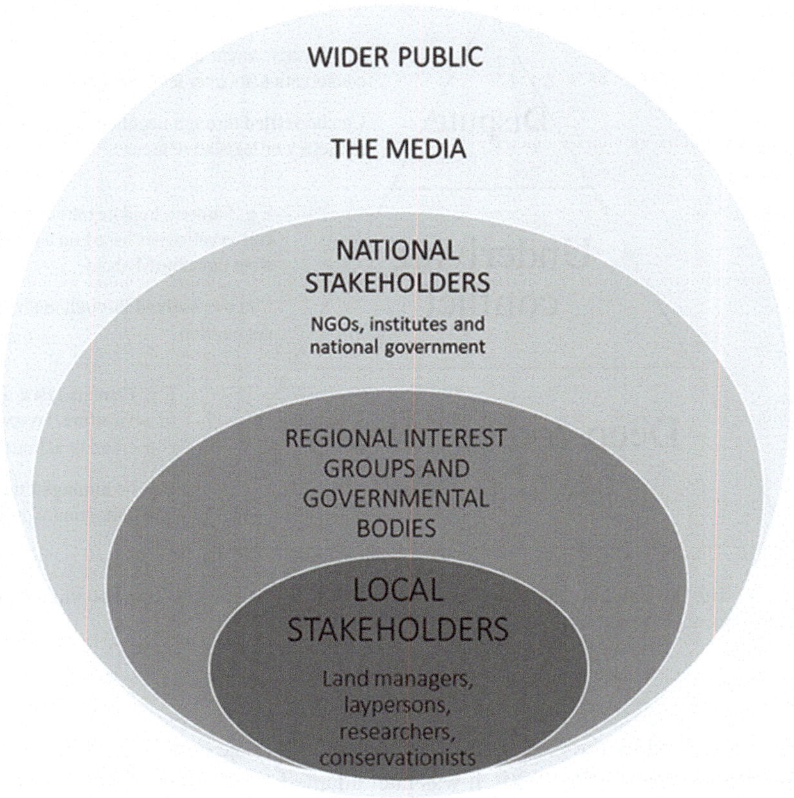

Fig. 2 Diagram illustrating the variety of actors and societal levels that may potentially be involved in a conservation conflict at any one time

migratory birds include hunters and activists, as well as government, non-governmental organisations (NGOs) and scientific bodies (Veríssimo and Campbell 2015). In addition, conflicts can span countries and geographical regions, especially when migratory species require management at an international scale (see Box 1) or large-scale natural features—such as river systems—provide key resources but cross national boundaries (Dresse et al. 2019).

Deep-rooted conflicts can be difficult to detect and distinguish from other levels. They can manifest as surface disputes (Madden and McQuinn 2014) and require more time and effort to fully understand and address (Baynham-Herd et al. 2018). Consequently, managers tend to focus on achieving relatively quick, 'win-win' outcomes to disputes and overlook more challenging, obstinate dimensions such as social and political tensions. However, if allowed to persist, deep-rooted identity conflicts may cause issues for conflict management further down the line. For example, even if negative human-nature impacts are lessened, or competing interests and values resolved, deep-rooted conflicts may simply lie dormant before emerging in response to new developments or events (Mathevet et al. 2015; Gerique et al. 2017). In recent years, these deeper-rooted elements of conservation conflicts have received more attention (Hodgson 2018; Hodgson et al. 2019; Harrison et al. 2019; Jani et al. 2019). Yet expansion and further integration of such knowledge into applied techniques remains a challenging endeavour.

Box 1 Waterbirds and Transboundary Conflicts

The United Nations Agreement on the Conservation of African-Eurasian Migratory Waterbirds (UN-AEWA) is an intergovernmental treaty working to bring countries and the wider conservation community together, to establish coordinated

(continued)

conservation and management of migratory waterbirds throughout their ranges (UNEP/AEWA 2018). One such species is the lesser white-fronted goose (LWfG; *Anser erythropus*), a long-distance migrant with variable and only partially known migration routes from Scandinavia and eastern Siberia to the Mediterranean and S.E. Asia, via key staging grounds in Central Asia and Europe (Marchant and Musgrove 2011). LWfG are globally threatened, with an estimated population of 16,000–27,000 adult individuals, and hunting is a key threat to this species (IUCN 2020a). In Kazakhstan—a key LWfG staging ground—hunting LWfG is illegal. To reduce accidental hunting of LWfG, which can happen when LWfG are flying in mixed flocks, specific conservation interventions to raise awareness of LWfG identification and of its protected status have been suggested (Jones et al. 2017). However, hunting and habitat loss occurs in other countries within the LWfG range, exacerbating LWfG population decline and reducing the efficacy of conservation interventions implemented. Not all countries along species' migration routes will be signatories to international agreements on species protection or may not have the resources to implement and enforce appropriate conservation measures (Runge et al. 2015). A difference in socio-economic status both within and between countries may also lead to different levels of legal and illegal hunting. Thus, conservation conflicts may develop across international boundaries, whereby there is an unequal distribution of costs associated with conservation interventions among nations. In such cases, species will not be fully protected across their migration route, undermining the efficacy of conservation interventions and creating transboundary conservation conflicts (Studds et al. 2017).

In the following sections, we describe approaches and tools that can be used to understand and manage conservation conflicts in their entirety. Due to the complex and multi-dimensional nature of conflicts, these approaches come from a range of fields and disciplines that include, but are not constrained to, the natural sciences. There are a number of disciplines exploring conservation conflict through a different lens and offering a multitude of tools, approaches and perspectives that may be utilised. This chapter serves to provide an overview and does not delve into the deeper theories and ideologies therein. Useful texts to expand this knowledge include Redpath et al. (2015b), Bunnefeld et al. (2017), Bennett et al. (2017a, b) and Hodgson et al. (2020).

Multi-disciplinary Approaches and Perspectives on Conservation Conflicts

A Natural Sciences Perspective

As conservation conflicts often centre around environmental issues, research and management is often subject to disciplinary and sectorial silos—predominantly with a bias towards the natural sciences and with expertise from ecology and conservation (Sandbrook et al. 2013; Bennett et al. 2017a). This is not to say that a natural science perspective is not integral to the process. Studies of animal behaviour, movements and habitat use are essential to the understanding of human-nature interactions, around which many disagreements centre. Box 2 illustrates how ecological modelling is helping to better predict and manage conflicts across Europe in relation to common cranes, *Grus grus*, a migratory waterbird that places pressure on agricultural interests (Nilsson et al. 2019). Using such models, negative impacts can be anticipated and potentially avoided or lessened. Further, technical projects—such as collecting data on animal physiology and behaviour, monitoring habitat use and movements or the testing of new technologies—

can also be useful tools with which to encourage collaboration and dialogue among conflicting stakeholders (Dresse et al. 2019; Duthie et al. 2018).

However, a bias towards this perspective risks narrowing the perception of what issues require the most attention and may limit what we can achieve when attempting to address conflicts (Redpath et al. 2015a; Moon et al. 2019a). For example, the natural sciences have a largely technical focus, traditionally relying on quantitative assessments, more rigid hypotheses and mechanistic perspectives. Moreover, for obvious reasons, its focus is usually on the animals or natural resource in question, with the human dimension (e.g. behaviour of stakeholders, resource users or interactions among them) often considered only in basic terms. Although this lends important insight into the natural resource aspect of conflict, human dimensions require a wider array of methodologies, the inclusion of more qualitative data and more open-ended research questions (White and Ward 2011). There are a wide variety of inter- and transdisciplinary perspectives which offer multiple lenses through which to view and understand conflict and provide a more holistic approach to managing them (Redpath et al. 2015b; Hodgson et al. 2020).

A Social Sciences Perspective

The social sciences encompass a vast diversity of theoretical and applied disciplines, including—but not limited to—anthropology, sociology, history, economics, ethnography, psychology, communication studies and law. These disciplines are vast, and within each exist innumerable paradigms, theories and concepts. In essence, the social sciences involve the analysis of a variety of social phenomena at all levels of society, from individual values to group dynamics, and further to wider societal patterns and trends (Tindall and Piggot 2015; Bennett et al. 2017a). They may assist in the understanding of not just human decision-making and behaviours but also how economic, political and historical factors—such as agency, governance and inequality—shape social events, structures and hierarchies (Hicks et al. 2016). Given that conflict is fundamentally a social phenomenon among humans (Brox 2000), the social sciences have much to offer in the way of understanding conservation conflicts.

Box 2 Understanding Staging Site Selection of Common Cranes Along the Western-European Flyway to Guide Crop Damage Prevention

Common cranes (*Grus grus*) have increased drastically along the Western-European flyway over the last decades, due to protection from the EU Birds Directive and wetland restorations, such as the European Natura 2000 network (Harris and Mirande 2013; EC 2020). The aim of the network is to support migratory and protected species, such as the common crane, by increasing supranational connectivity between protected areas (EC 2020). However, cranes now congregate in large numbers (occasionally up to 268,000 ind.) at wetland-agricultural sites along the flyway (LPO 2020). When foraging, cranes can have negative impacts on agricultural production, which fuels the reluctance of farmers to support wetland restorations and consequently encourages conflicts between conservation (i.e. species and wetland protection) and farming interests (i.e. maximising yields) (Salvi 2010; Montràs-Janer et al. 2019). Ecological studies of habitat and foraging patterns can be used to predict where species are likely to cause damage and thus guide management interventions (Fox et al. 2017; Nilsson et al. 2016). For example, location data derived from GPS transmitters demonstrate that cranes select Natura 2000 sites as wetland night roosts along their flyway, with a 97% probability that cranes will be present at these sites. However, the Natura 2000 sites do not fulfil their daily feeding

(continued)

requirements, which drives cranes to forage on the surrounding agricultural land. The probability of cranes utilising farmland is 63% for areas with close proximity to Natura 2000 sites; however, probability decreases to 27% with increasing distance from Natura 2000 sites (max 89.3 km; Nilsson et al. 2019).

The potential for crop damage on land that is close to protected areas indicates a high risk of conflicts between conservation and agricultural objectives. This in turn identifies a need for improved cross-boundary collaboration and policy development to reduce agricultural impacts and to manage conflicts when implementing protected areas (Nilsson et al. 2019). One key aspect is to decentralise management decisions to local stakeholders in order to implement effective compensation and damage prevention strategies within the vicinity of protected areas (Mason et al. 2018; Nilsson et al. 2019). Strategies could, for example, include providing undisturbed fields with attractive food (e.g. barley or wheat) for cranes within protected areas as a diversionary tactic and employing 'scaring' methods adjacent to growing crops (Nilsson et al. 2016, 2019).

As diverse and extensive as these disciplines are, so are the methodologies available to study social phenomena in the field. Quantitative approaches can involve questionnaires, surveys, lab and field experiments, choice experiments, demographic evaluations and cost-benefit analysis (Hanley et al. 2019; Bennett et al. 2017a). Such methods can offer large sample sizes capable of statistical analysis and thus are often used in deductive, hypothesis-driven studies, as well as to provide broad overviews of wide subject areas. For example, many large-scale attitudinal studies have compared levels of tolerance towards predators and management interventions in relation to demographic or geographical data (e.g. Mkonyi et al. 2017). However, where more in-depth, detailed and nuanced answers are needed, qualitative (non-numerical) data may be preferable (Rust et al. 2017). Methodologies include interviewing, focus groups, discourse analysis, ethnography, conservation analysis and analysis of oral and archival histories (Young et al. 2018; Hodgson et al. 2018; Bennett et al. 2017b). Typically, qualitative research takes a more inductive or grounded line of enquiry, where conclusions are more probable and open-ended, rather than certain answers to hypotheses-driven questions. It is not unusual for a mixed-methods approach to be used, where quantitative data is used to complement qualitative, and vice versa (Aramo-Immonen 2011; Schoonenboom and Johnson 2017). For example, in-depth debriefing interviews can be used to better understand the results from choice experiments and provide insight into decision-making processes. Extending beyond a theoretical understanding of conflict, the social sciences can also aid in the management of conflict in the field, facilitating the development and execution of participatory processes (Bennett et al. 2017a, b) and conservation planning (e.g. Ban et al. 2013).

The social sciences are increasingly applicable to conservation conflicts and to conservation in general. Within academia, more recent application of social science perspectives has extended to explore different values and meanings in conflict (St John et al. 2019), the motivations and reasoning behind wildlife crime (Von Essen et al. 2014; Von Essen and Allen 2017), ethics and social justice (Wright 2019; Brittain et al. 2020) and the role of discourse and social interaction in shaping conflict (Hodgson et al. 2018). However, there is still much potential to broaden the scope (Bennett et al. 2017a, b). For instance, Moon et al. (2019c) make a compelling argument that current research focuses on defining and quantifying social elements in conservation, when there are opportunities to further engage with different methodologies and philosophies. The authors refer to ethnographic approaches (methods that involve unstructured, open-ended interviews and participant observation) and advocate for plurality, openness and reflexivity in research. The next two sections describe in more detail two extensions of the social sciences that are highly relevant to conflict: socio-economics and political science.

Box 3 Using Ethnography to Gain a Deeper Understanding of Raptor-Grouse Conflicts in Scotland, UK

A contentious example of conservation conflict is illustrated in the competing interests of driven grouse shooting and the conservation of birds of prey (Thirgood and Redpath 2008). Driven grouse shooting is a recreational sport which traditionally takes place in upland areas. Rather than more conventional hunting methods—where hunters are on foot—in driven shooting, hunters remain stationary, targeting birds as they are 'driven' towards them by an advancing line of estate workers. Shooting estates, which are privately owned, rear artificially high populations of grouse in order to make a profit (Thompson et al. 2009). Predatory birds are viewed as a threat and have historically been persecuted on land managed for grouse shooting, driving some species to the brink of extinction. Protective legislation has slowed lethal control and allowed populations to recover; however, killing continues illegally. The ensuing conflict between grouse moor advocates and conservation interests has escalated, becoming entrenched due to the strong cultural aspects of grouse shooting and deep-rooted social and political elements (Hodgson et al. 2018). Many attempts to manage this conflict have failed, and parties have been unable to engage in meaningful dialogue. Until recently, research has focused on the surface dispute over the impacts of raptors and illegal wildlife crime (e.g. Thompson et al. 2009).

Hodgson et al. (2018) aimed to study the social and political elements of this conflict with the use of qualitative methodology. Given the highly sensitive nature of this situation and its associations with wildlife crime, an ethnographic approach was required whereby the researcher observed subjects in their day-to-day lives and held unrecorded interviews in a variety of settings. Once trust was established, semi-structured interviews—interviews with no formal questions, but guidelines to ensure key topics are covered (Young et al. 2018)—were conducted and recorded. Stakeholder perceptions were strongly influenced by trust, social relations, power dynamics and how they were perceived to be represented at a national level (Hodgson et al. 2018). These factors were major barriers to stakeholders engaging constructively with collaborative processes, causing them either to act antagonistically and reinforce their own interests or to feel powerless and disengage completely. This study demonstrated the importance of including social and political factors into conflict management processes, rather than the current focus on technical solutions and legislation.

A Socio-Economic Perspective

A socio-economic perspective can answer many questions relating to trade-offs and decision-making between different stakeholder groups. Although already used to some extent in conflict management (e.g. the use of compensation schemes; see Ravenelle and Nyhus 2017 for an overview), deeper exploration of the field can aid in the understanding and therefore more effective application of economic strategies in conservation (Hanley et al. 2019).

Financial incentives and compensation are popular tools used with which to alter stakeholder behaviour, commonly towards more pro-conservation or sustainable actions. Yet, their effectiveness in practice is questioned, as implementation has been challenging and, in many cases, difficult to monitor (Nyhus et al. 2005; Pozo et al. 2017). For example, in 2009, the Government of Botswana introduced a scheme aimed at reducing conflicts between local livelihoods and the conservation of protected wildlife (i.e. lions, elephants and crocodiles) in the eastern Okavango Delta Panhandle. Local villagers were asked to report any negative impacts caused by wildlife—such as damage to property or livestock loss—within 7 days of the incident occurring, following which a government official would visit the site and assess the level of impact before initiating a compensation process (Songhurst 2017; Noga et al. 2018). However, the scheme has encountered several issues, including transportation difficulties, delays to payment and claimants attempting to cheat the system, leading stakeholders to declare it inadequate (Noga et al. 2018).

Economics asks how and why stakeholders respond to various incentive-based schemes and the extent to which individuals are willing to commit to them. For example, contingent valuations or choice experiments are used to determine Willingness to Pay (WTP) or Willingness to Accept Compensation (WTAC) for certain assets, under different scenarios (Hanley et al. 2019). For example, choice experiments have been used to compare public WTP for actions to either protect or manage the hen harrier—a raptor at the centre of an entrenched political conflict in the UK (see Box 3; Hanley et al. 2010). Experimental games can also offer a low-cost and low-risk tool for testing the influence of different economic instruments on stakeholder behaviour (Redpath et al. 2018). Such methods have been used to understand farmer behaviour in elephant-related conflicts in Gabon (Box 4).

Box 4 Experimental Games to Understand Farmer Behaviour in Relation to Elephant Conflict Management Efforts in Gabon

Conservation conflicts related to elephants can impose considerable social and financial costs on farmers in Africa and Asia, and the retaliatory killing of elephants is common (Mackenzie and Ahabyona 2012). Experimental games with rural farmers in Gabon have helped to understand the impacts of subsidies and agglomeration payments on farmers' decision-making (Rakotonarivo et al. 2020). The games were framed around land-use management and played in groups of four, providing a relaxed atmosphere to explore local farmers' propensity to engage in lethal control.

The findings suggest that economic instruments were conducive to pro-conservation behaviour, with farmers opting to choose other methods over lethal control when a financial incentive was offered. However, results also implied that other factors influenced farmer decision-making. By combining game outcomes with household surveys, the study also shed light on the relationships between game decisions and key socio-economic and attitudinal factors, such as trust and equity attitudes. These factors have been identified as key determinants of farmer decision-making in other conflicts regarding wildlife and land management (Young et al. 2016b; Treves et al. 2017). This implies that addressing material, visible manifestations of conflict (such as wildlife impact) may not tackle underlying conflicts, and thus interventions must also seek ways and means of addressing issues like social equity.

Additionally, systematic economic evaluations of management actions—known as cost-benefit analysis (CBA)—can be used to map out the distribution of financial losses and gains, thereby helping to predict how different stakeholders will react to certain interventions (Hanley et al. 2019). For example, Mburu et al. (2003) analysed the transaction costs imposed on landowners by a collaborative management scheme in Kenya and were therefore able to recommend changes that could increase compliance.

A Political Sciences Perspective

Political science is also an extension of the social sciences, relating broadly to the mechanisms and structures that determine how society is governed and therefore studying the institutions, rules and norms that influence political activities, trends and behaviours (Chatturvedi 2005). Some major subfields include political theory, comparative politics and international relations. More recent developments include interdisciplinary studies that relate specifically to the environment and conservation, as their application to conservation conflicts can be extremely useful in the exploration of their underlying structural causes and dynamics.

One such field is political ecology. With roots in human geography, political ecology views conflict as socially mediated, with environmental factors providing the context, not the cause (LeBillon and Duffy 2018). Research interests lie within the history of conflicts, with particular focus on power theory and dynamics—for example, factors that foster resistance, collusion and repression (Koopman 2011) or stimulate cooperation and consent (Brock and Dunlap 2018). The potential application to conservation conflicts is therefore vast, as environmental injustices, power struggles and inequalities are argued as key drivers in their development and manifestation (Raik et al. 2008; Adams 2015; Von Essen and Allen 2017; Von Essen et al. 2014). Understanding the chain of political events that have resulted in conflict may help to explain current situations, as well as predict and prevent future occurrences (Mathevet et al. 2015). Key methodologies are ethnography (long-term immersion in knowledge and local cultures), oral or archival historical studies and analysis of policy narratives and discourses (LeBillon and Duffy 2018).

Another related—and extensive—field is peace and conflict studies. Much like political ecology, this field is vastly interdisciplinary, drawing on a wide range of theories from international relations, history, social sciences and politics (Rogers 2015). It could be argued that peace and conflict studies are perhaps the most applicable to conservation conflicts, given their focus on the more structural root causes (see Fig. 1) and interest in prevention and resolution (Rodríguez and Inturias 2018; LeBillon and Duffy 2018). However, application to conservation has been somewhat limited (Madden and McQuinn 2014).

From a peace research perspective, conflicts operate at multiple levels and stem from inequalities between stakeholders at these levels (Rodríguez and Inturias 2018). Therefore, subfields tend to have a more practical, applied framing—for example, conflict resolution, environmental peacebuilding and conflict transformation are all borne from peace studies (Lederach 2003; Madden and McQuinn 2014). Approaches range from quantitative and qualitative strategies, arbitration, mediation and facilitation (Rogers 2015) to addressing problematic power imbalances and injustices through, for instance, building local capacity and agency (Rodríguez and Inturias 2018). Environmental peacebuilding identifies different stages according to the level of polarisation between groups. 'Technical' solutions to environmental problems, for example, may be utilised as a starting point for dialogue between parties who are perhaps unable to engage in constructive mediation (Dresse et al. 2019). Further, peace studies commonly view conflict through a wider lens than political ecology, which typically concentrates on specific case studies (LeBillon and Duffy 2018). Peace studies search for global or more generalised trends and thus sometimes provide statistical analyses—which can be useful when thinking about conflict on a more global scale. Further, there is much we can learn from applied techniques and theories used in armed conflict (Box 5).

Box 5 Taking an Armed Conflicts Approach to Conservation Conflicts

Despite clear differences in the levels of violence involved, the development of conservation and armed conflicts share many similarities. Both involve the imposition of one, or several, interests over those of

(continued)

others, which results in situations of dominance, discord or power imbalance. Both are characterised by a combination of political discourse and concrete actions that contribute towards conflict escalation or de-escalation. Stakeholders in these conflicts may adopt antagonistic positions towards each other but may also act as third-party moderators or peacemakers. Lastly, like conservation conflicts, armed conflicts may show considerable variation in their historical or geo-political contexts. Yet, while research into the occurrence and characteristics of armed conflicts on a global scale has greatly benefitted from categorisations of conflict type and intensity (e.g. Metternich et al. 2019; Sundberg et al. 2012)—which, in turn, has enabled a better understanding of factors driving conflict dynamics (Hegre et al. 2019)—such frameworks are lacking for conservation conflicts.

Each conservation conflict will have its own unique historical and contextual characteristics. But focusing on the nature of human interactions may provide one way of generalising patterns across case studies. One potentially useful approach is the curve of conflict model (CCM), which is used in armed conflicts to track processes of escalation and de-escalation over time (Wallensteen 2018; Lund 1996; Crowley et al. 2017). The model was developed with the aim of guiding the prevention of armed conflicts and demonstrates how different conflict phases relate to one another, as well as to different kinds of third-party intervention. Figure 3 shows how the CCM can be adapted to conservation conflicts, thereby providing a classification for their intensity (Cusack et al. 2021).

Socio-Ecological Modelling and Game-Theoretic Approaches

Conservation conflicts are, by nature, multifaceted, involving dynamics of both natural resource populations and human decision-making and the interactions between these. Resolution often relies on the ability to accurately predict phenomena and their effects, such as changes in population dynamics and the outcome of management interventions. Yet until recently, techniques that achieve this successfully have been scarce. In part, this may be due to the fact that while quantitative modelling is common-place in some areas of research related to conflicts (e.g. natural sciences and economics), other fields—some of which we have already discussed—are more reliant on qualitative approaches that do not lend themselves as easily to numerical models (White and Ward 2011). Integrated models—which cover multiple aspects of socio-ecological systems—do exist, but have remained relatively niche.

Socio-ecological models, and in particular management strategy evaluation (MSE) approaches (see Fig. 4), provide increasingly promising means to integrate the perspectives from, and advances in, different fields, predicting the dynamics of different components of socio-ecological systems (Bunnefeld et al. 2017). In doing so, they provide a modelling framework to understand how changes in one component of a system (e.g. the behaviour of users) are likely to affect another (e.g. the natural resource). Due to the complex and integrated nature of conflicts, such models are unlikely to resolve them directly, but they do provide a systematic and transparent way to predict dynamics and outcomes.

MSE was originally a tool for aiding the management of harvesting systems (Smith et al. 1999). However, it can accommodate sub-models for user and manager dynamics (i.e. decision-making), as well as natural resource dynamics (Fig. 4). This ability to incorporate

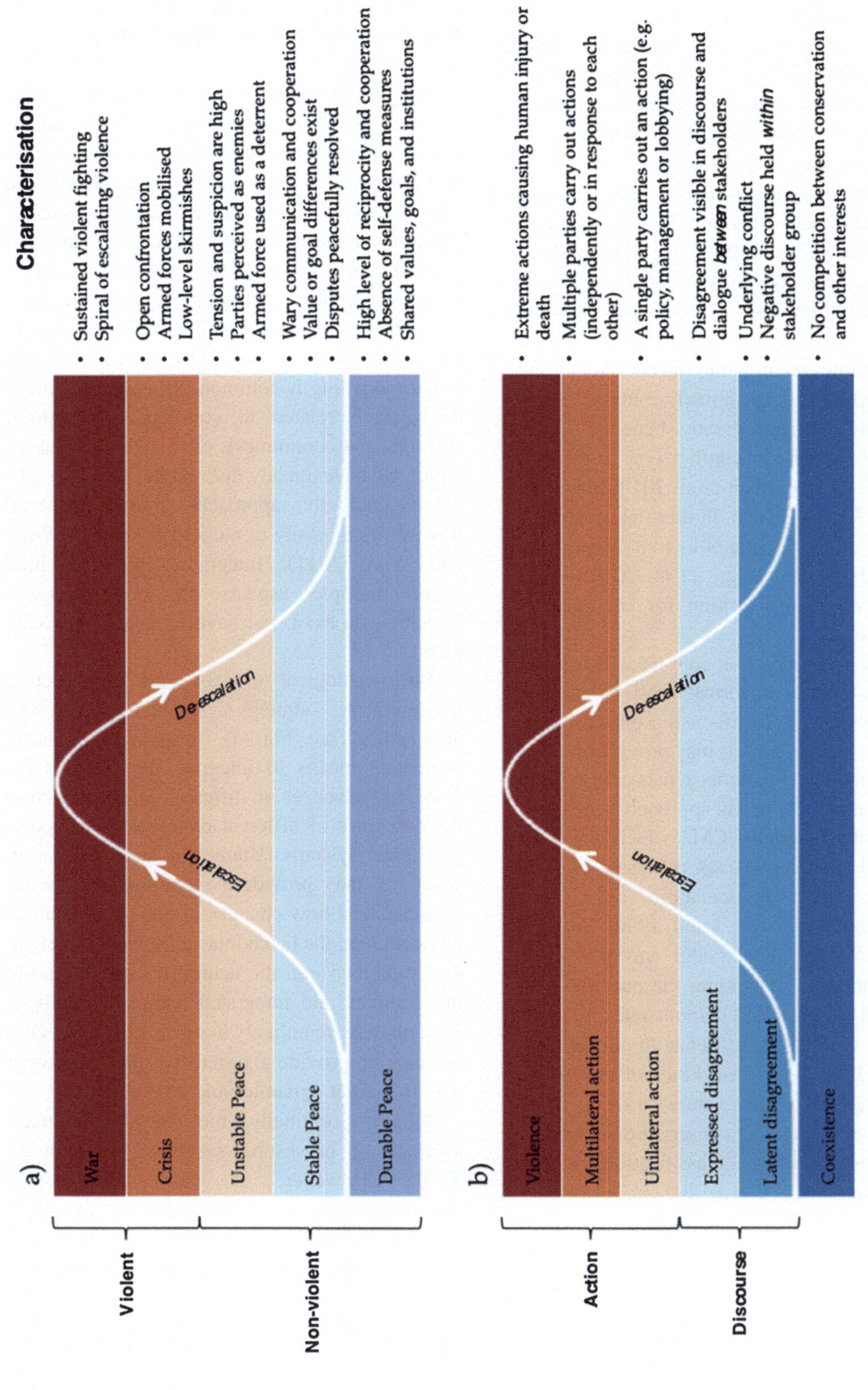

Fig. 3 Schematic representation of **(a)** Lund's (1996) curve of conflict and **(b)** the proposed conservation conflict curve, including the key characteristics relevant to each level. From Cusack et al., 2021

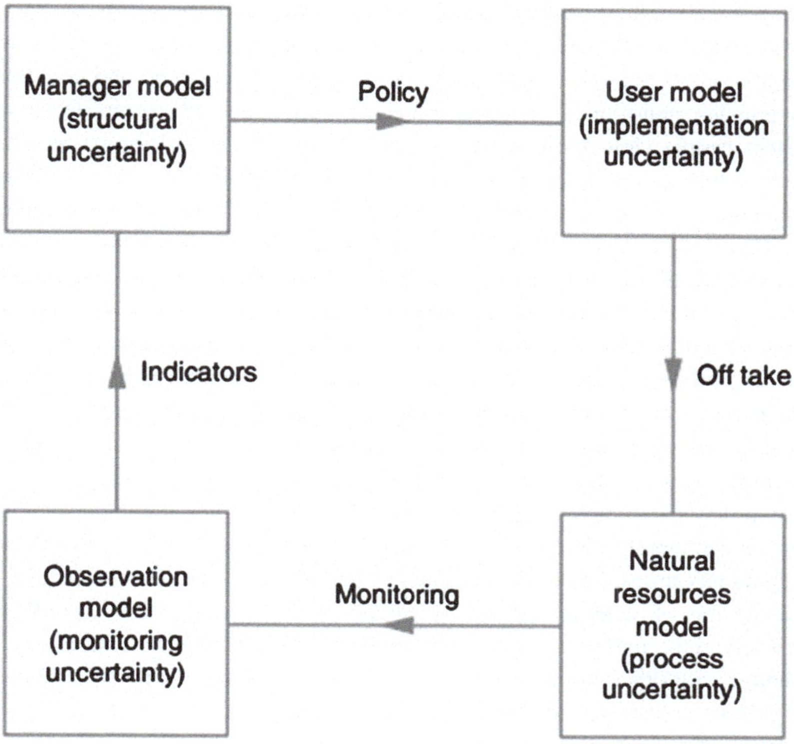

Fig. 4 Conceptual model representing management strategy evaluation (from Bunnefeld et al. 2017). This approach holistically models socio-ecological systems. Separate components model (1) the natural resource itself, (2) its observation, (3) management decisions and (4) users affecting the resource. While each of these can be modelled separately, their inputs and outputs are interdependent

social and ecological elements of systems makes it highly applicable to conservation conflicts (Bunnefeld et al. 2017). Indeed, the generalised MSE developed by Duthie et al. (2018) further extended this framework by integrating a game-theoretic approach to simulate user and manager decision-making in response to changes in, and impacts of, a natural resource population. Game theory is the mathematical study of strategic decision-making (Myerson 2013), where an individual's decision is dependent upon the decisions of others (the 'Prisoner's Dilemma' being a classic example). Incorporating this approach into the MSE framework allows for the quantitative modelling of decision-making of stakeholders (e.g. natural resource users and managers) in relation to specific goals, such as maximising yield, alongside models for the resource itself. Therefore, hypotheses relevant to conservation conflicts can be explicitly tested. For example, what are the consequences for the sustainability of natural resources when a manager's decision-making is influenced by either the interests of conservation or those of resource users, potentially through political acts such as lobbying (Cusack et al. 2020b)? More practical applications can include using the framework to provide decision support tools for real-world systems, where managers are frequently tasked with balancing trade-offs between resource exploitation and the protection of biodiversity (Cusack et al. 2020a).

In summary, holistic modelling frameworks that incorporate all aspects of socio-ecological systems are key in helping us understand the dynamics of conservation conflicts. They not only allow us to provide systematic and transparent predictions but are also an excellent means by which different perspectives and insights from different fields can be combined.

The Cognitive Sciences Perspective

As demonstrated in Fig. 1, conservation conflicts often forego surface disputes over species impact and have much deeper cognitive aspects (Redpath et al. 2013). These relate to how an individual shapes reality, their perceptions of the world and what happens within it (Hodgson et al. 2018). The cognitive sciences (i.e. philosophy, psychology, linguistics, neurology) tell us how and why different people view and construct reality and the factors that influence, or are influenced by, these perceptions. 'Cognition' refers to different ways of knowing, in that judgements, reasoning and awareness differ from person to person. For example, socio-psychological theory describes a cognitive hierarchy of certain factors that collectively influence an individual's worldview, consisting of values, beliefs, attitudes and norms, which ultimately result in behavioural actions (Vaske and Manfredo 2012). Examining the process of converting human thought into action can therefore help us to understand certain behaviours in conflict situations, such as the illegal killing of wildlife, refusal to comply with conservation measures and unwillingness to engage with other parties.

More recently, different theories and frameworks from the cognitive sciences have been applied to conservation conflicts. Barua et al. (2013) explored the 'hidden' impacts of conflicts related to wildlife on human well-being and found evidence that poor mental health, psychological trauma and severe stress are linked to such situations. Psychological theories (cognitive dissonance, reactance and motivation crowding) were used to understand historical shifts in attitudes towards wildlife by Maasai pastoralists of the Amboseli ecosystem in southern Kenya (Fernández-Llamazares et al. 2019). These findings demonstrated a history of cultural exclusion and marginalisation and highlighted the importance of building different meanings and conceptualisations of wildlife into conservation measures. Other explorations from a cognitive perspective have examined the concept of psychological 'ownership' of natural resources, and its importance in stakeholder cooperation (Matilainen et al. 2017), as well as the effect of different value orientations on stakeholder preferences for conflict management interventions (St John et al. 2019).

The mental model is an interesting technique, with multiple potential applications to conflict. In a basic sense, mental models are graphical representations of the way in which people interpret reality and how this influences the reasoning behind certain decisions and behaviours (Moon et al. 2019b). Using a variety of methodologies—from interviews with individuals to group role-playing and map building—mental models are constructed from individual or group knowledge, experiences, values and belief systems (Jones et al. 2014). Although their application to conservation conflicts is currently limited, they have been used in diverse ways to better understand and overcome issues in conservation and natural resources management, such as the suitability of possible management interventions (Biggs et al. 2011), the understanding of conflicts, barriers to progress and possible solutions (Mosimane et al. 2014). They can also be used in the practical management of conflicts, providing a joint activity around which trust and relationships can be built among stakeholders (Halbrendt et al. 2014).

Applying Multi-disciplinary Approaches to Conflicts in the Field: Asking the Right Questions

Throughout this chapter, we have described a number of different approaches to conflict research and management, all stemming from broad, interdisciplinary fields and perspectives. This complexity matches the undeniable convolution and 'wicked' nature of conflicts (DeFries and Nagendra 2017; Mason et al. 2018). However, to say that makes researching and tackling such problems is overwhelming, even for those experienced in conflict management, would be an understatement.

A common problem that stems from this inherent complexity is that 'conflict management' strategies—especially those focused on cognitive or structural elements—generally lack a cohesive framework to aid the development, implementation and monitoring of such processes (Redpath et al. 2013; Young et al. 2016a; Bunnefeld et al. 2017). Further, strategies are rarely evaluated—meaning there is often no concrete idea of what 'success' looks like and there is a lack of robust, empirical evidence to support future recommendations and improvements (Eklund et al. 2017). Many researchers and practitioners have advocated for a transformation in how both how conservation conflicts are perceived and approached, moving towards more long-term, adaptive and evaluative strategies that include opportunities for social learning (Redpath et al. 2013, 2015b; Eklund et al. 2017; Bunnefeld et al. 2017; Mason et al. 2018; Hodgson et al. 2020). However, integrating diverse perspectives and methodologies into an adaptive management process remains a significant challenge.

In recent years, stepwise frameworks have emerged, designed to aid in the design and implementation of management strategies (see Young et al. 2016a and Ainsworth et al. 2020). In agreement with a growing body of literature, such frameworks emphasise the fact that, before any strategy is implemented, key questions must first be addressed. For instance, what situation are we dealing with, and what is it about? Conflicts are not uncommon; however, it has been argued that the term is often used inappropriately (e.g. Peterson et al. 2013). For example, a scenario deemed a 'conflict' may in fact be a dispute over human-wildlife impacts, whereas the actual underlying conflict, which may consider social or political elements, may be unacknowledged (Young et al. 2010). This framing is vital, as it dictates which type of management intervention may be appropriate and can save valuable resources that may be better utilised elsewhere (Baynham-Herd et al. 2018). As previously discussed, a human-wildlife impact may be settled through technical, legislative, monetary or dialogic means; the human-human conflict may require more effort. Another important question is: who is involved? This question is particularly difficult, as some stakeholders involved in conflict may not be immediately obvious, or 'hidden'. Particularly in sensitive scenarios, such as problems where crime is involved, or in communities or cultures, where certain groups are traditionally hindered from participation, groups or individuals may be unable, or unwilling, to come forward. Furthermore, conflicts can occur at multiple different levels and within various sectors of society.

Once these questions have been established, it may be possible to ascertain relevant and appropriate management interventions (Young et al. 2016a), as well as how they may be implemented, and who should be responsible for their implementation. As indicated in this chapter, a wealth of techniques, strategies and approaches are available; however, their delivery and governance are of equal importance. For instance, community-based conservation schemes are growing increasingly popular, but can fail if governed by weak or centralised institutions, and/or are susceptible to corruption (Bluwstein et al. 2016). Closely linked is the query of how to evaluate and adapt strategies, i.e. are there methods in place to monitor conflict management and to address problems as they arise? Ideally, the process and the institutions that govern them must have an investment in the long-term, must be flexible and allow for adaptation to different scenarios. Conservation conflicts are by nature dynamic, i.e. constantly evolving, in relation to new developments, including management efforts (Redpath et al. 2013). Strategies must therefore reflect this uncertainty.

Uncertainty is a significant component to consider in future approaches to conflicts in conservation (Bunnefeld et al. 2017), due to the effects changes may have in such complex systems. For instance, it is well known that impacts of climate change will affect more adversely vulnerable and poorer agriculturalist communities, in particular women (UNDP 2013; Arora-Jonsson 2011; FAO 2017). These groups will be directly impacted by water scarcity, reductions in yields of forest biomass, land-use change and/or increased disease risk, all of which contribute to a perpetuation of poverty in vulnerable groups. Communities vulnerable to climate change are generally also more exposed to conflicts over natural resources, as they have a more direct dependence on them (FAO 2017; Lipper et al. 2014). Therefore, strategies aiming to create coexistence must be designed in order to adapt as well as to provide sustainable solutions for communities living in resilient environments.

In summary, conservation conflicts, in their entirety, are multi-dimensional, multi-levelled and dynamic, and this key concept should be reflected in the efforts made to manage and resolve them. A wealth of information, knowledge, tools and techniques exist—which we have merely brushed upon in this chapter—and there is no one way of utilising all of the tools in such a vast toolbox. However, acknowledging the breadth and complexity of conservation conflicts is a good starting point.

References

Adams WM (2015) The political ecology of conservation conflicts. In: Conflicts in conservation. Cambridge University Press, Cambridge, pp 64–75

Ainsworth GB, Redpath SM, Wilson M, Wernham C, Young JC (2020) Integrating scientific and local knowledge to address conservation conflicts: towards a practical framework based on lessons learned from a Scottish case study. Environ Sci Pol 107:46–55

Aramo-Immonen H (2011) Mixed methods research design. In: World summit on knowledge society. Springer, New York, pp 32–43

Arora-Jonsson S (2011) Virtue and vulnerability: discourses on women, gender and climate change. Glob Environ Chang 21(2):744–751

Ban NC, Mills M, Tam J, Hicks CC, Klain S, Stoeckl N, Bottrill MC, Levine J, Pressey RL, Satterfield T (2013) A social–ecological approach to conservation planning: embedding social considerations. Front Ecol Environ 11(4):194–202

Barua M, Bhagwat SA, Jadhav S (2013) The hidden dimensions of human–wildlife conflict: health impacts, opportunity and transaction costs. Biol Conserv 157:309–316

Bax V, Francesconi W, Delgado A (2019) Land-use conflicts between biodiversity conservation and extractive industries in the Peruvian Andes. J Environ Manag 232:1028–1036

Baynham-Herd Z, Redpath S, Bunnefeld N, Moloney T, Keane A (2018) Conservation conflicts: Behavioural threats, frames, and intervention recommendations. Biol Conserv 222:180–188

Bennett NJ, Roth R, Klain SC, Chan K, Christie P, Clark DA, Cullman G, Curran D, Durbin TJ, Epstein G, Greenberg A, Nelson MP, Sandlos J, Stedman R, Teel TL, Thomas R, Veríssimo D, Wyborn C (2017a) Conservation social science: understanding and integrating human dimensions to improve conservation. Biol Conserv 205:93–108

Bennett NJ, Roth R, Klain SC, Chan KM, Clark DA, Cullman G, Epstein G, Nelson MP, Stedman R, Teel TL (2017b) Mainstreaming the social sciences in conservation. Conserv Biol 31(1):56–66

Bhatia S, Redpath SM, Suryawanshi K, Mishra C (2019) Beyond conflict: exploring the spectrum of human–wildlife interactions and their underlying mechanisms. Oryx 2019:1–8

Biggs D, Abel N, Knight AT, Leitch A, Langston A, Ban NC (2011) The implementation crisis in conservation planning: could "mental models" help? Conserv Lett 4 (3):169–183

Bluwstein J, Moyo F, Kicheleri RP (2016) Austere conservation: understanding conflicts over resource governance in Tanzanian wildlife management areas. Conserv Soc 14(3):218–231

Brittain S, Ibbett H, de Lange E, Dorward L, Hoyte S, Marino A, Milner-Gulland E, Newth J, Rakotonarivo S, Veríssimo D (2020) Ethical considerations when conservation research involves people. Conserv Biol 34:925–933

Brock A, Dunlap A (2018) Normalising corporate counterinsurgency: engineering consent, managing resistance and greening destruction around the Hambach coal mine and beyond. Polit Geogr 62:33–47

Brox O (2000) Schismogenesis in the wilderness: the reintroduction of predators in Norwegian forests. Ethnos 65(3):387–404

Bunnefeld N, Nicholson E, Milner-Gulland EJ (2017) Decision-making in conservation and natural resource management: models for interdisciplinary approaches. Cambridge University Press, Cambridge, UK

Chatturvedi J (2005) Political governance: comparative politics. Gyan Publishing House, New Delhi

CICR (2000) Becoming a third-party neutral: resource guide. Ridgewood Foundation for Community-Based Conflict Resolution, Ottawa, Canada

COED (2011) Concise Oxford English Dictionary. Oxford University Press, Oxford, UK

Crowley SL, Hinchliffe S, McDonald RA (2017) Conflict in invasive species management. Front Ecol Environ 15(3):133–141

Cusack JJ, Kohl MT, Metz MC, Coulson T, Stahler DR, Smith DW, MacNulty DR (2020a) Weak spatiotemporal response of prey to predation risk in a freely interacting system. J Anim Ecol 89(1):120–131

Cusack JJ, Duthie AB, Minderman J, Jones IL, Pozo RA, Rakotonarivo OS, Redpath S, Bunnefeld N (2020b) Integrating conflict, lobbying, and compliance to predict the sustainability of natural resource use. Ecol Soc 25(2):13

Cusack JJ, Bradfer-Lawrence T, Baynham-Herd Z, Castelló y Tickell S, Duporge I, Hegre H, Moreno Zárate L, Naude V, Nijhawan S, Wilson J, Zambrano Cortes DG (2021) Measuring the intensity of conflicts in conservation. Conserv Lett:e12783 d'Harcourt E, Ratnayake R, Kim A (2017) How can the sustainable development goals improve the lives of people affected by conflict? Bull World Health Org 95(2):157

De Pourcq K, Thomas E, Elias M, Van Damme P (2019) Exploring park–people conflicts in Colombia through a social lens. Environ Conserv 46(2):103–110

DeFries R, Nagendra H (2017) Ecosystem management as a wicked problem. Science 356(6335):265–270

Dickman AJ, Hazzah L (2016) Money, myths and man-eaters: complexities of human–wildlife conflict. In: Problematic wildlife. Springer, New York, pp 339–356

Dresse A, Fischhendler I, Nielsen JØ, Zikos D (2019) Environmental peacebuilding: towards a theoretical framework. Cooperation Conflict 54(1):99–119

Duthie AB, Cusack JJ, Jones IL, Minderman J, Nilsen EB, Pozo RA, Rakotonarivo OS, Van Moorter B, Bunnefeld N (2018) GMSE: an R package for generalised management strategy evaluation. bioRxiv. https://doi.org/10.1101/221432

EC (2020) Natura 2000. https://ec.europa.eu/environment/nature/natura2000/index_en.htm. Accessed 26 Mar 2020

Eklund A, López-Bao JV, Tourani M, Chapron G, Frank J (2017) Limited evidence on the effectiveness of interventions to reduce livestock predation by large carnivores. Sci Rep 7(1):1–9

FAO (2017) Tackling climate change through rural women's empowerment. Food and Agriculture Organisation of the United Nations. http://www.fao.org/3/ca0178en/CA0178EN.pdf

Fernández-Llamazares Á, Western D, Galvin KA, McElwee P, Cabeza M (2019) Historical shifts in local attitudes towards wildlife by Maasai pastoralists of the Amboseli ecosystem (Kenya): insights from three conservation psychology theories. J Nat Conserv. https://doi.org/10.1016/j.jnc.2019.125763

Fox AD, Elmberg J, Tombre IM, Hessel R (2017) Agriculture and herbivorous waterfowl: a review of the scientific basis for improved management. Biol Rev 92(2):854–877

Gerique A, López MF, Pohle P (2017) Sitting on a ticking bomb? A political ecological analysis of conservation conflicts in the Alto Nangaritza Valley, Ecuador. DIE ERDE J Geogr Soc Berlin 148(2–3):134–149

Halbrendt J, Gray SA, Crow S, Radovich T, Kimura AH, Tamang BB (2014) Differences in farmer and expert beliefs and the perceived impacts of conservation agriculture. Glob Environ Change 28:50–62

Hanley N, Czajkowski M, Hanley-Nickolls R, Redpath S (2010) Economic values of species management options in human-wildlife conflicts hen harriers in Scotland. Ecol Econ 70(1):107–113

Hanley N, Shogren J, White B (2019) Introduction to environmental economics. Oxford University Press, Oxford

Harris J, Mirande C (2013) A global overview of cranes: status, threats and conservation priorities. Chin Birds 4 (3):189–209

Harrison HL, Kochalski S, Arlinghaus R, Aas Ø, Bailey M (2019) Do you care about the river? A critical discourse analysis and lessons for management of social conflict over Atlantic salmon (*Salmo salar*) conservation in the case of voluntary stocking in Wales. People Nat 1(4):507–523

Hegre H, Allansson M, Basedau M, Colaresi M, Croicu M, Fjelde H, Hoyles F, Hultman L, Högbladh S, Jansen R (2019) ViEWS: a political violence early-warning system. J Peace Res 56(2):155–174

Hicks CC, Levine A, Agrawal A, Basurto X, Breslow SJ, Carothers C, Charnley S, Coulthard S, Dolsak N, Donatuto J (2016) Engage key social concepts for sustainability. Science 352(6281):38–40

Hill CM e, Webber AD e, Priston NEC e (2017) Understanding conflicts about wildlife: a biosocial approach, 1st edn. Berghahn Books, New York

Hodgson ID (2018) A conflict with wings: understanding the narratives, relationships and hierarchies of conflicts over raptor conservation and grouse shooting in Scotland. Doctoral dissertation. University of Aberdeen

Hodgson ID, Redpath SM, Fischer A, Young J (2019) Who knows best? Understanding the use of research-based knowledge in conservation conflicts. J Environ Manag 231:1065–1075

Hodgson ID, Redpath S, Sandstrom C, Biggs D (2020) The state of knowledge and practice on human-wildlife conflicts. The Luc Hoffman Institute

IUCN (2020a) *Anser erythropus*. The IUCN Red List of threatened species. https://www.iucnredlist.org/species/22679886/23603064. Accessed 24 Mar 2020

IUCN (2020b) IUCN SSC human-wildlife conflict task force: what we do. http://www.hwctf.org/about/whatwedo. Accessed 23 Feb 2020

Jani V, De Wit AH, Webb NL (2019) Disputes, relationships, and identity: a 'levels of conflict' analysis of human-wildlife conflict as human-human conflict in the mid-Zambezi valley, northern Zimbabwe. S Afr Geogr J 102(1):59–76

Jones NA, Ross H, Lynam T, Perez P (2014) Eliciting mental models: a comparison of interview procedures in the context of natural resource management. Ecol Soc 19(1):1

Jones I, Whytock R, Bunnefeld N (2017) Assessing motivations for the Illegal Killing of Lesser White-fronted Geese at Key Sites in Kazakhstan

Koopman S (2011) Alter-geopolitics: other securities are happening. Geoforum 42(3):274–284

LeBillon P, Duffy RV (2018) Conflict ecologies: connecting political ecology and peace and conflict studies. J Polit Ecol 25(1):239–260

Lederach J (2003) Key issues in peace and reconciliation studies. Centre for the Study of Forgiveness and Reconciliation

Lipper L, Thornton P, Campbell BM, Baedeker T, Braimoh A, Bwalya M, Caron P, Cattaneo A, Garrity D, Henry K (2014) Climate-smart agriculture for food security. Nat Clim Change 4(12):1068–1072
LPO (2020) Informations about migration sites and common cranes. https://champagne-ardenne.lpo.fr/grue-cendree/grus-en. Accessed 26 Mar 2020
Lund MS (1996) Early warning and preventive diplomacy. In: Managing global chaos. United States Institute of Peace Press, Washington, DC, pp 379–402
Mackenzie CA, Ahabyona P (2012) Elephants in the garden: financial and social costs of crop raiding. Ecol Econ 75:72–82
Madden F, McQuinn B (2014) Conservation's blind spot: the case for conflict transformation in wildlife conservation. Biol Conserv 178:97–106
Marchant JH, Musgrove AJ (2011) Review of European flyways of the Lesser White-fronted Goose Anser erythropus. BTO Res Rep 595:2
Mason TH, Pollard CR, Chimalakonda D, Guerrero AM, Kerr-Smith C, Milheiras SA, Roberts M, Ngafack RP, Bunnefeld N (2018) Wicked conflict: using wicked problem thinking for holistic management of conservation conflict. Conserv Lett 11(6):e12460
Mathevet R, Peluso NL, Couespel A, Robbins P (2015) Using historical political ecology to understand the present: water, reeds, and biodiversity in the Camargue Biosphere Reserve, southern France. Ecol Soc 20 (4):17
Matilainen A, Pohja-Mykrä M, Lähdesmäki M, Kurki S (2017) "I feel it is mine!" – psychological ownership in relation to natural resources. J Environ Psychol 51:31–45
Mburu J, Birner R, Zeller M (2003) Relative importance and determinants of landowners' transaction costs in collaborative wildlife management in Kenya: an empirical analysis. Ecol Econ 45(1):59–73
Metternich NW, Çiflikli G, Ali A (2019, March 28) Predicting the severity of civil wars: an actor-centric approach. SocArXiv
Mkonyi FJ, Estes AB, Msuha MJ, Lichtenfeld LL, Durant SM (2017) Local attitudes and perceptions toward large carnivores in a human-dominated landscape of northern Tanzania. Hum Dimens Wildl 22(4):314–330
Montràs-Janer T, Knape J, Nilsson L, Tombre I, Pärt T, Månsson J (2019) Relating national levels of crop damage to the abundance of large grazing birds: implications for management. J Appl Ecol 56 (10):2286–2297
Moon K, Blackman DA, Adams VM, Colvin RM, Davila F, Evans MC, Januchowski-Hartley SR, Bennett NJ, Dickinson H, Sandbrook C, Sherren K, St. John FAV, van Kerkhoff L, Wyborn C, Ellison A (2019a) Expanding the role of social science in conservation through an engagement with philosophy, methodology, and methods. Methods Ecol Evol 10 (3):294–302
Moon K, Guerrero AM, Adams VM, Biggs D, Blackman DA, Craven L, Dickinson H, Ross H (2019b) Mental models for conservation research and practice. Conserv Lett 12(3):e12642
Moon K, Adams VM, Cooke B (2019c) Shared personal reflections on the need to broaden the scope of conservation social science. People Nat 1(4):426–434
Mosimane AW, McCool S, Brown P, Ingrebretson J (2014) Using mental models in the analysis of human–wildlife conflict from the perspective of a social–ecological system in Namibia. Oryx 48 (1):64–70
Myerson RB (2013) Game theory. Harvard University Press, Cambridge
Nilsson L, Bunnefeld N, Persson J, Månsson J (2016) Large grazing birds and agriculture—predicting field use of common cranes and implications for crop damage prevention. Agric Ecosyst Environ 219:163–170
Nilsson L, Bunnefeld N, Persson J, Žydelis R, Månsson J (2019) Conservation success or increased crop damage risk? The Natura 2000 network for a thriving migratory and protected bird. Biol Conserv 236:1–7
Noga SR, Kolawole OD, Thakadu OT, Masunga GS (2018) 'Wildlife officials only care about animals': farmers' perceptions of a Ministry-based extension delivery system in mitigating human-wildlife conflicts in the Okavango Delta, Botswana. J Rural Stud 61:216–226
Nyhus PJ, Osofsky SA, Ferraro P, Madden F, Fischer H (2005) Bearing the costs of human-wildlife conflict: the challenges of compensation schemes. Conserv Biol Ser 9:107
Peterson MN, Peterson MJ, Peterson TR, Leong K (2013) Why transforming biodiversity conservation conflict is essential and how to begin. Pac Conserv Biol 19 (2):94–103
Pozo RA, Coulson T, McCulloch G, Stronza AL, Songhurst AC (2017) Determining baselines for human-elephant conflict: a matter of time. PLoS One 12(6):e0178840
Raik DB, Wilson AL, Decker DJ (2008) Power in natural resources management: an application of theory. Soc Nat Resour 21(8):729–739
Rakotonarivo OS, Jones IL, Bell A, Duthie AB, Cusack J, Minderman J, Hogan J, Hodgson I, Bunnefeld N (2020) Experimental evidence for conservation conflict interventions: the importance of financial payments, community trust and equity attitudes. People Nat. https://doi.org/10.1002/pan3.10155
Ravenelle J, Nyhus PJ (2017) Global patterns and trends in human-wildlife conflict compensation. Conserv Biol 31(6):1247–1256
Redpath SM, Young J, Evely A, Adams WM, Sutherland WJ, Whitehouse A, Amar A, Lambert RA, Linnell JDC, Watt A, Gutierrez RJ (2013) Understanding and managing conservation conflicts. Trends Ecol Evol 28 (2):100–109
Redpath SM, Gutiérrez RJ, Wood KA, Young JC (2015a) Conflicts in conservation: navigating towards solutions. Cambridge University Press, Cambridge

Redpath SM, Bhatia S, Young J (2015b) Tilting at wildlife: reconsidering human-wildlife conflict. Oryx 49 (2):222–225

Redpath SM, Keane A, Andren H, Baynham-Herd Z, Bunnefeld N, Duthie AB, Frank J, Garcia CA, Mansson J, Nilsson L, Pollard CRJ, Rakotonarivo OS, Salk CF, Travers H (2018) Games as tools to address conservation conflicts. Trends Ecol Evol 33 (6):415–426

Rodríguez I, Inturias ML (2018) Conflict transformation in indigenous peoples territories: doing environmental justice with a 'decolonial turn'. Dev Stud Res 5 (1):90–105

Rogers P (2015) Peace research and conservation conflicts. In: Conflicts in conservation: navigating towards solutions. Cambridge University Press, Cambridge p 168

Runge CA, Watson JE, Butchart SH, Hanson JO, Possingham HP, Fuller RA (2015) Protected areas and global conservation of migratory birds. Science 350(6265):1255–1258

Rust NA, Abrams A, Challender DW, Chapron G, Ghoddousi A, Glikman JA, Gowan CH, Hughes C, Rastogi A, Said A (2017) Quantity does not always mean quality: the importance of qualitative social science in conservation research. Soc Nat Resour 30 (10):1304–1310

Salvi A (2010) Eurasian crane (*Grus grus*) and agriculture in France. In: Cranes, agriculture and climate change, Muraviovka Park, Russia, pp 65–70

Sandbrook C, Adams WM, Büscher B, Vira B (2013) Social research and biodiversity conservation. Conserv Biol 27(6):1487–1490

Schoonenboom J, Johnson RB (2017) How to construct a mixed methods research design. Kolner Z Soz Sozpsychol 69(2):107–131

Smith A, Sainsbury K, Stevens R (1999) Implementing effective fisheries-management systems–management strategy evaluation and the Australian partnership approach. ICES J Mar Sci 56(6):967–979

Songhurst A (2017) Measuring human–wildlife conflicts: comparing insights from different monitoring approaches. Wildl Soc Bull 41(2):351–361

St John FA, Steadman J, Austen G, Redpath SM (2019) Value diversity and conservation conflict: lessons from the management of red grouse and hen harriers in England. People Nat 1(1):6–17

Sterling EJ, Betley E, Sigouin A, Gomez A, Toomey A, Cullman G, Malone C, Pekor A, Arengo F, Blair M (2017) Assessing the evidence for stakeholder engagement in biodiversity conservation. Biol Conserv 209:159–171

Studds CE, Kendall BE, Murray NJ, Wilson HB, Rogers DI, Clemens RS, Gosbell K, Hassell CJ, Jessop R, Melville DS (2017) Rapid population decline in migratory shorebirds relying on Yellow Sea tidal mudflats as stopover sites. Nat Commun 8(1):1–7

Sundberg R, Eck K, Kreutz J (2012) Introducing the UCDP non-state conflict dataset. J Peace Res 49 (2):351–362

Thirgood S, Redpath S (2008) Hen harriers and red grouse: science, politics and human–wildlife conflict. J Appl Ecol 45(5):1550–1554

Thompson PS, Amar A, Hoccom DG, Knott J, Wilson JD (2009) Resolving the conflict between driven–grouse shooting and conservation of hen harriers. J Appl Ecol 46(5):950–954

Tindall D, Piggot G (2015) Influence of social ties to environmentalists on public climate change perceptions. Nat Clim Change 5(6):546–549

Treves A, Chapron G, López-Bao JV, Shoemaker C, Goeckner AR, Bruskotter JT (2017) Predators and the public trust. Biol Rev 92(1):248–270

UN, United Nations (2015) Transforming our world: the 2030 Agenda for Sustainable Development (A/RES/70/1)

UNDP (2013) Overview of linkages between gender and climate change. [Policy Brief], New York, U.S. https://www.undp.org/content/dam/undp/library/gender/Gender%20and%20Environment/PB1-AP-Overview-Gender-and-climate-change.pdf

UNEP/AEWA (2018) Agreement on the conservation of African-Eurasian Migratory Waterbirds (AEWA). UNEP/AEWA Secretariat, Bonn, Germany, pp 1–41

Vargas SP, Castro-Carrasco PJ, Rust NA (2019) Climate change contributing to conflicts between livestock farming and guanaco conservation in Central Chile: a subjective theories approach. Oryx:1–9

Vaske JJ, Manfredo MJ (2012) Social psychological considerations in wildlife management. Human dimensions of wildlife management, pp 43–57

Veríssimo D, Campbell B (2015) Understanding stakeholder conflict between conservation and hunting in Malta. Biol Conserv 191:812–818

Von Essen E, Allen MP (2017) Reconsidering illegal hunting as a crime of dissent: implication for justice and deliberative uptake. Crim Law Philos 11 (2):213–228

Von Essen E, Hansen HP, Nordström Källström H, Peterson MN, Peterson TR (2014) Deconstructing the poaching phenomenon: a review of typologies for understanding illegal hunting. Br J Criminol 54 (4):632–651

Wallensteen P (2018) Understanding conflict resolution. Sage, London

White PC, Ward AI (2011) Interdisciplinary approaches for the management of existing and emerging human–wildlife conflicts. Wildl Res 37(8):623–629

Wright JS (2019) Re-introducing life history methodology: an equitable social justice approach to research in education. In: Research methods for social justice and equity in education. springer, new york, pp 177–189

Young JC, Marzano M, White RM, McCracken DI, Redpath SM, Carss DN, Quine CP, Watt AD (2010)

The emergence of biodiversity conflicts from biodiversity impacts: characteristics and management strategies. Biodivers Conserv 19(14):3973–3990
Young JC, Thompson DB, Moore P, MacGugan A, Watt A, Redpath SM (2016a) A conflict management tool for conservation agencies. J Appl Ecol 53 (3):705–711
Young JC, Searle K, Butler A, Simmons P, Watt AD, Jordan A (2016b) The role of trust in the resolution of conservation conflicts. Biol Conserv 195:196–202
Young JC, Rose DC, Mumby HS, Benitez-Capistros F, Derrick CJ, Finch T, Garcia C, Home C, Marwaha E, Morgans C (2018) A methodological guide to using and reporting on interviews in conservation science research. Methods Ecol Evol 9(1):10–19

Transcending the Boundaries of Conservation and Community Development to Achieve Long-Term Sustainability for People and Planet

E. M. Naro and L. L. Lichtenfeld

Abstract

In the current epoch of the *Anthropocene*, developing communities must be a driving force for positive environmental change. This chapter focuses on overcoming the trade-offs between environmental efforts and human development. It aims to provide the necessary tools to transcend the boundaries of conservation and community development. We define and describe social-ecological systems (SES) and the functional mindset change that must take place for practitioners and environmental managers to imagine a new conservation paradigm. We outline several "stages" of community engagement and strategies to employ at each stage, recognizing that flexibility is a crucial aspect of adjusting these strategies to differing contexts. Lastly, we describe three categories of problems to be addressed in community conservation and how to appropriately diagnose these problems through a case study from African People & Wildlife.

Keywords

Conservation · Community · Social-ecological systems · Development · Community-based conservation · Tanzania

Introduction

The community of Loibor Siret in Manyara, Tanzania, is home to about 5000 people, many of whom lead the traditional pastoralist lifestyle of the Maasai tribe. Loibor Siret's communal pastures border the eastern side of Tarangire National Park, which is unfenced to allow for the natural migratory patterns of Tanzania's vast diversity of wildlife. This open corridor between the protected area and communal grazing lands, while ideal for conservation outcomes, leaves opportunity for conflict between people and wildlife. As in many pastoral cultures around the world, the Maasai living near Tarangire National Park often lose livestock to depredation by hyena (*Hyaenidae* sp.), leopard (*Panthera pardus*), lion (*Panthera leo*), African wild dog (*Lycaon pictus*), and cheetah (*Acinonyx jubatus*).

For pastoralists reliant on their livestock for food, income, and social status, conflict with predators is a significant problem (Mkonyi et al. 2017c). This problem is exacerbated by an increasing human population, cattle overstocking, climate change-induced shifts in seasonality, and competition between people and wildlife for limited water resources in the Tarangire-Manyara ecosystem (about 15,500 km^2). Historically, many pastoralists in Loibor Siret focused on techniques to reduce this conflict by fortifying their cattle corrals, or bomas, with thorn brush gathered from the surrounding area or retaliating

E. M. Naro · L. L. Lichtenfeld (✉)
African People & Wildlife, Arusha, Tanzania
e-mail: enaro@tanzaniapeoplewildlife.org; Llichtenfeld@africanpeoplewildlife.org; https://africanpeoplewildlife.org

S. C. Underkoffler, H. R. Adams (eds.), *Wildlife Biodiversity Conservation*,
https://doi.org/10.1007/978-3-030-64682-0_2

against predators by setting traps, poisoning carcasses, and spearing (Mkonyi et al. 2017b). While it is often young men who partake in retaliatory killings, women are primarily responsible for securing the boma and homestead against predators.

Frequent boma fortification causes a myriad of conservation, development, and women's empowerment challenges. Thorn brush bomas that require frequent maintenance can lead to conservation challenges due to the necessary shrub collection and tree cutting. Areas surrounding bomas demonstrate deforestation and degradation characteristics, fragmenting the landscape and requiring women to travel further from the boma to gather necessary maintenance materials (Mkonyi et al. 2017a). Further, women spend a significant amount of time gathering thorn brush and rebuilding gaps in the boma, thus resulting in limited time for other income-generating or social activities. And finally, when predators do attack a boma, blame is often put on the women responsible for boma maintenance, leading to potential domestic abuse or loss of social capital.

The scenarios above are far from unique for people living near protected areas and coexisting with predators. Similar cases can be found around the world, including outside Yellowstone National Park where cattle ranchers struggle to live with gray wolves (Nelson et al. 2016), near the Bandhavgarh Tiger Reserve in India where tigers are responsible for nearly 82% of livestock depredation events (Chouksey et al. 2018), and by Parque Nacional Queulat in Chile where farmers report frequent conflict with pumas (Rodriguez et al. 2019).

For these people, along with communities all around the world suffering from human-wildlife conflict (HWC), conservation and human development seem to be mutually exclusive goals. Tolerance of predators is sometimes low, and their value is often perceived to be primarily related to tourism (APW 2019). In conventional discourse on the topic, trade-offs between conservation and community development goals are inherent (Oldekop et al. 2016). Yet in a time of human population growth, climate change, and biodiversity loss, seeking to transcend these trade-offs and find solutions that benefit both people and wildlife is imperative. This chapter focuses on tools for pushing the boundaries of conservation and development, identifying areas where win-win solutions are possible, and imagining a new conservation paradigm in which human and natural systems are in balance.

Learning Objectives

Many of our existing environmental challenges today stem not from poor resource management at a local level, but rather from resource overuse in the most developed countries on the planet. The systems used to manage resources a thousand years ago may not be feasible or significant enough to solve our current problems in a world with nearly 8 billion humans and a rapidly changing climate. Thus, we recognize the need for formally protected areas, international regulations, and large-scale environmental action. But this does not detract from the necessity of community-level efforts, nor does it suggest that traditional systems are outdated or ineffective. Instead, it provides an opportunity to amplify the voices of local people, scale indigenous knowledge, and inspire adaptation and innovation.

Humanity's profound influence on every ecosystem on earth is such that small-scale conservation initiatives or those of protected areas in isolation—while very effective—are not sufficient on their own. These efforts must be grounded in a global movement toward sustainable living that does not merely move environmental destruction to less visible landscapes but removes it entirely. In the new epoch of the *Anthropocene*, developing communities must be a driving force for environmental change.

This chapter aims to provide the necessary tools to transcend the boundaries of conservation and community development. We begin with identifying some primary challenges faced in the past when working at the intersection of conservation and community development and those that remain to be addressed. We define and describe social-ecological systems (SES) and the functional mindset change that must take place for practitioners and environmental managers to imagine a new conservation paradigm. We outline several "stages" of community engagement and strategies to employ at each stage, recognizing that flexibility is a crucial aspect of adjusting these strategies to differing contexts. Lastly, we describe the three types of problems to be addressed in community conservation and how to appropriately diagnose these problems through a case study from African People & Wildlife.

Challenges to Past and Current Strategies

The history of community engagement in conservation provides modern practitioners with many lessons learned as well as pitfalls to avoid. In the nearly 50 years since conservationists have actively incorporated local people into environmental initiatives, four overarching critiques of community engagement strategies have emerged. This section, derived from *Community, Conservation, and Collaboration: A Framework for Success* (Lichtenfeld et al. 2019), explores each of these critiques and provides guidance to avoid them going forward. The full framework text can be found at africanpeoplewildlife.org/community-conservation-collaboration.

Romanticizing Community

In visiting the community of Loibor Siret, or any number of other Maasai communities, you may have a difficult time locating an individual who does not have a cell phone, seeing a shop without power, or finding a young adult without a social media presence. Despite these signs of globalization and development reaching extremely rural communities, the Maasai remain one of the most romanticized people on the planet today.

There is no barrier to the synergy between conservation and community development greater than that of romanticizing rural communities. The concept of *indigenous* has often been used synonymously with *primitive* and, by association, "ecologically noble" (Redford 1991). Where it exists, this perception further deepens the schism between community development and conservation goals. It suggests that indigenous people should not adopt modern technology or practices in the name of progress or to further their livelihoods. Cronon (1995) refers to the problematic wilderness ideology of romanticizing indigenous communities until they "do something unprimitive, modern, and unnatural, and thereby fall from environmental grace" as early as 1995, though it is still one of the most common pitfalls in community engagement today.

Homogenous Perspectives

From NGO websites to government reports (and even within this very chapter), the term "community" is often used vaguely to refer to local people. While this is sometimes for the sake of brevity, it can have the problematic consequence of oversimplifying the social system within that community. Oversimplifying community can lead to homogenizing the needs and perspectives of the individuals within it, thereby ignoring marginalized groups, further disenfranchising the vulnerable, and intensifying existing social imbalances (Agrawal and Gibson 1999; Waylen et al. 2013). Practitioners seeking to transcend the boundaries of conservation and community development must actively combat this homogenization by embracing complexity in social systems. Much of this chapter is dedicated to techniques and tools that can be used to assess different perspectives, understand social systems, and recognize complexity even in small, rural communities.

Imbalance of Benefits

As discussed above, conventional thought on conservation and community development is that the two are naturally at odds. Trade-offs are considered unfortunate (though necessary) management decisions for environmentalists and development practitioners alike (Alcorn 1993). Thus, the third overarching critique of community engagement strategies discussed here is the assumption that benefits cannot be balanced. Many programs throughout the history of conservation and development have claimed benefits for people and nature, but rarely can they claim that these benefits have been equitable (Oldekop et al. 2016). Conservation programs may marginally benefit a nearby community, and development programs may "do no harm" to the environment, but long-term sustainable benefits for both have been challenging to achieve.

Yet community-driven conservation programs are becoming more prevalent, and practitioners are becoming more knowledgeable about their social and ecological impact. Cases of win-win scenarios are emerging, and best practices to achieve these outcomes are being disseminated more and more frequently (Mkonyi et al. 2017a). This chapter seeks to contribute to the body of literature by continuing to demonstrate that community development and conservation need not contradict each other.

Illusion of Involvement

This last critique encompasses parts of the previous three. Many conservation initiatives in the past have sought to involve communities in some way. Yet some have either romanticized a community's history of natural resource management, homogenized their perspectives by reducing participation to only one subgroup in a community, or focused so heavily on the ecological outcomes that any planned social benefits were lost. This critique is often the result of well-intentioned conservationists seeking to engage communities but without the resources or experience to do so meaningfully.

The appearance of community participation in programs does not reflect legitimate investment in positive conservation outcomes. It does not reflect community ownership of programs or empowerment to begin new initiatives. Most importantly, it does not mean that such programs will be sustainable in the long term. Genuine community engagement requires much more than participation in meetings and a few local staff implementing activities. It requires long-term relationship building, mindset and behavior change, and a deep sense of trust and partnership between the community and the practitioners. In the long term, it requires that the community becomes the practitioner.

Social-Ecological Systems

In contrast to the problematic human-nature dualism that has dominated conservation thought for decades, social-ecological systems thinking provides a model for integrating social, human-centered systems with natural and ecological systems. Humans are, in fact, no less a part of the natural world than our closest primate relatives (Schmitz 2018). A social-ecological system is "the integration of both the natural and social sciences with systems theory, recognizing the interdependent interactions between biology, ecology, and social systems" (Lichtenfeld et al. 2019). Understanding the world through a social-ecological systems model allows for practitioners to see the complexity inherent in the system without becoming overwhelmed by it. Embracing this complexity is crucial for developing adaptable solutions to both conservation and development problems. Here, we parse out the varied elements in most social-ecological systems, providing examples of each and demonstrating the interaction effects between them.

Social Aspects

We begin with the social elements of a social-ecological system. Fundamentally, social systems revolve around individuals, groups, and institutions. Although groups and institutions are both comprised of individuals, the legal, economic, or political power of larger social structures can amplify or dilute the values of the component individuals. Thus, groups and institutions can be considered separate elements in social systems with values derived from, though not identical to, the values of individual members. This distinction between individuals and the larger structures they create can manifest in "mob mentality"—where the values of individuals within a group are compounded by other members and the group espouses a stronger form of those values than each individual member—or a "social contract," where individual instincts are weakened to conform to the values of the group or institution and thereby maintain social order.

More broadly, social systems incorporate the principles required for the system to function, the relationships between each element, and the values espoused at each level. The following section delves deeper into each of these concepts.

Principles Required for Social Systems to Function

Unsurprisingly, the principles paramount in functioning social systems reflect the principles conservation practitioners must also bear in mind when engaging with rural communities. Many lists of social system principles exist and are applicable in all societal levels from individual workplaces to small communities to international organizations. Most compilations of principles include variations of the following:

- Flexibility
  Social systems require flexibility to function in the long term. Social systems must remain adaptable to changing conditions and responsive to the needs of the individuals, groups, and institutions within them and adjust in a timely and relevant manner.
- Trust
  Social systems require a level of trust between individuals, groups, and institutions to function effectively. Each element in a social system must trust that inclusion in the system will provide more benefits than exclusion. Trust is strongest when the benefits for each element are transparent, equitable, and reciprocal.
- Participation
  Social systems are most functional when they incorporate a significant percentage of the individuals, groups, and institutions within them. Participation in the system should be appealing, while removal from the system should seem detrimental to the success of the individual, group, or institution.
- Respect
  Social systems function more effectively when each element in the system has value to add, recognizes the importance of the other elements, and seeks to uphold the system as a whole. When elements in a social system lose respect for other elements or for the system itself, the system is vulnerable to loss of participation or collapse.

Equality vs. Equity: Relationships Between Elements of a Social System

In all social systems, when these principles are not upheld, dissent and discord emerge. A lack of flexibility may reduce the level of trust that individuals, groups, and institutions have in the system, leading to decreased respect and participation. A common issue with inflexible social systems is the growth of inequity between individuals, groups, or institutions, often caused by the confounding of equity with equality.

Inequity and inequality are not interchangeable terms. Inequity refers to injustice or unfairness, whereas inequality refers to uneven distribution of services or resources. For example, a social program which provides the same financial support to all people regardless of their current financial needs, such as an economic stimulus, ensures equality but not equitability. On the other hand, a social program that specifically assists the most financially vulnerable people seeks to create equity, though the financial support is not equal. In short, equity implies fairness, while equality implies sameness.

Confusing the two terms and thereby the policies implemented to address them can exacerbate existing social and economic imbalances. If a social system is not flexible enough to shift between policies addressing inequality and those addressing inequity, the system is vulnerable.

When working in community development, it is critical for the components of a social system to address inequity. The institutions and groups in place to protect the values of individuals must respond to the inequitable distribution of resources. As these resources are often environmental or ecological, an SES model helps define the relationship between institutions, individuals, and natural resources.

Values in a Social System and the Governance Structures Upholding Them

People create governance institutions as a means to protect their base values, and differing base values among individuals are the foundation of different political groups, parties, and governance structures. Understanding the function of governance structures is a critical component of understanding broader social-ecological systems. Regardless of the type and structure of government in power, the social purpose and function of governance is to preserve and protect core human values (Clark and Wallace 2015).

These values, described by Lasswell and depicted in Fig. 1, exist in all individuals, though the ratio between these values reflects the different perspectives those individuals (and the groups or institutions they comprise) espouse (Lasswell 1970). For instance, some individuals may value wealth significantly more than skill, while others may value affection above all. Understanding individuals' core value ratios can provide unparalleled insight into their decision-making process. On a social system level, understanding group and institution core value ratios can highlight the priorities of the entire society. Where value ratios differ greatly between components of a social system, we can expect to see political or religious conflict at its worst, or healthy, compromising debate at its best.

As human-created institutions, governance structures are dependent on the individual to retain authority and command respect. If governance institutions fail to preserve the values of the individual (or some portion of individuals depending on the type of government), the individuals will seek a change in authority by means of a transition of power mechanism. Historically, these transitions of power can be anywhere on the spectrum from entirely peaceful to violent overthrow. Regardless of the mechanism through which individuals seek change in authority, principles of good governance and human values are intimately linked and interdependent (Lasswell 1970; Clark and Clark 2002; Lockwood 2010).

Economics of Natural Capital

The last element of a social system crucial to the comprehension of an SES is the economic element. Social systems and economic systems are intimately linked, influencing each other and co-evolving. A society's economic system impacts the mindset of the individuals within it and the policies implemented by its institutions. For instance, socialist economic systems influence the development of individuals within those systems from birth. Capitalist economic systems do likewise, affecting the values, goals, and priorities of the people educated and employed in those systems.

The economics of a social system depend heavily on the human and natural resources in that system. As stated above, the human resources are molded through membership in the system and are driven by core human values. The natural resources both are a driver of the economic system and can be impacted by that system in turn. To truly grasp the social aspects of an SES, we must understand the concept of natural capital.

Natural capital includes all of the world's natural assets, that is, natural resources and the ecosystem services they provide (Costanza and Daly 1992). Since natural capital is an input in all products humans use to survive and thrive, it must be included in economic models and considered in economic decisions (Fenichel et al. 2018). However, many economic systems in the developed world, and increasingly in the developing world, undervalue natural capital or fail to recognize its limits. All social systems today have institutions that allow individuals to buy, trade, or use products. However, many of these systems do not account for the raw natural capital inputs when valuing these products. For instance, the

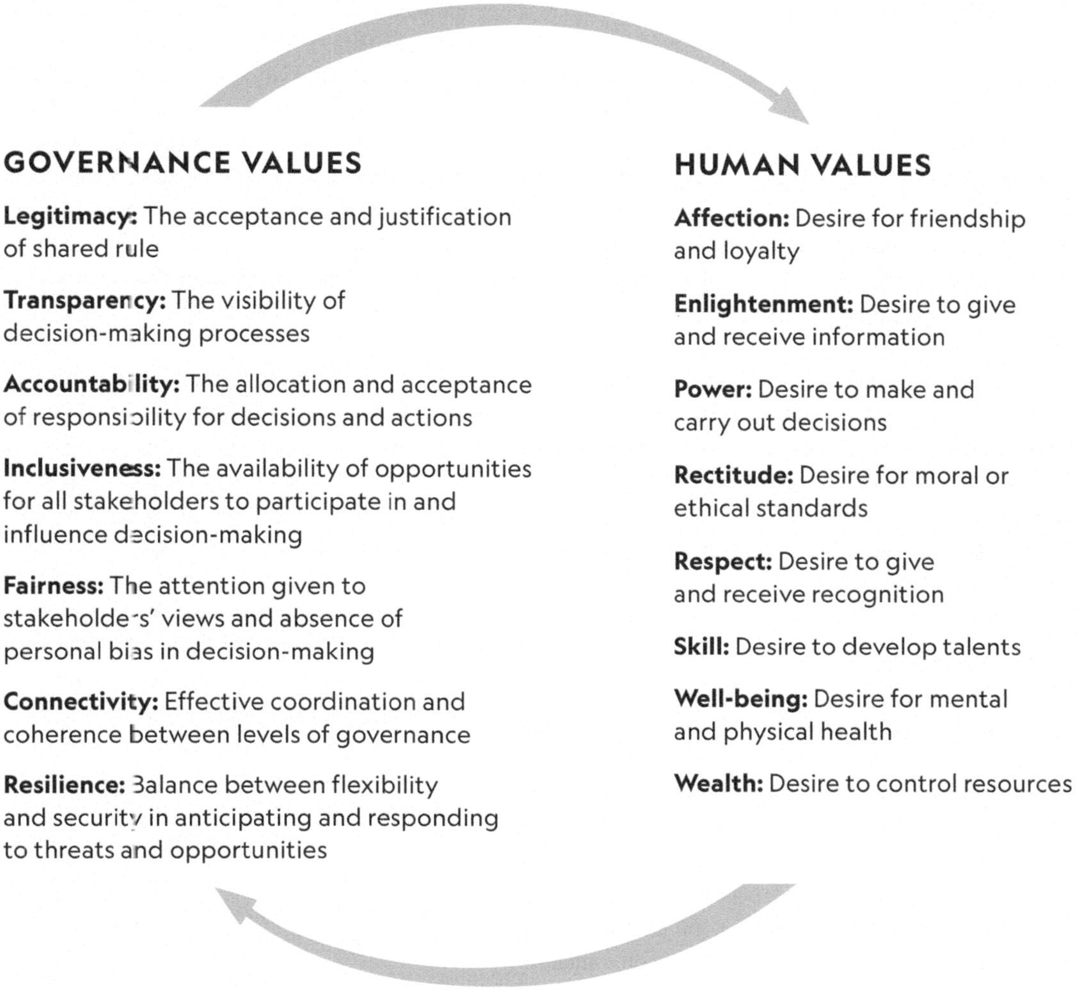

Fig. 1 Interdependent relationships between governance values and human values

cost of a piece of furniture may include the value of the timber, human labor, and associated trade fees. Yet it often overlooks the value of the ecosystem services lost in removing the timber-producing tree from an ecological system and the value of the soil that enabled the tree to grow in the first place. While this may be a small issue for individual products or communities, undervaluing natural capital on a global scale can create an economic system collapse in the long term.

Such non-optimized economic systems can lead to the inefficient or inequitable distribution of resources, the collapse of ecosystem services, and the degradation of critical natural stocks, ultimately causing significant losses in value and well-being (Fenichel et al. 2018). Thus, conservation practitioners seeking to understand the social-ecological systems in which they work must incorporate economic value modeling into their initial assessments. The discussion of the economics of natural capital provides a seamless transition between the social and ecological aspects of an SES. In the next section, we delve deeper into the ecological aspects of an SES, keeping in mind the natural capital value of ecosystem services and how their loss can affect social systems.

Ecological Aspects

As discussed above, the link between social and ecological aspects of an SES often lies in natural capital. Societies are dependent on the natural resources surrounding them, and these natural resources are significantly impacted by the people who use them. Generally, the ecological elements of an SES can be categorized into living resources, non-living resources, and the interactions between them. Importantly, the interactions between living and non-living elements of an ecological system constitute a complex adaptive system (CAS). A CAS is a system in which each component may change and have unpredictable effects on the whole (Lichtenfeld et al. 2019). Several examples of CAS are described below.

Living Resources

We begin with discussing living resources, recognizing that humans are both a living resource in ecological systems and also the primary element in social systems. Living resources also include much more than commonly referenced animal species; they include all extant plant life, fungi, algae, and bacteria. All living elements have some dependency on non-living elements and, in most cases, reliance on other living resources.

There is a wealth of diversity in the living elements in an ecological system, from variety in feeding patterns to locomotion to suitable habitat. Life on earth has been found in even the most unlikely places, like the thermophilic bacteria found in the hot springs of Yellowstone (Meyer-Dombard et al. 2005) or the tadpole shrimp (*Triops granarius*) found in the temporary pools of Qatar (Shama 1997). This diversity is both beautiful and functional. The result of millennia of evolution, functional diversity in living elements ensures the homeostasis of ecological systems. Further, each species' niche allows it some degree of distinction or competitive advantage in a certain context.

Within functional clades and intraspecies, particularly in biodiversity hotspots, negative density dependence is relatively common (Peters 2003; Johnson et al. 2012). This trait refers to the influence of species density on its survival. For instance, if many seedlings of a particular species or species with similar functions are densely located, they are likely to compete for the same limited resources and have a low survival rate. The great diversity in function and habitat requirements among living elements is necessary for the survival of many species, including humans.

Lastly, the diversity of living elements ensures whole ecosystem functioning. Living elements are commonly categorized by feeding type or trophic level. As a general though not perfect rule, each trophic level has an ecological efficiency of approximately 10% of the previous level. In other words, there are about ten times as much biomass in primary producers as there is in herbivores and about ten times as much biomass in herbivores as there is in carnivores in any given community (Pimm and Lawton 1977). This forms the basis of the relationship between trophic levels and highlights the importance of understanding trophic cascades. Trophic cascades represent a CAS in which a change in one trophic level can have an exponentially more significant (or cascading) effect on other levels (Schmitz 2003; Ripple et al. 2016). For example, if apex predators are removed from an ecosystem, hunting pressure on herbivores is reduced, which can lead to increased herbivory and, in turn, degraded grasslands or riparian areas. As these areas provide critical habitat for a myriad of other species, many of which are necessary pollinators and seed dispersers, their degradation has cascading effects on the entire ecological system. In short, the relationships between trophic levels are critical to the healthy functioning of ecological systems, and understanding them is a vital part of understanding SES.

Non-living Resources

Next, we look into the non-living resources in an ecological system. These include air, water, minerals, and sunlight, among others. As mentioned above, the sustainability of non-living resources is necessary for all life. Yet unlike living beings, non-living elements cannot migrate to ensure their sustainability or make decisions about their existence.

Although each non-living resource could have books dedicated specifically to it (and most do), we will focus on water as it exemplifies the complexity and variability in ecological systems. Water covers about 71% of the earth's surface, with oceans holding over 96.5% of that water (USGS 1984). The rest exists as water vapor in the air (0.001%), in rivers and lakes (0.007%), as ice in glaciers and the polar ice caps (1.74%), in groundwater and permafrost (1.71%), and of course in all living things (0.0001%) (Gleick 1993).

While the earth's water supply is constant, the state and location of that water are ever-changing. The water cycle includes many of the processes observable on a daily basis: precipitation, infiltration, evaporation, transpiration, sublimation, cloud transportation, and condensation (NASA 2019). It also includes less tangible processes such as percolation, plant uptake, groundwater flow, runoff and snowmelt (NASA 2019). These processes regulate the amount of water accessible to living things, and even small rate shifts can cause significant ecological change. For instance, changes in rainfall, sunlight, or soil infiltration can have extreme effects on living resources, such as grasslands including desertification, flooding, erosion, or general degradation. In turn, this affects the herbivores dependent on those grasses. These effects can further cascade through the trophic levels, impacting the entire system and even neighboring systems as living elements migrate to survive.

In this time of climate change, a focus on the non-living elements of an ecological system is more important than ever. Changes in weather patterns, the water cycle, soil composition, and even coastal boundaries threaten living elements in the system. Despite many species' adaptability, rapidly changing climatic conditions could cause mass extinction. Yet, many non-living elements are in abundance or renewable, even if not equally distributed (of course, this brings back the concept of equity in natural capital, since no part of an SES exists in a vacuum). So, these non-living resources can also serve as solutions to many of the challenges caused by their changing conditions. Solar and wind energy are some of the most common examples of this today. Still, living elements in ecological systems were engineering non-living resources to their benefit long before modern energy issues.

Interactions Between Ecological Elements

The last element of an ecological system critical to the understanding of SES is the interaction between living and non-living resources. Several examples have already been provided, including the relationship between water, sunlight, grasslands, and herbivores. However, no discussion of ecological systems is complete without incorporating the concept of ecosystem engineering.

Ecosystem engineering is the process by which living elements in an ecological system significantly modify their habitat through interactions with non-living or other living resources (Jones et al. 1994). These interactions are sometimes parasitic but often mutually beneficial and can result in coevolution over time. Ecosystem engineering is another example of a CAS, as the long-term and cascading effects of living elements on their environment are often unpredictable and interdependent.

A classic example of ecosystem engineering is the creation of natural dams by beavers (*Castor canadensis*). In much of North America, beaver dams drastically alter riparian ecosystems and create wetlands. The formation of wetlands as a result of beaver activity increases habitat heterogeneity, which in turn increases species richness on a landscape scale (Wright et al. 2002). This escalation in diversity facilitates a positive feedback loop; more plant diversity means more habitat options for a further increase in species richness. Thus, the interactions between beavers, water, and woody plants create whole ecosystems.

There are innumerable examples of ecosystem engineering, from frugivorous primates dispersing significant amounts of fruit tree seeds (Chapman et al. 2013) to plankton altering the light regimes in marine ecosystems and thereby controlling the depth at which other organisms can survive (Breitburg et al. 2010). But perhaps no species in history has had more impact on the environment than *Homo sapiens*. Humans have served as ecosystem engineers by drilling for oil to generate energy; setting aside specific areas of land for tourism, hunting, or conservation; and undertaking a variety of actions resulting in climate change. And with this recognition that humans are both the primary actors in social systems and extremely influential actors in ecological systems, we can begin to understand social-ecological systems in the Anthropocene.

Systems Thinking

The previous two sections have referenced social systems, ecological systems, and complex adaptive systems. But we have not yet explored the last aspect of SES—the systems thinking approach. Systems thinking requires a functional understanding of the components as described above, but more importantly, it requires analytical capacity and innovation. It is comparatively easy to recognize issues in a system when it has begun to show symptoms of corruption. It is much harder to predict challenges that have not yet demonstrated detectable signs, through simply reading the signals in the system.

Systems thinking provides a lens through which individuals can read these signals, anticipate unintended outcomes, and adapt as needed. It is a holistic approach to recognizing non-linear cause and effect models, interrelated and interdependent constituents, and how systems work in the context of larger systems (Schmitz 2018). Often, systems thinking results in a mechanism called adaptive management (Cundill et al. 2012). From savvy investors to conservation practitioners, the systems thinking approach is key to successful initiatives.

Scenario 1

Consider a hypothetical SES in a primarily pastoralist buffer zone of a protected area which houses high-profile predator species. Due to high unemployment rates in the area, alcohol dependence is high, and young people in a community often seek financial resources through whatever means available. Since the protected area is so close, the region has a significant tourism industry. To facilitate tourism and boost the economy of the region, the government partners with foreign investors to build infrastructure. In the foreign investor's home country, there is a high demand for illegal wildlife products, particularly the skins and bones of big cats. Poaching syndicates, therefore, target the vulnerable youth of the community for recruitment into poaching operations. As a result, over the course of a decade, the population of large felines in the region plummets. With fewer apex predators in the ecosystem, herbivore populations rise. The increased herbivory on grasslands along with a drying climate causes the buffer zone pastures shared by wildlife and livestock to degrade. Despite restoration efforts, overgrazing pressure threatens the pastoralists' livelihoods, as their livestock compete for limited pasture resources. Many pastoralists begin growing corn and beans, hoping to supplement their income with agriculture. As more land is converted from pasture to agriculture, crop-raiding from elephants increases. Frustration in the community rises as economic opportunities seem to disappear, and livelihoods and cultural values are threatened. This frustration leads to increased vulnerability among the youth to recruitment by poaching syndicates.

The system described in scenario 1 may be hypothetical, but it represents real events and interactions that occur in social-ecological systems around the globe. There are several noteworthy points to be considered in this example:

- The social aspects of the system include lifestyle and livelihood preferences of individuals, economic conditions outside the control of the individuals, global forces that impact local ecological systems, and human values that drive decision-making.
- The ecological aspects of the system include trophic cascades, top-down controls of predators, negative density dependence of wild herbivores and livestock, and changes in the water cycle and soil composition as a result of land allocation shifts.
- The systems thinking aspect of the case involves the unpredictable effect of improved infrastructure on organized crime, the unanticipated increase in human-wildlife conflict as a result of lifestyle change, and the feedback loops of unemployment-generated boredom increasing alcohol dependency and consequently increasing vulnerability of rural youth.

In such complex and unpredictable systems, conservation and development practitioners must seek solutions that address the social and ecological problems simultaneously. Understanding SES is a key first step in transcending the boundaries of community development and conservation. By viewing the challenges we face through an SES lens, we are more likely to anticipate unintended outcomes, recognize when serious issues are left unaddressed, and remain cognizant of the impact of programming on all life, human and wild.

Community Engagement Approach

Understanding the theory behind social-ecological systems is entirely different from putting that theory into practice. In this section, we provide actionable steps for practitioners seeking to transcend the boundaries of community development and conservation via African People & Wildlife's ACTIVE approach to community engagement, found in Fig. 2. While these are listed and described in a certain order, it is critical to remember that community engagement is a flexible process and no stepwise model can encompass its complexity. Therefore, these steps should be viewed more as guidelines of an approach rather than chronological actions.

Access

Accessing a community with which to partner for conservation initiatives requires time. This may mean spending months building relationships with key community members, attending community events, and listening to community needs before bringing up the topic of conservation programming. An initial assessment of the community and general interests can be formal or informal, but is undoubtedly a necessity.

Perhaps the most foundational step in accessing a community is to recognize the feature(s) that bind the community together. Communities share a geographic location, but usually, there are much deeper ties than locality. History, culture, language, religion, ethnicity, and race are just a few of the potential binding factors within a community. These factors often lead to similarities in core human values, as discussed previously.

Of course, to avoid the pitfall of homogenizing community perspectives, initial assessments should seek to identify how community subgroups differ, i.e., what factors bind subgroups together and how do the core human values of the various subgroups complement or contradict each other. Accepting and even embracing complexity in communities is key to understanding the social aspects of the SES in which the community exists.

Stakeholder analysis is a strong place to begin exploring the differential value systems in a society. There are a myriad of tools and resources for conducting stakeholder analyses, formally or informally. African People & Wildlife's community engagement toolkit can be found at https://africanpeoplewildlife.org/community-conservation-collaboration, with tools 5.1 and 5.2 providing guidance on stakeholder analysis. A thorough stakeholder analysis will include an assessment of all three components of a social

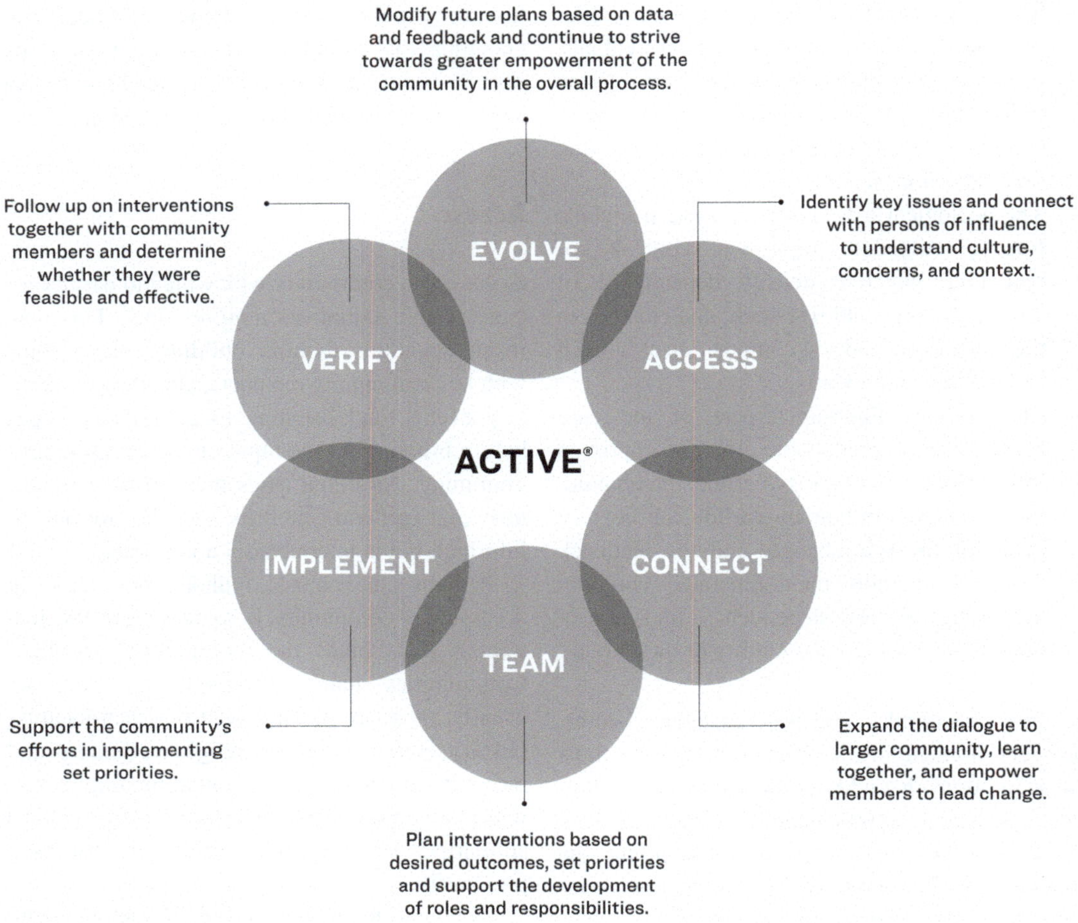

Fig. 2 African People & Wildlife ACTIVE approach to community engagement

system: individuals, groups, and institutions. By assessing multiple levels of a societal structure, we can better recognize the relationships between the components and thereby appreciate the complexity of the system.

Most stakeholder analysis guides recommend beginning with a brainstorming session to identify stakeholders for assessment. Throughout the process of assessment, however, more stakeholders may emerge, while others initially included may seem irrelevant. Maintaining flexibility is key.

Value assessment is perhaps the most useful form of stakeholder analysis for understanding the social aspect of an SES. For assessing individuals and groups (and to some extent institutions), Lasswell's eight core human values can be applied. Each stakeholder can be rated by how much weight it gives each of the eight values, described in Fig. 1. From these rating, value ratios can be determined and compared between stakeholders.

Power ranking is another form of stakeholder analysis that is particularly crucial for understanding the relationships between components of a social system. The power ranking model allows users to compare the political, economic, social, and cultural power of various stakeholders. While those four categories can be broadly applied in most contexts, other power dimensions can be added to fit the context of the social system (additional power dimensions could include legal, religious, financial, etc.). See Box 1 for more details on power ranking.

Box 1 Power Ranking and Radargrams

Power ranking tools often ask the assessor to rank each stakeholder on a scale of 1 to 5, with 1 demonstrating very little influence of the power dimension in question and 5 demonstrating significant influence in that dimension. Radargrams provide a useful way to visualize the results of a power ranking stakeholder analysis.

- *Political* power refers to a stakeholder's influence over local, national, or international governance, laws, and political systems.
- *Economic* power applies to the stakeholder's ability to influence funding as well as their power over local markets and economies.
- *Social* power relates to the stakeholder's influence on societal structures and behavior through thought leadership, perceived importance, or general respect within a community.
- *Cultural* power relates to the influence a stakeholder can have over the beliefs, practices, traditions, and values of a society as well as individuals.

To truly access a community, practitioners may require the support of community champions. A community champion is a member of the target community who is knowledgeable about community structure, has connections with community leadership, and is willing to facilitate introductions to other people. The community champion(s) can ensure that practitioners have access to the appropriate people to begin discussions about conservation programming.

Connect

With the facilitation of a community champion and the foundational understanding of community dynamics, practitioners can begin connecting with the broader community and leadership. Visioning workshops with communities can be beneficial for strengthening relationships and establishing a presence in the community. Similar to the strategies for inclusive participation discussed above, there are several strategies practitioners can use to connect with communities during visioning workshops and foster strong partnerships.

- Joint facilitation: Visioning workshops should be facilitated jointly by community leadership and the practitioner. This allows community members to be more connected to the process rather than observing the process happening around them.
- Language inclusion: Translators should be present to allow community members from different subgroups the opportunity to understand the proceedings and voice their perspectives.
- Disaggregation: Especially in communities with distinct subgroups and vulnerable or minority populations, offering separate visioning workshops for these groups can ensure they feel comfortable voicing their opinions. In particular, offering workshops specifically for women can be beneficial.
- Mirroring: Conduct visioning workshops in a manner similar to any regular community meeting. Mirroring the location, agenda, and protocol of existing structures can ensure that practitioners are genuinely connected to the community decision process and not distinct from it.

By the end of visioning workshops, of which there may be many, the practitioner and the community should have a clear sense of the other's values, goals, and objectives. Creating a shared goal document is a positive way to begin a partnership. The practitioner may have conservation goals beyond those of the community, and the community may have development goals beyond those of the practitioner. However, many goals will likely overlap or complement one another. Focusing on these shared goals, which address both conservation and development needs, is a key initial step in transcending the boundaries between conservation and development.

Box 2 Ensuring Inclusivity

For practitioners attempting to transcend the boundaries of conservation and community development, including the perspectives of marginalized groups is a necessity. Yet while physical inclusion is frequent via community meetings or stakeholder engagement forums, real and active participation of vulnerable populations is more challenging.

Several strategies can be used to ensure that all relevant components of a social system are included substantially in conservation and community development initiatives.

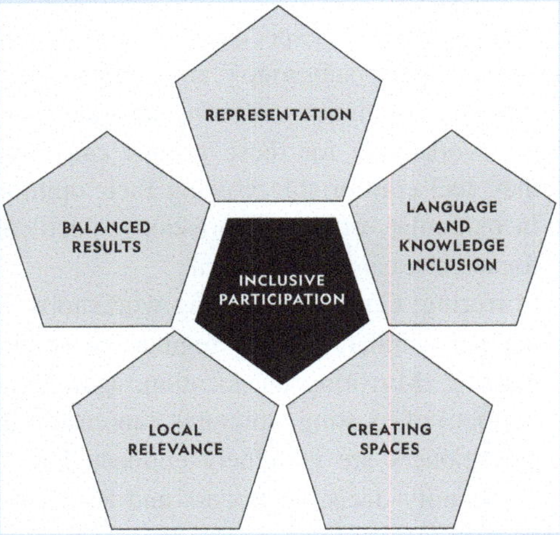

1. Practitioners should reach out to minority members of a social system (individuals, groups, or institutions) and create opportunities for those members to access initiatives. This may include providing transportation, relocating events to more rural areas, or offering separate gender-disaggregated meetings.
2. Practitioners should ensure that translators are present at initiative meetings or activities and encourage marginalized social groups to contribute traditional knowledge. Allow for this knowledge to be shared via music, dance, story, etc., rather than just through formal writings or records.
3. Practitioners should be available frequently and actively present in a social system. Both informal and formal spaces for feedback, idea sharing, and debate should be accessible for marginalized or vulnerable populations.
4. Practitioners should assure that conservation initiatives are co-designed with broad representation from members of the social system, particularly to link conservation goals with community development needs. Initiatives are more successful when they are relevant to both majority and minority groups in a system.
5. Practitioners should aim to balance the results of programming so that marginalized members are not excluded. This relates to the distinction between equal and equitable benefit distribution. For instance, initiatives explicitly aimed at women or minorities can strengthen the trust those members have in the social system and its benefits.

Team

Practitioners should aim to team with communities, rather than implement programs at or for them. This begins with defining shared goals, as described above, and continues with jointly assessing strengths, weaknesses, opportunities, and threats. SWOT analysis was designed initially for business ventures but can easily be applied to conservation and development work, organizational capacity, and community engagement. This commonly used strategic planning technique helps committees or other planning teams to understand their assets, capacities, needs, and challenges.

The following questions can be used by a facilitator to lead the SWOT process. This process should include community leadership, practitioners, and other community stakeholders who may be involved with joint programming.

- *S*trengths: What ecological, social, political, and economic assets do we have as a community and as individual subgroups?
- *W*eaknesses: What ecological, social, political, and economic capacities do we lack, which may limit our ability to achieve our goals? This area of discussion could culminate in highlighting potential partnerships that could fill these gaps.
- *O*pportunities: What external ecological, social, political, and economic factors can we use to achieve our goals?
- *T*hreats: What external ecological, social, political, and economic challenges may threaten programming and the achievement of our goals?

In addition to identifying key factors for potential program success, conducting SWOT analyses as a team can deepen the trust communities have in the practitioner and vice versa. Results from a SWOT analysis are most valuable when shared transparently with community members and used to inform program design.

Co-designing activities with communities is a natural next step after conducting a SWOT analysis. Many practitioners may already have activities in mind, based on values, past programs, or donor requirements. However, implementing pre-determined programs in a community is a sure way to strengthen the barriers between conservation and community development. Pre-determined programs are likely to overlook community needs and values, not align with existing governance structures, and ultimately lose the support of community members and leadership. Instead, practitioners and their community partners must undergo the process of backward mapping to truly co-design activities. See Box 3 for more detail on backward mapping.

Box 3 Co-designing Activities Through Backward Mapping

Backward mapping is the process by which we create a theory of change. A theory of change is a cause-and-effect hypothesis, or model, for how a suite of activities will achieve their goals. Backward mapping begins with the shared goals determined during visioning workshops and, using the assets and capacities identified during SWOT analysis, generates ideas for how to achieve those goals.

Usually led by a facilitator, this process works backward from shared goals through cause-and-effect logic until arriving at implementable activities. Often, theories of change are visual depictions of this logic, with building blocks, or boxes, describing the inputs, outputs, outcomes, and eventual impact of activities.

Developing a theory of change jointly with a community is a necessary aspect of a community-driven conservation model. It ensures that both community and practitioner interests are considered and hence that both community development and conservation goals are addressed. A theory of change can be used to visualize the extent to which a program will transcend the boundaries of community development and conservation. By coloring community development outcome boxes in black, for instance, and coloring conservation outcome boxes in white, you can easily determine if programs are balancing benefits or are focused heavily on one side. Of course, some activities may be inherently social and others may be inherently ecological, but the program as a whole should strive for balance. This may mean discussing additional activities that address the gaps in holistic programming.

See Section "Recognizing Technical, Systemic, and Constitutive Problems in Community-Driven Conservation Initiatives" for real-world examples of how African People & Wildlife co-designs holistic programming with rural communities in Tanzania to transcend the boundaries of conservation and community development.

Implement

Implementing activities jointly with communities can be an extremely rewarding process for both the practitioner and the community members involved. To ensure that the community is invested in project success, activities should be led by community members whenever possible. Local capacity building should be an active part of every project, including the capacity of local people to lead and implement projects. Project and team management is a skill in itself; thus, it is critical for local staff members to be given the resources and mentorship they need to succeed in management roles.

Work plans for individual activities will differ substantially depending on the activity and the community context. However, some features of a work plan will remain constant. For instance, an activity timeline should be included in a work plan along with a section defining the roles and responsibilities of both practitioner and community team members. A useful example of a role definition tool is the RACI chart. A RACI chart outlines who is responsible, accountable, consulted, or informed of all activity tasks and decisions.

- *R*esponsible: The people performing the task
- *A*ccountable: The project manager or other entity (usually reporting back to donors)
- *C*onsulted: Community members or groups who may have knowledge or expertise about the activity and its implementation
- *I*nformed: Community members or groups who may be affected by the activity

To ensure that activities are community-driven and not merely tolerated by the community, practitioners should play a role of support, guidance, and advocacy. Community members should more often act as implementers, project managers, and technical leads. When done well, this model of community-driven conservation can create positive change for the environment while simultaneously developing the skills of the community. Having community members implement activities and benefit from them builds enthusiasm for joint conservation and community programs and ensures long-term sustainability of initiatives.

Verify

Early in this chapter, we discussed the challenge of balancing benefits for the community with benefits for the environment. This challenge forms the base of the question this chapter seeks to answer: how do we transcend the boundaries of community development and conservation?

The next "step" in the ACTIVE community engagement approach—verify—is critical to addressing this question. Thus far, the practitioner and the community leadership have defined shared goals, co-designed activities, and begun implementing projects aimed at providing benefits for both people and planet. But these good intentions do not inherently create balanced benefits and positive change. The logic in a theory of change must be verified as the project is implemented.

Verifying a theory of change requires routine monitoring. For communities to be genuinely invested in project success, they must see measurable change in their livelihoods and their environment. Thus, involving community members in the process of monitoring program outcomes is critical. Of course, not every building block of a theory of change can be monitored. There is too much complexity in a theory of change for most projects and not enough resources to accurately monitor all outcomes. Rather, it is helpful to focus monitoring efforts on parts of a theory of change that are new, are innovative, or have inherent assumptions.

This is particularly important when attempting to balance community development and conservation outcomes. In many projects, the logic leading to social benefits is tried and true (e.g., income-generating activities). In other projects, the logic leading to conservation benefits is clear and measurable (e.g., demarcating protected areas). Yet when projects are intended to have balanced benefits, it is useful to test the assumptions that these benefits are accruing. For instance, an environmental enterprise program may have obvious social benefits as an income-generating activity, but the environmental benefits may be questionable. Thus, practitioners should develop indicators specifically to monitor and test the environmental effects of the program.

In the long run, being able to demonstrate the positive effects of a program for both people and planet can ensure the sustainability and scalability of that program. This makes verification through monitoring one of the most essential factors in transcending the boundaries of conservation and community development.

Evolve

Last, but certainly not least, is the process of evolving with a community. Program evolution could occur organically or be the result of decisions informed by program evaluation. Whereas regular monitoring allows practitioners and communities to test their output and outcome-level assumptions, program evaluation can provide answers to higher-level questions about program effectiveness, impact, and sustainability.

Through both monitoring and evaluation efforts, practitioners and community partners should seek to learn and adapt. Conservation and community development have ever-shifting priorities and goals. Likewise, the social and ecological aspects of an SES are always changing due to political, economic, and ecological events outside of our control. To ensure the long-term sustainability of program benefits for both people and nature, programs must be flexible, open to evaluative critique, and responsive to the changing needs of communities and the environments on which they depend.

This "step" in community engagement recalls the need for practitioners to maintain a systems thinking approach to natural resource management and be open to system-level change. This means both understanding the social-ecological systems in which we work and also recognizing how program interventions affect these systems. Interventions and their effects on an SES constitute components of another complex adaptive system (CAS). Hence, an understanding of the individual components of a CAS—for instance, understanding how an intervention works—does not imply an understanding of the whole system due to the interdisciplinary and interdependent relationships between the components.

The component of a CAS over which we have control, that is, the intervention, must be flexible and adaptable to unpredictable changes in the rest of the system. Importantly, practitioners must seek to accurately diagnose the cause of these changes or barriers to effectiveness. The next section focuses on problem diagnosis and recognizing when problems are truly technical or if they are embedded in deeper systemic and constitutive challenges.

Recognizing Technical, Systemic, and Constitutive Problems in Community-Driven Conservation Initiatives

One of the most significant barriers to transcending the boundaries of conservation and community development is the misdiagnosis of problems. There are three types of problems commonly recognized in the policy sciences (Clark and Clark 2002):

- *Technical:* focus on the ground-level logistical or ecological issues
- *Systemic:* focus on the broader governance challenges at an institutional level
- *Constitutive:* focus on the roots of the human condition based on historical, cultural, and psychological factors

For effective problem-solving in the conservation field, we must have a better method for accurate problem definition. Often, problems are misdiagnosed as technical when, in reality, there are deeper systemic and constitutive issues underpinning the technical symptoms. Without addressing these governance and cultural challenges, the technical problems will continue to surface. Addressing only the technical problems is similar to treating the symptoms of an illness but not the root cause.

To ensure the long-term sustainability of program outcomes, problems must be explored from the surface level down to the root causes. When working with rural communities in conservation, reaching to these constitutive causes can take years. Short-term projects to address technical issues must be layered with medium-term programs to address systemic challenges and long-term initiatives invested in mindset and behavior change. The following section describes African People & Wildlife's model and demonstrates how technical, systemic, and constitutive issues can be addressed simultaneously to achieve sustainable impact for communities and conservation.

African People & Wildlife Case Study

Let's explore the case study of African People & Wildlife's work in Tanzania introduced at the beginning of this chapter with a problem orientation lens.

Technical Problems

The case from the beginning of this chapter demonstrates a technical challenge that can be addressed with technical solutions. This easily identifiable technical challenge is human-wildlife conflict. This challenge is commonly discussed in village meetings, pastoralists will describe their experiences openly, and a technical solution is feasible: reduce depredation of livestock.

Since 2005, African People & Wildlife has been working closely with rural communities in Tanzania to develop a technical solution to human-wildlife conflict. In 2008, the first Living Wall was built in Loibor Siret village. A Living Wall is an environmentally friendly corral that keeps livestock safe from predators. To build a living wall, community members plant a circle of trees to serve as posts for chain-link fencing. As the trees grow, they add height to the wall and create an impenetrable barrier. As of 2019, over 1100 Living Walls have been constructed across 35 villages in northern Tanzania. Depredation at the boma has decreased by 90% in some of these villages, and retaliatory killings have subsequently plummeted (Lichtenfeld et al. 2015).

Living Walls have proved to be an effective technical solution to a technical problem. And yet Living Walls alone cannot ensure sustainable livelihoods for the people of Loibor Siret nor ensure the long-term persistence of big cat populations. There are deeper challenges at play that threaten coexistence between people and wildlife. Thus, after years of relationship building with the community of Loibor Siret, and generating trust and respect through technical problem-solving, African People & Wildlife began exploring the systemic challenges.

Systemic Problems

Systemic, or governance, challenges can be much more difficult to identify than technical challenges. Their solutions are proportionally more challenging. Yet with stakeholder analysis, in-depth discussions with communities, and a history of trust and mutual respect, practitioners can explore and seek to resolve systemic issues.

In Loibor Siret, and in many other pastoral communities, a growing population and shifting land use priorities result in limited pasture resources for both livestock and wildlife. The fragmentation and loss of habitat due to increased settlement and agricultural expansion force wildlife and humans into closer proximity, increasing opportunities for conflict. Lack of pasture resource monitoring or effective land use planning exacerbates these issues, and in turn, political conflict proliferates. At a higher level of systemic issues, pastoral communities may not have the legal tenure over their rangeland to manage it appropriately. If these issues are simply rendered as technical, human-wildlife conflict may be addressed at the surface, but the deeper problems of habitat loss and natural resource management capacity may be ignored.

African People & Wildlife seeks to address the systemic issues at play through a sustainable natural resource monitoring and management program, which is implemented jointly between the village of Loibor Siret and the NGO. This program seeks to strengthen the existing governance structures in the village by building their capacity to make evidence-based decisions about natural resource management and land use planning. Working closely with African People & Wildlife and additional partners, community-level governance institutions implement land use plans that allow for income-generating agriculture while preserving communal pastures for both livestock and wildlife. Further, the program supports community members to monitor these pastures through a combination of indigenous knowledge and modern technology. Monitoring the rangeland health and productivity allows land use committees at the village level to use real-time data to allocate rangeland for livestock at strategic times of the year. In this way, limited pasture resources can be preserved for livestock, and vital wildlife habitat is protected.

Several other NGOs in the region focus on land tenure issues through existing legal mechanisms at a higher level of governance. By first ensuring that communities have rights over their natural resources and then assisting with the sustainable management of those resources, the systemic challenges around land use and habitat loss can be mitigated.

Constitutive Problems

Lastly, and most challenging to identify and address, are cultural or constitutive problems. These types of problems underpin all other systemic and technical challenges. They frame the way people think, behave, and relate to each other. They are the basis of people's value systems and priorities. Accurately identifying constitutive problems requires deep trust-building and ethnography and can often take years. Addressing these problems can take generations.

At the heart of most constitutive problems related to the environment is the human relationship with nature. The distinction between people and animals, developed and wild, modern and primitive, has permeated human cultures since time immemorial. Yet this dualism forms the foundation of all the environmental challenges we face today. Of course, this human-nature dichotomy has been intensified in much of the developed world and is historically the basis of the conservation field. As a result, many rural communities today, particularly those living in and around protected areas, have a mistrust of conservation work and environmental institutions. It is these constitutive challenges that must be addressed to transcend the boundaries of human development and conservation.

African People & Wildlife's organizational model is focused on addressing these constitutive problems. Just as understanding constitutive problems requires long-term and deep engagement with a community, solving these problems requires more than an individual program. It requires that an entire organizational philosophy be aimed at altering the current conservation paradigm. Thus, all of African People & Wildlife's programs follow a *community-driven* conservation model, not just a community-based model.

To begin addressing the mistrust of conservation initiatives and the imbalanced power dynamic between rural communities and environmental institutions in northern Tanzania, African People & Wildlife's headquarters is based in the village of Loibor Siret. This gives all staff, many of whom are from the community, an opportunity to personally connect with the pastoral communities living among predators, grazing their livestock on limited resources, and raising their children in coexistence with wildlife. All African People & Wildlife programs are co-designed and co-implemented with community members and leadership. And African People & Wildlife seeks to support community-driven environmental initiatives, such as watershed management or environmental enterprise, through small grants provision, capacity enhancement, and mentorship. Through this type of holistic engagement, practitioners can begin to repair relationships with those who have suffered environmental injustices.

The even deeper-seated constitutive problem of the human-nature relationship requires a different type of engagement—the type that lasts a lifetime. Hence, African People & Wildlife focuses on mindset and behavior change programs specifically tailored to youth. With a long-term impact objective of creating a community-level conservation ethic, African People & Wildlife supports wildlife clubs in primary and secondary schools, hosts environmental summer camps, and brings students to the nearby protected area, Tarangire National Park, to instill an appreciation for the natural world that reaches beyond the fear of predators.

For decades, the conservation field has implemented education programs that focus on the value of nature and wildlife. Yet this value has often been expressed in terms of wildlife as an economic asset to the country. While this is important information to impart on rural communities, it strengthens the divide between people and nature, demonstrating how wildlife have economic value when tourists visit national parks but little inherent value outside of those government-managed protected areas. Since those protected areas are rarely accessed by local people, they are seen as areas for tourists and the wildlife within them as tourist attractions. Thus, even with people who historically have an environmental ethic (such as the Maasai), the conservation education model of the last 30 years has furthered the divide between humans and the environment.

African People & Wildlife's youth environmental education program seeks to change this model by demonstrating the non-economic value of wildlife and reinvigorating the traditional value of coexistence. For instance, environmental camp activities with youth are aimed at educating students about the ecological roles of apex predators and the top-down trophic controls they impart on the landscape. Particularly in pastoralist communities, where natural controls on wild herbivores are necessary to maintain healthy pastures for livestock, these lessons can shift the way youth view predators. Rather than wanting predators to remain in the national parks where they can only be seen by tourists, young pastoralists can grow with a greater tolerance for their presence in community rangelands and an appreciation of their ecological role. Combining this constitutive strategy with a technical one to reduce human-wildlife conflict can greatly decrease the fear of predators while rebuilding the relationship between people and nature as one of interdependence and stewardship.

African People & Wildlife's holistic approach to remedying constitutive problems includes another long-term program—scholarships. The cost of both private and government secondary school education in Tanzania can be restrictive to many rural students, especially girls, for whom education is not considered a priority. While thousands of non-profit programs exist throughout Africa to provide funding for schools and schoolchildren, rarely do these programs invest long term in the same students—instead providing one-off funding for a semester or year of schooling with no guarantee of continuity. African People & Wildlife's scholarship program, started in 2009, provides selected students with a full 6-year tuition guarantee, enough for them to graduate from secondary school. In 2017, the program also began supporting university tuition for students who wished to continue onto higher education.

This long-term investment in student's education addresses constitutive problems in several ways: (1) it instills in students a conservation ethic and understanding that humans are part of our ecosystem, breaking down the human-nature dualism from a young age; (2) it supports continuity in education for girls, thereby counteracting deep-seated societal biases which threaten women's empowerment; (3) it does not restrict beneficiary students to higher education in an environmental field (i.e., one of African People & Wildlife's university scholars recently graduated from medical school), ensuring that future leaders in different disciplines have a strong foundation in environmental management; and (4) it provides an environmentally non-consumptive alternative to lion spearing as a method of gaining social status for young men (i.e., completing secondary school is a significant achievement for Maasai and can fulfill young men's desire for higher social status).

Jointly, African People & Wildlife's youth environmental education program and general philosophy of community-driven conservation work to address constitutive problems in rural communities in Tanzania. This model can be adopted by other practitioners around the globe and adapted to fit their community context. By not rendering all problems technical and instead actively seeking to address the root constitutive challenges in a system, practitioners can begin to transcend the boundaries of community development and conservation.

Conclusion

In this chapter, we have discussed social-ecological systems theory, explored how that theory can be put into practice, and provided real-world examples of practitioners effecting positive change. From recognizing the elements in a system to accurately diagnosing problem types, conservation practitioners must maintain a holistic SES mindset. All three types of problems—technical, systemic, and constitutive—must be addressed in all components of an SES to achieve lasting change.

Of course, conservation practitioners alone cannot realistically accomplish this. Here we make a case for interdisciplinary programming based on strong partnerships between scientists and practitioners in conservation, health, economics, ecology, development, and education. Sustainable, equitable results require a system overhaul. Human and environmental well-being are so intimately linked that no development program should be without a conservation element and no conservation program can succeed in the long term without embracing the values of environmental justice for impacted people. To truly shift the conventional paradigm of human-nature dualism to one of integrated human and environmental welfare, both a conservation ethic and a human dignity approach to programs must be instilled in all aspects of society.

We have reason to hope. Poverty and hunger have decreased significantly in the past few decades, and environmental enterprises around the globe are bringing new economic opportunities to rural communities. Innovative technological solutions to mitigate climate change are being implemented in some of the most developed countries in the world, while traditional natural resource management practices are making their way into the national policies of many developing countries.

Scenario 2

Consider a hypothetical SES in a primarily pastoralist buffer zone of a protected area which houses high-profile predator species. Due to the innovation of local women's groups, thousands of beehives have been hung in trees growing in degraded communal pastures. Since the protected area is so close, the region has a significant tourism industry. Environmentally conscious tourists seek out markets to purchase locally produced honey and share ideas with the women's groups about the environmental enterprises in their home countries. The local business expands to produce other honey products, and the women's groups hang more beehives in degraded pastures. As bees colonize the hives in these pastures, they pollinate native species, and the grasslands begin to regenerate. Pastoralists monitoring the pasture health notice the improvement and provide evidence to the local government of the successful restoration techniques. The government committees responsible for land use planning make evidence-based decisions about land allocation. As pastures regenerate, pastoralists feel confident in their livelihoods and teach their children about their cultural value of environmental stewardship. The children grow to respect the importance of predators in an ecosystem

(continued)

and take measures like building living walls to protect their livestock from depredation. Some of these youth grow to become leaders in the community, while others attend school to become the future managers of their nearby protected area.

The system described in scenario 2 may be hypothetical, but it does not have to be. It represents real events and interactions that can occur in social-ecological systems around the globe—so long as we support them.

References

Agrawal A, Gibson CC (1999) Enchantment and disenchantment: the role of community in natural resource conservation. World Dev 27:629–649

Alcorn JB (1993) Indigenous peoples and conservation. Conserv Biol 7:424–426

APW (2019) Perceptions of human wildlife conflict baseline survey. Unpublished Data, African People & Wildlife

Breitburg DL, Crump BC, Dabiri JO, Gallegos CL (2010) Ecosystem engineers in the pelagic realm: alteration of habitat by species ranging from microbes to jellyfish. Integr Comp Biol 50:188–200

Chapman CA, Bonnell TR, Gogarten JF, Lambert JE, Omeja PA, Twinomugisha D, Wasserman MD, Rothman JM (2013) Are primates ecosystem engineers? Int J Primatol 34:1–14

Chouksey S, Singh S, Pandey R, Tomer VS (2018) Monitoring the status of human-wildlife conflict and its impact on community based conservation in Bandhavgarh tiger reserve, Madhya Pradesh, India. J Appl Nat Sci 10:710–715

Clark TW, Clark SG (2002) The policy process: a practical guide for natural resources professionals. Yale University Press, London

Clark SG, Wallace RL (2015) Integration and interdisciplinarity: concepts, frameworks, and education. Policy Sci 48:233–255

Costanza R, Daly HE (1992) Natural capital and sustainable development. Conserv Biol 6:37–46

Cronon W. 1995. The Trouble with Wilderness; or. Getting Back to the Wrong Nature.

Cundill G, Cumming G, Biggs D, Fabricius C (2012) Soft systems thinking and social learning for adaptive management. Conserv Biol 26:13–20

Fenichel EP, Abbott JK, Do Yun S (2018) The nature of natural capital and ecosystem income. In: Handbook of environmental economics. Elsevier, New York, pp 85–142

Gleick PH (1993) Water in crisis. Pacific Institute for Studies in Development, Environment & Security. Stockholm Environment Institute, Oxford University Press, Oxford, 473p

Johnson DJ, Beaulieu WT, Bever JD, Clay K (2012) Conspecific negative density dependence and forest diversity. Science 336:904–907

Jones CG, Lawton JH, Shachak M (1994) Organisms as ecosystem engineers. In: Ecosystem management. Springer, New York, pp 130–147

Lasswell HD (1970) The emerging conception of the policy sciences. Policy Sci 1:3–14

Lichtenfeld LL, Trout C, Kisimir EL (2015) Evidence-based conservation: predator-proof bomas protect livestock and lions. Biodivers Conserv 24:483–491

Lichtenfeld LL, Naro EM, Snowden E (2019) Community, conservation, and collaboration: a framework for success. National Geographic Society, Washington, DC, United States and African People & Wildlife, Arusha, Tanzania

Lockwood M (2010) Good governance for terrestrial protected areas: a framework, principles and performance outcomes. J Environ Manag 91:754–766

Meyer-Dombard D, Shock E, Amend J (2005) Archaeal and bacterial communities in geochemically diverse hot springs of Yellowstone National Park, USA. Geobiology 3:211–227

Mkonyi FJ, Estes AB, Msuha MJ, Lichtenfeld LL, Durant SM (2017a) Fortified Bomas and vigilant herding are perceived to reduce livestock depredation by large carnivores in the Tarangire-Simanjiro ecosystem, Tanzania. Hum Ecol 45:513–523

Mkonyi FJ, Estes AB, Msuha MJ, Lichtenfeld LL, Durant SM (2017b) Local attitudes and perceptions toward large carnivores in a human-dominated landscape of northern Tanzania. Hum Dimens Wildl 22:314–330

Mkonyi FJ, Estes AB, Msuha MJ, Lichtenfeld LL, Durant SM (2017c) Socio-economic correlates and management implications of livestock depredation by large carnivores in the Tarangire ecosystem, northern Tanzania. Int J Biodivers Sci Ecosyst Serv Manag 13:248–263

NASA (2019) A multi-phased journey in observatory

Nelson AA, Kauffman MJ, Middleton AD, Jimenez MD, McWhirter DE, Gerow K (2016) Native prey distribution and migration mediates wolf (*Canis lupus*) predation on domestic livestock in the greater Yellowstone ecosystem. Can J Zool 94:291–299

Oldekop J, Holmes G, Harris W, Evans K (2016) A global assessment of the social and conservation outcomes of protected areas. Conserv Biol 30:133–141

Peters HA (2003) Neighbour-regulated mortality: the influence of positive and negative density dependence on tree populations in species-rich tropical forests. Ecol Lett 6:757–765

Pimm S, Lawton J (1977) Number of trophic levels in ecological communities. Nature 268:329

Redford KH (1991) The ecologically noble savage. Cult Survival Q 15:46–48

Ripple WJ, Estes JA, Schmitz OJ, Constant V, Kaylor MJ, Lenz A, Motley JL, Self KE, Taylor DS, Wolf C (2016) What is a trophic cascade? Trends Ecol Evol 31:842–849

Rodriguez V, Poo-Muñoz DA, Escobar LE, Astorga F, Medina-Vogel G (2019) Carnivore-livestock conflicts in Chile: evidence and methods for mitigation. Hum Wildl Interact 13:10

Schmitz OJ (2003) Top predator control of plant biodiversity and productivity in an old-field ecosystem. Ecol Lett 6:156–163

Schmitz OJ (2018) The new ecology: rethinking a science for the Anthropocene. Princeton University Press, Princeton, NJ

Shama A (1997) Density and diversity of the desert Arthropoda of Qatar

USGS (1984) The hydrologic cycle in survey

Waylen KA, Fischer A, McGowan PJ, Milner-Gulland EJ (2013) Deconstructing community for conservation: why simple assumptions are not sufficient. Hum Ecol 41:575–585

Wright JP, Jones CG, Flecker AS (2002) An ecosystem engineer, the beaver, increases species richness at the landscape scale. Oecologia 132:96–101

The Challenges of Conserving Biodiversity: A Spotlight on Southeast Asia

Kathryn Strang and Nathan Rusli

Abstract

Biodiversity is being lost at a rapid pace, mainly due to anthropogenic pressures from a growing human population. Southeast Asia is a biodiversity hotspot with high species endemism; however, it is also a region undergoing a biodiversity crisis. Unregulated wildlife trade, high rates of deforestation, and increasing human-wildlife conflict are threatening many Southeast Asian species. Mitigating many of these threats is difficult because the level of poverty of forest-neighbouring communities is a main driver to these activities. This, coupled with demand for wild-harvested animal products and fertile land for agriculture, has led to rapid biodiversity loss. Conservation work not only needs to mitigate these threats through applied conservation actions (e.g. restoration, protection, or reintroductions) but also needs to address these social drivers. This chapter outlines the complexities of biodiversity conservation in a Southeast Asian context and describes how a multidisciplinary approach is necessary for biodiversity conservation.

Keywords

Conservation · Habitat conversion · Poaching · Community education · Habitat restoration · Wildlife trade · Human-wildlife conflict · Habitat loss

Introduction

The world is in the midst of a biodiversity crisis, where the rate of species lost has increased to alarming levels, so much so that Earth may be undergoing a sixth mass extinction event (Thomas et al. 2004; Wake and Vredenburg 2008; Barnosky et al. 2011; Ceballos et al. 2015; McCallum 2015; Ceballos et al. 2017; Cowie et al. 2017). Many threats to biodiversity have arisen due to increased human populations and the subsequent increase in pressure on land use and natural resources. A region that encompasses many of these biodiversity challenges is Southeast Asia. With almost 9% of the world's human population, and only 3% of the world's land area (World Population Review 2020), Southeast Asia faces its own biodiversity crisis. Though it is home to over 22,000 endemic plant species and over 4000 endemic vertebrate species, less than 8.3% of the original vegetation cover remains (Myers et al. 2000).

There are many threats to the native fauna and flora of Southeast Asia. An online survey was given to individuals working in biodiversity

K. Strang (✉)
Stay Wild Tiger Protection Trust, Lankat, North Sumatra, Indonesia

N. Rusli
Indonesia Herpetofauna Foundation, Bogor, Indonesia

S. C. Underkoffler, H. R. Adams (eds.), *Wildlife Biodiversity Conservation*,
https://doi.org/10.1007/978-3-030-64682-0_3

conservation within the Southeast Asia region, with a series of questions to help identify the main issues, causes, and necessary mitigating actions in biodiversity conservation. There were 32 participants, and multiple answers were allowed for each question. Over-harvesting (including poaching and the wildlife trade) and habitat loss were identified as main threats to biodiversity conservation in Southeast Asia (Fig. 1a). Respondents also listed some of the causes for these and other concerns, such as lack of education and awareness, weak policy and enforcement, and unregulated development. The most prevalent conservation issues reported were development and a growing human population, corruption and a lack of government support for conservation programmes, and a lack of education and awareness (Fig. 1b). When asked about the efforts put in place to resolve these issues, the answer most provided was establishing education and awareness programmes (Fig. 1c).

The primary threats to conservation and causes of biodiversity decline in Southeast Asia identified in the survey are interrelated, and many of the conflicts are driven by the socio-economic background of local communities. The following discussion aims to highlight the challenges of biodiversity conservation, particularly within Southeast Asia, by discussing some of the emerging and pressing threats to species in this region.

Wildlife Trade

Two of the primary threats to biodiversity in Southeast Asia are the wildlife trade and the unsustainable harvesting of wildlife. Wildlife is exported from the region at a massive scale, with Nijman (2010) estimating that over 35 million Convention of International Trade in Endangered Species of Wild Fauna and Flora (CITES)-listed animals were exported over a 10-year period (1998–2007), 30 million of which originated from free-ranging populations. Harvesting of wildlife is not a new activity, with households using hunting as a form of sustenance in the past and present (Rao et al. 2011; Pangau-Adam et al. 2012). However, in recent years, there has been an increase in the demand for wildlife products. Subsequently, harvesting rates of free-ranging wildlife have risen beyond that which could be considered sustainable (Krishnasamy and Zavagli 2020).

Wildlife trade and poaching has mostly been associated in the media with high-profile species in the past, such as rhinoceros and elephants, despite it being a large threat to many different groups of taxa. Scheffers et al. (2019) found that almost 20% of mammals, birds, reptiles, and amphibians are affected by trade. The demand for wildlife varies, with some species being used for medicine, food, ornaments, or pets. The use of different species also varies by country. For example, Nekaris et al. (2010) found that in Cambodia, slender and slow lorises (*Nycticebus* and *Loris*) are used for traditional medicine, while in Indonesia those species are traded as pets. Knowledge of the uses and demand for different taxa may help conservationists understand the drivers behind wildlife trade, as well as to devise potential actions to mitigate against these pressures.

Demand for a particular species can lead to a rapid increase in harvesting rates (beyond sustainable levels) and push the species towards the brink of extinction, even for those considered to be common throughout their range. This is the case for three species of songbirds endemic to Indonesia: black-winged myna (*Acridotheres melanopterus*), grey-backed myna (*A. tricolor*), and grey-rumped myna (*A. tertius*) (Nijman et al. 2018). Over 20 years, these species have undergone high levels of harvesting for pet markets (Nijman et al. 2018), and while two of the *Acridotheres* species were common until recent years, all three are now Critically Endangered (Birdlife International 2018a, b, c). Three of the four pangolin (*Manis*) species residing in Asia are listed as Critically Endangered due to their populations having undergone, or expecting to undergo, an 80% decline in three generations (approximately 21 years) (Challender et al. 2019a, b; Schoppe et al. 2019). Pangolins are now the most trafficked mammal in the world, driven by demand for their scales, which are used

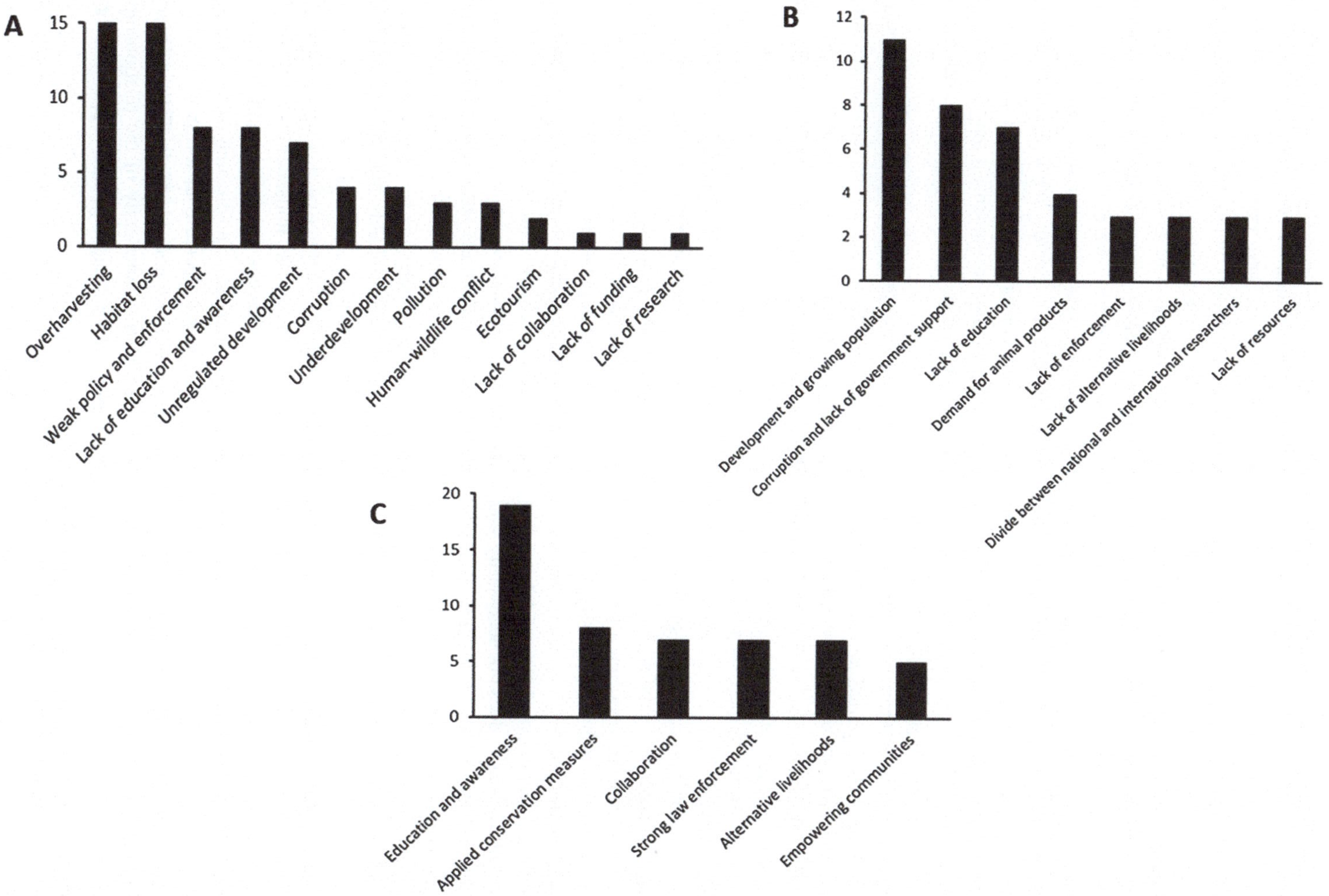

Fig. 1 Results of the survey from the 32 respondents identifying (**a**) the main threats to biodiversity conservation in Southeast Asia, (**b**) the causes of conservation issues, and (**c**) the efforts that have been put in place to resolve conservation issues.

in traditional medicine, and for their meat (Aisher 2016; Harrington et al. 2018). High harvesting pressures for the wildlife trade, often coupled with other threats such as habitat loss, have contributed to rapid reductions in free-ranging populations of wildlife.

There are various trade demands for wildlife, the most publicized being wildlife used as medicinal products (particularly tigers and rhinoceros). Some species are targeted as bushmeat, with the meat either eaten by the hunter's family or sold for income. Sandalj et al. (2016) investigated wild meat consumption in Vietnam and found that 85% of respondents had consumed wild meat at some point. Only 23% of respondents had consumed wild meat that was farmed (Sandalj et al. 2016). Interestingly, of those who did not try farmed wildlife, 14% reported an expected preference for the taste of wild over farmed meat (Sandalj et al. 2016). Dutton et al. (2011) found that wild bear bile (used for medicinal purposes) was preferred over farmed bear bile, with respondents willing to pay higher prices for wild bear bile. Dutton et al. (2011) also found that prices were dependent on the supply of farmed bear bile, with lower prices offered for wild bear bile when farmed bear bile was also supplied. The studies of Sandalj et al. (2016) and Dutton et al. (2011) highlight an important point—substituting farmed wildlife products does not necessarily lead to a lower demand of wild-harvested products. This suggests that, in order to combat the trade in wildlife, it is necessary to understand the drivers behind consumer demand for desired products.

The wildlife trade is not a simple market based on supply and demand, but is intertwined with socio-economic background, culture, and beliefs. Biggs et al. (2017) developed a theory of change framework to mitigate the drivers of wildlife trade, which highlighted the importance of strong policy and implementation and building community capacity. Although CITES (an international agreement) exists to ensure the trade of fauna and flora does not threaten their survival (CITES 2020), the actual protection of species depends on both national laws and local policy implementation. Application of conservation-based laws may be weak due to authorities having little knowledge of conservation regulations and/or a lack of resources (Lee et al. 2005). Enforcement is further complicated by corruption within government agencies, and interestingly, countries with high levels of illegal wildlife trade also tend to have poor governance (Biggs et al. 2017). In addition, Lee et al. (2005) stated that protecting species targeted for wildlife trade requires community awareness programmes, as many local communities are unaware of laws governing wildlife species. While it can be challenging for conservationists to elicit positive behavioural change in humans, it may be accomplished through well-developed outreach programmes.

Lack of law enforcement is a driver of increased wildlife trade, and poverty drives the supply of wild-harvested animals and poaching. With high levels of poverty and the lack of employment opportunities for additional income being a contributing factor to poaching (TRAFFIC 2008; Rao et al. 2011; Duffy and St John 2013), organizations have been introducing alternative livelihood projects (such as agricultural practices) that may alleviate harvesting pressure on free-ranging populations of wildlife. However, this may not necessarily translate to a reduction in poaching in all circumstances. A review of international and national experts' opinions on Southeast Asian wildlife trade stated that 80% of the time when interventions were implemented to alleviate poverty there was no significant change to the number of households engaged in poaching activities (TRAFFIC 2008). However, Wilfred and MacColl (2010) reported that a predictor of wildlife poaching for households in western Tanzania was income from crop and livestock sales. An increase in income from these livelihoods saw a decrease in levels of wildlife poaching (Wilfred and MacColl 2010). These studies show the complexity of the relationship between poverty, poaching, and supply for the wildlife trade, reinforcing the call from Duffy and St John (2013) for additional research to improve our understanding of this relationship. Nonetheless, a potential mitigating action for alleviating the pressure of poaching and unsustainable harvesting on free-ranging wildlife populations could be the introduction of alternative livelihood projects and income.

The increase in wildlife trade has seen the establishment of non-governmental organizations (NGOs), such as TRAFFIC and Free the Bears, focused on creating strong policy to combat overharvesting for trade, as well as aiding wildlife confiscations. This has led to a demand for rehabilitation centres that rescue wildlife destined for trade and rehabilitate and reintroduce wildlife into their natural habitat whenever possible. Reintroduction of confiscated wildlife is difficult and requires a high level of species-specific knowledge within an in situ and ex situ environment, as well as extensive planning (Cheyne 2009). There is also an ethical debate on rehabilitation and release of confiscated animals and their conservation value (Palmer 2018); however, until the wildlife trade has been reduced or eliminated, there will always be a need for these rescue centres.

Success in reducing both the wildlife trade and poaching must not only focus on lowering demand for animal products but must also focus on addressing the socio-economic backgrounds of citizens of rural villages living close to threatened ecosystems and work to increase community engagement. A multidisciplinary approach is needed to address policy formation and implementation, product demand, education, and alternative livelihoods. This approach would require a high level of engagement between conservationists, policy-makers, law enforcement, and local communities.

Box 1 Poaching as Income

Interviews with poachers were conducted by the NGO Stay Wild Tiger Protection Trust in North Sumatra, Indonesia. Fifteen different interviews were conducted to gather initial information on the motives for and methods of poaching in the area. The majority (80%) used poaching as supplemental income, while three poachers used hunting as their sole income. Most of the animals poached were sold within the local community, and five people reported that their families sometimes consumed the meat. Five respondents used a gun as their method of hunting, two used only snares, and seven used a combination of both guns and snares. Between five and twelve snares were set by each poacher (Fig. 2). Poachers reported catching between two and ten animals per month, with 40% catching more than four animals per month. A wide range of animals were targeted (Table 1), though deer, wild pigs, and monkeys were the most common.

While most of the animals caught were mammals, there were also two bird species. One was the white-rumped shama (*Kittacincla malabarica*), a song bird that, while currently listed by the IUCN as Least Concern, has seen population declines, likely due to increases in the caged-bird trade (Leupen et al. 2019). One poacher reported targeting hornbills (family Bucerotidae), a group of birds poached for their casques, which are used to make ornaments (Beastall et al. 2016; Phassaraudomsak et al. 2019). The casques are often referred to as "red ivory", with the price per weight exceeding that of elephant ivory (Beastall et al. 2016). Phassaraudomsak et al. (2019) reported on hornbill products sold through online portals and found that the majority (73%) of casques were from helmeted hornbill (*Rhinoplax vigil*), while the rest of the products had been harvested from eight different species, over half of which are threatened with extinction. Due to intense hunting pressures exerted on the helmeted hornbill, the species' status changed following reassessment in 2015 from Near Threatened to Critically Endangered on the IUCN Red List (Birdlife International 2019).

Respondents reported a lack of money and inability to find alternative means of supplementary income as their primary reasons for poaching. Rao et al. (2011) reported that households living in poverty

(continued)

Fig. 2 Several different snare traps used by poachers to capture animals (left and right). One snare had captured a Javan mongoose (*Herpestes javanicus*) that was able to be released into the wild (middle). Photos from Stay Wild Tiger Protection Trust

Table 1 Summary of the animals that the interviewed poachers ($n = 15$) targeted and the local price that was obtained for the animals

Animal common name	No. of respondents poaching animal	Unit	Price (IDR)	Price (USD)
Deer	10	Kilogram	70,000	$4.90
Monkey	6	Animal	20,000–30,000	$1.40–$2.10
Wild pig[a]	5	Kilogram	30,000–40,000	$2.10–$2.80
Snakes	3	Metre	50,000	$3.50
White-rumped shama[b]	2	Animal	1,000,000	$70.40
Sun bear[c]	2	Animal	1,000,000–2,000,000	$70.40–$140.80
Porcupine[d]	2	Kilogram	25,000	$1.80
Hornbill	1	Gram of casque	80,000	$5.60
Wild chicken	1	Animal	150,000	$10.60
Mouse deer[e]	1	Kilogram	25,000	$1.80
Thomas leaf monkey[f]	1	Animal	50,000	$3.50
Slow loris[g]	1	Animal	70,000–100,000	$4.90–$7.00
Leopard cat[h]	1	Animal	50,000	$3.50

The price per unit is given in both Indonesian rupiah (IDR) and converted to US dollar (USD). Species or genus names for those that could be identified: [a]*Sus scrofa*, [b]*Kittacincla malabarica*, [c]*Helarctos malayanus*, [d]*Hystrix brachyura*, [e]*Tragulus* spp., [f]*Presbytis thomasi*, [g]*Nycticebus* spp., [h]*Prionailurus javanensis* with few alternative livelihood sources relied upon hunting as a significantly higher source of income as compared to other activities such as farming. TRAFFIC (2008) reported that hunting was more likely to be conducted by low-income households as compared to middle-income or wealthy households. Nijman (2010) found that the wildlife trade was largely driven by economic forces, by generating high revenue nationally as well as by providing income to locals with few alternatives for income generation. While all of these studies, along with the interviews conducted by Stay Wild Tiger Protection Trust, demonstrate that poaching is often attributed to poverty, it is a complex issue requiring more research to understand the mechanisms behind it (Duffy and St John 2013).

Box 2 The Role of Captive Institutes in Breeding and Reintroductions

The confiscation of wild caught animals for the pet and wildlife trade has increased the demand for captive rescue and rehabilitation facilities. Captive institutes play a role in breeding captive wildlife and release of rehabilitated individuals into free-ranging populations, though reintroduction attempts are difficult, with high failure rates (Kleiman 1989; Mallinson 1995; Bowkett 2009). Captive populations are also used to safeguard the species from a genetic standpoint (Mallinson 1995).

The Cikananga Conservation Breeding Center (CCBC) in West Java was founded in 2007, with a focus on breeding and releasing species on the brink of extinction (Owen et al. 2014; Vernia et al. 2018). Much of its work is focused on endemic Indonesian songbirds, as many songbird species are threatened by increased poaching for the caged-bird trade (Nijman et al. 2018). The Javan green magpie (*Cissa thalassina*) (Fig. 3) is highly sought after by collectors due to their ability to mimic the song of other species. This has exacerbated the harvesting pressure on free-ranging populations, leading to rapid decline of the species (Eaton et al. 2015). There are less than 250 *C. thalassina* individuals estimated in the wild (Birdlife International 2018d). CCBC houses over half the global captive population of *C. thalassina* and utilizes captive breeding in an effort to safeguard the species (B. Ferns, pers. comm.). Additionally, CCBC has translocated captive individuals to European zoological collections, to further strengthen captive populations (B. Ferns, pers. comm.). Due to the presence of the ongoing threats within the species' natural habitat, captive individuals cannot yet be reintroduced to the wild.

CCBC also has a captive population of black-winged myna (*Acridotheres melanopterus*) that they breed and release into the wild (B. Ferns, pers. comm.) (Fig. 3). *A. melanopterus* are Critically Endangered with few individuals persisting throughout West Java (Shepherd et al. 2016; Birdlife International 2018a). CCBC has successfully bred this species in captivity (a difficult task for some species) and subsequently coordinated multiple reintroduction attempts. One population was reintroduced to Halimun Salak National Park, West Java, and was successful at reproducing in the wild (B. Ferns, pers. comm.).

Captive breeding and reintroduction are resource-intensive, and in order to ensure the persistence of the reintroduced population, threats to the population, such as poaching, must be reduced or removed (Robinson et al. 2020). Education and community engagement also contribute to the success of reintroduction programmes (Robinson et al. 2020). With many species

(continued)

Fig. 3 The black-winged myna (*Acridotheres melanopterus*), image supplied by Simon James (left), and the Javan green magpie (*Cissa thalassina*), image supplied by CCBC (right), are threatened with over-harvesting for the songbird trade

throughout Southeast Asia declining, and wild populations becoming smaller, it is likely that reintroductions will become even more important in the future for species conservation.

Habitat Loss

Southeast Asia holds the highest number of threatened and endemic species in the world (Koh and Sodhi 2010). The region also boasts the highest rate of deforestation of any tropical region and is predicted to lose three quarters of its original forests and 42% of its biodiversity by 2100 (Sodhi et al. 2004). Singapore has already seen large-scale deforestation (>95% habitat loss) that resulted in high local extinction rates of species and at least a 28% loss in biodiversity (881 recorded species) (Castelletta et al. 2000; Brook et al. 2003). In Singapore, residual species are now dependent on protected reserves (Brook et al. 2003). Habitat loss is not restricted to forests, with an estimated 30% loss of mangrove ecosystems within Asia (Richards and Friess 2016). If current rates of deforestation and habitat conversion are sustained within Southeast Asia, there will be mass extinctions and high biodiversity losses (Brooks et al. 2002).

Much of the habitat destruction within Southeast Asia has been attributed to land conversion for agriculture (Fig. 4), with two types of plantations at the forefront—rubber and oil palm. Southeast Asia is one of the largest cultivators of rubber trees (*Hevea brasiliensis*), at one point holding 84% of the global rubber plantation area (Warren-Thomas et al. 2015). Although many rubber plantations are being converted to the more profitable oil palm, given the increased global demand for natural rubber, it is expected that rubber plantations will continue to expand (Warren-Thomas et al. 2015). African oil palm (*Elaeis guineensis*) plantations are used to produce palm oil, a highly traded vegetable oil and a crop highly publicized as a driver of deforestation in tropical countries. Both Vijay et al. (2016) and Koh and Wilcove (2008) found that 45–59% of surveyed oil palm plantations in Southeast Asia were developed in areas that were forested up to 30 years previously. Fitzherbert et al. (2008) noted that due to a lack of accurate records, it is difficult to ascertain whether oil palm plantations were the drivers behind deforestation or whether they were planted following land clearance for other reasons. However, with illegal oil palm plantations encroaching on protected reserves (Fitzherbert et al. 2008), it cannot be denied that oil palm plantations play a role in habitat loss in Southeast Asia.

Fig. 4 Forest that has been recently cleared and burned on the forest edge (left) and a plot of land that was previously forest but was cleared for a new plantation (right). Photos from Stay Wild Tiger Protection Trust

The conversion of forest to plantations results not only in habitat loss due to land clearance and a reduction in the overall amount of habitat available to species but also in a loss of biodiversity. Fitzherbert et al. (2008) found that oil palm plantations held less than half the number of vertebrate species supported by primary forest ecosystems and had lower species richness than disturbed (secondary or logged) forests. Aratrakorn et al. (2006) also demonstrated that the conversion of lowland forest to oil palm or rubber plantations resulted in a 60% reduction in bird species richness. In addition, while the conversion of native forest to oil palm plantations results in large biodiversity losses, there is also a loss in biodiversity with the conversion of rubber plantations to oil palm (Koh and Wilcove 2008).

Logging is a large driver of deforestation and habitat degradation in tropical countries, with protected reserves succumbing to illegal logging at times (Wilcove et al. 2013; Edwards et al. 2014). Selective logging may not result in high impacts on biodiversity as compared to forest conversion, but it can affect species abundance (Berry et al. 2010). Furthermore, logged forests are more likely to be converted to agricultural land and plantations due to their low conservation value. However, although logged forests reduce biodiversity and species richness when compared to primary forests, logged forests support higher biodiversity levels than forest conversion (Edwards et al. 2011, 2014; Konopik et al. 2015). Thus, the preservation of forest, whether primary, secondary, or logged, is beneficial for biodiversity conservation.

There are additional drivers to deforestation and habitat loss in Southeast Asia, including mining, fire (Hughes 2017; Sonter et al. 2018), and infrastructure development, which can be a major threat to endemic species (Alamgir et al. 2019). One such endemic species are orangutans—the only non-human great ape found in Asia. The Bornean orangutan, *Pongo pygmaeus*, inhabits the island of Borneo, while the Sumatran orangutan (*Pongo abelii*) inhabits most of Sumatra. One Sumatran population is quite special, however, as it constitutes a distinct and recently described species—the Tapanuli orangutan, *Pongo tapanuliensis* (Nater et al. 2017). Its range is restricted to Batang Toru, an area only a tenth the size of Sydney, Australia (Nater et al. 2017). This newly discovered species is threatened by a multibillion dollar hydropower and road-building scheme, which is ongoing despite the damage it will cause towards biodiversity (Wich et al. 2019; Laurance et al. 2020). Loss of native forest restricts surviving orangutans to small pockets of forest, and they are more likely to leave the forest in search of food sources (Meijaard et al. 2011). This increases human-wildlife conflict—they are viewed as pests and often killed, and their young sold as illegal pets, despite being protected by law in Indonesia (Meijaard et al. 2011).

Multiple studies have demonstrated the linkage between habitat loss, fragmentation, deforestation, and a resulting loss of biodiversity and species abundance, along with increasingly isolated populations of native wildlife. As a result, many species action and recovery plans focus on preserving remaining habitat in an effort to reduce further losses. However, the success of forest protection is dependent on strong policy and local law enforcement, which as discussed previously is often weak (Lee et al. 2005). The needs of forest-border communities also must be considered, and if access to the forest is restricted, alternatives must be provided for local people such as planting additional community forests.

To counteract the effects of habitat loss on free-ranging wildlife, degraded or deforested land can be restored to expand habitat so that sustainable population sizes can be reached and maintained. However, this method is limited, as Turner et al. (1997) reported that after 100 years of being planted on intensively farmed land, secondary forest exhibits only 60% of tree species richness as compared to primary forest. This supports the idea that conserving existing habitat is better (in terms of biodiversity conservation) than trying to create new forest. Deforestation may also result in patches of forest within an open or uninhabitable matrix, leading to genetically isolated populations that cannot be sustained naturally. Corridors are used to reconnect fragmented habitats and allow movement of animals between otherwise isolated forest patches. Though there has been some scepticism to their practicality (Beier and Noss 1998; Christie and Knowles 2015), an increasing amount of research has demonstrated the positive effects of corridors on animal movements, population connectivity, and species richness within connected habitat patches (Aars and Ims 1999; Tewksbury et al. 2002; Haddad et al. 2003; Damschen et al. 2006; Dixon et al. 2006; LaPoint et al. 2013). However, the use of corridors should be proposed with caution, as it may inadvertently exacerbate habitat destruction and could aid the spread of invasive species, disease, or fire (Jain et al. 2014; Resasco et al. 2014). Haddad et al. (2014) noted such potential negative effects can be managed and are small compared to the positive effects of corridors.

Human-Wildlife Conflict

Human-wildlife conflict (HWC) has been a long-standing problem in tropical regions, where humans share a landmass with a vast diversity of wildlife. Simply put, HWC is a situation in which the requirements of wildlife overlap with the human population, creating costs to both human residents and wild animals (Distefano 2005). This has resulted in various wild animals being persecuted in Southeast Asia, especially those considered as pests or potentially dangerous to human life. HWC has contributed to the decline of many species, such as Sumatran tigers (*Panthera tigris sumatrae*) (Nyhus and Tilson 2004), and is therefore considered a major threat to biodiversity in this region.

One of the main driving factors of HWC is the ever-increasing human population. To fulfill the requirements of modern humans, there has been mass conversion of natural landscapes into agricultural, residential, or industrial areas. This has caused a great deal of habitat loss, degradation, and fragmentation, pushing humans and wild animals even closer together, as mentioned above.

In some cases, the increasingly easy access to nature reserves can also contribute to HWC. One example of this is the population of long-tailed macaques (*Macaca fascicularis*) in Muara Angke Nature Reserve, North Jakarta (Entoh 2011). The reserve is situated in the city and is very easily accessible. Because of this, the behaviour of macaques, which tend to avoid humans, has been altered due to regular feedings by tourists and local residents (Entoh 2011) (Fig. 5). Over time this population lost their fear of people and began causing problems for the residents of North Jakarta by raiding houses and biting visitors (Entoh 2011). Human-primate conflict is an issue in many parts of Southeast Asia. On Sulawesi, the moor macaques (*Macaca maura*)

Fig. 5 Long-tailed macaques feast on the food items thrown at them from moving vehicles in North Jakarta, Indonesia. Image supplied by Nathan Rusli

are involved with conflict due to crop raiding (Zak and Riley 2017), and on Java there are similar cases with the Javan gibbon (*Hylobates moloch*) (Lappan et al. 2020) (R. Oktaviani pers. comm.).

Climatic factors and stochastic events also play a role in human-wildlife conflict. An example of this are reticulated pythons (*Malayopython reticulatus*) which are highly adaptable to urban environments (De Lang 2017; Low 2018; Rusli 2020). In the metropolitan city of Jakarta, Indonesia, the pythons live in drains and sewer systems, preying on rats and other animals which inhabit the same environment. During the monsoon season, flooding often occurs, flushing the pythons out of their hiding places and greatly increasing the risk of conflict with humans (Rusli 2016; Rusli and Rini 2020).

Mitigating actions for HWC in Southeast Asia involve community outreach—educating local people about wildlife and how to coexist with it and minimizing the risk to both humans and nature (Tangley 1997; Rusli and Rini 2020).

Box 3 Human-Tiger Conflict with Forest-Edge Communities

Human-wildlife conflict has been a threat to tiger populations throughout the twentieth century, with the extinctions of the Javan tiger (*Panthera tigris sondaica*) and Bali tiger (*P.t. balica*) attributed to human-tiger conflict. This struggle has also contributed to the decline of Sumatran tiger (*P.t. sumatrae*) populations (Nyhus and Tilson 2004). While human-tiger conflict has decreased over recent decades (due mainly to a decreasing free-ranging tiger population), there are fears it may increase as tiger numbers rise (Nyhus and Tilson 2004; Goodrich 2010). This could lead to more conflict and retaliatory killings, which would undermine conservation efforts.

The issues between humans and tigers include not only the threat of people losing their lives but also the loss of income, such as through livestock attacks. One such case was recorded by Stay Wild Tiger Protection Trust, working with forest-edge communities near Gunung Leuser National Park, North Sumatra, Indonesia. Local farmers graze their cattle on the edges of the National Park as well as the karst forest that borders the protected area. In May 2020, two tigers in two different incidents killed adult cattle grazing near forest edges (Fig. 6).

(continued)

Fig. 6 Camera trap still of a Sumatran tiger returning to its kill (a cow carcass) on the forest edge. Image supplied by Stay Wild Tiger Protection Trust

In a previous study, Miller et al. (2016) found that most livestock attacks by carnivores occurred near the forest edge, which are areas regularly used for livestock grazing. Human-tiger conflict is more common with old, sick, or wounded tigers that have been driven into human-dominated landscapes (Goodrich 2010). However, in a case study by Stay Wild Tiger Protection Trust, both tigers were in good condition and returned over several days to feed on the cattle carcasses. This may indicate that tigers left the protected forest in search of prey. Hunting for bushmeat reduces tiger prey populations, which may lead to increased human-tiger conflict. This highlights not only how habitat loss and encroachment into habitat can increase human-wildlife conflict but also the hunting of prey species.

To reduce the risk of further livestock attacks by the tigers, rangers from local NGOs collaborated with government rangers to monitor the tigers as they returned each night to feed on cattle carcasses. Non-lethal methods such as noise deterrents were used to try to drive the tigers back towards the protected forest. To prevent the possibility of retaliatory killings, NGOs worked with local farmers and provided a compensation scheme, a technique often employed to reduce the effects of human-wildlife conflict (Nyhus and Tilson 2004; Goodrich 2010; Miller et al. 2016).

Box 4 Human-Snake Conflict in Indonesia

Indonesia is home to over 300 species of snakes, many of them endemic (Uetz et al. 2020). There are over 70 species of medically significant (highly venomous) snakes in Indonesia, and snakebite has been a neglected issue for many years (Indonesia Toxinology Society, 2020 (unpublished data)) (Adiwinata and Nelwan 2015; Warrell 2017). With the human population constantly on the rise, humans and snakes are pushed closer together, causing potential problems.

Snakes have developed several adaptations to defend themselves from predators. Many species use cryptic colouration to avoid detection. One example of this is the Malayan pit viper (*Calloselasma rhodostoma*), a common species in lowland forest and plantations

(continued)

in Southeast Asia. This venomous snake is brown in colour, with light and dark markings that resemble leaf litter. Unfortunately, due to its impeccable camouflage, people often step on them by accident. In some cases, this will prompt the serpent to bite in self-defence and inject its lethal venom.

Events such as this and other envenomations by venomous species have caused a deep-rooted fear of snakes for many Indonesians. This has resulted in the persecution of snakes. It is common practice to kill a snake on sight in Indonesia (Rusli 2016; Rusli and Rini 2020). Sadly, the act of killing a snake (usually by hitting it with a hard object, i.e. a stick or machete) can sometimes go wrong. When cornered and unable to flee, the final line of defence for most snakes is to bite. In the case of venomous snakes, this can be potentially life-threatening.

This is especially apparent on the island of Java, which has the highest population of humans in Indonesia (Badan Pusat Statistik 2019). Out of the 89 species of snakes found here, several species such as the reticulated python (*Malayopython reticulatus*) (Fig. 7) and Indonesian spitting cobra (*Naja sputatrix*) have adapted to live in urban and rural areas (Das 2010; De Lang 2017; Rusli 2020). Reticulated pythons are very successful in cities, living in drainage systems and preying on other urban fauna, just as they do in Jakarta. They are often involved in conflict with humans, especially during the monsoon season. This is also the breeding season for spitting cobras, when they are most active, increasing the likelihood of human encounters (Rusli and Rini 2020).

The Indonesia Herpetofauna Foundation is a local NGO focused on the conservation of amphibians and reptiles in Indonesia. As part of their conflict mitigation programme, they work with governments, NGOs, and local communities to reduce human-snake conflict in Indonesia. This is done through outreach and education, raising awareness about the ecological importance of snakes, common snakes in the area, what to do in a snake encounter, how to prevent snakebite, and the correct first aid for snakebite. Removal and relocation of conflict animals is another way to reduce human-snake conflict, which is also undertaken by several herpetologists and reptile enthusiasts in Indonesia.

Fig. 7 The reticulated python (*Malayopython reticulatus*) is threatened with human-wildlife conflict because they often inhabit urban areas and come into contact with people. Image supplied by Nathan Rusli

Threats to Biodiversity in Aquatic Ecosystems

While this chapter has focused mainly on terrestrial ecosystems, many of the threats discussed are also causing biodiversity losses within marine and freshwater ecosystems. Not only is this important in terms of biodiversity conservation, but fishing is a source of income and food security within Southeast Asia. Pomeroy et al. (2016) reported that fish are relied upon as a primary protein source and income provider in Southeast Asia, more so than in any other part of the world. However, habitat destruction, plastic pollution, and climate change threaten marine and freshwater ecosystems (Reid et al. 2019). Plastic from single-use packaging and microplastics are polluting marine ecosystems, but efforts are being made to educate coastal communities and provide alternatives. The NGO Philippine Reef and Rainforest Conservation Foundation Inc. (PRRCFI) engages local communities and governments to reduce plastic pollution on Danjugan Island and adjacent coastal communities in Southern Negros, Philippines, to help conserve marine ecosystems (Fig. 8). Urbanization along coastal zones results in sedimentation and pollutant discharge into marine ecosystems, which can lead to loss of coral reefs and essential ecosystem functions (Heery et al. 2018). Overfishing, illegal catches, and shark finning also pose threats to sustainable fishery management (Pomeroy et al. 2016; Dulvy et al. 2017; de Mitcheson et al. 2020). Degradation of coastal nesting sites, fishing mortalities, hunting, and egg exploitation have led to population declines in turtle species (Chan 2006; Hitipeuw et al. 2007). Lastly, some ornamental tropical fish may be targeted for wildlife trade (Ng and Tan 1997). All of these threats are similar to those faced by terrestrial ecosystems and are equally difficult to manage, relying on strong policy, law enforcement, public awareness and engagement, management systems, and an assessment of community needs.

Additional Threats

Although this chapter focuses on the primary and immediate threats to biodiversity in Southeast Asia, there are additional risks to consider. Climate change is predicted to have adverse effects on native fauna and flora, and Bickford et al. (2010) suggest severe consequences for reptiles and amphibians in Southeast Asia. Rising global temperature is predicted to negatively impact temperature-dependent sex determination, metabolic rates, and water availability for herpetofauna (Bickford et al. 2010). Additionally, with continued land clearance and urbanization, there is an increased risk of disease spread by domestic animals to free-ranging populations of wildlife and vice versa (Daszak et al. 2000; Cook

Fig. 8 Danjugan Island in the Philippines (left) and volunteers of PRRCFI cleaning up plastic washed ashore (right). Images supplied by PRRCFI

and Karesh 2012; Beineke et al. 2015). There is also the threat of zoonoses being passed from wild animals through wildlife trade and possible subsequent persecution of wildlife (Daszak et al. 2000; Karesh et al. 2005; Karesh and Noble 2009; MacFarlane and Rocha 2020). Further, invasive species pose a threat through predation, competition, hybridization, and environmental disturbance, particularly within insular ecosystems (Peh 2010; Doherty et al. 2016).

There are inherent challenges in conservation, namely, those of ensuring efficient implementation of conservation measures, and proper controls for conservation management. While ecotourism can increase the value of wildlife to local communities and can provide an alternative income and benefit some species (Buckley et al. 2016), it must be monitored to ensure sustainability and to ensure it does not pose a danger to free-ranging wildlife. Corruption within the government or within organizations tasked with protecting the environment also undermines policies that safeguard nature (Irland 2008; Wyatt et al. 2018). In addition, conservation work is underfunded, which limits the ability to utilize interdisciplinary collaboration and reduces overall capacity for conservation projects. Waldron et al. (2013) assessed countries based on threatened biodiversity and amount of conservation funding. Based on their model, Malaysia and Indonesia are in the top 40 most underfunded countries for biodiversity conservation (Waldron et al. 2013).

Another impediment to the implementation and efficiency of conservation programmes is a lack of appropriate research. Monitoring and analysis of wildlife and ecosystems in Southeast Asia is important to conduct not only for quantifying the rate of decline and loss of biodiversity but also for identifying the main issues and potential mitigating actions that will help to conserve biodiversity. There are many species in Southeast Asia at risk of going extinct even before they are formally described (Giam et al. 2010), meaning that rates of biodiversity decline may be underestimated. Conservation programmes should be evidence-based and formed based on previous research that identifies primary threats to the survival of the species, as well as on alleviating actions that have been shown to reduce, or are likely to reduce the decline of the species. Any mitigating action implemented should be monitored to demonstrate its effectiveness. This should also be the case for any actions taken to reduce social drivers, such as education and awareness campaigns, increased policy enforcement, and implementation of alternative livelihood programmes.

Conclusion

There are many different threats to biodiversity in Southeast Asia which are intertwined with both social development and governance within their respective countries. Poverty and underdevelopment may restrict both educational opportunities and income sources and may increase the dependence on natural resources or illegal activity for survival. A growing human population coupled with development for economic growth puts additional pressure on fertile land and increases land clearance. Protection laws and policies are undermined by corruption within authority organizations and under-resourced law enforcement. These are often the issues that drive the major threats to biodiversity, such as wildlife trade, habitat loss, and human-wildlife conflict. All of these aspects contribute to the decline of biodiversity in Southeast Asia.

The information in this chapter was provided not to present an in-depth examination of all causes of biodiversity loss, but instead to outline the complexities of conservation. As many of the threats to biodiversity are anthropogenic in nature, it is no surprise that the approach taken by conservationists has changed over the years. While protection status and the formation of national parks were used frequently in the past, there has been a movement to incorporate the needs of forest-neighbouring communities and to take a community-based approach. This is emphasized by the online survey results presented here, where education and community engagement, alternative livelihoods, and community empowerment were listed as important for conservation work. This approach, coupled with applied conservation measures (such as research, policy, and applied conservation actions), could enhance the effectiveness of biodiversity conservation.

Acknowledgements The authors would first like to thank all the anonymous conservationists throughout Southeast Asia who have participated in our survey to gather valuable information for this chapter. We would also like to thank Vanessa Rowe and the Stay Wild Tiger Protection Trust team for providing photographs and case studies in regard to tigers and poaching. The team at Ciliwung Herpetarium, as well as Dr. dr. Tri Maharani Msi SpEM from Indonesia Toxinology Society have been very helpful in providing information regarding snake conflict. We would also like to thank Bertie Ferns (Cikananga Conservation Breeding Centre), Dave Gumban Albao (Philippine Reef and Rainforest Conservation Foundation Inc.), and Rahayu Oktaviani (Javan Gibbon Research and Conservation Project) for providing their firsthand insight on conservation issues and efforts in Southeast Asia. Finally, we would like to thank Simon James, PRRCFI, and Stay Wild Tiger Protection Trust for providing photographs for this chapter.

References

Aars J, Ims RA (1999) The effect of habitat corridors on rates of transfer and interbreeding between vole demes. Ecology 80(5):1648–1655. https://doi.org/10.1890/0012-9658(1999)080[1648:TEOHCO]2.0.CO;2

Adiwinata R, Nelwan EJ (2015) Snakebite in Indonesia. Acta Med Indones 47(4):358

Aisher A (2016) Scarcity, alterity and value: decline of the pangolin, the world's most trafficked mammal. Conserv Soc 14(4):317–329

Alamgir M, Campbell MJ, Sloan S, Suhardiman A, Supriatna J, Laurance WF (2019) High-risk infrastructure projects pose imminent threats to forests in Indonesian Borneo. Sci Rep 9(1):140. https://doi.org/10.1038/s41598-018-36594-8

Aratrakorn S, Thunhikorn S, Donald PF (2006) Changes in bird communities following conversion of lowland forest to oil palm and rubber plantations in southern Thailand. Bird Conserv Int 16(1):71–82. https://doi.org/10.1017/S0959270906000062

Badan Pusat Statistik (2019) Statistical year book of Indonesia: BPS

Barnosky AD, Matzke N, Tomiya S, Wogan GOU, Swartz B, Quental TB et al (2011) Has the Earth's sixth mass extinction already arrived? Nature 471(7336):51–57. https://doi.org/10.1038/nature09678

Beastall C, Shepherd CR, Hadiprakarsa Y, Martyr D (2016) Trade in the helmeted hornbill *Rhinoplax vigil*: the 'ivory hornbill'. Bird Conserv Int 26(2):137–146. https://doi.org/10.1017/S0959270916000010

Beier P, Noss RF (1998) Do habitat corridors provide connectivity? Conserv Biol 12(6):1241–1252. https://doi.org/10.1111/j.1523-1739.1998.98036.x

Beineke A, Baumgärtner W, Wohlsein P (2015) Cross-species transmission of canine distemper virus: an update. One Health 1:49–59. https://doi.org/10.1016/j.onehlt.2015.09.002

Berry NJ, Phillips OL, Lewis SL, Hill JK, Edwards DP, Tawatao NB et al (2010) The high value of logged tropical forests: lessons from northern Borneo. Biodivers Conserv 19(4):985–997. https://doi.org/10.1007/s10531-010-9779-z

Bickford D, Howard SD, Ng DJJ, Sheridan JA (2010) Impacts of climate change on the amphibians and reptiles of Southeast Asia. Biodivers Conserv 19(4):1043–1062. https://doi.org/10.1007/s10531-010-9782-4

Biggs D, Cooney R, Roe D, Dublin HT, Allan JR, Challender DWS, Skinner D (2017) Developing a theory of change for a community-based response to illegal wildlife trade. Conserv Biol 31(1):5–12. https://doi.org/10.1111/cobi.12796

Birdlife International (2018a) *Acridotheres melanopterus*. The IUCN red list of threatened species. https://doi.org/10.2305/IUCN.UK.2018-2.RLTS.T103870843A131892465.en

Birdlife International (2018b) *Acridotheres tertius*. The IUCN red list of threatened species. https://doi.org/10.2305/IUCN.UK.2018-2.RLTS.T103871084A134212031.en

Birdlife International (2018c) *Acridotheres tricolor*. The IUCN red list of threatened species. https://doi.org/10.2305/IUCN.UK.2018-2.RLTS.T103870973A134212688.en

Birdlife International (2018d) *Cissa thalassina*. The IUCN red list of threatened species. https://doi.org/10.2305/IUCN.UK.2018-2.RLTS.T22724821A134213647.en

Birdlife International (2019) *Rhinoplax vigil* (amended version of 2018 assessment). The international red list of threatened species 2019. https://doi.org/10.2305/IUCN.UK.2019-3.RLTS.T22682464A155467793.en

Bowkett AE (2009) Recent captive-breeding proposals and the return of the ark concept to global species conservation. Conserv Biol 23(3):773–776

Brook BW, Sodhi NS, Ng PKL (2003) Catastrophic extinctions follow deforestation in Singapore. Nature 424(6947):420–423. https://doi.org/10.1038/nature01795

Brooks TM, Mittermeier RA, Mittermeier CG, Da Fonseca GAB, Rylands AB, Konstant WR et al (2002) Habitat loss and extinction in the hotspots of biodiversity. Conserv Biol 16(4):909–923. https://doi.org/10.1046/j.1523-1739.2002.00530.x

Buckley RC, Morrison C, Castley JG (2016) Net effects of ecotourism on threatened species survival. PLoS One 11(2):e0147988

Castelletta M, Sodhi NS, Subaraj R (2000) Heavy extinctions of forest avifauna in Singapore: lessons for biodiversity conservation in Southeast Asia. Conserv Biol 14(6):1870–1880

Ceballos G, Ehrlich PR, Barnosky AD, García A, Pringle RM, Palmer TM (2015) Accelerated modern human–induced species losses: entering the sixth mass extinction. Sci Adv 1(5):e1400253. https://doi.org/10.1126/sciadv.1400253

Ceballos G, Ehrlich PR, Dirzo R (2017) Biological annihilation via the ongoing sixth mass extinction signaled by vertebrate population losses and declines. Proc Natl Acad Sci 114(30):E6089. https://doi.org/10.1073/pnas.1704949114

Challender D, Willcox DHA, Panjang E, Lim N, Nash H, Heinrich S, Chong J (2019a). *Manis javanica*. The IUCN red list of threatened species. https://doi.org/10.2305/IUCN.UK.2019-3.RLTS.T12763A123584856.en

Challender D, Wu S, Kaspal P, Khatiwada A, Ghose A, Ching-Min Sun N, et al. (2019b) *Manis pentadactyla* (errata version published in 2020). The IUCN red list of threatened species. https://doi.org/10.2305/IUCN.UK.2019-3.RLTS.T12764A168392151.en

Chan E-H (2006) Marine turtles in Malaysia: on the verge of extinction? Aquat Ecosyst Health Manage 9 (2):175–184. https://doi.org/10.1080/14634980600701559

Cheyne SM (2009) The role of reintroduction in gibbon conservation: opportunities and challenges. In: Whittaker D, Lappan S (eds) The gibbons: new perspectives on small ape socioecology and population biology. Springer, New York, pp 477–496. https://doi.org/10.1007/978-0-387-88604-6_23

Christie MR, Knowles LL (2015) Habitat corridors facilitate genetic resilience irrespective of species dispersal abilities or population sizes. Evol Appl 8(5):454–463. https://doi.org/10.1111/eva.12255

CITES (2020) Convention on international trade in endangered species of wild fauna and flora. Retrieved from https://cites.org/eng

Cook RA, Karesh WB (2012) Emerging diseases at the interface of people, domestic animals, and wildlife. Fowler's zoo and wild animal medicine, p 136

Cowie RH, Regnier C, Fontaine B, Bouchet P (2017) Measuring the sixth extinction: what do mollusks tell us. Nautilus 131(1):3–41

Damschen EI, Haddad NM, Orrock JL, Tewksbury JJ, Levey DJ (2006) Corridors increase plant species richness at large scales. Science 313(5791):1284. https://doi.org/10.1126/science.1130098

Das I (2010) A field guide to the reptiles of South East Asia. New Holland Publishers, London

Daszak P, Cunningham AA, Hyatt AD (2000) Emerging infectious diseases of wildlife: threats to biodiversity and human health. Science 287(5452):443. https://doi.org/10.1126/science.287.5452.443

De Lang R (2017) The snakes of Java. Edition Chimaira, Bali and surrounding islands

de Mitcheson YJS, Linardich C, Barreiros JP, Ralph GM, Aguilar-Perera A, Afonso P et al (2020) Valuable but vulnerable: over-fishing and under-management continue to threaten groupers so what now? Mar Policy 116:103909

Distefano E (2005) Human-wildlife conflict worldwide: collection of case studies, analysis of management strategies and good practices. Food and Agricultural Organization of the United Nations, Rome

Dixon JD, Oli MK, Wooten MC, Eason TH, McCown JW, Paetkau D (2006) Effectiveness of a regional corridor in connecting two Florida black bear populations. Conserv Biol 20(1):155–162. https://doi.org/10.1111/j.1523-1739.2005.00292.x

Doherty TS, Glen AS, Nimmo DG, Ritchie EG, Dickman CR (2016) Invasive predators and global biodiversity loss. Proc Natl Acad Sci 113(40):11261. https://doi.org/10.1073/pnas.1602480113

Duffy R, St John F (2013) Poverty, poaching and trafficking: what are the links? London. https://doi.org/10.12774/eod_hd059.jun2013.duffy

Dulvy NK, Simpfendorfer CA, Davidson LNK, Fordham SV, Bräutigam A, Sant G, Welch DJ (2017) Challenges and priorities in shark and ray conservation. Curr Biol 27(11):R565–R572. https://doi.org/10.1016/j.cub.2017.04.038

Dutton AJ, Hepburn C, Macdonald DW (2011) A stated preference investigation into the Chinese demand for farmed vs. wild bear bile. PLoS One 6(7):e21243. https://doi.org/10.1371/journal.pone.0021243

Eaton J, Shepherd C, Rheindt F, Harris J, Van Balen S, Wilcove D, Collar N (2015) Trade-driven extinctions and near-extinctions of avian taxa in Sundaic Indonesia. Forktail 31:1–12

Edwards DP, Gilroy JJ, Woodcock P, Edwards FA, Larsen TH, Andrews DJR et al (2014) Land-sharing versus land-sparing logging: reconciling timber extraction with biodiversity conservation. Glob Chang Biol 20 (1):183–191. https://doi.org/10.1111/gcb.12353

Edwards, D. P., Larsen, T. H., Docherty, T. D., Ansell, F. A., Hsu, W. W., Derhé, M. A., . . . Wilcove, D. S. (2011). Degraded lands worth protecting: the biological importance of Southeast Asia's repeatedly logged forests. Proc R Soc B Biol Sci, 278(1702), 82–90

Entoh AE (2011) Laporan Program Penanganan Konflik Monyet ekor panjang (Macaca fascicularis) di Wilayah Kelompok Hutan Angke Kapuk dan Sekitarnya Juni 2011

Fitzherbert EB, Struebig MJ, Morel A, Danielsen F, Brühl CA, Donald PF, Phalan B (2008) How will oil palm expansion affect biodiversity? Trends Ecol Evol 23 (10):538–545. https://doi.org/10.1016/j.tree.2008.06.012

Giam X, Ng TH, Yap VB, Tan HTW (2010) The extent of undiscovered species in Southeast Asia. Biodivers Conserv 19(4):943–954. https://doi.org/10.1007/s10531-010-9792-2

Goodrich JM (2010) Human–tiger conflict: a review and call for comprehensive plans. Integrative Zoology 5 (4):300–312

Haddad NM, Bowne DR, Cunningham A, Danielson BJ, Levey DJ, Sargent S, Spira T (2003) Corridor use by diverse taxa. Ecology 84(3):609–615. https://doi.org/10.1890/0012-9658(2003)084[0609:CUBDT]2.0.CO;2

Haddad NM, Brudvig LA, Damschen EI, Evans DM, Johnson BL, Levey DJ et al (2014) Potential negative ecological effects of corridors. Conserv Biol 28 (5):1178–1187. https://doi.org/10.1111/cobi.12323

Harrington LA, D'Cruze N, Macdonald D (2018) Rise to fame: events, media activity and public interest in pangolins and pangolin trade, 2005–2016. Nat Conserv 30:107

Heery EC, Hoeksema BW, Browne NK, Reimer JD, Ang PO, Huang D et al (2018) Urban coral reefs: degradation and resilience of hard coral assemblages in coastal cities of east and Southeast Asia. Mar Pollut Bull 135:654–681

Hitipeuw C, Dutton PH, Benson S, Thebu J, Bakarbessy J (2007) Population status and internesting movement of leatherback turtles, *Dermochelys coriacea*, nesting on the northwest coast of Papua, Indonesia. Chelonian Conserv Biol 6(1):28–36. https://doi.org/10.2744/1071-8443(2007)6[28:PSAIMO]2.0.CO;2

Hughes AC (2017) Understanding the drivers of southeast Asian biodiversity loss. Ecosphere 8(1):e01624. https://doi.org/10.1002/ecs2.1624

Irland LC (2008) State failure, corruption, and warfare: challenges for forest policy. J Sustain For 27 (3):189–223. https://doi.org/10.1080/10549810802219963

Jain A, Chong KY, Chua MAH, Clements GR (2014) Moving away from paper corridors in Southeast Asia. Conserv Biol 28(4):889–891. https://doi.org/10.1111/cobi.12313

Karesh WB, Cook RA, Bennett EL, Newcomb J (2005) Wildlife trade and global disease emergence. Emerg Infect Dis 11(7):1000

Karesh WB, Noble E (2009) The bushmeat trade: increased opportunities for transmission of zoonotic disease. Mount Sinai J Med 76(5):429–434. https://doi.org/10.1002/msj.20139

Kleiman DG (1989) Reintroduction of captive mammals for conservation. Bioscience 39(3):152–161

Koh LP, Sodhi NS (2010) Conserving Southeast Asia's imperiled biodiversity: scientific, management, and policy challenges. Biodivers Conserv 19(4):913–917

Koh LP, Wilcove DS (2008) Is oil palm agriculture really destroying tropical biodiversity? Conserv Lett 1 (2):60–64. https://doi.org/10.1111/j.1755-263X.2008.00011.x

Konopik O, Steffan-Dewenter I, Grafe TU (2015) Effects of logging and oil palm expansion on stream frog communities on Borneo, Southeast Asia. Biotropica 47(5):636–643

Krishnasamy K, Zavagli M (2020) Southeast Asia: at the heart of wildlife trade. TRAFFIC, Southeast Asia Regional Office, Petaling Jaya

LaPoint S, Gallery P, Wikelski M, Kays R (2013) Animal behavior, cost-based corridor models, and real corridors. Landsc Ecol 28(8):1615–1630. https://doi.org/10.1007/s10980-013-9910-0

Lappan S, Malaivijitnond S, Radhakrishna S, Riley EP, Ruppert N (2020) The human–primate interface in the new normal: challenges and opportunities for primatologists in the COVID-19 era and beyond. Am J Primatol 82(8):e23176

Laurance WF, Wich SA, Onrizal O, Fredriksson G, Usher G, Santika T et al (2020) Tapanuli orangutan endangered by Sumatran hydropower scheme. Nat Ecol Evol:1–2

Lee, R. J., Gorog, A. J., Dwiyahreni, A., Siwu, S., Riley, J., Alexander, H., . . . Ramono, W. (2005). Wildlife trade and implications for law enforcement in Indonesia: a case study from North Sulawesi. Biol Conserv, 123(4), 477–488. https://doi.org/10.1016/j.biocon.2005.01.009

Leupen B, Krishnasamy K, Shepherd C, Chng S, Bergin D, Eaton J et al (2019) Trade in white-rumped shamas *Kittacincla malabarica* demands strong national and international responses. Forktail 34:1–8. https://doi.org/10.13140/RG.2.2.17829.04320

Low M-R (2018) Rescue, rehabilitation and release of reticulated pythons in Singapore (no. 2831719011). IUCN/SSC Reintroduction Specialist Group, Abu Dhabi

MacFarlane D, Rocha R (2020) Guidelines for communicating about bats to prevent persecution in the time of COVID-19. Biol Conserv 248:108650. https://doi.org/10.1016/j.biocon.2020.108650

Mallinson JJ (1995) Conservation breeding programmes: an important ingredient for species survival. Biodivers Conserv 4(6):617–635

McCallum ML (2015) Vertebrate biodiversity losses point to a sixth mass extinction. Biodivers Conserv 24 (10):2497–2519. https://doi.org/10.1007/s10531-015-0940-6

Meijaard E, Buchori D, Hadiprakarsa Y, Utami-Atmoko SS, Nurcahyo A, Tjiu A et al (2011) Quantifying killing of orangutans and human-orangutan conflict in Kalimantan, Indonesia. PLoS One 6(11):e27491. https://doi.org/10.1371/journal.pone.0027491

Miller JRB, Jhala YV, Jena J (2016) Livestock losses and hotspots of attack from tigers and leopards in Kanha Tiger Reserve, Central India. Reg Environ Change 16 (1):17–29. https://doi.org/10.1007/s10113-015-0871-5

Myers N, Mittermeier RA, Mittermeier CG, Da Fonseca GA, Kent J (2000) Biodiversity hotspots for conservation priorities. Nature 403(6772):853–858

Nater A, Mattle-Greminger MP, Nurcahyo A, Nowak MG, de Manuel M, Desai T (2017) Morphometric, behavioral, and genomic evidence for a new orangutan species. Curr Biol 27(22):3487–3498.e3410. https://doi.org/10.1016/j.cub.2017.09.047

Nekaris KAI, Shepherd CR, Starr CR, Nijman V (2010) Exploring cultural drivers for wildlife trade via an ethnoprimatological approach: a case study of slender and slow lorises (*Loris* and *Nycticebus*) in South and Southeast Asia. Am J Primatol 72(10):877–886. https://doi.org/10.1002/ajp.20842

Ng PKL, Tan HH (1997) Freshwater fishes of Southeast Asia: potential for the aquarium fish trade and conservation issues. Aquar Sci Conserv 1(2):79–90. https://doi.org/10.1023/A:1018335617835

Nijman V (2010) An overview of international wildlife trade from Southeast Asia. Biodivers Conserv 19 (4):1101–1114. https://doi.org/10.1007/s10531-009-9758-4

Nijman V, Langgeng A, Birot H, Imron MA, Nekaris KAI (2018) Wildlife trade, captive breeding and the imminent extinction of a songbird. Global Ecol Conserv 15: e00425. https://doi.org/10.1016/j.gecco.2018.e00425

Nyhus PJ, Tilson R (2004) Characterizing human-tiger conflict in Sumatra, Indonesia: implications for conservation. Oryx 38(1):68–74

Owen A, Wilkinson R, Sözer R (2014) In situ conservation breeding and the role of zoological institutions and private breeders in the recovery of highly endangered Indonesian passerine birds. Int Zoo Yearb 48 (1):199–211

Palmer A (2018) Kill, incarcerate, or liberate? Ethics and alternatives to orangutan rehabilitation. Biol Conserv 227:181–188

Pangau-Adam M, Noske R, Muehlenberg M (2012) Wildmeat or bushmeat? Subsistence hunting and commercial harvesting in Papua (West New Guinea), Indonesia. Hum Ecol 40(4):611–621

Peh KSH (2010) Invasive species in Southeast Asia: the knowledge so far. Biodivers Conserv 19 (4):1083–1099. https://doi.org/10.1007/s10531-009-9755-7

Phassaraudomsak M, Krishnasamy K, Chng SCL (2019) Trading faces: online trade of helmeted and other hornbill species on Facebook in Thailand. TRAFFIC, Petaling Jaya, Malaysia

Pomeroy R, Parks J, Courtney K, Mattich N (2016) Improving marine fisheries management in Southeast Asia: results of a regional fisheries stakeholder analysis. Mar Policy 65:20–29. https://doi.org/10.1016/j.marpol.2015.12.002

Rao M, Zaw T, Htun S, Myint TJEM (2011) Hunting for a living: wildlife trade, rural livelihoods and declining wildlife in the Hkakaborazi National Park, North Myanmar. Environ Manag 48(1):158–167

Reid AJ, Carlson AK, Creed IF, Eliason EJ, Gell PA, Johnson PTJ et al (2019) Emerging threats and persistent conservation challenges for freshwater biodiversity. Biol Rev 94(3):849–873. https://doi.org/10.1111/brv.12480

Resasco J, Haddad NM, Orrock JL, Shoemaker D, Brudvig LA, Damschen EI et al (2014) Landscape corridors can increase invasion by an exotic species and reduce diversity of native species. Ecology 95 (8):2033–2039. https://doi.org/10.1890/14-0169.1

Richards DR, Friess DA (2016) Rates and drivers of mangrove deforestation in Southeast Asia, 2000–2012. Proc Natl Acad Sci 113(2):344. https://doi.org/10.1073/pnas.1510272113

Robinson NM, Dexter N, Brewster R, Maple D, MacGregor C, Rose K et al (2020) Be nimble with threat mitigation: lessons learned from the reintroduction of an endangered species. Restor Ecol 28 (1):29–38

Rusli N (2016) Snakes of Jakarta and its surroundings: bypass

Rusli N (2020) A photographic guide to the snakes of Java. Indonesia Herpetofauna Foundation

Rusli N, Rini CP (2020) Ular Disekitar Kita - Pulau Jawa. Indonesia Herpetofauna Foundation

Sandalj M, Treydte AC, Ziegler S (2016) Is wild meat luxury? Quantifying wild meat demand and availability in Hue, Vietnam. Biol Conserv 194:105–112. https://doi.org/10.1016/j.biocon.2015.12.018

Scheffers BR, Oliveira BF, Lamb I, Edwards DP (2019) Global wildlife trade across the tree of life. Science 366 (6461):71. https://doi.org/10.1126/science.aav5327

Schoppe S, Katsis L, Lagrada L (2019) *Manis culionensis*. The IUCN red list of threatened species. https://doi.org/10.2305/IUCN.UK.2019-3.RLTS.T136497A123586862.en

Shepherd CR, Nijman V, Krishnasamy K, Eaton JA, Chng SC (2016) Illegal trade pushing the critically endangered black-winged myna *Acridotheres melanopterus* towards imminent extinction. Bird Conserv Int 26(2):147

Sodhi NS, Koh LP, Brook BW, Ng PKL (2004) Southeast Asian biodiversity: an impending disaster. Trends Ecol Evol 19(12):654–660

Sonter LJ, Ali SH, Watson JE (2018) Mining and biodiversity: key issues and research needs in conservation science. Proc R Soc B 285(1892):20181926

Tangley L (1997) The importance of communicating with the public. In: Principles of conservation biology, 2nd edn. Sinauer Associates, Sunderland

Tewksbury JJ, Levey DJ, Haddad NM, Sargent S, Orrock JL, Weldon A et al (2002) Corridors affect plants, animals, and their interactions in fragmented landscapes. Proc Natl Acad Sci 99(20):12923. https://doi.org/10.1073/pnas.202242699

Thomas, J. A., Telfer, M. G., Roy, D. B., Preston, C. D., Greenwood, J. J. D., Asher, J., . . . Lawton, J. H. (2004). Comparative losses of British butterflies, birds, and plants and the global extinction crisis. Science, 303(5665), 1879. https://doi.org/10.1126/science.1095046

TRAFFIC (2008) What's driving the wildlife trade? A review of expert opinion on economic and social drivers of the wildlife trade and trade control efforts in Cambodia, Indonesia, Lao PDR and Vietnam. East Asia and Pacific Region Sustainable Development Department, Washington, DC

Turner I, Wong Y, Chew P, bin Ibrahim A (1997) Tree species richness in primary and old secondary tropical forest in Singapore. Biodivers Conserv 6(4):537–543

Uetz P, Freed P, Hosek J (2020) The reptile. Database. Retrieved from http://www.reptile-database.org

Vernia R, Tritto A, Abinawanto A, Winarni N, Mayasari A, Sedayu A, Bowolaksono A (2018) Differentiating hybrid and juvenile of black-winged myna (*Acridotheres melanopterus*) by using morphometric data. In: Paper presented at the IOP Conference Series: Materials Science and Engineering

Vijay V, Pimm SL, Jenkins CN, Smith SJ (2016) The impacts of oil palm on recent deforestation and biodiversity loss. PLoS One 11(7):e0159668. https://doi.org/10.1371/journal.pone.0159668

Wake DB, Vredenburg VT (2008) Are we in the midst of the sixth mass extinction? A view from the world of amphibians. Proc Natl Acad Sci 105(Supplement 1):11466. https://doi.org/10.1073/pnas.0801921105

Waldron, A., Mooers, A. O., Miller, D. C., Nibbelink, N., Redding, D., Kuhn, T. S., . . . Gittleman, J. L. (2013). Targeting global conservation funding to limit immediate biodiversity declines. Proc Natl Acad Sci, 110 (29), 12144. https://doi.org/10.1073/pnas.1221370110

Warrell DA (2017) Clinical toxicology of snakebite in Asia. In: Handbook of clinical toxicology of animal venoms and poisons. CRC Press, Boca Raton, FL, pp 493–594

Warren-Thomas E, Dolman PM, Edwards DP (2015) Increasing demand for natural rubber necessitates a robust sustainability initiative to mitigate impacts on tropical biodiversity. Conserv Lett 8(4):230–241. https://doi.org/10.1111/conl.12170

Wich SA, Fredriksson G, Usher G, Kühl HS, Nowak MG (2019) The Tapanuli orangutan: status, threats, and steps for improved conservation. Conservation Science and Practice

Wilcove DS, Giam X, Edwards DP, Fisher B, Koh LP (2013) Navjot's nightmare revisited: logging, agriculture, and biodiversity in Southeast Asia. Trends Ecol Evol 28(9):531–540

Wilfred P, MacColl AD (2010) Income sources and their relation to wildlife poaching in Ugalla ecosystem, Western Tanzania. Afr J Environ Sci Technol 4 (12):886–896

World Population Review (2020) South eastern Asia population 2020. Retrieved from https://worldpopulationreview.com/continents/south-eastern-asia-population/

Wyatt T, Johnson K, Hunter L, George R, Gunter R (2018) Corruption and wildlife trafficking: three case studies involving Asia. Asian J Criminol 13(1):35–55

Zak AA, Riley EP (2017) Comparing the use of camera traps and farmer reports to study crop feeding behavior of moor macaques (*Macaca maura*). Int J Primatol 38 (2):224–242

Governance and Challenges of Wildlife Conservation and Management in Kenya

George E. Otianga-Owiti, Joseph John L. Okori, Stephen Nyamasyo, and Dorothy A. Amwata

Abstract

Wildlife in Kenya is both a national resource and a key source of revenue for the government. Wildlife and tourism are interdependent and essential sectors in Kenya's socio-economic development agenda. This chapter reviews the contribution of wildlife to tourism, wildlife management approaches, policy and legal framework, stakeholder involvement, as well as the challenges facing wildlife conservation and management. The insights and approaches illustrated may be used to formulate and implement solutions to enhance wildlife conservation and management for the benefit of all stakeholders. Kenya is at a crossroads with wildlife management. It is recommended that Kenya embrace a more holistic management approach that integrates effective political and related governance frameworks. This chapter proposes a novel vision of conservation in Kenya that includes additional space for wildlife, the adoption of a zero-tolerance policy on corruption and wildlife crime, substantial stakeholder participation, and a community-based approach to conservation.

Keywords

Governance · Challenges · Wildlife · Conservation · Management · Tourism · Kenya

Introduction

Kenya lies on the eastern coast of the African continent with the equator bisecting the country into two nearly equal parts. Kenya is bordered by the Indian Ocean, Uganda, Tanzania, Ethiopia, South Sudan, and Somalia. It has a total area mass of 582,646 km^{2,} with a land mass of 571,466 km^2. Approximately 20% of this mass is arable land, while 80% consists of arid and semi-arid lands (ASAL). Kenya has 7 unique ecosystems, including savannah, forest, woodland, mountain, fresh water, marine-coastal, and urban-cropland, and ranks among the world's richest biodiversity nations with over 25,000 species of animals and 7004 species of plants (Groombridge 1992; Rathbun 2009). Each ecosystem supports a diverse array of biodiversity that is of significant scientific, intrinsic, and

G. E. Otianga-Owiti (✉)
Department of Veterinary Anatomy & Physiology, Faculty of Veterinary Medicine CAVS, University of Nairobi, Kangemi, Nairobi, Kenya
e-mail: gowiti@uonbi.ac.ke

J. J. L. Okori
Enviro Wildlife Consult, Kampala, Uganda

S. Nyamasyo
Kenya Wildlife Service Training Institute, Naivasha, Uganda

D. A. Amwata
Department of Agricultural Sciences, South Eastern Kenya University, Kitui, Kenya

S. C. Underkoffler, H. R. Adams (eds.), *Wildlife Biodiversity Conservation*,
https://doi.org/10.1007/978-3-030-64682-0_4

economic value and has a considerable extent of wildlife habitat, with 10%–12% designated for biodiversity conservation (Government of Kenya 2009; Ojwang et al. 2017). Wildlife is currently hosted and managed in a variety of conservation areas through a landownership tenure system. Land in Kenya, according to Article 61 (2) of the Constitution 2010, is classified as public, or reserved for public use or environmental protection; community, or held by communities on basis of ethnicity, culture, or similar community interest; and private, or held by natural or legal persons (Republic of Kenya 2010a). Wildlife inhabit all categories, with the majority of wildlife inhabiting community and private lands (Waithaka and Western 2005; Western et al. 2009).

Wildlife tourism is a key contributor to Kenya's socio-economic development. Its contribution to Kenya's gross domestic product (GDP) and employment, together with other tourism-related sectors, will be discussed in later sections of this chapter. Wildlife and tourism are, by design, interlinked, with policies and legislative frameworks developed along cohesive lines. This cohesive approach promotes a strategy of connectivity between stakeholders, which are critical in ensuring that wildlife development does not go to naught due to social, economic, or political entanglements.

Wildlife and its habitat in Kenya face a suite of chronic and emerging threats, including human population pressure and associated pressure on resources, land use and land cover changes (LULCCs), poverty, climate change impacts, insufficient policies, and political influence or interference. The resulting impact of such threats on wildlife and their habitats includes habitat loss, land degradation, over-utilization of natural resources, poaching and illegal wildlife trade, pollution and invasive species, siltation and over-abstraction of water bodies, and human-wildlife conflict. Over the last four decades, significant negative impacts on wildlife species have resulted in a decline in wildlife populations, as well as a severe degradation of native habitat Grunblatt et al. 1996; Ottichilo et al. 2000; Ottichilo et al. 2001; Reid et al. 2008; Nyamasyo 2016; Ogutu et al 2016).

Significance of Tourism and Wildlife Sectors in Kenya

Tourism and wildlife sectors are closely interlinked in Kenya, as the tourism industry is primarily wildlife-based, with terrestrial wildlife as the primary draw. Over the past 15 years, tourism has contributed an average of 10.51%–11.26% toward Kenya's national gross domestic product (GDP). Kenya's Vision 2030 is the official developmental blueprint and recognizes tourism as an important sector in attaining the anticipated national GDP growth of 10% per annum (Government of Republic of Kenya 2007). As depicted in Fig. 1, the total contribution to GDP from tourism and travel has been fluctuating for the past 15 years. The sector's share of GDP has steadily declined from 13.84% in 2005 to 9.96% in 2018, although the contribution to socio-economic development in Kenya continues to remain significant. This trend appears to be replicated in the percentage share of total employees in Kenya (Fig. 2). The percentage contributions to the GDP and total workforce are mostly attributed to the numbers and earnings received, as depicted in Fig. 3. Despite the declining GDP contribution, the tourist numbers and earnings have been growing, with fluctuation occurring in 2007–2008. The key cause for the fluctuations in 2007–2008 was the post-election violence, with subsequent security concerns and negative travel advisories from some European source markets that may have influenced tourism to shift elsewhere. However, even with such figures, tourism continues to remain the second largest source of foreign exchange revenue in Kenya following agriculture. In 2018, 2.027 million people visited Kenya, with a revenue of 157.4 billion shillings generated, and tourism

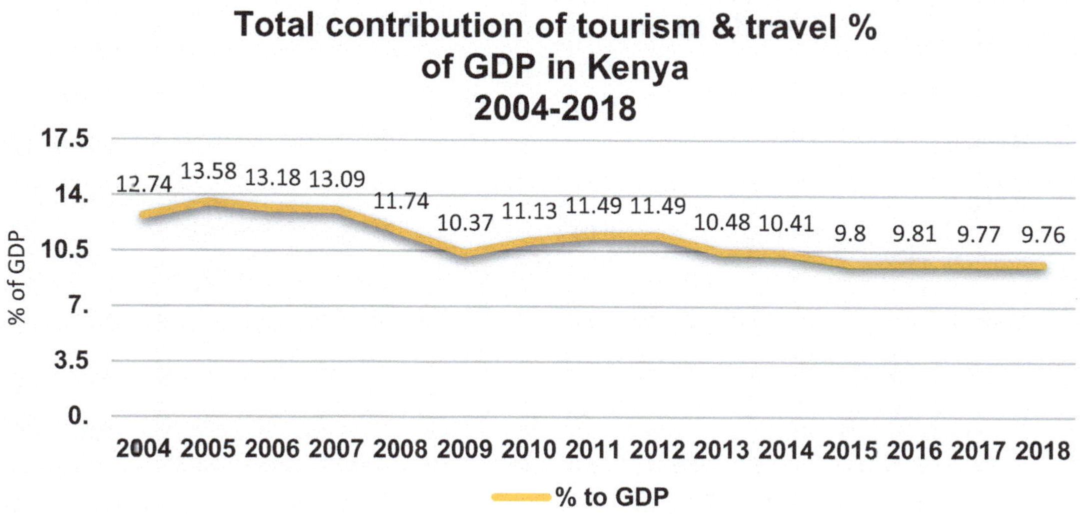

Fig. 1 Total contribution of tourism and travel % of GDP in Kenya 2004–2018 (Developed from data in Price, R.A. 2017)

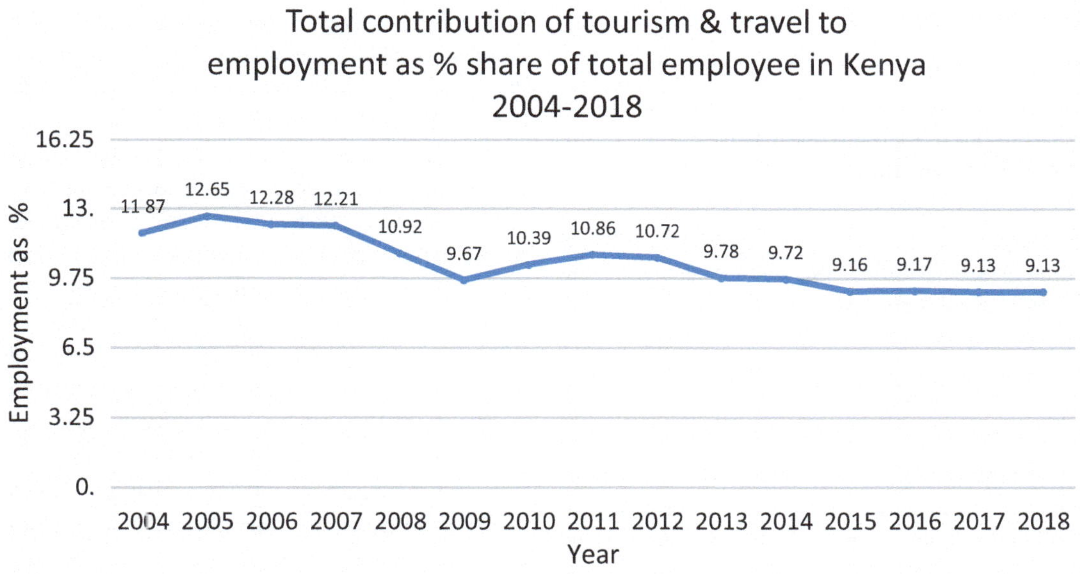

Fig. 2 Total contribution of tourism and travel to employment as % share of total employees in Kenya 2004–2018 (Developed from data in Price, R.A. 2017)

contributed 9.76% to the country's GDP. The United Kingdom has been the leading source market for Kenyan tourism, followed by the United States India, Germany, the United Arab Emirates (UAE), Italy, China, Canada, France, and the Netherlands (Okello 2014; Price 2017). In Vision 2030, Kenya aims to be among the top ten long-haul tourist destinations in the world offering a high-end, diverse, and distinctive visitor experience (Government of Republic of

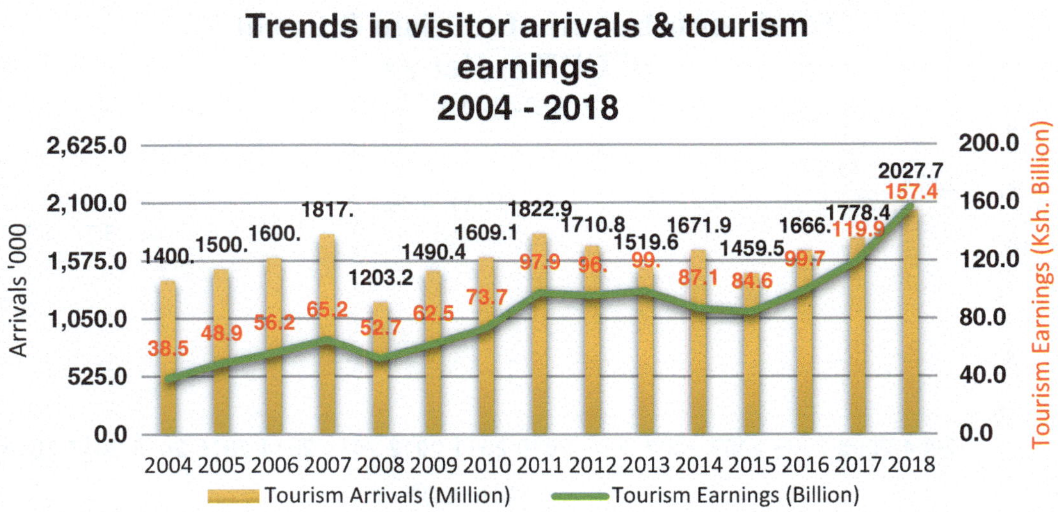

Fig. 3 International visitor arrivals and tourism earnings from 2004 to 2018 (Developed from Economic Survey Reports 2004–2018)

Kenya 2007; Ministry of Tourism and Wildlife 2018).

Conservation Management and Approaches in Kenya

This section looks at the development of different strategies of wildlife conservation and management in Kenya and examines some of the factors or conditions that lead to emergence and subsequent adoption of various management approaches. Wildlife conservation in Kenya still bears the scars of the colonial era, when colonialists deliberately excluded indigenous people from their land and hunting rights, utilizing wildlife exclusively for exploitation and recreation. As wildlife resources dwindled, colonial bureaucrats liberally employed statecraft, in which policies of social control for safeguarding flora and fauna that favored a neocolonial premise centered on conquest and land acquisition for government, the elites, and their hobbies. These included the exertion of draconian retribution for locals caught flouting conservation laws and regulations, as well as practices that limited indigenous access to land, wildlife, and other key commercial and subsistence resources, effectively exerting political and societal control over the indigenous people of Kenya (Neumann 2001; Neumann 2004a; Neumann 2004b; Waithaka 2012; Mwaura 2016; Kamau and Sluyter 2018; Cockerill and Hagerman 2020).

Kenya has been slow in adapting to the changing models of conservation and management of wildlife that is practiced in other parts of the world. After gaining independence in 1963, the Government of Kenya continued the legacy of its colonial masters by retaining game reserves or converting conservation areas into fully protected national parks, as a means of income generation for the central government and those with political influence. There are three wildlife conservation and management models which have been used in Kenya, namely, the informal conservation and management model; the protected area model, combined with an outreach approach; and the community-based conservation or community wildlife conservancy model. These models are discussed below.

Informal Conservation Model

Kenyan communities in pre-colonial times (prior to 1800) lived among and utilized wildlife resources without formally recognized policies and legislation, during which time wildlife thrived in abundance and diversity. Local communities ensured conservation of wildlife resources through proven cultural and social bonds, traditional customs, rules, taboos, beliefs, and practices. This model was used by various ethnic groups, and it enabled them to derive income and livelihoods from wildlife and natural resources, thus offering an important mechanism for maintaining balance in the ecosystem. Sacred beliefs were often tied to wildlife species, thus ensuring conservation principles as a way of life. Those who broke the law were punished by the ancestral spirits. In addition, recreational sport hunting was an abomination in most ethnic groups in Kenya and was disrespectful to the gods who provided the resource. This type of indigenous conservation practice persisted until the advent of colonialism. During colonization, this model was viewed as sub-standard, and western models were adopted. Western laws prohibited local communities from venturing into game reserves, as wildlife and their protected habitats were viewed as property of the white settlers (Neumann 2004a; Neumann 2004b; Waithaka 2012; Kamau and Sluyter 2018).

Protected Area (PA) Model

The PA model is a fortress conservation model adopted from the Yellowstone National Park model in the United States and forms the cornerstones of virtually all national and international conservation approaches in many countries of the world (Borrini-Feyerabend 1996; Mburu 2004; Jones 2006). It is considered as a top-down, "fence and fines approach," which is at times punitive to local human populations and was introduced during the colonial era. In Kenya, like in most parts of the developing world, the justification for the creation of PAs during the colonial era was often based on the grounds of preserving wildlife resources, including marine species and their habitats, for the benefit of those in power and/or the wealthy and privileged of society. Such policies of preservation resulted in a form of social control and spatial segregation between indigenous populations and wildlife, with fixed boundaries between nature, culture, and ultimately society, i.e., to "civilize the local population." The state protected areas currently stand at approximately 12.34% of the country's area, all of which have a focus on wildlife conservation. The areas are comprised of 23 national parks, 29,357 km^2 (5.2%); 28 national reserves, 18,042 km^2 (2.8%); 4 national sanctuaries, 37 km^2 (0.01%); 6 marine national reserves, 1063 km^2 (0.12%); 4 marine national parks, 76.3 km^2 (0.01%); and forest reserves, 18,979 (3.33%) (Republic of Kenya 1976; Ministry of Tourism and Wildlife 2018). These are administered as conservation areas (Fig. 4). This system has been posited as arguably one of the best in Africa. However, it must be stated that in setting up the PAs, the colonialist and post-independence government actions of annexation of land were associated with a prodigious number of displacements of local communities, the most controversial being the Tsavo National Park (Kassam and Bashuna 2004; Kamau and Sluyter 2018). Although at the moment this model appears to be one of hope for managing wildlife, in particular for threatened species, it has been considered by some scientists as a "conservation against the people" approach.

Tsavo National Park was carved out from an area of land considered unsuitable for agriculture or domestic livestock farming and unlikely to be required for any other form of land use in the foreseeable future in 1949. The colonial government deemed it viable land for elephant conservation, and as a result, human inhabitation was banned (Sheldrick 1973; Ayeni 1974; Schauer 2015). Local communities were forcibly evicted and further prevented from using the land as a livelihood resource. All hunting, livestock grazing, and land utilization were forbidden. Over the next few decades, displaced ethnic groups, whose livelihoods were closely tied to sustainable land use, suffered a loss of cultural

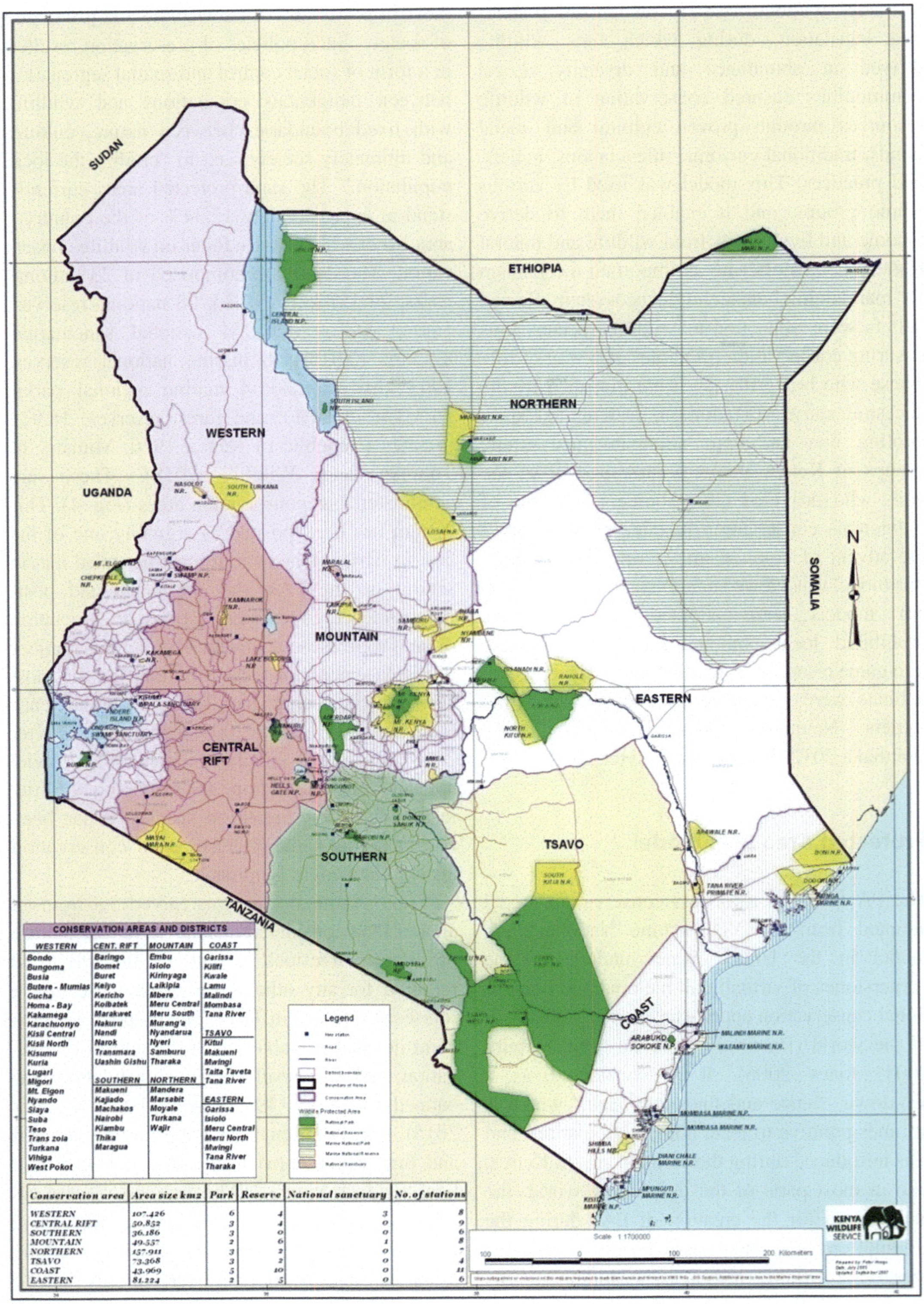

Fig. 4 Kenya Wildlife Service conservation areas (Source: KWS)

identity and became largely assimilated into the greater culture of Kenya (Bashuna 1993; Kassam and Bashuna 2004; Kamau and Sluyter 2018).

In order to enhance the biological integrity of the parks, a modified PA model was introduced in some areas, adding a more community-based approach through the provision of amenities, such as dispensaries, schools, water, and roads, for communities at the interface with the parks. In turn, this community focus helps to increase local support for conservation.

Community-Based Conservation Approach (CBCA)

The long-term goals of the CBCA are to empower local communities to be the overseers and beneficiaries of their diverse environments, whether that be through tourism, payment for ecosystem services, or revenue-sharing schemes, as they create space for wildlife and other components of biodiversity. In Kenya, it is estimated that approximately 65–75% of Kenya's wildlife is found outside of protected areas at any given time. With this realization in the early 1990s, the Kenya Wildlife Service adopted a form of CBCA in areas adjacent to PAs and adopted the philosophy "when wildlife pays, wildlife stays." CBCAs require that any derived benefits must be non-consumptive. Arising from this conceptual framework, landowners formed associations that would address key wildlife conservation challenges outside national parks. Some of the well-known associations include the Northern Rangelands Trust (NRT), the Laikipia Wildlife Forum, the South Rift Association of Landowners (SORALO), the Maasai Mara Management Association, and the Amboseli Ecosystem Trust. NRT appears to be the largest and currently works with 39 community conservancies across northern and coastal Kenya, managing an area just over 45,000 km^2 (Ministry of Tourism and Wildlife 2018). At the time of publication, there are 160 conservancies covering an area of over 63,600 km^2, with numbers expected to rise (King 2014; King et al. 2015; Ministry of Tourism and Wildlife 2018). It is apparent that most CBCAs involve public-private or private-community partnership (PPPs/PCPs) (Godfrey 2016). In these arrangements, communities partner with local, national, or external groups and organizations, including hospitality and tourism operators. Such groups take the form of partnership facilitators, managerial bodies, or direct investors. Only a few have complete community ownership and management. Opinions as to the success of this model are divided, with some promoting their success and pointing to an increase in the number of conservancies as a testimony of local community engagement (Thompson and Homewood 2002; African Wildlife Foundation 2016). However, others contend that some CBCAs are riddled with complications related to equitable benefit-sharing, pastoralist rights, favoritism, sustainability, and unfair partnership deals from governments (national/county) or foreign investors (Sibanda 1995; Rutten 2002; Mburu et al. 2003; Ondicho 2012; Godfrey 2016; Cockerill and Hagerman 2020).

Land managed under this model constitutes approximately 11% of the total country land mass, more than the total land area contained in Kenya's national parks, and is still expanding (African Wildlife Foundation 2016; KWS, personal communication 2017). This enables and promotes bridging the gap between the protected area and local stakeholders. It is a governance system that combines state/county control with local, decentralized decision-making and accountability.

In general, these models appear to be practiced in different areas of the country and in most situations run in tandem or alongside each other (Mburu et al. 2003; Mburu 2004). While all these conservation models have made a remarkable contribution to wildlife conservation management, more remains to be done to consolidate their efforts into a coherent, inclusive, and practical approach that would promote sustainable conservation in terms of biodiversity, ecosystem services, and socio-economic integrity. It is also important to mention that the creation of community wildlife areas and the empowering of local communities to sustainably manage natural resources are most often reliant on, or even established by, external support from either the government, non-governmental organizations (NGOs), or financial donors.

Stakeholder Analysis and Involvement in Wildlife Conservation and Management in Kenya

There is significant conflict in Kenya between various stakeholders on the issues of management and ownership of wildlife. In Kenya, wildlife is considered to be a government-owned resource, whether on public or private lands. This has created tension between stakeholders who desire greater autonomy and more benefits for keeping wildlife on their lands. In recent years, the government has made compromises in order to reach agreements among stakeholders, communities, landowners, businesses, international agencies, and non-governmental organizations. International agencies often play a key role in influencing which species to prioritize. This section will therefore look at the various stakeholders involved in the decision-making process.

Stakeholder Analysis and Involvement

There are a variety of stakeholders in Kenyan wildlife conservation and management, which have been placed into four categories, shown in Table 1.

Policy and Legal Framework of Wildlife Conservation and Management in Kenya

This section provides an overview of the policies and legislative framework that supports wildlife conservation and management in the country. It begins by providing an overview of evolution of wildlife policies and legislation in Kenya from pre-colonial, colonial, and post-colonial periods. It further highlights the different national and sectoral policies that have relevance to wildlife. Kenya's wildlife policy follows the theory of conservation through protection, where the conservation and management of wildlife are administered through a system of PAs that excludes local communities from active participation in their management. This is because the Government of Kenya, as is the case with most developing countries, follows western guidelines and philosophies of nature conservation. In this regard, wildlife conservation in Kenya thus continues to emphasize law enforcement to protect the wildlife resources.

Evolution of Wildlife Policies and Legislation

Policy and legislation regarding wildlife dates to 1898, after Kenya became a British protectorate and laws were enacted to control hunting and the trade of wildlife. In 1900, the East African Game Regulations were enacted, followed by the establishment of the Kenya Game Department in 1907 to manage game reserves. In 1945, the Royal National Parks of Kenya Ordinance was promulgated to provide for the establishment of national parks, which led to the creation of Nairobi and Tsavo National Parks. The Ordinance was further altered to become the National Park of Kenya Act. This Act led to the establishment of 56 protected areas (26 national parks and 30 reserves) (Republic of Kenya 1976). The Act created two institutions to administer wildlife policies, namely, the Kenya National Park organization and the Game Department.

The first wildlife policy in Kenya is the Sessional Paper No. 3 of 1975, entitled *A Statement on Future Wildlife Management Policy in Kenya* (Republic of Kenya 1975). Its role was to optimize the returns from this resource, taking into account the returns from other forms of land use. This policy focused on direct negotiations between the newly created Wildlife Conservation and Management Department (WCMD) and local communities on the future of wildlife in dispersal areas. It also led to the establishment of the Kenya Wildlife Service Training Institute (KWSTI) to train manpower for the wildlife industry.

Table 1 Key stakeholders involved in wildlife conservation and management in Kenya

Stakeholder	Level of action	Roles
Ministry of Tourism and Wildlife	National	• Policy formulation • Facilitation of good governance for tourism and conservation of wildlife • Marketing of Kenya as wildlife-based tourism destination
Kenya Wildlife Service (KWS)	National	• Conservation and management of wildlife resources in Kenya within and outside protected area • Establishment of networks and support for wildlife conservation with stakeholders/communities
The Judiciary	National	• Administer justice to all • Hearing and determination of wildlife-related cases
Ministry of Environment and Natural Resources, including the National Environmental Management Authority (NEMA) and Kenya Forest Service	National	• Policy formulation on environmental matters • Conserve, develop, and sustainably manage environmental resources including forest resources and environmental impact assessment and audit
County government	National	• Conservation of wildlife at county level • Enforcement of wildlife regulation
Kenya Tourist Board (KTB)	National	• Marketing Kenya as a tourist destination both locally and internationally
National Environmental Management Authority (NEMA)	National	• Environmental impact assessment and auditing
National Museum of Kenya (NMK)	National	• Conservation of wildlife specimens • Development of a wildlife database • Coordination of research and monitoring of wildlife
United States Agency for International Development (USAID)	International	• Funding conservation and management of wildlife • Combating wildlife trafficking
World Wide Fund	International	• Protecting the future of nature • Funding, lobbying, and advocacy
Wetlands International	International	• Protection of wetland and wetland resources • Funding, lobbying, and advocacy
Nature Kenya	International	• Conservation of birds and their habitats • Mapping of important bird areas (IBAs)
International Union for Conservation and Nature	International	• Conservation status of wildlife species • Protecting the future of nature
African Wildlife Foundation	Regional	• Conserving land, protecting species, and empowering people
International Fund for Animal Welfare (IFAW)	International	• Rescue of individual animals, safeguarding of populations, preservation of habitat, and advocacy for greater protections
BirdLife International	Regional	• Conservation of birds
African Conservation Centre (ACC)	Local	• Integration of knowledge, environment, and livelihoods in resolving problems facing biodiversity conservation in East Africa
Laikipia Wildlife Forum (LWF)	Local	• Forest management, conservation enterprise, human-wildlife conflict management, and environmental education

(continued)

Table 1 (continued)

Stakeholder	Level of action	Roles
Northern Rangeland Trust	Local	• Improving security for people and wildlife • Promoting trade and tourism • Conservation of wildlife
Save the Elephants	Local	• Elephant research
Kenya Wildlife Conservancies Association (KWCA)	National	• Wildlife governance • Lobbying and advocacy • Conservation of wildlife
Local communities	Local	• Establishment of community conservancies • Provision of more land for conservation
Community Forest Associations	Local	• Conservation of forest as wildlife habitats • Anti-poaching operations
Coastal Beach Management Unit (BMU)	Local	• Conservation of marine ecosystems • Beach operations • Cleaning of the beach
Academic and research institutions (universities, colleges)	National	• Provision of information for conservation • Education and awareness • Research
Wildlife Clubs of Kenya (WCK)	National	• Education and awareness
Media	National and local	• Marketing • Education and awareness
Private ranches and conservancies	Local	• Hold a significant percentage of wildlife on their land • Provide land for conservation of wildlife • Promote and invest in wildlife conservation

In 1976, the Wildlife (Conservation and Management) Act was enacted. This Act amalgamated the Game Department and the Kenya National Parks to form a single agency, the WCMD. Despite its positive impacts, there were concerns of corruption and inefficiency within the department, resulting in a reduction of relative effectiveness of national park management operations in PAs and beyond. Both the Wildlife Policy and the Act failed to tackle poaching, human-wildlife conflict (HWC), and loss of biodiversity, due to inadequate legal framework, political and bureaucratic interference, and massive corruption. Further, local communities had no access to wildlife benefits, despite coexisting with wildlife on communal lands (Honey 1999; Waithaka 2012).

In 1989, the Wildlife (Conservation and Management) Act was amended (Act No. 16 of 1989) to create the Kenya Wildlife Service (KWS), which would replace the WCMD. The legislative establishment of KWS was followed in 1990 by the elaboration of a comprehensive framework of policy and implementation strategies known informally as the "Zebra Books" and more formally as "KWS Policy Framework and Development Programme 1991–1996." Through this framework, the Community Wildlife Service (CWS) was created to forge co-management or partnerships with communities outside the PAs, which enabled them to derive direct cash benefits from wildlife on their land, as well as a reduction in poaching and human-wildlife conflict (KWS 1995; KWS 1996; Rutten 2002; Mburu 2004; Waithaka 2012).

The Wildlife Conservation and Management Act of 2013 became operational on 10 January 2014, after the repealing of the Wildlife Conservation and Management Act Cap 376. The implementation of the Act was guided by the principles of devolution on conservation and management of wildlife to landowners and managers in areas where wildlife occurs, the recognition of wildlife conservation as a form of land use, better access to benefits from wildlife conservation, and adherence to the principles of sustainable utilization. The new Act has also enhanced penalties for wildlife crimes and reviewed wildlife compensation, which has been a thorn to those living with wildlife. However, the Statute Law (Miscellaneous Amendments) Act No. 18 of 2018 substantially amended the Wildlife Conservation and Management Act of 2013. This amendment introduced new offenses, higher penalties and expansion of some offenses, and no compensation for death and injury caused by poisonous snakes, sharks, stonefish, whales, stingrays, and wild pigs, as well as damage to crops, livestock, and property caused by snakes, zebras, wildebeest, wild dogs, and eland.

Current Legal Frameworks Supporting Wildlife Conservation and Management in Kenya

Over the years, Kenya, like most African countries, has struggled to find and implement appropriate legal and policy instruments that can protect as well as allow sustainable management of its natural resources, especially wildlife and its habitats. This section highlights the diverse legal and policy frameworks currently employed in Kenya to manage and conserve its biological resource with a focus on wildlife (see Table 2).

Challenges of Wildlife Conservation and Management in Kenya

Wildlife conservation management in Kenya faces several drivers and threats/pressures, namely, a rapidly expanding human population, land use changes, poverty, climate change, poaching, limited human expertise, inadequate financial resources, conflicting policies, land transformation, and encroachment of wildlife habitats, leading to a decline in wildlife numbers. As outlined earlier in the introduction, Kenya's wildlife numbers for both fauna and flora have declined in the past few decades. This decline in terms of wildlife numbers can be attributed mostly to numerous anthropogenic activities, including climate change (Grunblatt et al. 1996; Noe 2003; Lamprey and Reid 2004; Reid et al. 2008; Norton-Griffiths and Said 2010; Okello and Kioko 2010; Ogutu et al. 2011, 2014, 2016; Ojwang et al. 2017). Studies and reports including the Economic Survey of 2005–2018 demonstrate that the population of popular wildlife species such as zebras have declined by 21%, giraffes (25%), elands (32%), and buffaloes (27%) between 2005 and 2018 (Ogutu et al. 2014, 2016; Economic Survey Report 2018; Aduma et al. 2018). This rapid decline has affected carnivore species (leopards, cheetahs, lions, hyenas, and wild dogs) that depend on such herbivores. However, the numbers of charismatic species like elephants and rhino have increased, and this can be attributed to the level of protection conferred to them. Currently, 325 wildlife species (including fish and plants) in Kenya are listed as threatened. Of these, 34 are classified as critically endangered, 93 are endangered, and 198 are vulnerable (Weru 2016).

The problems of managing wildlife are in many ways akin to the problems of managing any common-pooled resources. The distinctive features of wildlife as a common-pooled resource are comprised of low ownership, mobility of the resource, non-recognition of user rights, criminalization of its use, difficulty of monitoring the resource, and low barriers to its entry in the exploitation of the resource. All of these features imbue wildlife with the characteristics of common-pooled resources and usually encourage "free rider" behavior (free access situation without any restriction or control), whereas those who exploit the resource have little ability or incentive to manage it sustainably. With these factors in mind, the challenges facing wildlife conservation and management potentially compromise the sustainability (and to an extent viability) of wildlife in Kenya. The multifaceted challenges are comprised of diverse drivers that change with time. The survival of Kenya's wildlife is a race against these drivers and pressures. With the passage of each decade, the task of protecting wildlife and its habitats becomes increasingly difficult, as drivers and various pressures acting

Table 2 The current legal frameworks supporting wildlife conservation and management in Kenya Modified from Government of Kenya 2007; 2012; 2013; 2015; Republic of Kenya, 1976; 1989; 2009; 2010a; 2010b; 2010c; 2012; 2017)

Legal framework	Type of framework	Objective
The Constitution of Kenya 2010	Constitution	• Provides grounds for the formulation of wildlife management legislation, policies, and strategies on sustainable management of the environment and natural resources. Article 69 emphasizes the duties of state organs and its people to protect and conserve the environment and natural resources including wildlife
Tourism Act No. 28 of 2011	Act of Parliament	Enacted to provide for the development, management, marketing, and regulation of sustainable tourism and tourism-related services, in concert with relevant stakeholders (lead agencies and communities), and realization of the economic blueprint, as laid out in Vision 2030
National Policy on Arid and Semi-Arid Lands (*Sessional Paper No. 8 of 2012*)	Sectoral Policy 1	• This is a Sessional Paper No. 8 of 2012 on the National Policy for the Sustainable Development of Northern Kenya and other Arid Lands. Policy acknowledges pastoralism as a legitimate and productive livelihood and gives attention to wildlife, which is one of the drivers of pastoral livelihoods, through tourism and employment
National Water Policy of 2012 (NWP, 2012)	Sectorial Policy	• Developed in line with the mandate, vision, and mission of the ministry responsible for water affairs in Kenya
Wildlife Conservation and Management Act, 2013	Act of Parliament	• Provides for protection, conservation, and management of wildlife and related matters in Kenya. It applies to all wildlife resources on public, community, and private land and Kenya territorial waters
Environmental Management and Coordination Act, 2009, and Environmental Management Coordination (Amendment) Act 2015	Act of Parliament	• This is the principal instrument of government for the management of the environment and provides for the relevant institutional framework for the coordination of environmental management • Imposes restrictions necessary for protection from environmental degradation and guarantees to all citizens the right to a clean and healthy environment
National Spatial Plan—2015–2045	Strategy	• This plan provides a national spatial structure that defines how the national space is utilized to ensure optimal and sustainable use of land, including wildlife conservation • Environmental protection and conservation zones with major environmentally sensitive areas included water towers, flood plains, indigenous forests, marine parks, wetlands, and national parks/reserves/conservancies
Water Act 2016	Act of Parliament	• Provides for the regulation, management, and development of water resources (lakes, aquifers, and rivers) and water and sewerage services in line with the Constitution
Forest Conservation and Management Act of 2016	Act of Parliament	• Makes provision for the conservation and management of public, community, and private

(continued)

Table 2 (continued)

Legal framework	Type of framework	Objective
		forests and areas of forest land that require special protection, defines the rights in forests, and prescribes rules for the use of forest land, including the protection of wildlife and flora populations
Land Act of 2012 and Land Amendment Act of 2016	Act of Parliament	• Governs the management and administration of public, private, and community land, as outlined in the Constitution • Conservation of land-based natural resources, especially where land is situated within a PA • The Amendment Act of 2016 has outlined the role the National Land Commission (NLC) responsible for managing public land on behalf of national and county governments, in which a number of conservation areas lie
Fisheries Management and Development Act of 2016	Act of Parliament	• Provides for the conservation, management, and development of fisheries and other aquatic resources to enhance the livelihood of communities dependent on fishing and key fisheries institutions • It also implements obligations under international law concerning fisheries
Mining Act 12 of 2016	Act of Parliament	• Provide for prospecting, mining, processing, refining, treatment, transport, and any dealings in minerals, as well as for related purposes, including the rights to mining required in land falling with protected conservation areas
Community Land Act of 2016	Act of Parliament	• Deals with the recognition, protection, and registration of community land rights and the management and administration of community land, including the special rights and entitlements associated with community land
Natural Resources (Classes of Transactions Subject to Ratification), Act of 2016	Act of Parliament	• Involves the grant of a right or concession for the exploitation of any natural resource in Kenya subject to ratification by Parliament • Classes are set out in the Schedule to this Act and include among others wildlife (export and re-export of endangered wildlife species, as well as the extraction of oil, gas, and minerals within a wildlife protection area)
Protection of Traditional Knowledge and Traditional Cultural Expressions Act of 2016	Act of Parliament	• Provide a framework for the protection and promotion of traditional knowledge and cultural expressions, including intellectual property right (IPR), held by community
Climate Change Act of 2016	Act of Parliament	• Provides a regulatory framework for enhanced response to climate change and offers mechanisms and measures to improve resilience to climate change and promote low-carbon development
National Land Use Policy of 2017 (*Sessional Paper No. 1 of 2017*)	Sectoral Policy	• Provides legal, administrative, institutional, and technological framework for optimal utilization and productivity of land-related resources in a sustainable and desirable manner at national, county, and community levels

(continued)

Table 2 (continued)

Legal framework	Type of framework	Objective
		• The Policy focuses on conservation and sustainable management of land-based natural resources, mapping, identification, and gazettement of biodiversity areas. In addition, it provides incentives for community participation in conservation of natural resources and the environment
Vision 2030	Strategy	• This is the country's developmental blueprint which aims to achieve a clean, secure, and sustainable environment by 2030 • Wildlife conservation management and tourism is expected to contribute to the economic and social pillars of Vision 2030
National Wildlife Strategy 2030	Strategy	• Provides a mechanism to coordinate the wildlife sector and implement the Wildlife Conservation and Management Act (2013), as well as to bring Kenyans together through a shared vision for wildlife as a cornerstone of our social, cultural, environmental, and economic development
National Wildlife Conservation and Management Policy of 2017	*This document is still in progress, with annual updates. Kenya's wildlife policy is still embodied in the Sessional Paper No 3 of 1975* National Wildlife Conservation and Management Policy of 2017 *Sessional Paper No. 01 of 2020 on Wildlife Policy - June 2020 is currently in circulated for Stakeholders contribution*	

on the resources build up and become more complicated. This section reviews drivers/threats/pressures that have led to the decline of wildlife population and its habitats.

The strategic approach to wildlife conservation shows that the main causes of wildlife decline in Kenya, as in most parts of modern Africa, can be categorized into three drivers or threats, namely, *proximate*, *ultimate*, and *social* (see Table 3). The *proximate drivers* are those threats that account immediately for wildlife decline; they can be addressed effectively by a combination of national investment, conservation-development projects, and international agreements and action. *Ultimate drivers* refer to the wider changes in society that ultimately bring about the proximate threats; however, not much can be done to address these drivers. The last category is the *social drivers*, encompassing the socio-economic, political, and institutional weaknesses within society that undermine conservation efforts.

Looking at the various drivers and threats facing wildlife conservation and management in Kenya as outlined in Table 3, it can be gathered that they are varied and dependent upon the conservation area or wild species being reviewed. The task of managing existing wildlife habitats and establishing new natural areas has become increasingly difficult. The key drivers that make it difficult are variable, but the outcomes include a plethora of negative consequences, decline in wildlife numbers, reduction and modification of wildlife areas or habitats, and conflicts. Numerous studies have examined the causes of decline of wildlife populations in different parts of Kenya, with the biggest threats being centered mostly on human population growth, land tenure reforms and policies, severe economic stress and rising poverty, conflict, and climate change.

Table 3 Drivers of wildlife and habitat loss

Category of threat	Drivers of wildlife and habitat loss
Proximate drivers	• Demand-driven illegal and unsustainable off-take of wildlife (poaching)—elephant, rhino, and bushmeat trade of other wild species
	• Settlement and accompanied development lead to fragmentation and loss of habitat and species' range through alteration and conversion of natural ecosystems
	• Functional failure in protected and other non-protected areas (inadequate coverage, lack of investment in management, encroachment/excision, insecure tenure to land and illegal allocation, poaching) • Livestock incursion
	• Failure in governance of wildlife industry
	• Human-wildlife conflict in the wider sense (pesticides, pollution, roadkill, farm-wildlife conflicts)
	• Invasive alien species-ASALs and aquatic ecosystems
Ultimate drivers	• Human population growth in Kenya
	• Rising demand for land and natural resources leading to loss of habitat and increased poaching
	• Climate change
Social drivers	• Political indifference to wildlife issues and conflicts
	• Economic stress and poverty
	• Legal and policy frameworks that promote "fortress management" of protected areas
	• Competition with livestock
	• Financial constraints and underfunding of parks and reserves
	• Lack of conservation policy that is embedded in African society and alienation or inadequate involvement of locals
	• Inadequate incentives (for communities and landowners) to adopt land use practices compatible with wildlife conservation and management

Source: Modified from EU (2014)

Causes of Loss and Fragmentation of Wildlife Habitat in Kenya

Human Population Growth

As of 2019, the total population of Kenya stands at 47.6 million people and has exponentially increased over the last 57 years from 8.6 million people in 1962 (KNBS 2019). Kenya's current population growth rate, at 2.2% per annum, is one of the highest in the world (UN 2017; KNBS 2019). It is expected to reach nearly 55–60 million in 2030 and 77 million by 2050 (NCPD 2013). The recent population growth saturation in urban areas of Kenya, together with escalating poverty, has forced immigration of humans into more arid lands with lower potential for sustainable agriculture. In the last 15 years, pastoralists have been transforming to more sedentary lifestyles in the ASALs of Kenya. The transformation from pastoralist to agrarian society has placed additional pressure at the human-wildlife interface. There are numerous examples where ASALs have been converted to settlement and metropolis cities. To date, population growth around the parks and reserves has continued to be a major setback to wildlife conservation and management (Western et al. 2009; Elliot et al. 2013; Bhandari 2014; Ogutu et al. 2014).

The rapid rise in human population has increased demand and competition for resources, resulting in an augmented exploitation of resources at the highest level, beyond the capacity of available resources (Scholte 2011; Kideghesho et al. 2013; Nyamasyo 2016). The demands are associated with wildlife and habitat destruction, including land for settlements, cultivation and livestock grazing, wood products, and water points for livestock and domestic use (Kideghesho et al. 2013). Settlements are expanding more rapidly nearer to the PAs and associated wildlife dispersal areas and migratory corridors as a result of enhanced anthropogenic activities (Lamprey and Reid 2004; Western et al. 2009; Kideghesho et al. 2013; Ojwang et al. 2017; Mukeka et al. 2018). Human population growth is driving requirements for more food, hence expanding agricultural activities. Expansion of agriculture destroys natural habitats, alters landscapes and ecosystem services, and fuels human-wildlife conflicts and subsequently reduces local support for conservation. Increased agriculture results in fencing off farms, thereby disrupting wildlife movement and migration, causing a loss of tourism due to decreased aesthetic appeal (Reid et al 2008; Elliot et al. 2013; Bhandari 2014; Ogutu et al. 2014, 2016).

The role of human population growth and the changing lifestyles for those living adjacent to protected areas generates conflicts that can be summarized in three categories: (i) disruption of ecological processes and functions that are essential in maintaining wildlife and related biodiversity, (ii) increased illegal and unsustainable off-take of wildlife and their products, and (iii) increased pressure from local people to open protected lands for community use, some of which may not be compatible with wildlife management. The latter may include regular uncontrolled burning (which sometimes emanates from areas outside of the park boundaries) and arbitrary extension of boundaries (Kideghesho et al. 2013).

Land Tenure Reforms and Policies

Tenure systems may contain many categories of rights (e.g., rights to ownership, right to use, right to access, right to control, and right to transfer), and in Kenya, it has been a source of conflict and political debate (Kameri-Mbote and Kindiki 2008; Doshi et al. 2014). The new Constitution of Kenya has given a clear direction on matters related to landownership, use, and management. As outlined in Doshi et al. 2014 and Kameri-Mbote (2019), Chapter 5 of the Constitution of Kenya 2010 Article 62 states that "all land in Kenya belongs to the people of Kenya collectively as a nation, as communities, and as individuals." Thus, land is classified as public land, private land, and community land (Republic of Kenya 2010a). Land is the most sought-after resource in Kenya. Land has deep cultural importance for Kenyans and, in the current economy, represents the only livelihood option for many.

Several case studies have shown that land use changes are driven by a combination of resource scarcity, changing opportunities created by markets, inappropriate policy intervention, loss of adaptive capacity and increased vulnerability, and changes in social organization, resource access, and attitudes (Lambin et al. 2003). Land use and land cover changes (LULCCs) are manifested through conversion and modification, which are caused by interactions between climatic and anthropogenic forces owing to its inherently complex nature. The main drivers of LULCCs can be divided into seven factors: multiple causes, natural variability, economic and technological, demographic, institutional, cultural, and globalization (Lambin et al. 2003). Land use changes are driven by human actions, and subsequently alterations limit availability of products and services for humans, livestock, and wildlife, and this can further undermine environmental health and biodiversity distribution. All these drivers are present in the Kenyan land system.

With Kenya's human population at 46.7 million people, and with a prediction of 60 million by 2030, the demand for land is on the rise. For example, ASAL population is ever-increasing, placing more pressure on wildlife habitat. Land is one of the most significant resources in Kenya, as it is the foundation for activities such as agriculture, wildlife conservation, urban development, human settlement, and infrastructure development. Wildlife conservation was historically excluded as a recognized form of land use in Kenya, which was exacerbated by the lack of adequate and effective national land use policy and planning. In recent years, the negative impact of other land use types such as agriculture or rural and urban development on wildlife conservation has been recognized in several ecosystems such as Maasai Mara, Amboseli, and Laikipia (Worden et al. 2003; Georgiadis et al. 2007; Reid et al. 2008; Kioko et al. 2008; Kioko and Okello 2010; Ogutu et al. 2014; Nyamasyo and Kihima 2014). Despite an overall decline of 70% in wildlife populations in the Maasai Mara System, it remains one of the richest and most diverse landscapes in Africa. It is host to more than 95 species of mammals, over 550 species of birds, and thousands of insect species. The seasonal movements of hundreds of thousands of ungulates, such as zebras, gazelles, and, in particular, wildebeest (known as "the great migration"), contribute to its preference as a spectacular and highly popular tourist attraction (Ottichilo et al. 2000; Ottichilo et al. 2001; Nelson 2012; Elliot et al. 2013).

Land use in Kenya is changing rapidly, as much of the land is being transformed into farmlands, grazing lands, human settlements, and urban centers, at the expense of native ecosystems. The deliberate policy of subdividing land traditionally held as communal, and the provision of unrestricted access to resources by pastoralists, wildlife, and smallholder farmers, contributes to habitat fragmentation and restricted access by key stakeholder groups. Population growth and land use change alter the interactions of people and animals in terms of animal numbers and species diversity (Campbell et al. 2000; Maitima et al. 2009).

As the land use changes in favor of human activities (expansion of agriculture, settlements, fences, infrastructure, and demand for fuel wood), the fragmentation and degradation of natural habitats lower the numbers of large mammals that such spaces can support and further accelerate the local extirpation of wildlife populations (Ogutu et al. 2011, 2014, 2016). Additionally, fragmentation and incompatible land use practices impact ecosystem resilience, reducing support mechanisms for maintaining biodiversity. In some instances, habitat destruction may interfere with wildlife migratory corridors and dispersal areas and thus decrease predator-prey interactions and other ecological factors (Okello and D'Amour 2008; Okello and Kioko 2010; Fynn and Bonyongo 2010; Ojwang et al. 2017).

Human-Wildlife Conflict (HWC)

Human-wildlife conflict (HWC) is frequently defined as conflict that occurs between people and wildlife (Woodroffe et al. 2005). It encompasses actions by humans or wildlife that have an adverse effect on one another; threats posed by wildlife to human life, economic security, or recreation; or the perception that wildlife threatens human safety, health, food, and property (Treves and Karanth 2003; Peterson et al. 2011; Redpath et al. 2013; Fisher 2016; Nyhus 2016). In Kenya, the primary cause of HWC is competition for finite natural resources and space. Kenya's HWC can be categorized into two primary areas: (i) true problems between animals and humans and (ii) interpersonal conflicts over wildlife and its habitats that occur between individuals and stakeholders. These types of conflict are evident between communities, landowners, wildlife agencies, and personalities that manage such resources. As in most parts of the world, in Kenya, conflicts between humans and wild animals occur when either the need or behavior of wildlife impacts negatively on human livelihoods or when humans pursue goals that impact negatively on the needs of wildlife.

HWC escalated with the establishment of PAs, which were believed to be the most feasible strategy of maintaining biodiversity. Most of the PAs in Kenya are situated in the rangelands within the ASALs, with a few close to agricultural areas. However, given the multiple uses of the rangelands, decisions to allocate lands for conservation have often faced resistance. This type of land use is perceived as an infringement on the rights of other local communities. Such is the case when the conservation process involves evicting people from these areas and/or denying them access to critical livelihood resources. With regard to policy in Kenya, priority has historically been given to wildlife over local communities. Some examples include the eviction of indigenous communities like the Taitas, Waata, Ndorobos, and Maasai, among others, in order to provide room for wildlife conservation, which has taken place in almost all PAs of Kenya, justified by expansion of national parks and creation of game reserves. The eviction has, over time, worsened the conflicts between these parks and surrounding local communities.

In this section, we shall restrict the discussion to wildlife behavior that is usually perceived to negatively impact social, economic, or cultural aspects of human life or species of conservation concern, i.e., “human-wildlife impacts” (Redpath et al. 2013; Snijders et al. 2019). Conflict between humans and wildlife is one of the most widespread and intractable issues facing conservation agencies today in Kenya and Africa as a whole. To many who live in Kenya, wildlife is a threat and a liability. Conflicts between humans, livestock, and wildlife in Kenya are heightened by the expansion of human and livestock populations, cultivation, infrastructural developments, barriers, and settlements that reduce space and other resources for wildlife. HWC encompasses a huge diversity of situations and species, from grain-eating rodents, to man-eating lions, and to the largest mammal of terrestrial soil, the elephant. Living alongside such species can impose a variety of significant costs upon local communities. Livestock depredation, crop damage, human attack, disease transmission to livestock, loss of livelihoods, and fueling of poverty are major examples of HWC consequences that communities face (Musyoki et al. 2012; Ogada and Nyingi 2013; Mukeka et al. 2018, 2019). However, for others, wildlife can be a source of income, such as through tourism, which places a priority on areas with higher biodiversity and an abundance of fauna for wildlife viewing (Okello 2014). HWCs are escalating as human resource demands increase. In Kenya, it has become a persistent problem and a major threat to wildlife conservation and management efforts. Each year, it is fueled by the changes in increased human population growth, land use changes (including blockage of wildlife migratory corridors and dispersal areas), high livestock and wildlife population densities, climatic change, and changing perceptions by some communities living with wildlife (Makindi et al. 2014; Ojwang et al. 2017; Ministry of Tourism and Wildlife 2018). Figure 5 depicts the key hotspots of HWC, primarily in the counties of Lamu, Taita-Taveta, Laikipia, Nyeri, Narok, Kajiado, and Baringo. In general, HWC exhibits seasonal and annual fluctuations, reflecting underlying precipitation variations in most areas. Elephants are the most problematic species due to crop damage, property destruction, and human attacks, some of which result in death. Carnivores lead in livestock depredation in areas where livestock keeping is the main livelihood in these counties.

Crop raiding is most acute where maize, tomatoes, vegetables, beans, and wheat are grown at small and large scales (*personal observation*; KWS, personal communication; Long et al. 2019). Crop raiding species, in order of damage inflicted, are elephants (*Loxodonta africana*), baboons (*Papio* spp.), buffalo (*Syncerus caffer*), hippos (*Hippopotamus amphibius*), vervet monkeys (*Cercopithecus* spp.), and zebras (*Equus quagga*). Birds are a menace where wheat is grown (KWS, personal communication). In most areas, crop raiding usually peaks in the late rainy season when crops have matured. Areas with highest crop destruction by wildlife in the past decade include Lamu, Taita-Taveta, Narok, Laikipia, Meru, and Chyulu areas (personal observation; Makindi et al. 2014; Mukeka et al. 2018, 2019; KWS, personal communication). Livestock predation occurrence also varies depending on the area of the country and season, with greater predation occurring when the natural prey density is lowest or where there are a high density of livestock and poor protective measures (Mukeka et al. 2018, 2019). Lions appear to prefer predating on cattle, while leopards and spotted hyena (*Crocuta crocuta*) kill mostly shoats. Attacks on humans are predominantly limited to elephants, buffaloes, lions (*Panthera leo*), hippos, crocodiles (*Crocodylus niloticus*), hyenas, and snakes (Serpentes suborder) (personal observation; KWS personal communication 2019; Mukeka et al. 2018, 2019; Long et al. 2019).

Illegal and Unsustainable Off-Take of Wildlife and the Bushmeat Trade

This encompasses poaching and overexploitation (uncontrolled harvesting) of different plant and animal species (bushmeat), both of which are considered to be key contributors to the current conservation crisis and major contributors to the decline of biodiversity in Kenya and other East African savannah areas (Lindsey et al. 2013).

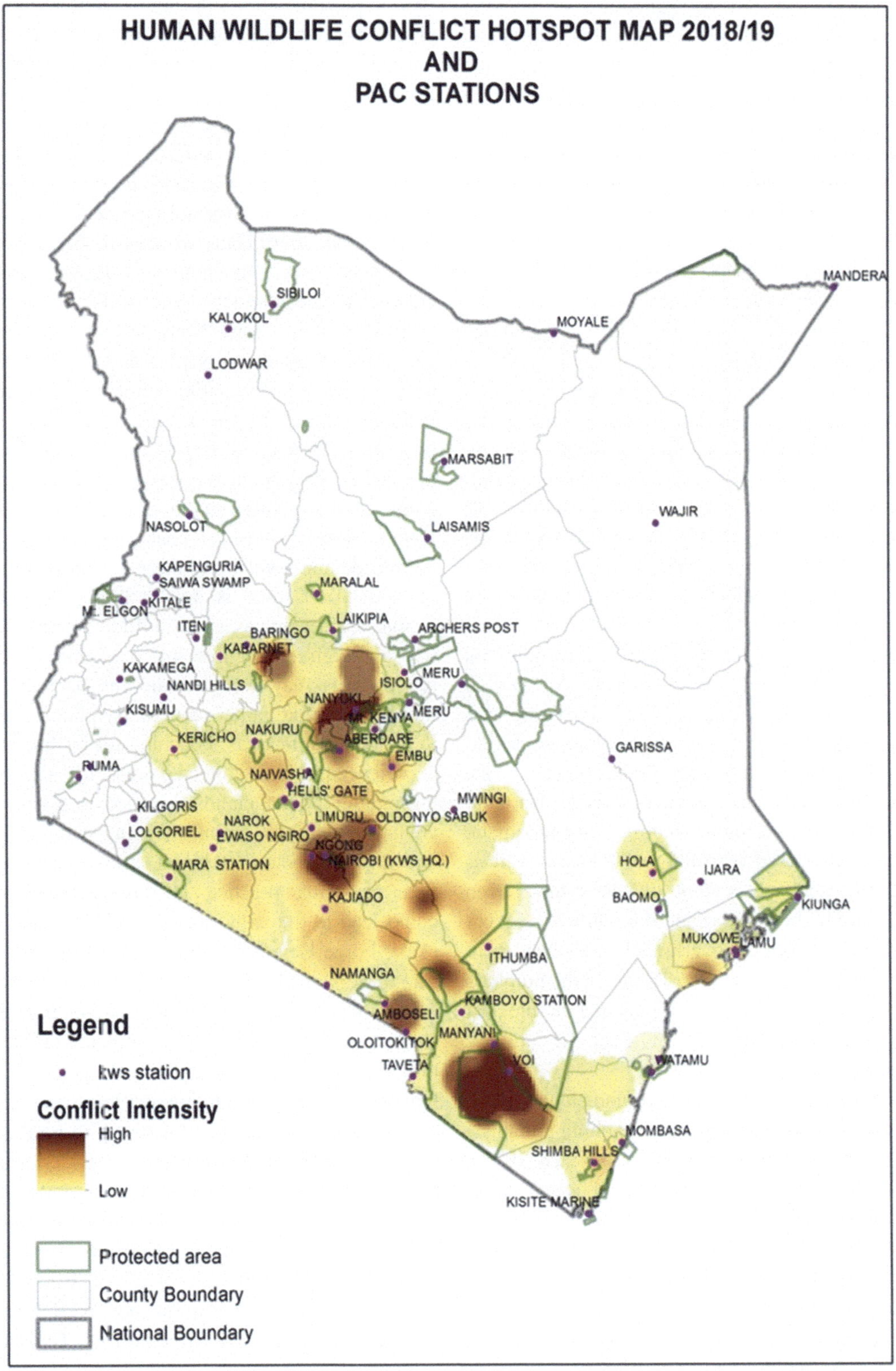

Fig. 5 Human-wildlife conflict hotspot areas in Kenya (Source: KWS)

Poaching and overexploitation of natural resources are driven by several factors comprising poverty, availability of lucrative markets, and lack of institutional capacity in implementing anti-poaching laws. Some of the critically endangered species in Kenya that are poached or exploited for bushmeat or trophies include elephant, lion (*Panthera leo*), both black and white rhinos (*Diceros bicornis* and *Ceratotherium simum simum*, respectively), Grévy's zebra (*Equus grevyi*), cheetah (*Acinonyx jubatus*), leopard (*Acinonyx jubatus*), hirola antelope (*Beatragus hunteri*), Eastern red colobus (*Procolobus rufomitratus*), Sokoke scops owl (*Otus ireneae*), roan antelope (*Hippotragus equinus*), Rothschild's giraffe (*Giraffa camelopardalis rothschildi*), and East African sandalwood (*Osyris lanceolata*) (Wildlife & Conservation Act, 2013; Weru, 2016). While biological diversity within some of the PAs remains high, incidents of illegal extraction are common (Weru 2016).

Kenya is a signatory to the Convention on International Trade in Endangered Species of Wild Fauna and Flora (CITES). Poaching was historically a critical issue for elephant conservation in Kenya. In recent years, however, the poaching crisis has declined. KWS efforts to stem poaching through monitoring of charismatic species, increased patrolling of PAs, as well as arresting and prosecuting poachers have been scaled up in the last 6 years. For national reserves managed by county governments, however, there is a need for improvement. Trends in the numbers of the rhino and elephant poached in the last 15 years are depicted in Figs. 6 and 7, respectively.

Though the country's rhinos and elephants are kept under close surveillance, poaching remains a serious threat to their survival. In the early 1970s, Kenya's population of black and white rhino numbered around 20,000. Between 1980 and 1999, the numbers greatly declined and have only recently increased from a low of approximately 350 in 1983/1974 to 1367 in 2018, due in part to the efforts of the KWS and support from other government agencies like the National Police Service and National Intelligence Agency (KWS, personal communication). With regard to elephants, the international ivory trade ban enacted in 1989 has helped in recovering elephant populations across Kenya, and as of 2018, Kenya had 33,136 elephants (KWS, personal communication). Kenya has not suffered the onslaught of ivory poaching as witnessed in other African countries such as Mozambique and Tanzania (Chase et al. 2016; Hauenstein et al. 2019); however, when such incidents occur, interdiction efforts have been mostly successful.

Bushmeat has long been part of local consumption in many parts of Kenya, and recent trends indicate an escalating number of poaching incidents linked to the killing of wildlife for bushmeat (Task Force Report on Wildlife Security 2014). This practice poses a significant challenge to both conservation and ecotourism in Kenya. Subsistence bushmeat poaching has hit unprecedented levels, and the growing commercial bushmeat trade is now a highly lucrative business, emerging as a multimillion dollar industry, although no figures are available in this report. This may well explain the decrease in numbers of wild game, particularly plains game, in major wildlife areas. Poaching for bushmeat is experienced in both protected areas and non-PAs throughout the rangelands. All species of wildlife are harvested indiscriminately using snares, bows and arrows, spears, clubbing, and occasionally firearms. This practice is unsustainable and could lead to the extermination of many species (Task Force Report on Wildlife Security 2014). The drivers of illegal hunting stem directly from local consumption (subsistence) and/or immediate local community trade to commercial trade in urban centers or even international markets. The key drivers of illegal hunting and bushmeat trade are increasing demand for bushmeat in rural and urban areas. However, when one considers studies done in Africa and the results specific to each country or region, illegal hunting for bushmeat was considered to be the most serious threat facing wildlife in protected areas in Botswana, Malawi, Mozambique, Zambia, and West Africa; the second most serious issue in Tanzania, Zimbabwe, and Central Africa; and a less serious

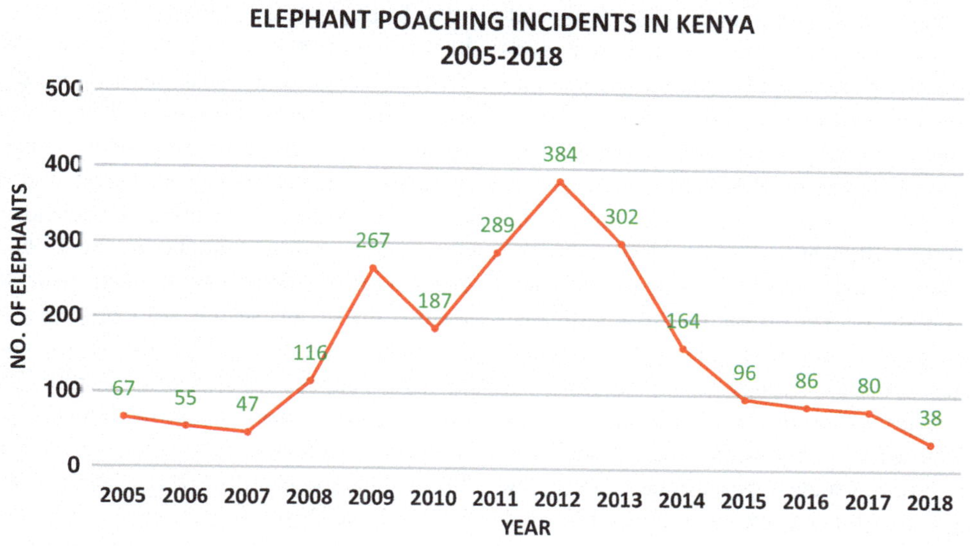

Fig. 6 Number of elephant poached between 2005 and 2018 (Source: KWS, personal communication)

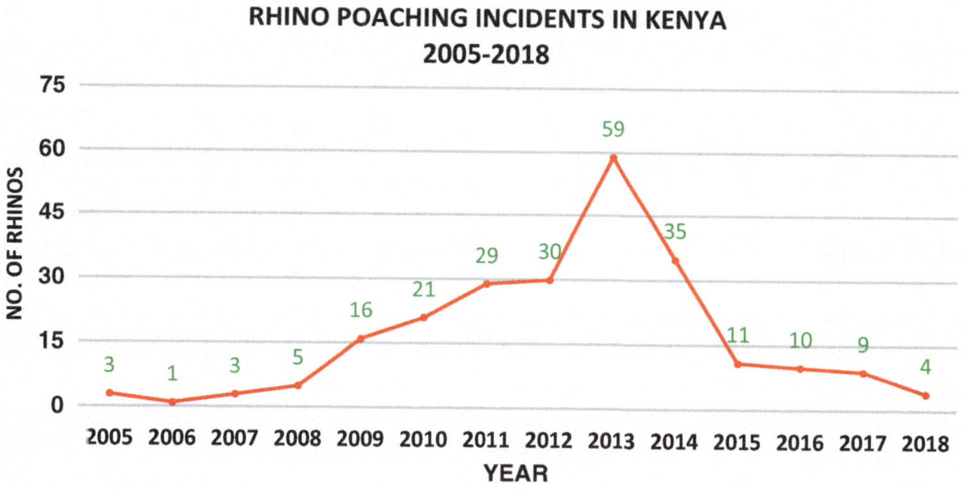

Fig. 7 Number of rhinos poached between 2005 and 2018 (Source: KWS, personal communication)

issue in South Africa, Kenya, and Namibia (Lindsey et al. 2015).

Human encroachment into wildlife areas occurs as a result of the following: inadequate enforcement of the penal system and poor law enforcement; lack of alternative livelihoods; insufficient alternative food sources; lack of clear rights over wildlife or land and/or inadequate benefits from legal use of wildlife; political instability, corruption, and poor governance; demand for wildlife body parts for traditional medicine and ceremonies; and abundant supplies of trapping devices. As a result of these drivers, illegal hunting is a big problem in some parts of Kenya, around Narok, Naivasha, Isiolo, Samburu, Machakos, Kitengela, Namanga, and Tsavo areas, the extent of which is higher than previously estimated (*personal observation*; Task Force Report on Wildlife Security 2014; KWS, personal communication).

The illegal acquisition and exchange of wild meat is receiving little attention in Kenya, perhaps due to a misconception that bushmeat hunting is a low-impact subsistence activity when compared to hunting of the large trophy species (Task Force Report on Wildlife Security 2014). Though data on impacts are scarce, indications are that the bushmeat trade is a widespread problem in the country, with severe impacts on wildlife populations. The impacts of the bushmeat trade in Kenya vary from edge effects around protected areas to disproportionate declines of some species, to severe wildlife declines in areas with inadequate anti-poaching operations. The illegal bushmeat trade appears to be increasingly commercialized due to elevated demand in rural areas, urban centers like Nairobi, and even overseas cities (Task Force Report on Wildlife Security 2014). Other drivers for the trade include human encroachment of wildlife areas, poverty and food insecurity, and inadequate legal frameworks enabling communities to benefit legally from wildlife as a food resource (Lindsey et al. 2013; Task Force Report on Wildlife Security 2014; Lindsey et al. 2015).

Climate Change

Climate change is currently acknowledged as a global crisis threatening human existence and biological resources. Numerous studies, especially those dealing with increasing temperatures, have predicted that there will be significant impacts on the world's physical, biological, and human systems and it is expected to become more severe in the future if not mitigated (Pimm 2009; Sinclair et al 2010; Monzón et al. 2011; Kideghesho et al. 2013; Sintayehu 2018). Studies suggest that many plants and animals are unlikely to survive the long-term impacts of anthropogenic climate change (Pimm 2009; Román-Palacios and Wiens 2020). By 2050, climate change will lead to the extinction of 15–37% of a total sample of 1103 land plants and animals (Thomas et al. 2004; Pimm 2009; Román-Palacios and Wiens 2020). Changes and variations in climate pose serious threats to biodiversity in Kenya, in both aquatic and terrestrial ecosystems, and the impacts of climate change have been felt in virtually all ecosystems, including rangelands. For example, the severe droughts in the 1990s and 2000s forced pastoralists to shift their herds toward PAs in search of pastures, with devastating effects such as the destruction of habitats, reduced biodiversity, destruction of water sources, and increased human-wildlife conflicts (Lovett et al. 2005; Otiangá-Owiti et al. 2011; Mango et al. 2011; Ongugo et al. 2014; Mbote 2016; Aduma et al. 2018). In the long run, such climate change impacts will affect various tourism destinations, which are major contributors to the nation's economy, as most of them are wildlife-based (Nyamwange 2016).

The rise of temperature and change of rainfall patterns in ASAL areas provide further illustration of the impacts of climate change on biodiversity. Climate change poses a serious threat to wildlife and national security, as it may cause drastic ecosystem changes that could alter the reservoirs for emerging infectious disease, contribute to food and water scarcity, and accelerate conflict between stakeholders over resources in many parts of Kenya (Githeko and Ndegwa 2001; Zhou et al. 2004; Otiangá-Owiti et al. 2011; Aduma et al. 2018). Climate change coupled with poverty may force communities to adopt coping strategies that are destructive to biodiversity, such as encroachment and illegal hunting, both natural and human-induced wildfires, and an increase in human-wildlife conflict (Aduma et al. 2018). Other destructive practices such as the cutting of trees in water catchment areas augment soil erosion and siltation of water bodies that eventually become prone to eutrophication, thereby negatively impacting both aquatic and terrestrial wildlife. Examples of this effect have been found in lakes such as Jipe, Naivasha,

Nakuru, and Baringo (Otianga-Owiti et al. 2011; Mbote 2015).

For example, Lake Naivasha is a wetland of national and international importance. However, it is under constant anthropogenic pressure, including the quest for socio-economic development within the lake ecosystem, as well as other human-related activities within the catchment and basin areas. In addition to climate change effects, the basin is additionally threatened by an increasing reduction of lake levels, deterioration of lake and river water quality, deforestation, increased soil erosion and siltation of rivers, increased lake sedimentation, fish mortality and decreasing fish yields, increased land conversion, encroachment and transformation of the lakeshore riparian zone, encroachment and transformation of the riverine buffer zones in the catchment areas, increasing population and unplanned human settlements, poor waste management in the urban areas, inaccessibility to the lake by pastoralists fishermen and general public, and lake infestation by invasive species (Otiang'a-Owiti and Oswe 2007; Ministry of Environment, Water and Natural Resources 2014).

Invasive Alien Plant Species

Invasive alien plant species are a major threat to wildlife resources, particularly in ASALs and aquatic ecosystems in Kenya. They have been shown to transform the structure and species composition of ecosystems by repressing or excluding native species, either directly (competition) or indirectly (altering ecosystem nutrient cycles).

Many habitats in Kenya, including national parks and other forms of PAs, are not immune to infestation by invasive species. As a result, invasive species have now been recognized in conservation agendas countrywide. Aquatic and wetland biodiversity is seriously compromised by alien invasive species. The most significant areas infested by invasive species include Lake Nakuru National Park, Tsavo East and West National Parks, Hells Gate National Park, Lake Bogoria National Reserve, and Amboseli National Park. Invasive plant species in Kenya include *Datura stramonium*, *Solanum incanum*, *Lippia javanica*, *Psiadia punctulata*, *Sida tenuicarpa*, *Tagetes minuta*, *Opuntia excelsa*, *Prosopis juliflora* (commonly known as 'mathenge'), *Parthenium hysterophorus*, *Momosa pigra*, *Chromolaena odorata* and *Eichhornia crassipes* (Kedera and Kuria 2003; Kanga et al. 2013). Among the more prevalent species causing havoc in aquatic ecosystems is the water hyacinth, *Eichhornia crassipes.* For terrestrial ecosystems, the most destructive invasive species is the tick berry, *Lantana camara.* The primary impact of invasive species is the general disruption of the ecosystem, which has a ripple effect extending to multiple native species within the ecosystem.

Insights into Workable Solutions to Wildlife Conservation and Management in Kenya

Introduction

The survival of wildlife in Kenya is a race against mounting social and economic development. With the passage of each decade, the task of protecting wildlife and its habitats is becoming increasingly political, challenging, and complex. The future of Kenya's wildlife depends on the ability to conserve wildlife while balancing the needs of the people, including their economic expectations and political affiliations. A conservation crisis is looming in Kenya. Kenya's wildlife policies must address and tackle the root causes of the existing and future conservation problems. These are centered on the rapidly expanding human population in Kenya and a lack of appropriate social support for those living at the interface with wildlife. Policies must ensure effective legal frameworks and deterrents for those harming the environment, among other requirements listed below. These problems must be addressed urgently and aggressively, however "unpalatable" that may be. Wildlife management in Kenya will be best achieved when or if it is combined with the social and economic uplift of local people.

This section focuses on some of the mechanisms that can be used to promote sustainable wildlife conservation in Kenya and takes into consideration the challenges and opportunities for sustainable management of the wildlife industry.

Opportunities for Strengthening Wildlife Conservation and Management in Kenya

Create More Space for Wildlife

Pressure from growing human populations and a commensurate demand for land for development has put open space for wildlife at great risk of being converted to other uses, especially in the ASALs and related areas. Wildlife habitat fragmentation is on the increase, with land being subdivided, fenced, cultivated, or developed for human settlement. This trend is of great concern to the survival of wildlife in Kenya, as PAs alone are not sufficient for wildlife to prosper. Land outside protected areas is essential for large mammals to migrate between protected areas and seasonal grazing areas. Loss of open space, therefore, presents a threat to the survival of wildlife species that are the basis of the tourism industry in the country. Additionally, there is a lack of appropriate tools, mechanisms, and benefits for landowners and communities to encourage them to keep their land open, as they believe wildlife does not pay.

There is a need for the government, players in wildlife industry, and other stakeholders to come up with enabling policies and fiscal resources to ensure that more space is created for wildlife. Currently there are efforts being carried out by the government for acquiring space for wildlife conservation outside protected areas. Such efforts include securing dispersal areas and wildlife migratory corridors, a Vision 2030 flagship project. It aims at the identification and mapping of landscapes and resources used by key species, with a view of connecting wildlife to key resources of pasture, water, breeding sites, and other ecological requirements. These habitats have been severely compromised by human activities, and thus the need exists to restore them by improving ecosystem resilience and, by extension, wildlife habitat.

Other related and critical initiatives center around the development of community and private wildlife conservancies and the use of environmental easements. Additionally, the human dimensions of conservation such as the livelihoods of local communities must be taken into consideration.

The National Spatial Plan (2015–2045) should complement the above efforts through the implementation of effective land use plans through zones, based on their potential for posterity. Further, this plan should be cognizant of county government needs. Kenyans over the years have developed a mentality that the acquisition of land is a symbol of wealth. This cultural perception must be addressed, as it influences the amount of land available for wildlife. The political elites and their cohorts have capitalized on policy weaknesses in order to take hold of critical wildlife corridors, dispersal areas, or buffer zones.

Social Support for Conservation

Successful environmental conservation depends upon the involvement and participation of local communities. Communities living in and around a protected or conservation area (especially those without barriers or fencing) can determine its fate, either by overexploiting its resources or by supporting its boundaries and laws. Furthermore, there is a direct link between poverty and wildlife conservation, and thus strategies of payment to encourage coexistence must be employed in order to facilitate wildlife conservation and alleviate local poverty.

Harness Local Communities' Goodwill

Most wildlife-related policies and legislation in Kenya have failed to consider the wildlife resource user rights of surrounding communities and pastoralists. The establishment of PAs has been biased to the exclusion of local communities. This controlled environment denies and hampers other land uses for locals and brings tension between community landowners and wildlife conservation. Therefore, sustainable frameworks that support such groups must be implemented.

Participatory or Collaborative Wildlife Management Approach

The wildlife management and conservation approach adopted by the Government of Kenya emphasizes the protection of ecosystems, with less emphasis on the local communities that rely on these ecosystems for their livelihoods. This implies a general limited involvement and participation of locals in matters of policy formulation, implementation, and evaluation of state conservation programs. It is critical to note that local communities have a right to benefit from the ecosystem as much as the animals. This can only be achieved through a participatory approach in line with a national policy for wildlife conservation and management, which only exists on paper and is rarely effectively implemented. While participatory approaches such as community-based conservation are ripe with potential, the social, economic, and political entanglements of stakeholders in conservation are thus far proving to be barriers to tangible community benefit in Kenya.

Socio-economic Empowerment of the Rural Communities

Local communities, especially pastoralist communities, historically evolved in a harmonious relationship with their natural environment, including wildlife. In more recent times, they have failed to fully benefit from wildlife. Stakeholders, in particular the government, have recognized the impracticality of managing wildlife in PAs alone. It is therefore necessary to ensure that developments on land adjacent to wildlife PAs do not interfere with wildlife conservation and PAs are not managed as islands in a sea of humanity, but rather as an integral part of the national land use process. Presently, the Community Land Act of 2016 empowers local communities and pastoralists to take control on land use matters and offers great potential for minimization of resource-use conflicts, for a win-win situation.

Equitable and Effective Wildlife Benefit-Sharing Mechanisms Among Partners

Currently there are no well-functioning mechanisms in Kenya for the sharing of wildlife revenue obtained from state PAs among stakeholders, especially local people living around these conservation areas. There is a need to devise an innovative model on the revenue and benefits accruing from wildlife and tourism that can be shared between the government, park authorities, and the local communities. Such benefits will reduce negative attitudes toward wildlife, hence improving conservation and development. Where financial benefits are shared between local communities and park agencies, local communities develop a strong sense of ownership and commitment to wildlife conservation. Recently, in 2018, Kenya's Minister for Tourism and Wildlife established a task force to investigate how wildlife may be used in a way that works for both people and wildlife. The report has been released and in general concluded that Kenya was not yet ready to implement consumptive wildlife utilization (CWU) (Ministry of Tourism and Wildlife 2019). In brief, some of the reasons outlined are as follows:

- Serious challenges such as lack of education and awareness of wildlife matters, especially among the communities that host wildlife on their land.
- Fear that CWU would open up hunting in Kenya, leading to extinction of its wild animals; some understood CWU to mean meat in restaurants.
- Corruption which is still prevalent in Kenya would wipe out our wildlife and that there was

no framework to monitor, regulate, and enforce CWU.

In the report, most stakeholders felt that the government should instead implement some of the options outlined in the WCMA 2013 that offer benefits to communities, including compensation for death, injury, and crop damage. Furthermore, they indicated that the government should develop regulations to implement this Act, the national strategy, and develop a wildlife policy before effecting the CWU.

Promotion of Community-Based Wildlife Management Models

There is a need to promote participation and consultation of communities on wildlife conservation through a bottom-up approach, as opposed to top-down strategies, or the "fortress conservation model." This primarily entails involving the public in the decision-making process and the overall management of wildlife. The government could create incentives in community-based conservation models which employ sustainable livelihoods and economic development practices. Introduction of economic promise projects, through either ecotourism or other non-consumptive utilization ventures, must be encouraged. A good example is the Rukinga Carbon Project in the coast of Kenya, where PAs provide carbon credit for mitigation of climate change. Providing protection and improvement on wildlife habitats promotes their ability to act as carbon sinks to sequester carbon. In addition, it helps maintain ecosystem integrity, thus promoting provision of ecosystem goods and services. Such models could be replicated in other PAs to diversify livelihood sources in communities adjacent to PAs.

Ensuring Public Participation in the Establishment of PAs

The provisions in Kenya's legislation regarding the powers of the government to establish PAs neither oblige the government to consult the local communities nor make provision for public participation. This is an anomaly, as in a democratic state the role of public participation in decision-making, especially on a public resource such as wildlife, need not be overemphasized. The laws should be amended to provide mechanisms for such consultations and negotiations with stakeholders.

Ensuring Effective Legal Frameworks and Deterrents

The Government of Kenya, in partnership with international agencies and donors, must:

- Strengthen the legal framework and facilitate law enforcement to combat poaching and other illegal wildlife trade and assist prosecution and the imposition of penalties that are an effective deterrent
- Strengthen the ability to achieve successful prosecutions and deterrent sanctions by raising awareness in the judicial sector about the seriousness, impact, and potential profits of wildlife crime
- Adopt a zero-tolerance policy on corruption associated with wildlife crime and consider illegal wildlife trade as an economic crime with significant consequences
- Invest in capacity-building to strengthen law enforcement in order to protect key populations of species threatened by poaching and the bushmeat trade
- Improve intelligence by working with local communities and establish monitoring and law enforcement networks in conservation areas where wildlife exists
- Improve capacity-building to strengthen law enforcement, awareness, and education and enhance training in investigative techniques for law enforcement, including identification techniques of wildlife products and the use of forensics
- Strengthen cross-border and regional cooperation through better coordination and through full support for regional wildlife law enforcement networks like Lusaka Task Force and the

newly created Horn of African Wildlife Enforcement Network, among others

- Provide dedicated leadership at high levels as critical for wildlife conservation that can educate local communities and build a shared ethic of conservation across Kenyan society

Other Opportunities

Strengthening of the Science-Policy-Practice Interface

There is a need for adequate, accurate, and coherent research to aid in the formulation of policies and legal framework for the conservation and management of wildlife in Kenya. In order to facilitate effective decision-making, KWS and relevant institutions must harmonize research methodologies and tools to allow for monitoring and evaluation over time. It is extremely important to gather the most comprehensive and reliable data on wildlife population trends and habitat status that would allow managers and other relevant stakeholders to make appropriate decisions to the benefit of wildlife conservation and management, even if it may not fully support the hypotheses and theories of those providing funding. It is also imperative that to acquire good and accurate data, we must enhance the capacity of local scientists, practitioners, and communities, particularly in advanced research skills and monitoring techniques of biological resources. Additionally, the requirement to develop and implement conservation and management strategies of specific or key species is critical for the future of wildlife in Kenya.

Harmonization of Existing Wildlife and Other Relevant Sectoral Policies and Legislation

There have been efforts to harmonize the Wildlife Act and other relevant sectoral policies and legislation. However, the implementation has not occurred because critical intersectoral planning and implementation has been slowed through political interests.

Decentralization of Wildlife Conservation Operations

Devolution is still in its infant stages, with just 4 years of operation, but already there is recognition of inclusive contribution of county governments to local communities. Even though KWS, an authority in wildlife conservation and management, has stations countrywide, coordination is still centralized at the national level, making service delivery rather slow. Therefore, wildlife conservation should be decentralized to county levels in order to empower communities and other stakeholders to participate effectively in the conservation planning, implementation, and decision-making processes. Effective national-level policies that ensure conservation's benefits and are shared with local people are critical. These policies should clearly define the quantity and the process of benefit-sharing.

Conservation Education

Transforming wildlife resources from a liability into an asset that communities will value, and thus be motivated to conserve through attitudinal and behavioral change, is a key conservation goal. Local communities must be actively involved in both conservation planning and decision-making. Conservation education for both urban and rural communities should be emphasized. Conservation strategy implementation failures have occurred as a result of limited awareness by the people of their role in biodiversity conservation, ecosystem management, as well as human health implications. Other compounding factors include limited financial and human resources. Transforming communities into conservators requires a clear understanding of the value that nature offers, as well as the consequences of having non-functional ecosystems. Poverty, one of the root causes of wildlife loss, must be tackled through provision of incentives that promote alternative livelihoods to wildlife dependence.

Climate

Climate change impacts include changes in physical conditions, weather patterns, and ecosystem functioning (WWF 2006; Monzón et al. 2011; Otianga-Owiti et al. 2011; Kideghesho et al. 2013; Muoria et al. 2015; Sintayehu 2018). Wildlife conservation will be severely impacted by the effects of climate change, unless we manage to cope through decisive planning and action. However, it is fair to mention that confronting the climate crisis requires that we address the underlying causes of climate change and simultaneously prepare for and adapt to current and future impacts. There is a need to integrate programs to build resilience to climate change in most parts of the country, and adaptive capacity of vulnerable communities living around conservation areas, and adopt a climate-smart wildlife concept in Kenya. This could entail designing and carrying out conservation in the face of a rapidly changing climate as outlined in Climate-Smart Conservation (Stein et al. 2014), Climate Change Act of 2016, and National Climate Change Action Plan (Government of Kenya 2018).

The whole rationale in these strategies is to ensure the mechanisms and measures to achieve low-carbon climate-resilient development in a manner that prioritizes adaptation, pays particular attention to increasing wildlife habitat cover through reforestation and restoration of habitat, rehabilitates degraded lands, and increases resilience of wildlife. These can be achieved through the following overarching themes:

- Act with intentionality
- Manage for change, not just persistence
- Link actions to climate impacts by ensuring that conservation strategies and actions are designed specifically to address the impacts of climate change
- Embrace forward-thinking conservation goals that look to the future rather than the past
- Consider ground actions designed in the context of broader geographic scales to account for likely shifts in species distributions, that sustain ecological processes
- Employagile and informed conservation planning and resource management with dynamic adjustment to accommodate uncertainty, taking advantage of new knowledge, that cope with rapid shifts in climatic, ecological, and socio-economic conditions
- Minimize carbon footprint with strategies and projects that minimize energy use and greenhouse gas emissions and sustain the natural ability of ecosystems to cycle, sequester, and store carbon
- Safeguard people and nature by adopting strategies and actions that would enhance the capacity of ecosystems to protect human communities from climate change impacts in ways that also sustain and benefit wildlife
- Avoid maladaptation by ensuring that actions taken to address climate change impacts on human communities or natural systems do not exacerbate other climate-related vulnerabilities or undermine conservation goals and broader ecosystem sustainability

Conclusion

Wildlife conservation in Kenya is at a crossroads, and we must implement strategies that will meet the needs of both people and wildlife among the other requirements of the country's Vision 2030. However, as a country, there must be goodwill from key stakeholders for the implementation and enforcement of relevant legislation and policies that support wildlife conservation. The Government of Kenya must subscribe to a zero-tolerance attitude toward wildlife crime, whereby justice is swift and firm to those involved. We must remember that wildlife conservation is not a business to be compared with other land uses such as farming or real estate development, but must be considered as a social investment that requires adequate subvention from government and external partners, and thus is valued for its posterity. In order for wildlife conservation and management to work, it is important that we "put nature on the balance sheet." However, it is also important to remember that wildlife conservation and management efforts will still need external partners, as it is a very expensive affair. Kenya's unique diversity and wealth of wildlife species is a heritage, both for Kenya and for the world. Finally, all Kenyans must be decisive and committed, for wildlife's sake.

Acknowledgments We acknowledge the support in terms of availing relevant data, reports, and literature from the following: Prof. Charles M. Musyoki, Messrs. Linus Kariuki, Moses Maloba, Asher Odhiambo, James Wachira, and Eric Wamari Owiti of the Kenya Wildlife Service.

References

Aduma MM, Ouma G, Said MY, Wayumba GO, Omondi PA, Njino LW (2018) Potential impacts of temperature projections on selected large herbivores in savanna ecosystem of Kenya. Am J Clim Chang 7:5–26

African Wildlife Foundation (2016) African conservancies volume: towards best practices, African conservancies, vol 1. African Wildlife Foundation, Nairobi

Ayeni JS (1974) Observation on waterhole utilization by wild game in Tsavo National Park (east). A thesis submitted in part fulfilment of the Degree of Master of Science (Biology of conservation) - The University of Nairobi

Bashuna AB (1993) The Waata, hunter-gatherers of northern Kenya. Past Present 25:36–38

Bhandari M (2014) Is tourism always beneficial? A case study from Maasai Mara National Reserve, Narok, Kenya. Pac J Sci Technol 15(1):456–483

Borrini-Feyerabend G (1996) Collaborative management of protected areas: tailoring the approach to the context. Issues in social policy. IUCN, Gland, Switzerland

Campbell DJ, Gichohi H, Mwangi A, Chege L (2000) Land use conflict in Kajiado District, Kenya. Land Use Policy 17:337–348

Cockerill KA, Hagerman SM (2020) Historical insights for understanding the emergence of community-based conservation in Kenya: international agendas, colonial legacies, and contested worldviews. Ecol Soc 25 (2):15–34. https://doi.org/10.5751/ES-11409-250215

Chase MJ, Schlossberg S, Griffin CR, Bouché PJC, Djene SW, Elkan PW, Ferreira S, Grossman F, Kohi EM, Landen K, Omondi P, Peltier A, Selier SAJ, Sutcliffe R (2016) Continent-wide survey reveals massive decline in African savannah elephants. PeerJ 4:e2354. https://doi.org/10.7717/peerj.2354

Doshi M, Kago CW, Kamunde-Aquino N, Kiguatha L, Idun YNA, Chapman S (2014) Land tenure classification in Kenya. REDD+ Law Project. https://www.4cmr.group.cam.ac.uk/filecab/redd-law-project/20140821%20BP%20Land%20Tenure%20Classifications%20in%20Kenya.pdf

Economic Survey (2004-2018) Kenya National Bureau of Statistics, 9966-767-47-9

Elliot J, Gibbons H, King D, King A, Leménager T (2013) "Greater than the sum of their parts": exploring the environmental complementarity of different types of protected areas in Kenya. Report financed by AFD, commissioned to AWF, IIED and UNEP-WCMC

EU (2014) Framework Contract COM 2011 – Lot 1 Request for Services 2013/328436 - Version 2 Inputs for the design of an EU strategic approach to wildlife conservation in Africa Final Report December 2014

Fisher M (2016) Whose conflict is it anyway? Mobilizing research to save lives. Oryx 50:377–378. https://doi.org/10.1017/S0030605316000673

Fynn RWS, Bonyongo MC (2010) Functional conservation areas and the future of Africa's wildlife. Afr J Ecol 49:175–188

Georgiadis NJ, Olwero JGN, Ojwang G, Romanach SS (2007) Savannah herbivore dynamics in a livestock-dominated landscape: I. Dependence on land use rainfall density and time Biol Cons 137:461–472

Githeko AK, Ndegwa W (2001) Predicting malaria epidemics in the Kenyan highlands using climate data: a tool for decision-makers. Global Change Human Health 2:54–63. https://doi.org/10.1023/A:1011943131643

Godfrey K (2016) Approaches to community-based conservation in Kenya: case Studies from Amboseli, Maasai Mara, and Laikipia. https://www.icanconserve.org/wp-content/.../Godfrey2016CBC-Approaches-Study.pdf

Government of Kenya (2009) National Wildlife Policy 2009. Government Printer, Nairobi

Government of Kenya (2012) National tourism strategy 2013-2018. Department of Tourism, Ministry of East Africa, Commerce and Tourism, Nairobi, Kenya

Government of Kenya (2013) National Climate Change Action Plan 2013-2017: Executive Summary. Retrieved from http://www.kccap.info/

Government of Kenya (2015) Kenya National Spatial Plan 2015-2045: an integrated spatial plan for balanced and sustainable national development. https://lands.go.ke/wp-content/uploads/2018/03/National-Spatial-Plan.pdf

Government of Kenya (2018) National Climate Change Action Plan (Kenya): 2018-2022. Ministry of Environment and Forestry, Nairobi

Government of Republic of Kenya (2007) Kenya Vision 2030 Popular Version. https://www.ieakenya.or.ke/publications/other-documents/kenya-vision-2030-popular-version

Groombridge B (1992) Global biodiversity: status of the Earth's living resources. WMC, New York

Grunblatt JM, Said M, Wargute P (1996) National Rangelands report. Summary of population estimates of wildlife and livestock (1977–1994). Department of Resource Surveys and Remote Sensing, Ministry of Planning and National Development, Nairobi, Kenya

Hauenstein S, Kshatriya M, Blanc F, Dormann CF, Beale C (2019) African elephant poaching rates correlate with local poverty, national corruption and global ivory price. Nat Commun 10:1–9

Honey M (1999) Ecotourism and sustainable development: who owns paradise? Island Press, Washington, DC

Jones S (2006) A political ecology of wildlife conservation in Africa. Rev Afr Polit Econ 33(109):483–495. https://doi.org/10.1080/03056240601000911

Kamau PN, Sluyter A (2018) Challenges of elephant conservation: insights from oral histories of colonialism and landscape in Tsavo, Kenya. Geogr Rev 108 (4):523–544. https://doi.org/10.1111/gere.12288

Kameri-Mbote P (2019) Wildlife conservation and land rights in Kenya: competing or complementary agendas? Chapter 9. In: Cullet P, Koonan S (eds) Research handbook on law, Environment and the global south, pp 169–189. Handbook ISBN: 9781784717452

Kameri-Mbote P, Kindiki K (2008) Trouble in Eden: how and why unresolved land issues landed 'peaceful Kenya' in trouble in 2008. Forum for Development Studies (2):167–193

Kanga EM, Kenana L, Ngoru B, Lala FO (eds) (2013) National strategy and action plan for the management of invasive species in Kenya's protected areas. Kenya wildlife service (KWS). Nairobi, Kenya

Kassam A, Bashuna AB (2004) Marginalisation of the Waata Oromo hunter–gatherers of Kenya: insider and outsider perspectives. Afr J Int Afr Inst 74(2):194–216

Kedera C, Kuria B (2003) Invasive alien species in Kenya: Status and management. IPPC Secretariat. Identification of risks and management of invasive alien species using the IPPC framework. In: Proceedings of the workshop on invasive alien species and the international plant protection convention. Braunschweig, Germany, pp 22–26

Kenya National Bureau of Statistics (KNBS) (2019) Kenya population and housing census Volume I: population by county and sub-county. https://www.knbs.or.ke/?wpdmpro=2019-kenya-population-and-housing-census-volume-i-population-by-county-and-sub-county

Kenya National Spatial Plan 2015-2045: an integrated spatial plan for balanced and sustainable national development

Kideghesho JR, Rija AA, Mwamende KA, Selemani IS (2013) Emerging issues and challenges in conservation of biodiversity in the rangelands of Tanzania. Nat Conserv 6:1–29

King J (2014) A review of Wildlife Conservation and Management Act, 2013, Northern Rangeland Trust (NRT) Community Conservancies

King J, Kaela D, Buzzard B, Warigia G (2015) Establishing a wildlife conservancy in Kenya: a guide for private land-owners and communities. Kenya Wildlife Conservancies Association

Kioko J, Muruthi P, Omondi P, Chiyo PI (2008) The performance of electric fences as elephant barriers in Amboseli, Kenya African. J Wildl Res 38:52–58

Kioko J, Okello MM (2010) Land use cover and environmental changes in a semi-arid rangeland, Southern Kenya. J Biogeogr Rangeland Plan 3:322–326

KWS (1995) Kenya wildlife service: report of the five person review committee on wildlife-human conflict. Nairobi, Kenya

KWS (1996) Wildlife policy 1996. Kenya Wildlife Service, Nairobi

Lambin EF, Geist HJ, Lepers E (2003) Dynamics of land-use and land-cover change in tropical regions. Annu Rev Environ Resour 28:205–241

Lamprey RH, Reid RS (2004) Expansion of human settlement in Kenya's Maasai Mara: what future for pastoralism and wildlife? J Biogeogr 31:997–1032

Lindsey PA, Balme G, Becker M, Begg C, Bento C, Bocchino C, Dickman A, Diggle RW, Eves H, Henschel P, Lewis D, Marnewick K, Mattheus J, Weldon McNutt J, McRobb R, Midlane N, Milanzi J, Morley R, Murphree M, Opyene V, Phadima J, Purchase G, Rentsch D, Roche C, Shaw J, Westhuizen HVD, Vliet NV, Zisadza-Gandiwa P (2013) The bushmeat trade in African savannas: impacts, drivers, and possible solutions. Biol Conserv 160:80–96

Lindsey P, Taylor WA, Nyirenda V, Barnes L (2015) Bushmeat, wildlife-based economies, food security and conservation: insights into the ecological and social impacts of the bushmeat trade in African savannahs. FAO/Panthera/Zoological Society of London/SULI Report, Harare. 58 p

Long H, Mojo D, Fu C, Wang G, Kanga E, Oduor AMO, Zhang L (2019) Patterns of human-wildlife conflict and management implications in Kenya: a national perspective. Hum Dimens Wildl. https://doi.org/10.1080/10871209.2019.1695984

Lovett JC, Midgely GF, Barnard PB (2005) Climate change and ecology in Africa. Afr J Ecol 43:279–281

Maitima JM, Mugatha SM, Robin S, Reid RS, Lyaruu H, Pomery D, Gachimbi LN, Mathai S, Mugisha S (2009) The linkages between land use change, land degradation and biodiversity across East Africa. Afr J Environ Sci Technol 3(10):310–325

Makindi S, Mutinda M, Olekaikai NKW, Olelebo WL, Aboud AA (2014) Human-wildlife conflicts: causes and mitigation measures in Tsavo conservation area, Kenya. Int J Sci Res 3(6):1025–1031

Mango LM, Melesse AM, McClain ME, Gann D, Setegn SG (2011) Land use and climate change impacts on the hydrology of the upper Mara River Basin, Kenya: results of a modeling study to support better resource management. Hydrol Earth Syst Sci 15:2245–2258. https://doi.org/10.5194/hess-15-2245-2011

Mbote BW (2016) Assessing the impacts of climate variability and climate change on biodiversity in Lake Nakuru, Kenya. A project submitted in partial fulfilment of the requirements for award of the Degree of Master of Science in Climate Change of the University of Nairobi

Mburu J (2004) Wildlife conservation and management in Kenya: towards a co-management approach, Nota Di Lavoro, p 47

Mburu J, Birner R, Zeller M (2003) Relative importance and determinants of landowners' transaction costs in collaborative wildlife management in Kenya: an empirical analysis. Ecol Econ Transdiscip J Int Soc Ecol Econ 45(1):59–73

Ministry of Environment, Water and Natural Resources (2014) Lake Naivasha Basin Integrated Management Plan 2012-2022

Ministry of Tourism and Wildlife (2018) National Wildlife Strategy 2030

Ministry of Tourism and Wildlife (2019) Report of the task force on consumptive wildlife utilization in Kenya. Nairobi, Kenya. vi + 101pp

Monzon J, Moyer-Horner L, Palama MB (2011) Climate change and species range dynamics in protected areas. Bioscience 61:752–761. https://doi.org/10.1525/bio.2011.61.10.5

Mukeka JM, Ogutu JO, Kanga E, Røskaft E (2018) Characteristics of human-wildlife conflicts in Kenya: examples of Tsavo and Maasai Mara regions. Environ Nat Resour Res 8(3). https://doi.org/10.5539/enrr.v8n3p148

Mukeka JM, Ogutu JO, Kanga E, Røskaft E (2019) Human-wildlife conflicts and their correlates in Narok County, Kenya. Global Ecol Conserv 18:1–22

Muoria P, Matiku P, Ng'weno F, Munguti S, Barasa F, Ayiemba W (2015) Coping with climate change in Kenya. Policy guide for ecosystem based adaptation for national and county government. Nature Kenya, Nairobi

Musyoki C, Andanje S, Said M, Chege M, Anyona G, Lukaria L, Kuloba B (2012) Challenges and opportunities for conserving some threatened species in Kenya. George Wright Forum 29(1):81–89

Mwaura F (2016) Wildlife heritage ownership and utilization in Kenya-the past, present and future. In Anne-Marie Deisser, Mugwima Njuguna (eds), Conservation of natural and cultural heritage in Kenya, a cross-disciplinary approach, UCL Press, University College London. www.ucl.ac.uk/ucl-press

National Council for Population and Development (NCPD) (2013) Kenya population situation analysis

Nelson F (2012) Recognition and support of ICCAs in Kenya. In: Kothari A, et al. (eds) Recognizing and supporting territories and areas conserved by indigenous peoples and local communities: global overview and national case studies. Secretariat of the Convention on Biological Diversity, ICCA Consortium, Kalpavriksh, and Natural Justice, Montreal, Canada, Technical Series no. 64

Neumann RP (2001) Africa's 'Last wilderness': reordering space for political and economic control in colonial Tanzania. Africa 71(4):641–665

Neumann R (2004a) Moral and discursive geographies in the war for biodiversity in Africa. Polit Geogr 23 (7):813–837

Neumann RP (2004b) Nature-state-territory: toward a critical theorization of conservation enclosures. In: Peet D, Watts C (eds) Liberation ecologies: environment, development, social movements, 2nd edn. Routledge, London, pp 179–200

Noe C (2003) The dynamics of land use changes and their impacts on the wildlife corridor between Mt. Kilimanjaro and Amboseli National Parks. LUCID working paper No. 31.International Livestock Research Institute. www.lucideastafrica.org

Norton-Griffiths M, Said MY (2010) The future for wildlife on Kenya's rangelands: an economic perspective. In: du Toit JT, Kock R, Deutsch JC (eds) Wild rangelands: conserving wildlife while maintaining livestock in semi-arid ecosystems. Wiley, Chichester, UK

Nyamasyo S (2016) Effects of land use change on wildlife conservation in Kenya: changing land use patterns and their impact on wild ungulates in Kimana wetland ecosystem, Kajiado County. LAP LAMBERT Academic Publishing, Kenya

Nyamasyo ST, Kihima BO (2014) Changing land use patterns and their impacts on wild ungulates in Kimana Wetland Ecosystem, Kenya. Int J Biodivers 2014:1–10. https://doi.org/10.1155/2014/486727

Nyamwange M (2016) Impacts of climate change on tourism in Kenya. J Geogr Earth Sci 4(2):1–10. http://jgesnet.com/journals/jges/Vol_4_No_2_December_2016/1.pdf

Nyhus PJ (2016) Human–wildlife conflict and coexistence. Annu Rev Environ Resour 41:143–171

Ogada MO, Nyingi DW (2013) The Management of Wildlife and Fisheries Resources in Kenya: origins, present challenges and future perspectives. In: Paron P, Olago D, Omuto CT (eds) Developments in earth surface processes, vol 16. Elsevier, Amsterdam, The Netherlands, pp 219–236

Ogutu JO, Owen-Smith N, Piepho HP, Said MY (2011) Continuing wildlife population declines and range contraction in the Mara region of Kenya during 1977–2009. J Zool 285:99–109. https://doi.org/10.1111/j.1469-7998.2011.00818.x

Ogutu JO, Piepho HP, Said MY, Kifugo SC (2014) Herbivore dynamics and range contraction in Kajiado County Kenya: climate and land use changes, population pressures, governance, policy and human-wildlife conflicts. Open Ecol J 7:9–31

Ogutu JO, Piepho H-P, Said MY, Ojwang GO, Njino LW, Kifugo SC, Wargute PW (2016) Extreme wildlife declines and concurrent increase in livestock numbers in Kenya: what are the causes? PLoS One 11(9): e0163249. https://doi.org/10.1371/journal.pone.0163249

Ojwang GO, Wargute PW, Said MY, Worden JS, Davidson Z, Muruthi P, Kanga E, Ihwagi F, Okita-Ouma B (2017). Wildlife migratory corridors and dispersal areas: Kenya Rangelands and Coastal Terrestrial Ecosystems

Okello MM (2014) Economic contribution, challenges and way forward for wildlife-based tourism industry in Eastern African Countries. J Tour Hosp 3:122. https://doi.org/10.4172/2167-0269.1000122

Okello MM, D'Amour DE (2008) Agricultural expansion within Kimana electric fences and implications for natural resource conservation around Amboseli National Park, Kenya. J Arid Environ 72:2179–2192

Okello MM, Kioko JM (2010) Contraction of wildlife dispersal area in Olgulului-Ololorashi Group Ranch around Amboseli National Park, Kenya

Ondicho TG (2012) Local communities and ecotourism development in Kimana, Kenya. J Tour XIII(1):41–60

Ongugo PO, Langat D, Oeba VO, Kimondo JM, Owuor B, Njuguna J, Okwaro G, Russell AJM (2014) A review of Kenya's national policies relevant to climate change adaptation and mitigation: insights from Mount Elgon. Working paper 155. CIFOR, Bogor, Indonesia

Otiang'a-Owiti GE, Oswe IA (2007) Human impact on lake ecosystems: the case of Lake Naivasha, Kenya. Afr J Aquat Sci 32(1):79–88. https://doi.org/10.2989/AJAS.2007.32.1.11.148

Otianga-Owiti NS, Malel E, Onyuro R (2011) impacts of climate change on human wildlife conflict in East Africa. Kenya Vet 35(2):15

Ottichilo WK, De Leeuw J, Skidmore AK, Prins HHT, Said MY (2000) Population trends of large non-migratory wild herbivores and livestock in the Masai Mara ecosystem Kenya between 1977 and 1997. Afr J Ecol 38:202–216

Ottichilo WK, de Leeuw J, Prins HHT (2001) Population trends of resident wildebeest [*Connochaetes taurinus hecki* (Neumann)] and factors influencing them in the Maasai Mara ecosystem, Kenya. Biol Conserv 97:271–282. https://doi.org/10.1016/S00063207(00)00090-2.

Peterson MN, Birckhead JL, Leong K, Peterson MJ, Peterson TR (2011) Rearticulating the myth of human–wildlife conflict. Conserv Lett 3:74–82

Pimm SL (2009) Climate disruption and biodiversity. Curr Biol 19(14):R595–R601

Price RA (2017) The contribution of wildlife to the economies of sub Saharan Africa. K4D helpdesk report. Institute of Development Studies, Brighton, UK

Rathbun G (2009) Why is there discordant diversity in Sengi (Mammalia: Afrotheria: Macroscelidea) taxonomy and ecology? Afr J Ecol 47:1–13

Redpath SM, Young J, Evely A, Adams WM, Sutherland WJ, Whitehouse A, Amar A, Lambert R, Linnell J, Watt AD (2013) (2013). Understanding and managing conservation conflicts. Trends Ecol Evol 28:100–109

Reid RS, Gichohi H, Said MY, Nkedianye D, Ogutu JO, Kshatriya M, Kristjanston P, Kifugo SC, Agatsiva JL, Adanje SA, Bagine R (2008) Fragmentation of a peri-urban savanna, Athi-Kaputiei plains, Kenya. In: Galvin KA, Reid RS, Behnke RH, Hobbs NT (eds) Fragmentation in semi-arid and arid landscapes; consequences for human and natural systems. Springer, New York, pp 195–224

Republic of Kenya (1975) Sessional paper no. 3 of 1975: statement on the future of Wildlife Management Policy in Kenya. Government Printer, Nairobi

Republic of Kenya (1976) Wildlife (Conservation and Management) Act, 1976 (Cap. 376)

Republic of Kenya (1989) The Wildlife (Conservation and Management) (Amendment) Act of 1989 (Act No. 16 of 1989), Kenya Wildlife Service, Nairobi, Kenya

Republic of Kenya (2009) Kenya Vision 2030. The popular version

Republic of Kenya (2010a) Constitution of Kenya 2010, Article 60(2). Article 61(2)

Republic of Kenya (2010b) National Climate Change Response Strategy (NCCRSAP), 2013-2017. Ministry of Environment, Water and Natural Resources, Nairobi, Kenya

Republic of Kenya (2010c) Population and housing census report 2009. Ministry of State for Planning, National Development and Vision 2030, Nairobi, Kenya

Republic of Kenya (2012) Sessional paper no. 8 of 2012 on the National Policy for the sustainable development of northern Kenya and other arid lands. Ministry of State for Development of Northern Kenya and other Arid Lands, Nairobi

Republic of Kenya (2017) Sessional paper, no. 1 of 2017 on National Land use Policy: Ministry of Lands and Physical Planning

Román-Palacios C, Wiens JL (2020) Recent responses to climate change reveal the drivers of species extinction and survival. Proc Natl Acad Sci 117(8):4211–4217. https://doi.org/10.1073/pnas.1913007117

Rutten M (2002) Parks beyond parks: genuine community-based wildlife eco-tourism or just another loss of land for Maasai pastoralists in Kenya? IIED Drylands Programme Issues. Paper No. 111. Institute for International Environment and Development, London

Schauer J (2015) The elephant problem: science, bureaucracy, and Kenya's National Parks, 1955 to 1975. Afr Stud Rev 58(1):177–198

Scholte P (2011) Towards understanding large mammal population declines in Africa's protected areas: a west-central African perspective. Trop Conserv Sci (1):1–11

Sheldrick D (1973) The Tsavo story. Harvill Press, London

Sibanda B (1995) Wildlife conservation in Kenya: wildlife or local communities at the crossroads? Environ Policy Pract 5(1):35–45

Sinclair S, White M, Newell G (2010) How useful are species distribution models for managing biodiversity under future climates? Ecol Soc 15:8

Sintayehu DW (2018) Impact of climate change on biodiversity and associated key ecosystem services in Africa: a systematic review. Ecosyst Health Sustain 4 (9):225–239. https://doi.org/10.1080/20964129.2018.1530054

Snijders L, Greggor AL, Hilderink F, Doran C (2019) Effectiveness of animal conditioning interventions in reducing human–wildlife conflict: a systematic map protocol. Environ Evid 8:10. https://doi.org/10.1186/s13750-019-0153-7

Stein BA, Glick P, Edelson N, Staudt A (eds) (2014) Climate-smart conservation: putting adaptation principles into practice. National Wildlife Federation, Washington, DC. www.nwf.org/ClimateSmartGuide

Task Force Report on Wildlife Security (2014) Lift the siege - securing Kenya's wildlife. Ministry of Environment, Water and Natural Resources Publication

Thomas CD, Cameron A, Green RE, Bakkenes M, Beaumont LJ, Collingham YC, Erasmus BF, De Siqueira MF, Grainger A, Hannah L, Hughes L, Huntley B, Van Jaarsveld AS, Midgley GF, Miles L, Ortega-Huerta MA, Peterson AT, Phillips OL, Williams SE (2004) Extinction risk from climate change. Nature 427(6970):145–148

Thompson M, Homewood K (2002) Entrepreneurs, elites and exclusion in Maasailand: trends in wildlife conservation and pastoralist development. Hum Ecol 30:107–138

Treves A, Karanth KU (2003) Human-carnivore conflict and perspectives on carnivore management worldwide. Conserv Biol 17:1491–1499

United States Agency for International Development-Country Profile (2017) Kenya–land tenure and property rights profile. https://www.land-links.org/wp-content/uploads/2016/09/USAID_Kenya_Country_Profile_For_Posting_2.21.17-1.pdf

Waithaka J (2012) Historical factors that shaped wildlife conservation in Kenya. George Wright Forum 29 (1):21–29

Weru S (2016) Wildlife protection and trafficking assessment in Kenya: drivers and trends of transnational wildlife crime in Kenya and its role as a transit point for trafficked species in East Africa, TRAFFIC

Western D, Waithaka J (2005) Policies for reducing human-wildlife conflict - the Kenya case with wider application. In: Woodroffe R, Thirgood S, Rabinowitz A (eds) People and wildlife – conflict or coexistence? Conservation biology no. 9. Cambridge University Press, Cambridge, UK

Western D, Russell S, Cuthill I (2009) The status of wildlife in protected areas compared to non-protected areas of Kenya. PLoS One 4:e6140

Wildlife conservation and management act (2013) Government Printer. Kenya, Nairobi

Woodroffe R, Thirgood S, Rabinowitz A (2005) The impact of human-wildlife conflict on natural systems. People wildlife, conflict or coexistence? Cambridge University Press, Cambridge, pp 1–12. https://doi.org/10.1017/S0030605306000202

Worden J, Reid R, Gichohi H (2003) Land use impacts on large wildlife and livestock in the swamps of the greater Amboseli ecosystem, Kajiado District, Kenya. LUCID Working Paper Series No. 27. ILRI, Nairobi, Kenya. www.lucideastafrica.org

WWF (2006) Impacts climate change on East Africa a review of the scientific literature. https://www.wwf.or.jp/activities/lib/pdf_climate/l.pdf/. www.pnas.org/lookup/suppl/Doi:10.1073/pnas.1500664112

Zhou G, Minakawa N, Githeko AK, Yan G (2004) Association between climate variability and malaria epidemics in the east African highlands. PNAS 101:2375–2380

Wildlife Conservation Law

Margaret E. Cooper

Abstract

Legislation is a significant element of wildlife conservation and management. Wildlife conservation law is found at international, regional and national levels. The relevant legislation is listed and sources and relevant literature are provided.

Keywords

Wildlife · Species · Animal · Law · Legislation · International · Convention · Enforcement

Introduction

Throughout the world wildlife is suffering from a wide variety of depredations. Populations of animals, plants and other organisms, both rare and common, are being depleted at an ever-increasing rate, together with their habitat and wider environment.

Numerous measures, some discussed in other chapters in this book, are used to conserve what remains. These include conservation practices and management, politics, economics, traditional culture, community relations and education plus anything else that might make a difference. In addition, the law is seen as an essential, if not necessarily particularly effective, instrument to protect and conserve wildlife, habitats, the environment and biodiversity.

Law can be used in a variety of ways to aid wildlife conservation. It can provide the administrative structure for the management of species and their habitats. It can establish offences, rules and regulations regarding wildlife that can be enforced through the judicial system.

Ideally, conservation legislation should be fit for purpose and properly enforced. Unfortunately, while wildlife legislation can be found in most countries, the extent and quality can vary considerably from old-fashioned hunting controls to comprehensive, contemporary provisions tailored to the needs of species and their habitat. Judicial systems and law enforcement in some countries are of high quality, imposing substantial penalties. However, elsewhere they are inadequate and under-used.

Legislation emanates from many sources, both within a country or internationally. National laws provide the essential authority and framework for wildlife conservation and management. International legislation is of considerable importance in wildlife conservation given the need for collaboration between nations arising from the extensive movement of non-domesticed species around the world as a natural occurrence, for trade and for other purposes.

No law is fully effective unless it is enforced. Where this involves punishing offences by

M. E. Cooper (✉)
Durrell Institute of Conservation and Ecology, University of Kent, Canterbury, United Kingdom

S. C. Underkoffler, H. R. Adams (eds.), *Wildlife Biodiversity Conservation*,
https://doi.org/10.1007/978-3-030-64682-0_5

prosecution through the judicial system, evidence must be provided. In the case of wildlife offences, forensic science is often called upon to produce sound evidence. Evidence, based on forensic science, is frequently required for the pursuit of claims involving wildlife in other fields of law, for example, for compensation for injury or damage to property or persons or for a breach of contract.

In addition to legislation, wildlife management and habitat protection are at times underpinned by "soft" law, such as agreements, statements of principles, memoranda of understanding, guidelines and standards. These are not legally enforceable but enable the development of co-operation and good practice in circumstances where authorities are not able to make a legally binding agreement (Birnie et al. 2009).

Customary laws and practices in use within a given community are sometimes applied as part of good conservation practices, see, for example, Kahler, A (2018) https://www.nomos-elibrary.de/10.5771/9783845294360-912.pdf.

General Information and Definitions

In both science and law, definitions and terminology play an important role in providing clarity and precision in the production of sound information, documents and interpretation. Words often have assumed meanings that are used without further thought. Prime examples are the terms "wildlife" and "wild". These words may intentionally refer to species that are truly free-living. Alternatively, they are often used for species that are of a kind commonly found in the wild but are actually held in captivity or under constraint. Thus, many zoological collections are described as a "wildlife park" or "wild animal attraction". We speak readily of the wildlife trade or "the trade in wild animals", but in fact an animal in trade is, of necessity, under some form of captivity or control and not "in the wild". On closer examination, further questions arise. Does "wildlife" refer to animals? To invertebrates (Fig. 1)? To plants? To other organisms? To biodiversity in general?

Fig. 1 Invertebrates, such as this East African hesperiid butterfly, form 90 percent of living creatures. Image credit: Mrs. Margaret E Cooper

It may be useful to use the term "wild" or "wildlife" in a general sense when it is clear from the context. However, in legal or scientific writing, particularly in respect of conservation, it may be important to make it clear whether the subject refers to free-living specimens or to those of a species that are most commonly found in the wild but are actually in captivity. In applying legislation, the laws or requirements often differ according to whether a species is free-living or captive. For example, an elephant, African or Asian, in most circumstances is assumed to be free-living, but, in fact, some are displayed in enclosures in zoological collections. In certain countries, Asian elephants are kept as working transport or ceremonial animals or live in holy temples. Conservation laws apply to the free-living elephants, but those in captivity are entitled to the benefits of any welfare legislation and may be subject to property laws. There may be grey situations where species are under partial control, for example, when captive-bred non-domesticated species are on "soft release" as part of a translocation or repopulation procedure.

In many day-to-day situations, the context will provide an insight into the intended meaning. However, when writing academic, scientific or forensic documents, it is important to examine critically the usage and to provide definitions appropriate in the circumstance. Décory (2019) looks in depth at definitions of animals.

In this chapter the terms will be used in a generic sense with more specific terminology for clarification where necessary. Some other key terms (primarily used in respect of animals) that should be considered carefully are also listed in Table 1 below.

Types of Law

There are many types or fields of law that can be relevant to wildlife conservation. For the most part, the basis and authority for wildlife conservation are parliamentary statutes (acts) or subsidiary (administrative) legislation such as regulations that provide the detailed implementation of the statute. The law is most commonly couched in terms of criminal law in that offences are provided as a means to enforce the laws. However, some civil law rights regarding land and property are used, particularly in environmental issues.

The types of law listed in Table 2 are based on the legislation of the "common law" countries, historically derived from the British legal system, that relies both on parliamentary legislation and on precedent, i.e. decisions of the courts. Other countries may have "civil law" systems based on Roman law, with codes of laws as the primary legal authority. Islamic law forms the basis of some countries' legal systems and may be mixed with other regimes. For further information on the nature of legal systems, and the status of individual countries, refer to the University of Ottawa's website JuriGlobe (https://juri-globe.ca/en/). For a brief summary of the legislative framework of EU member states, see N_LEX (https://n-lex.europa.eu/n-lex/), a database providing access to the national laws of European Union (EU) countries.

Wildlife Conservation Law

Levels and Sources

The purpose of wildlife conservation law is to provide measures to conserve or manage wildlife in its many forms: species (of animal, plant and other organisms) and the habitat (locations, landscapes and other areas of land or water) upon which species (and people) depend for survival.

This chapter will focus on legislation relating to animals. However, past experience and present principles and practice of conservation management make it clear that, without the appropriate habitat, no species will survive in the wild (Fig. 2). It is acknowledged, therefore, that biodiversity, habitat and environmental law are equally important.

Wildlife conservation legislation should also include the means to implement and enforce such laws. Legislation is found in various levels and sources of law; see Table 3:

The Development of Wildlife Laws

Wildlife laws existed at least a thousand years ago in England. Their early purpose was largely to manage wildlife and forests in order to preserve hunting, economic and other rights of the sovereign and other landowners (Law Commission 2012). In the USA, the right to hunt free-living animals for food, sport or pest control was, and is, fiercely defended. The late nineteenth century saw the beginnings of the legal protection of land and species for conservation purposes in various countries. In the UK, the Wild Birds Protection Act 1880 was one of the first species protection statutes. In the USA the passenger pigeon received legal protection by some states in the 1880s prior to its extinction. The federal Lacey Act was passed in 1900 to make it an offence to trade, in the widest sense, with illegally acquired, owned or transported wildlife, whether in violation of US or other countries' laws.

National laws directed at wildlife conservation expanded during the next century in many parts of the world (see Table 7). The multiplicity of laws and lack of harmonisation lead to wide variation in the quality of protection and enforcement. The Council of Europe drew up the Berne Convention (see Table 6), and, subsequently, the European Union introduced the Habitats

Table 1 Definitions

Term	Definitions	Comments
Wildlife	Species (animals, plants or organisms) that have not been domesticated Species that are commonly found free-living (free-ranging), living not under the control of people	Note: specimens of "wildlife" or "wild" animals or plants may also exist in captivity, e.g. in collections or research facilities. Some aspects of wildlife conservation law may apply to captive specimens as do animal welfare and animal health laws
Wild	Indicates variously "free-living" or "free-ranging", i.e. not under control or constraint Non-domesticated species Non-cultivated plants Often used to describe species that are *of a kind* that is normally free-living but are actually under control	The term should be carefully defined in documents, research and evidence
Animal (scientific context)	Member of the scientific kingdom "Animalia" (see below)	Animals include both vertebrates and invertebrates
Animal (legal context)	The term "animal" has a different meaning in almost every piece of legislation, in accordance with its purpose	The definition is usually provided in the interpretation section of a law or regulation
Species Scientific name	Every known animal, plant and other organism has a designated scientific name The International Commission on Zoological Nomenclature (ICZN) advises and arbitrates on the naming of species through the International Code of Zoological Nomenclature (https://www.iczn.org/). The Royal Botanic Gardens, Kew, maintains the International Plant Names Index (IPNI) of the scientific names of plants (https://www.ipni.org/)	Commonly used in modern wildlife legislation to define species to which the law applies Sometime the scientific name is revised, but the legislation continues to use the old nomenclature. This may need to be explained in forensic or other documents
Species Common name	Non-scientific name There may be multiple or local names for a single species	Used in old wildlife legislation. Explanation may be needed.
Species Vernacular name	Name used in the language of a particular group of people or in an area	Occasionally appears in old wildlife legislation This may need to be explained in forensic or other documents
Non-domesticated species	Species of animal not established as domesticated	
Free-living Free-ranging	Not under the control of people	
Feral	Free-living domesticated species	
Hybrid	Bred from more than a single species	In legislation (e.g. CITES) an offspring of a captive and a free-living specimen of a species may be referred to as "hybrid"
Domesticated species	An animal that breeds under human control, provides a product or service useful to humans, is tame and has been selected away from the wild type (Mason 1984)	Usually involves day-to-day control; dependent on humans for survival and welfare See also Décory (2019)
Captive animals	Living under the control or care of people May be domesticated or non-domesticated	Non-domesticated species are kept in captivity for, e.g. pets, zoos, hobbies, rehabilitation, breeding, sport, research and breeding programmes for commercial or conservation purposes

(continued)

Table 1 (continued)

Term	Definitions	Comments
Livestock	Domesticated species kept by people for food, other products or traction	"Wildlife"/non-domesticated species are kept for production, e.g. crocodile, ostrich or snail farming
Vertebrates/ invertebrates	Animals with/without backbones	In legislation the definition of "animal" may include or exclude invertebrates
Alien invasive species	IUCN: "species that are introduced, accidentally or intentionally, outside of their natural geographic range and that become problematic" EU "'alien species' means any live specimen of a species, subspecies or lower taxon of animals, plants, fungi or micro-organisms introduced outside its natural range; it includes any part, gametes, seeds, eggs or propagules of such species, as well as any hybrids, varieties or breeds that might survive and subsequently reproduce" EU 'invasive alien species' means an alien species whose introduction or spread has been found to threaten or adversely impact upon biodiversity and related ecosystem services'	International Union for Conservation of Nature (IUCN) https://www.iucn.org/regions/europe/our-work/biodiversity-conservation/invasive-alien-species Definition EU Regulation (EU) No 1143/2014 https://eur-lex.europa.eu/legal-content/EN/TXT/?uri=CELEX%3A32014R1143
Endangered species	Term used in the IUCN Red List (IUCN 2020) Term used generically for animals at risk of extinction	The IUCN Red List of Threatened Species. Version 2020-2. https://www.iucnredlist.org The Red List categories describe a scientific, not a legal, status Also used as a specific designation in legislation, e.g. The Endangered Species Act, 1973 (USA) (Vucetich et al. 2006; https://www.jstor.org/stable/3879130?seq=1 Gleaves et al. 1992 https://scholarship.law.umt.edu/cgi/viewcontent.cgi?article=1231&context=plrlr)
Threatened species	The level of risk of extinction of a species is assessed in the IUCN Red List (IUCN 2020) There are 9 categories including "threatened" "critically endangered", "endangered", and "vulnerable" Also used as a specific designation in legislation, e.g. The Endangered Species Act, 1973 (USA)	The IUCN Red List of Threatened Species. Version 2020-2. https://www.iucnredlist.org. The Red List categories describe a scientific, not a legal, status

Directive and the EU Convention on International Trade in Endangered Species of Wild Fauna and Flora (CITES) Regulations (see Table 6) to provide a level of uniformity in the standards and provisions of wildlife legislation amongst their member countries. Regional treaties have also been drawn up in other parts of the world for the same purpose (see Table 6).

Legislation is made at different levels (see Table 3), and this has led to multiple designations of protected species and habits. For example, the mountain gorilla (*Gorilla beringei beringei*) (Fig. 3) is listed on CITES and the Convention on Migratory Species (CMS) and in national legislation, and large parts of its habitat are designated as World Heritage sites (one area also as a Ramsar site) as well as being national parks (see Table 4).

Even a World Heritage site that is designated for its cultural value can nevertheless provide

Table 2 Types and sources of law relevant to wildlife

Law	Description	Comments
Statute law	Legislation made by a parliament/legislature	Can apply to the fields of criminal, civil and other law
Criminal law	Laws that rely on offences and punishments by way of enforcing them	
Subsidiary law (UK) Administrive law (USA)	Regulations that implement the main statutory law	
Civil law	Applies to disputes between individuals, corporations and institutions Includes the law of nuisance, negligence, contract, property	Aspects of these forms of law may be used in claims relating to damage to the environment, e.g. claims for compensation or injunctions for personal injury, for damage to property or causing a "nuisance" such as excessive noise or the unregulated discharge of waste
Common law, case law	Law arising from decisions made by the courts. Often used in the interpretation and application of subsequent cases (precedent)	Interpretation of existing statute and subsidiary law or case law
Public law	Constitutional, administrative and criminal law	
European Union law	Regulations take direct effect in the law of member states Directives must be given legal effect by member states	https://ec.europa.eu/info/law/law-making-process/types-eu-law_en
International law	Treaties and other instruments made between nations and other recognised international entities	Global or multi-lateral conventions, treaties, protocols, agreements, memoranda of agreement
Unwritten/customary law	Laws of an ethnic or other community or a locality; supervised by elders or leaders	Incorporating traditional management of natural resources to conservation practices
"Soft" law	Non-binding agreements, statements of principles, memoranda of understanding, guidelines and standards. These are not legally enforceable but enable the development of co-operation and good practice when legally binding agreement is not feasible (Birnie et al. 2009)	Widely used in multi-lateral wildlife negotiations

Fig. 2 This chimpanzee (*Pan troglodytes schweinfurthii*) in Kibale National Park, Uganda, depends on its habitat for survival. Image credit: Nigel Harcourt-Brown

Table 3 Levels and sources of wildlife conservation legislation

Level of law	Examples	Comments
International law	Global treaties, conventions, protocols	Made between governments Open to all countries They have to be implemented and enforced by measures in the national law of the participating countries
	Multi-lateral treaties Bilateral treaties	Made by specified governments for a common purpose
Regional		
European Union 27 member states	Directives Regulations	Legislation applies in all EU member states
Council of Europe 49 member states	Conventions	https://www.coe.int/en/web/conventions/full-list/-/conventions/treaty/104
Treaties for Africa, ASEAN countries, Central America	See Table 6	
Regional Seas Conventions, Protocols and Agreements	See Table 6	Protection of coastal marine areas and areas beyond national jurisdiction
National law	Made by individual countries	Applicable within that country's jurisdiction
Unitary countries	E.g. Bangladesh, Kenya, New Zealand, Uganda	Laws apply throughout the country
Regionalised unitary countries	E.g. China, France, Indonesia, UK	Some powers devolved to regional governments
Federal countries Federal law	E.g. Australia, Canada, Germany, India, Mexico, Nigeria, USA	Federal laws apply and are enforced throughout the country International treaty conservation implementation and enforcement are managed thorough federal laws.
Federal countries State law	Law that applies within an individual state (province, lande, keshetra) that is a component of a federal country	The law applies and is enforced only within the jurisdiction of that state. Many conservation laws are made and enforced at state level
Local law, municipal law	Law applied within a city, county or other lower administrative jurisdiction	Usually deals with local concerns, e.g. licencing the keeping of non-domesticated animals and public health
Customary law (see above)	Unwritten law that is accepted by a community and supervised by elders or leaders	E.g. traditional rules of managing or sharing wildlife resources or recognising property rights May be recognised or applied in the course of wildlife conservation and management

some protection for the species and habitat within the protected area (Fig. 4).

International Wildlife Conservation Legislation

The Main Conventions

International conservation law began at the opening of the twentieth century with the London Convention of 1900 (Convention for the Preservation of Wild Animals, Birds, and Fish in Africa) as a response to the loss of wildlife in Africa caused by hunting. While early treaties had little impact, the modern conservation conventions dealing with trade, habitat protection and cultural heritage and migratory species emerged alongside such policies as the Stockholm Declaration of 1972, World Conservation Strategy of 1980, World Charter for Nature 1982, and Report of the Brundtland Commission

Fig. 3 The mountain gorilla and its habitat are protected by multiple legal designations. Image credit: Nigel Harcourt-Brown

(1987) on the need for sustainable development. This body of soft law (i.e. non-binding agreements, statements and other documents) recognising the need for conservation to be coupled with the sustainable use of natural resources, the recognition of the role played by human beings, their right to share the benefits of natural resources and the need to conserve the environment and ecology as a whole, led to the Convention on Biological Diversity (CBD) 1993. While the first four conventions dealt with single issues of species and habitat, the CBD addresses the need to conserve biodiversity in conjunction with the sustainable use and the fair sharing of natural resources. The five conventions, although developed separately, now collaborate extensively on common and overlapping issues. Authors who discuss the international wildlife legislation in detail include Birnie et al. (2009), Bowman et al. (2010) and Sands et al. (2018). A useful list of current international conservation laws and soft law agreements with a discussion of their value to conservation is provided by Trouwborst et al. (2017). The convention websites have the full legal and administrative documents and extensive information about the conventions and allied legislation.

The five conservation conventions are listed below in Table 4, together with sources of further information.

Sectoral and Single Species Treaties

Some treaties and agreements are restricted in their application. They may apply to a specific area, species or purpose and may not relate directly or solely to wildlife conservation but may, nevertheless, be beneficial to conservation. Some are related to the management of stocks rather than species conservation. Some of these instruments are listed in Table 5.

Regional Wildlife Conservation Legislation

In addition to treaties that are open to all nations, there are many regional and special purpose conventions and agreements. They may be directed at co-operation in a particular part of the world or may relate to a particular species or purpose in a region.

An example of the former is the Council of Europe Convention on the Conservation of European Wildlife and Natural Habitats (1979), known as the Berne Convention. It was the first conservation treaty to include provisions for both species and their habitats. It served to raise the standards of conservation and relevant legislation amongst its 49 member states, the EU and five African countries. Subsequently, the European Union brought in the Habitats Directive 1992 which has to be transposed into (implemented in) all 27 member states' national legislation.

There are regional conservation treaties for the Western Hemisphere, Africa, the South Pacific, the ASEAN countries and Central America that promote co-operation on conservation amongst countries of the particular areas (see Table 6).

Transboundary protected areas, such as the Great Limpopo Transfrontier Park in Mozambique, South Africa, and Zimbabwe are set up and managed under regional multi-lateral treaties between the concerned countries. The

Table 4 The main conventions

Treaties/conventions	Purpose	Comments and definitions	Websites and references
Convention on Wetlands (Ramsar) 1971 171 parties	Designation and development of the wise use of wetland areas of international importance and international co-operation	"Wise use" means the conservation and sustainable use of wetlands	Convention website https://www.ramsar.org/ Full definition of "wise use": https://www.ramsar.org/about/the-wise-use-of-wetlands
World Heritage Convention (WHC) 1972 193 parties	Designation, protection and conservation of sites of cultural or natural heritage	Includes areas of habitat important to threatened species or areas of exceptional value to conservation	Convention website https://whc.unesco.org/en/convention/
Convention on International Trade in Endangered Species of Wild Fauna and Flora (CITES) 1973 183 parties	Regulation of the inter-country movement of threatened species (or their derivatives) listed on the Appendices, whether or not for commercial purposes	"Trade" means export, re-export, import and introduction from the sea https://cites.org/eng/disc/text.php#I Biological samples—simplified permits https://cites.org/sites/default/files/document/E-Res-12-03-R18.pdf	Convention website https://cites.org/ Appendices https://cites.org/eng/app/appendices.php
Convention on the Conservation of Migratory Species of Wild Animals (CMS or Bonn) 1979 130 parties	Framework for the conservation and sustainable use of migratory species listed on Appendices	CMS definition of migratory species: "cyclically and predictably cross one or more national jurisdictional boundaries" https://www.cms.int/sites/default/files/instrument/CMS-text.en_.PDF Numerous Agreements and Action Plans to promote co-operation between "range states" in conserving migratory species, such as birds, gorillas and marine turtles	Convention website https://www.cms.int/
Convention on Biological Diversity (CBD) 1992 196 parties	Promotes the conservation and sustainable use of biological diversity and the fair and equitable sharing of benefits derived from using natural resources	Cartagena Protocol To ensure the safe handling, transport and use of living modified organisms http://bch.cbd.int/protocol Nagoya Protocol Provides a legal framework: the fair and equitable sharing of benefits derived from using genetic resources. https://www.cbd.int/abs/about/default.shtml/	Convention website https://www.cbd.int/intro/ Common portal for the conservation conventions https://www.cbd.int/cooperation/related-conventions/guide.shtml

Fig. 4 Gedi Ruins is a tentative World Heritage site and a designated national monument and museum. As such, it provides some protection for the species and habitat at the location. Image credit: Mrs. Margaret E Cooper

ensuing co-operation is also useful for dealing with wildlife crime. These "peace parks" occur in many parts of the world for the purpose of conservation of species, migratory patterns, habitats, landscapes and marine areas. Many protected areas are listed in the IUCN World Database on Protected Areas (https://www.iucn.org/theme/protected-areas/our-work/world-database-protected-areas).

Numerous multi-lateral treaties and agreements have been made under the Convention on Migratory Species in order to encourage co-operation in the conservation of the areas along the migratory routes of species listed on the Appendices to the CMS. These include agreements relating to various species of bird, gorillas and marine turtles.

The Antarctic Treaty serves to protect the continent in restricting its use to peaceful purposes and scientific research (htps://www.ats.aq/e/antarctictreaty.html). The Convention on the Conservation of Antarctic Marine Living Resources provides measures for the management of marine resources and fishing stocks in the region (https://www.ccamlr.org/en/organisation/convention). There are conservation laws for Antarctic polar bears (*Ursus maritimus)* and six species of Antarctic seals. Together, these treaties comprise the Antarctic Treaty System (https://www.scar.org/policy/antarctic-treaty-system/). A number of books have been written on the Antarctic laws and are available at https://newbooksinpolitics.com/political/the-antarctic-treaty-regime/.

The UNEP Regional Seas Programme promotes the conservation of coastal seas and high seas and their natural resources. This includes treaties, agreements and plans of action for improving conservation legislation and species protection and preventing pollution in marine areas. The Regional Seas Programme provides a framework for inter-state collaboration in protection and conserving the environment of most of the world's oceans and coastal regions (https://www.unenvironment.org/explore-topics/oceans-seas/what-we-do/working-regional-seas/why-does-working-regional-seas-matter).

National Wildlife Conservation Legislation

Most countries have national wildlife legislation. The provisions vary greatly between nations in content, quality and development. Some examples of individual countries' legislation are

Table 5 Sectoral or single species treaties

Treaty	Purpose	Websites
International Convention for the Regulation of Whaling 1946 89 members	The conservation of whale stocks and the management of catch limits	https://iwc.int/convention
Convention on the Conservation of Antarctic Marine Living Resources (CCLAMR) 26 members	Management of marine resources and fish stocks in southern Antarctic seas	https://www.ccamlr.org/en/organisation/convention
Agreement on the Conservation of Polar Bears 1973 5 range states	Range states recognise the need to conserve the polar bear	https://polarbearagreement.org/about-us/1973-agreement
Convention for the Conservation of Antarctic Seals 1972 17 parties	Protection, scientific study and rational use of Antarctic seals, maintaining ecological balance	https://iea.uoregon.edu/treaty-text/2808
Convention on the Conservation and Management of the Vicuna 1979 5 signatories	Conservation and management, including economic use, of vicuna (*Vicugna vicugna*)	https://www.ecolex.org/details/treaty/convention-for-the-conservation-and-management-of-the-vicuna-tre-000102/

listed in Table 7, and the portals to collections of national laws are listed in Table 8.

National governments give effect to international treaties, such as the conservation conventions and other legislation mentioned earlier, by putting the provisions into their national law and providing powers to enforce them. For example, the US federal law, the Endangered Species Act of 1973 implements the CITES Convention. The EU CITES Regulations do the same for the 27 EU member states (https://ec.europa.eu/environment/cites/legislation_en.htm).

National wildlife legislation can include species protection for both plants and animals and habitat protection for both land and water. Habitat conservation is based on the designation of protected areas. There are so many different categories, from those internationally recognised to small local areas managed by private organisations, that the IUCN classifies them according to their management needs and governance types (Dudley et al. 2013). The protection accorded thereby also provides benefits (intentionally or otherwise) for the species located in the area. The conservation value of both species and habitat is unequal as those that are most at risk receive stricter protection.

Despite inevitable variabilities, there are some core topics that are commonly included in species protection, as follows:

- Protected species are usually listed according to varying degrees of protection.
- Killing, taking, injuring and disturbing protected species are made offences (Fig. 5).
- Possession of wild animals may require authorisation, especially if they are protected species.
- Sites, nests, young and eggs are protected during the breeding season.
- Trade, i.e. sale or other commercial use, is an offence or controlled. CITES implementation may be included or put in separate legislation.
- Hunting may be illegal or regulated. There may be restrictions on the equipment used and methods of hunting.
- Close season restrictions are set for game species so that they are not hunted in the breeding season.
- Sustainable use of some species may be allowed by permit (see below).

Table 6 Regional legislation

Regional	Regional conventions	Protection of coastal marine areas and areas beyond national jurisdiction
European Union	Birds Directive Habitats Directive CITES Regulations	Legislation applies in all EU member states
Council of Europe	Convention on the Conservation of European Wildlife and Natural Habitats (Berne Convention)	https://www.coe.int/en/web/bern-convention https://www.coe.int/en/web/conventions/full-list/-/conventions/treaty/104
Africa	African Convention on the Conservation of Nature and Natural Resources (revised 2006)	https://portals.iucn.org/library/sites/library/files/documents/EPLP-056.pdf
Southeast Asia	ASEAN Agreement on the Conservation of Nature and Natural Resources	http://agreement.asean.org/media/download/20161129035620.pdf
UNEP Regional Seas programme Other regional seas programmes	Conventions, protocols and agreements on the conservation of coastal marine areas and the high seas	https://www.unenvironment.org/explore-topics/oceans-seas/what-we-do/working-regional-seas/why-does-working-regional-seas-matter
The Antarctic Treaty	For the management of the Antarctic for peaceful and scientific purposes	htps://www.ats.aq/e/antarctictreaty.html

Some national legislation is made for the protection of an individual species that is particularly at risk or of special national significance, for example, the Philippine Eagle (*Pithecophaga jefferyi*) (https://www.chanrobles.com/republicacts/republicactno6147.html) (between 1970 and 2001) (see Table 7 and Krupa (1989)), bald (*Haliaeetus leucocephalus*) and golden eagles (*Aquila chrysaetos canadensis*) in the USA (https://www.fws.gov/birds/policies-and-regulations/laws-legislations/bald-and-golden-eagle-protection-act.php) and the European badger (*Meles meles*)in the UK.

Strict regulation usually requires some scope for flexibility, and authorisation may be given for the use of protected species or activities that may be acceptable as sustainable use or necessary control, such as:

- Captive breeding, farming or ranching, particularly when this reduces the pressure on wild populations
- Wildlife rescue, rehabilitation and release—in many countries, this requires training and authorisation
- Taxidermy of dead animals, legally acquired
- Display of protected species for education and entertainment
- Scientific research
- Sport, such as falconry
- Farming/ranching
- Protection of property
- Protection of crops
- Public health

Wildlife legislation often includes restrictions on the release into the wild of non-indigenous animals and plants outside their natural range. This has become an issue of serious concern due to the propensity of alien species of animal and plant to damage or to displace native species or habitat. Release may be illegal altogether or have to be authorised. By way of example, in the EU where there are estimated to be 12,000 alien species, 15% of which are invasive, the EU regulation on invasive alien species makes it illegal to bring them into the EU, breed or transport or release them. There are also measures to deal with accidental introduction and for their control and eradication (https://ec.europa.eu/environment/nature/invasivealien/index_en.htm).

The variability of national wildlife conservation laws is such that it is essential to examine the precise law relating to any species, habitat or particular situation in a given country. The laws of a number of countries have been listed in

Table 7 Examples of national wildlife legislation and where to find it

Country	Legislation and sources
Africa, West and Central	See LAGA Legal Regional Library in Table 8
Australia states and territories	Wildlife legislation within each state Database for federal and state legislation Australasian Legal Information Institute http://www.austlii.edu.au/ Environmental Crime in Australia, Samantha Bricknell, Australian Institute of Criminology 2010 Canberra https://core.ac.uk/download/pdf/30682032.pdf
Australia federal law	Environment Protection and Biodiversity Conservation Act 1999 (amnd 2001); implementing obligations under international treaties https://www.environment.gov.au/biodiversity/wildlife-trade/law#
Cameroon	LOI N°94/01 du 20 janvier 1994 portant régime des forêts, de la faune et de la pêche See LAGA database in Table 8 Djeukam (2012), Nkoke et al. (2018)
Canada	Wild Animal and Plant Protection and Regulation of International and Interprovincial Trade Act (S.C.1992, c. 52) https://laws-lois.justice.gc.ca/eng/acts/W-8.5/ Wildlife Area Regulations (C.R.C., c. 1609) https://laws-lois.justice.gc.ca/eng/regulations/C.R.C.,_c._1609/index.html
	Migratory Birds Regulations (C.R.C., c. 1035) https://laws-lois.justice.gc.ca/eng/regulations/C.R.C.,_c._1035/index.html
	Canada Wildlife Act (R.S.C., 1985, c. W-9) https://laws-lois.justice.gc.ca/eng/acts/W-9/wildlife
	National Parks Wildlife Regulations (SOR/81-401) https://laws-lois.justice.gc.ca/eng/regulations/SOR-81-401/
	Canada National Parks Act (S.C. 2000, c. 32) https://laws-lois.justice.gc.ca/eng/acts/n-14.01/
	National Parks Wildlife Regulations (SOR/81-401) https://laws-lois.justice.gc.ca/eng/regulations/SOR-81-401/
India	The Wild Life (Protection) Act, 1972 (amnd 2003) http://legislative.gov.in/sites/default/files/A1972-53_0.pdf Forest (Conservation) Act, 1980 (amnd 1988) The Environmental (Protection) Act 1986 No. 29 OF 1986 [23rd May, 1986.] No. 29 OF 1986 [23rd May, 1986]
	Sinha (2010) Handbook on Wildlife Law Enforcement in India. https://www.traffic.org/site/assets/files/6284/handbook-wildlife-law-enforcement-india.pdf Wildlife Conservation Laws in India An Overview http://awsassets.wwfindia.org/downloads/lecture_notes2_session_11_1.pdf Legal Framework for Wildlife Conservation in India https://www.conservationindia.org/resources/the-legal-framework-for-wildlife-conservation-in-india-2
Ireland	Wildlife Acts 1976 to 2018 https://www.npws.ie/legislation
Jamaica	The Wildlife Protection Act 1945 https://moj.gov.jm/sites/default/files/laws/The%20Wild%20Life%20Protection%20Act.pdf
Kenya	The Wildlife (Conservation and Management) Act 2009 http://kenyalaw.org/kl/fileadmin/pdfdownloads/Acts/Wildlife__Conservation_and_Management_Act___Cap_376_.pdf
Malaysia	Wildlife Conservation Act 2010 https://sherloc.unodc.org/cld/uploads/res/document/wildlife-conservation-act-2010_html/Wildlife_Conservation_Act_2010.pdf
Monserrat	Forestry, Wildlife, National Parks and Protected Areas Act (revised 2008) http://agc.gov.ms/wp-content/uploads/2011/10/Forestry-Wildlife-National-Parks-Act.pdf

(continued)

Table 7 (continued)

Country	Legislation and sources
New Zealand	Wildlife Act 1953 https://www.doc.govt.nz/about-us/our-role/legislation/wildlife-act/
Pakistan	Wildlife Trade Control Act 2012 http://extwprlegs1.fao.org/docs/pdf/pak164599.pdf Rules of the Pakistan Wildlife Trade Control Act of Fauna and Flora 2012
Papua New Guinea	Fauna (Protection and Control) Act 1966 http://www.paclii.org/pg/legis/consol_act/faca1966290.pdf
Philippines	Republic Act No. 9147 on the Wildlife Resources Conservation and Protection Act of 2001 https://www.lawphil.net/statutes/repacts/ra2001/ra_9147_2001.html Republic Act No. 6147—an act declaring the *Pithecophaga jefferyi* . . . as a protected bird in the Philippines (and other laws) (Krupa (1989)). Repealed in 2001 by the Republic Act 9147
Republic of China	Wildlife Conservation Act https://law.moj.gov.tw/ENG/LawClass/LawAll.aspx?pcode=M0120001 Laws and regulations database of the Republic of China
Tanzania	Wildlife Conservation Act 2009 https://tanzlii.org/tz/legislation/act/2009/5-0 https://www.loc.gov/law/help/wildlife-poaching/tanzania.php#_ftn16 Wildlife cases https://www.iucn.org/sites/dev/files/content/documents/wildlife_cases_in_tanzanian_courts_1_0.pdf
Trinidad	Conservation of Wildlife Act 1958 as amended https://rgd.legalaffairs.gov.tt/Laws2/Alphabetical_List/lawspdfs/67.01.pdf
Uganda	Uganda Wildlife Act 2019 Act No. 17 of 2019.pdf (https://www.ugandawildlife.org)
UK	Wildlife and Countryside Act 1981 as amended and other legislation https://www.legislation.gov.uk/ukpga/1981/69/contents Wildlife-related laws (Cooper, 2016)
USA—all US law	Guide to Law online (all US law) https://www.loc.gov/law/help/guide.php US Code (of all federal statutes) https://uscode.house.gov/browse/prelim@title16&edition=prelim
USA—federal wildlife law	https://www.fws.gov/le/laws-regulations.html
Code of Federal Regulations	https://www.fws.gov/international/laws-treaties-agreements/regulations.html Fish and wildlife: https://ecfr.io/Title-50/
US state laws	https://www.loc.gov/law/help/guide/states.php
US state wildlife law	A general summary of state laws is provided by the organisation Born Free https://www.bornfreeusa.org/campaigns/animals-in-captivity/summary-state-laws-exotic-animals/

Table 7, and, for a wider range of legislation, Table 8 provides databases and portals that give access to a further range of national wildlife conservation laws.

National wildlife legislation normally includes provisions to prohibit or control the trade in wildlife species that occur within the country. The cross-border trade in CITES-listed species is regulated by the Convention provisions

The 193 parties to the CITES Convention are required to incorporate the terms of the Convention in their national law. Again, the final national provisions will vary; nevertheless, the fundamental framework is common to all countries and is summarised below.

The Convention on the International Trade in Endangered Species of Wild Fauna and Flora (CITES) regulates the "trade" in endangered species. The basic provisions are as follows:

- "Trade" means any cross-frontier movement of species of animals or plants that are listed on the Appendices to CITES.

Table 8 List of databases and other sources providing access to national legislation

EUR-LEX	Database of all EU law https://eur-lex.europa.eu/homepage.html
European Union Official Journal	Database of all EU legislation, notices and information as published https://eur-lex.europa.eu/oj/direct-access.html
United Kingdom	http://www.legislation.gov.uk https://www.legislation.gov.uk/all?theme=environment https://www.cps.gov.uk/legal-guidance/wildlife-offences Wild Birds and the Law. England and Wales https://www.rspb.org.uk/globalassets/downloads/documents/positions/wild-birds-and-the-law/wild-birds-and-the-law%2D%2D-a-plain-guide-to-bird-protection-today.pdf Wild Birds and The Law: Scotland https://www.rspb.org.uk/globalassets/downloads/documents/positions/wild-birds-and-the-law/wild-birds-and-the-law%2D%2Dscotland.pdf
EUR-LEX National Transposition	Searchable database of national laws of member states that implement EU laws https://eur-lex.europa.eu/collection/n-law/mne.html
N-LEX	Database of access to national laws of EU member states Includes summary of the country's legislation framework https://n-lex.europa.eu/n-lex/
LEGIS-PALOP + TL	The juridical official database of the Portuguese-speaking African Countries +Timor Leste https://www.legis-palop.org/
ECO-LEX	Portal to environmental law. Treaties, agreements, soft law, national laws, judicial decisions and policy. Compiled from the resources of the Food and Agriculture Organization, IUCN Environmental Law Centre and United Nations Environment Programme (UNEP) https://www.ecolex.org/
FAOLEX	Database of national laws, mainly agricultural and environmental http://www.fao.org/faolex/associated-databases/en/
ELAWS	List of Caribbean environmental laws https://home.elaw.org/caribbean/laws
International Environmental Law Research Guide	Georgetown University International Environmental Law database https://guides.ll.georgetown.edu/c.php?g=273374&p=1824812
LAGA Legal Regional Library	Last Great Ape Organisation. Links to the wildlife laws of 11 West and Central African countries https://www.laga-enforcement.org/en/legal-regional-library
Wildlife Law Africa	Website featuring legislative, enforcement and judicial aspects of wildlife law of African countries https://wildlifelawafrica.com/
Uganda Wildlife Crime Database	Under construction https://uganda.wcs.org/Initiatives/Law-Enforcement/Wildlife-Crime-Database.aspx
WILDLEX	New database currently containing wildlife laws and case reports from Tanzania and Cameroon. https://www.wildlex.org/
Free Access to Law Movement	Portals to the legislation of a variety of counties in Africa, Asia, the Commonwealth and elsewhere. Provided by organisations committed to free access to legal information http://www.fatlm.org/members/current/
Reports	Country Reports to the United Nations (UN) or Convention Secretariats Publications by UNEP, IUCN and other international agencies Report on national conservation strategy or policy These often contain summaries of relevant national laws
Gorillas	A survey of legislation relating to gorillas (Cooper 2017)

Fig. 5 This letterbox is closed annually while a small passerine bird produces young. In accordance with UK law, it must not be disturbed while nesting. Image credit: Mrs. Margaret E Cooper

- Such a movement must be authorised by the appropriate CITES permit.
- "Trade" includes movement to another country for any purpose (commercial or non-commercial). Consequently, transfer for scientific purposes requires a CITES permit.
- This rule applies to the movement of live animals and plants, dead specimens and any "derivatives", i.e. anything that can be identified, by any means, as belonging to a CITES-listed species.
- The Appendices:
  - Appendix I: Species of animals and plants at risk of extinction
  - Appendix II: Species not immediately at risk of extinction but are likely to become so if trade is not closely monitored (Fig. 6)
  - Appendix III: Species listed by a country that requests help from other countries in controlling the trade in it
- Permits are required for import and export and are issued by the CITES Management Authority on the advice of the Scientific Authority of each country
- CITES permits:
  - Appendix I:
    Permits are not given for primarily commercial purposes.
  - Appendix II:
    Permits for commercial purposes are issued but the trade monitored.
  - Appendix III:
    For species listed as requested by parties.
- Export permits may only be given for legally obtained specimens and must not be detrimental to the survival of the species. Live specimens must be transported to minimise the risk of injury, damage to health or ill treatment

There are extensive provisions beyond these basic elements. The CITES website provides the documentation and information regarding the Convention (https://cites.org/eng), and a detailed account of its functioning is provided in Wijnstekers (2018).

As mentioned above, CITES allows parties to impose measures that are stricter than the actual Convention requirements. For example, the European Union has raised the status of a good number of Appendix II species to the equivalent of Appendix I within the EU. It also has stricter control on the commercial use of CITES species but provides for a regulated trade in captive-bred Annex A (equivalent of Appendix I) species. The convention is implemented by EU Regulations that take direct effect in all member states. The latter have only to provide national enforcement legislation. Information on, and links to, the EU CITES legislation is provided at http://ec.europa.eu/environment/cites/legislation_en.htm, and the implementation is explained in the Reference Guide to EU Regulations (http://ec.europa.eu/environment/cites/legis_refguide_en.htm). Links to CITES relevant information for individual member states are provided at https://ec.europa.eu/environment/cites/links_national_en.htm, and N-LEX is a portal to all the national legislation of the member states (https://ec.europa.eu/environment/cites/links_national_en.htm).

Fig. 6 The giraffe (*Giraffa camelopardalis*) was added to CITES Appendix II in 2019 to enable trade, particularly in derivatives, to be monitored. Image credit: Mrs. Margaret E Cooper

The UK left the EU on 31 January 2020, but its CITES legislation will be comparable to that of the EU. Specific information and guidance on UK CITES law is available at https://www.gov.uk/guidance/cites-imports-and-exports, and the UK legislation database is at https://www.legislation.gov.uk/.

CITES is implemented in the United States of America (USA) by the Endangered Species Act and managed and enforced by the United States Fish and Wildlife Service (USFWS); see https://www.fws.gov/endangered/laws-policies/ and https://www.fws.gov/international/cites/index.html. The permit procedures are described in https://www.fws.gov/international/pdf/factsheet-cites-permits-and-certificates-2013.pdf.

Many countries now maintain a public database of their national legislation on the Internet, and this is likely to include their CITES legislation. If this is accessible, information may be available from the country's CITES Management Authority as it is the national body responsible for issuing permits.

Wildlife Law Enforcement

Wildlife is traded at every level of society and commerce. In legitimate trade, animals and plants are in demand for breeding or propagation, exhibition, private collections, pets and scientific study. Plants and timber also have widespread uses. TRAFFIC estimates that the legal trade in wildlife is worth Euro 13 billion per year in the EU alone and takes the view that a sustainable legal trade can be beneficial. CITES regulates (see above) the trade in some 5950 species of animals and 32,800 species of plants. In parts of the world where electricity is scarce, charcoal is widely used for cooking. The long haul, sometimes cross-boundary, trade in charcoal, whether or not permitted, has depleted landscapes of trees to the detriment of the environment (Fig. 7).

In addition, large numbers of animals and plants, of all degrees of endangerment, are moved around the world illegally. The illegal trade in wildlife ranks with drug and human trafficking as a substantial cause of serious and organised national and international crime. It attracts corruption at every stage, and cybercrime and dark Internet markets are also implicated. The sheer numbers of species involved, and the quantity of each, put their wild populations at risk of extinction (Fig. 8).

There are extensive efforts at national and international levels by some countries to apply enforcement measures to curtail the illegal trade in wildlife; on the other hand, there are parts of

Fig. 7 In this case, charcoal-burning is illegal, and a witness is indicating the evidence. Image credit: Mrs. Margaret E Cooper

Fig. 8 The trade in ivory puts free-living elephant (*Loxodonta africana*) populations at risk of extinction. Image credit: Mrs. Margaret E Cooper

the world where illegal trade is rife and benefits from poor governance, legislation and enforcement. The UK government states that the illegal trade is worth £17 billion per year, and Clifton and Rastogi (2016) suggest US$19 billion or more, depending on the authority cited. INTERPOL reported that more than 300 tonnes of wood and timber, 3.9 tonnes of pangolin scales, 25 tonnes of animal parts and 22,000 derivatives and processed products, amongst other wildlife, were seized in a single 92-country enforcement operation (INTERPOL 2018). Ivory is one of the largest and best-known categories of illegal trade; with the inevitable effect on elephant populations. The Wildlife Conservation Society states that "some 100,000 elephants across Africa [were] killed illegally between 2010 and 2012" (https://www.wcs.org/our-work/species/african-elephants), and there is a similar impact on pangolins, rhinos and many other species of animals and plants (INTERPOL 2018).

Countless efforts are being made to counter poaching and other aspects of illegal trade from on-the-ground support for wildlife rangers in the field, reports and analysis based the collection of in situ market data (https://www.traffic.org/about-us/legal-wildlife-trade/) and identification services (The Royal Botanic Gardens Kew https://www kew.org/read-and-watch/tackling-illegal-wildlife-partnership) (RGB Kew 2017) to multi-disciplinary theoretical studies of consumer patterns and dark networks based on theories and modelling (The Oxford Martin Programme on the Illegal Wildlife Trade https://www.illegalwildlifetrade.net/). TRAFFIC launched, in April 2020, the Wildlife Trade Portal, an open-source database of wildlife trade incidents and seizures (https://www.wildlifetradeportal.org/).

Many countries that are rich in rare species have long suffered from weak laws and enforcement procedures. In recent years, increasing recognition has been accorded to the need for improvement in these areas and is included in strategy and planning and by NGOs, funding and other agencies, both private and public and governmental and international. It has led to a steady increase in training and capacity building for enforcement authorities and for the judiciary—covering investigation and forensic facilities, prosecution, the legislative framework and legislation, judicial procedure, advocacy, sentencing and the recording of decided cases (The International Consortium on Combating Wildlife Crime ICCWC https://cites.org/eng/prog/iccwc.php; INTERPOL 2018; UNODOC 2016, 2020). Recent examples of capacity building are the setting up of a wildlife crime task force in Uganda (https://uganda.wcs.org/About-Us/News/ID/13876/Government-of-Uganda-Inaugurates-a-National-Wildlife-Crime-Task-Force.aspx) and training in the analysis of wildlife prosecutions in Tanzania and Cameroon (https://www.wildlex.org/). In West and Central Africa, the Last Great Ape Organisation (LAGA) has developed a comprehensive model to improve the effectiveness of wildlife law enforcement, first established in Cameroon and now extended to other countries in the region (https://www.laga-enforcement.org/en). A guide to law enforcement in Cameroon is provided by Nkoke et al. (2018).

There have been several multi-national statements on illegal trade, culminating in the London Declaration of 2018, made at the London Conference on the Illegal Wildlife Trade and signed by 73 countries. The Declaration affirms that these countries recognise the social and economic damage and risks to security caused by wildlife crime and undertake to strengthen, inter alia, the capacity of legal and justice systems to combat wildlife crime (https://www.gov.uk/government/publications/declaration-london-conference-on-the-illegal-wildlife-trade-2018/london-conference-on-the-illegal-wildlife-trade-october-2018-declaration).

Governments of high-income countries have made funds available for such projects, for example, the UK's Illegal Wildlife Trade Challenge Project Funding is aimed specifically at projects that include the "strengthening of law enforcement" and "ensuring effective legal frameworks" (https://assets.publishing.service.gov.uk/government/uploads/system/uploads/attachment_data/file/891874/iwt-challenge-fund-list.pdf).

The US Department of State has established a task force pulling together federal departments and agencies to combat wildlife crime by "strengthening law enforcement, reducing demand, and building international cooperation" under the Eliminate, Neutralize, and Disrupt Wildlife Trafficking Act of 2016 (https://www.state.gov/2019-end-wildlife-trafficking-strategic-review/ and https://www.state.gov/2019-end-wildlife-trafficking-report/).

CITES has a role in wildlife law enforcement. Parties are required to incorporate the CITES convention's provisions in their national law and to include provisions for enforcement. Compliance with this is still in progress for many countries, according to the CITES Secretariat's National Legislation Project (https://cites.org/eng/legislation/National_Legislation_Project). It also provides accounts of wildlife seizure and prosecutions in its twice-yearly bulletins and annual illegal trade reports (https://cites.org/eng/resources/reports/Annual_Illegal_trade_report). TRACE Wildlife Forensics Network (https://www.tracenetwork.org/) supports capacity-building in DNA forensics in Africa and Asia. The provision of sound evidence is essential to the success of a prosecution, and a number of initiatives have produced written guidance on forensics, for example, FWG/PAW (2014) on forensic techniques (Baker et al. (2020)) and UNODC/IWCCWC (2014) on ivory identification. For a broader approach to forensic investigation, see Cooper and Cooper (2013) and Huffman and Wallace (2012).

International wildlife trade is one of the major categories of cross-border serious and organised crime and can only be effectively combatted with collaborative action. To this end a number of regional Wildlife Enforcement Networks (WENs) have been established, such as the Lusaka Agreement Task Force (https://www.peacepalacelibrary.nl/ebooks/files/C08-0009-Lusaka-Mrema.pdf) and the International Consortium on Combating Wildlife Crime and the South Asia Wildlife Enforcement Network (ASEAN-WEN) to share information and take joint action on wildlife crime in their region (https://cites.org/eng/prog/iccwc.php/Action/report_second_global_meeting_WENs).

In addition, there are overarching collaborative entities that bring together major international institutions. The International Consortium on Combating Wildlife Crime brings together INTEPOL, UNODOC, the World Bank and World Customs Union with CITES to take concerted wildlife enforcement initiatives, including strengthening wildlife legislation and building capacity in enforcement and judicial systems (https://cites.org/eng/prog/iccwc.php). The merits of international collaboration in wildlife crime are discussed by Pink and White (2016).

Without the implementation of improved wildlife conservation legislation and enforcement, species, their habitat, biodiversity and environment are likely to contemplate a bleak future (Fig. 9).

Fig. 9 This Kibale chimpanzee appears to contemplate the future with gravity. Image credit: Nigel Harcourt-Brown

Conclusion

Legislation is an essential tool in wildlife conservation. It provides the framework that permits authorities to implement conservation policy management and other actions. It includes enforcement powers to support such activities and to combat illegal use, damage or destruction of species and their habitat. It is important to know what legislation exists and to be able to locate and examine the current version. Despite the convenience of accessible law databases and the immediacy of Internet-based information, deeper consideration of the legislation and its implications is available in published literature. This chapter provides a guide to the structure, sources and discussion of wildlife conservation legislation.

Acknowledgements The author would like to express her appreciation of the help and support of Susan Underkoffler, senior editor of this book, for the invitation and encouragement to contribute, for her help in editing this contribution and for her generous approach to submission deadlines. Nigel Harcourt-Brown kindly provided the splendid great ape images in this chapter, and I would like to thank him for raising the standards of photography in the chapter. The remaining images were taken by the author, and she is also grateful to Dr. Dino Martins for the opportunities to take photographs at Mpala Research Centre, Laikipia, Kenya, some of which are included in this chapter.

References

Baker B, Jacobs R, Mann M, Espinoza E, Grein G (2020) CITES identification guide for ivory and ivory substitutes (4th edn, Allan C (ed)), World Wildlife Fund Inc., Washington, DC. Commissioned by CITES Secretariat, Geneva, Switzerland. https://cites.org/sites/default/files/common/resources/The_Evolution_of_CITES_2018.pdf

Birnie P, Boyle A, Redgwell C (2009) International law and the environment, Oxford University Press, Oxford UK (4th edn due 2021)

Bowman M, Davies P, Redgwell C (2010) Lyster's International Wildlife Law, 2nd edn. Cambridge University Press, Cambridge, UK

Clifton K, Rastogi A (2016) Curbing illegal wildlife trade: the role of social network analysis working paper 5 (July 2015). IUCN, Washington. https://www.iucn.org/sites/dev/files/pdf_final_wildlife_crime.pdf

Cooper ME (2016) Law affecting British wildlife casualties. In: Mullineaux E, Keeble E (eds) BSAVA manual of wildlife casualties, 2nd edn. British Small Animal Veterinary Association, Quedgeley, UK

Cooper ME (2017) Legal considerations. In: Cooper JE, Hull G (eds) Gorilla pathology and health. Academic Press, Elsevier, New York

Cooper JE, Cooper ME (2013) Wildlife Forensic Investigation. Principles and practice. CRC Press, Boca Raton, FL

Décory MSM (2019) A universal definition of 'domestication' to unleash global animal welfare progress, dA. Derecho Animal (Forum of Animal Law Studies) 10/2. https://doi.org/10.5565/rev/da.424. https://ddd.uab.cat/pub/da/da_a2019v10n2/da_a2019v10n2p39.pdf

Djeukam R (2012) The wildlife law as a tool for protecting threatened species in Cameroon. Ministry of Forestry and Wildlife (MINFOF) Department of Wildlife and Protected Areas and the Last Great Ape Organisation (LAGA). https://www.laga-enforcement.org/media/legal_library/Cameroon/legal%20book%20English.pdf

Dudley N, Shadie P, Stolton S (2013) Guidelines for applying protected area management categories including IUCN WCPA best practice guidance on recognising protected areas and assigning management categories and governance types. IUCN, Gland, Switzerland

FWG/PAW (2014) Wildlife crime. A guide to the use of forensic and specialist techniques in the investigation of wildlife crime. Forensic Working Group (FWG), Partnership for Action Against Wildlife Crime (PAW), London, UK. https://assets.publishing.service.gov.uk/government/uploads/system/uploads/attachment_data/file/324558/wildlife-crime-forensics-guide-April-2014.pdf

Gleaves K, Kurue K, Montanio P (1992) The meaning of "species" under the endangered species act. 13 Publ Land Law Rev 2. https://scholarship.law.umt.edu/cgi/viewcontent.cgi?article=1231&context=plrlr

Huffman JE, Wallace JR (2012) Wildlife forensics methods and applications. Wiley-Blackwell, Oxford, UK

INTERPOL 2018 Global Wildlife Enforcement. Strengthening Law Enforcement Cooperation Against Wildlife Crime, INTERPOL General Secretariat, France. https://www.interpol.int/en/Crimes/Environmental-crime/Wildlife-crime

IUCN 2020) The IUCN Red List of Threatened Species. Version 2020-2. International Union for Conservation of Nature. https://www.iucnredlist.org

Krupa RE (1989) Legislation for the Philippine Eagle *Pithecophaga jefferyi*. In: Meyburg B-U, Chancellor RD (eds) Raptors in the modern world WWGBP: Berlin, London & Paris. http://www.raptors-international.org/book/raptors_in_the_modern_world/Krupa_1989_597-598.pdf

Law Commission (2012) Wildlife Law. Law Commission Consultation Paper No 206. Law Commission, London, UK. http://www.lawcom.gov.uk/app/uploads/2015/03/LCCP206_Wildlife_law_consultation_paper_for_web.pdf

Mason IL (1984) Evolution of domesticated animals. Longman, London, UK

Nkoke SC, Nya FA, Ononino AB (2018) Guide to wildlife law enforcement, Cameroon: competences, attributions, duties, and responsibilities of the different law enforcement agencies. TRAFFIC, Yaoundé, Cameroon and Cambridge, UK

Pink G, White R (2016) Environmental crime and collaborative state intervention. Palgrave Macmillan, Basingstoke, UK. https://link.springer.com/chapter/10.1007/978-1-137-56257-9_5

RBG Kew (2017) Plant conservation policies and international trade (Chapter 12), State of the Worlds Plants. Royal Botanic Gardens Kew. http://stateoftheworldsplants.org/2017/report/SOTWP_2017_12_plant_conservation_policies_and_international_trade.pdf

Sands P, Peel J, Fabra A, MacKenzie R (2018) Principles of international environmental, 4th edn. Cambridge University Press, Cambridge, UK

Sinha S (2010) Handbook on wildlife law enforcement in India. TRAFFIC India, New Delhi. https://www.traffic.org/site/assets/files/6284/handbook-wildlife-law-enforcement-india.pdf

Trouwborst A, Blackmore A, Boitani L et al (2017) International wildlife law: understanding and enhancing its role in conservation. Bioscience 67(9):784–790. https://doi.org/10.1093/biosci/bx086

UNODC/IWCCWC (2014) Guidelines on Methods and Procedures for Ivory Sampling and Laboratory Analysis. United Nations Office, Vienna, Austria

UNODC (2016) World Wildlife Crime Report Trafficking in Protected Species. United Nations Office on Drugs and Crime. Austria, Vienna. https://globalinitiative.net/wp-content/uploads/2017/12/UNODC-World_Wildlife_Crime_Report_2016_final.pdf

UNODC (2020) World Wildlife Crime Report Trafficking in Protected Species. United Nations Office on Drugs and Crime. Austria, Vienna. https://www.unodc.org/documents/data-and-analysis/wildlife/2020/World_Wildlife_Report_2020_9July.pdf

Vucetich JA, Nelson MP, Phillips MK (2006) The normative dimension and legal meaning of endangered and recovery in the U.S. Endangered Species Act. Conserv Biol 20(5):1383–1390. https://www.jstor.org/stable/3879130?seq=1

Wijnstekers W (2018) The evolution of CITES, 11th edn. International Council for Game and Wildlife Conservation, Budakeszi, Hungary. https://cites.org/sites/default/files/common/resources/The_Evolution_of_CITES_2018.pdf

Part II

Applying Forensic Science and Integrative Disciplines to Conservation

Conservation Forensics: The Intersection of Wildlife Crime, Forensics, and Conservation

Aaron M. Haines, Stephen L. Webb, and John R. Wallace

Abstract

Poaching and the illegal wildlife trade (i.e., wildlife crime) are a multibillion-dollar global industry. The commercialization and overexploitation of wildlife caused by wildlife crime threaten biodiversity, particularly many of the species already on the cusp of extinction. Wildlife crime also leads to ecosystem collapse and loss of government revenues and threatens the strength and economic aspiration of developing nations. Efforts from wildlife law enforcement to prevent wildlife crime are a conservation necessity. The purpose of this chapter is to introduce the field of conservation forensics. Conservation forensics is an applied field of conservation crime science that fits within the broader frameworks of green and conservation criminology. This field of study applies hard science techniques used to gather wildlife crime data such as genetics, chemical analysis, geographical analysis, statistics, artificial intelligence, and computational modeling toward techniques that can directly benefit the efforts of law enforcement personnel involved in protecting imperiled wildlife. This chapter identifies and reviews tools and techniques that can help achieve the goals of conservation forensics: the prosecution of wildlife criminals and the prevention of wildlife crime to conserve biodiversity.

Keywords

Conservation forensics · Illegal wildlife trade · Technology · Wildlife crime · Wildlife law enforcement

Introduction

The term wildlife trade refers to the trade of live animals and plants, including a diverse collection of animal- or plant-based products from the wild (TRAFFIC 2014). The long human history of wildlife use and trade is ingrained in cultures around the world (Kahler and Gore 2017). More recently, many cases of wildlife trade have led to overexploitation, as evidenced by significant reductions in population sizes of species due to fishing (e.g., sea cucumbers and sharks), illegal wildlife trade (e.g., pangolins [*Manis* spp.]), and extinction of species due to overharvest (e.g., passenger pigeon [*Ectopistes migratorius*]) (Wyatt 2013a; McLellan 2014; Primack 2014; Eriksson et al. 2015; Maxwell et al. 2016). In response, many countries have implemented national and international policies and regulations

A. M. Haines (✉) · J. R. Wallace
Department of Biology, Millersville University, Millersville, PA, USA
e-mail: aaron.haines@millersville.edu; john.wallace@millersville.edu

S. L. Webb
Noble Research Institute, LLC, Ardmore, OK, USA
e-mail: slwebb@noble.org

S. C. Underkoffler, H. R. Adams (eds.), *Wildlife Biodiversity Conservation*,
https://doi.org/10.1007/978-3-030-64682-0_6

to maintain wildlife populations by making specific wildlife markets illegal and others more sustainable (Gore 2017). Wildlife crime involves the breaking of these policies and regulations, such as the illegal take or killing of wildlife (i.e., poaching), the alteration of wildlife into products, smuggling wildlife products within or between countries, or the selling of these products (Kahler and Gore 2017).

Globally, nearly one out of five terrestrial vertebrates is traded in wildlife markets, with traded species more likely to be threatened or vulnerable to extinction than those not bought or sold (Scheffers et al. 2019). The commercialization of wildlife is threatening species already on the cusp of extinction. Efforts from wildlife law enforcement to prevent illegal harvest and poaching of wildlife are a conservation necessity because reducing wildlife crime makes for effective wildlife conservation (Haines et al. 2015). The purpose of this chapter is to introduce the field of conservation forensics and provide a brief review of the latest in wildlife forensics and current technologies to help wildlife law enforcement efforts. The goal of this chapter is to identify tools and techniques that can help achieve the goals of conservation forensics: the prosecution of wildlife criminals and the prevention of wildlife crime to conserve biodiversity.

Causes and Scale of Wildlife Crime

Wildlife criminals are motivated by the scale at which benefits are received to the individual, family, or society. The benefits received at these levels can include economic, social, and political rationales (see Table 1). Forsyth and Forsyth (2018) hypothesized that wildlife crime is culturally passed down, like hereditary behavior, from family to family (or individual to individual), and tied to historical motivations (Table 1).

The current scale and intensity of both poaching and the illegal wildlife trade are global, to the point where wildlife trafficking ranks as one of the most profitable crimes in the world, making it an intricate and diverse multibillion-dollar industry involving a range of species, products, illegal organizations, and countries (Wyatt 2013a; Brashares et al. 2014; Gore 2017). The illegal trade of wildlife has become as lucrative as the sale of illegal drugs, weapons, and human trafficking, with profit estimates ranging from a total of $7 to $23 billion dollars, based on estimates from the Organisation for Economic Co-operation and Development (OECD), United Nations Office on Drugs and Crime (UNODC), United Nations Environment Programme (UNEP), and International Criminal Police Organization (INTERPOL) (Nellemann et al. 2014). Additionally, the estimated number of confiscations and seizures reported may only be one-tenth of the volume of wildlife smuggled (Wyatt 2013a).

Illegal wildlife smuggling operations range from a single individual to extensive transnational crime syndicates with organizational and logistical resources to move large volumes of wildlife or parts taken illegally to consumer markets around the globe (Nellemann et al. 2016; Shelley and

Table 1 Annotated list of what motivates individuals to commit wildlife crime

Motivation	Level of scale
Recreation satisfaction	Individual level[a]
Thrill killing	Individual level[a]
Commercial gain	Individual/family level[a,b,c,d]
Household consumption	Individual/family level[a]
Protection of self/property	Individual/family level[a,b]
Traditional rights	Societal level[a,b,c]
Regulation disagreement	Societal level[a,c]
Rebellion/political unrest	Societal level[a,c,d]

[a]Muth and Bowe (1998)
[b]Treves et al. (2017)
[c]Warchol (2018)
[d]Passas (1999)

Kinnard 2018). The illegal movement and trade of wildlife products have expanded to established trade routes used for many other illegal products (Shelley and Kinnard 2018) (Fig. 1). These trade routes and associated facilitators are often protected or given immunity by corrupt government officials. Affluent individuals are the primary purchasers of illegal wildlife imports, with most of these illegal products exported to Asia, the United States, and Europe. The increased value of illegal wildlife products has further fueled government corruption, whereby organized crime syndicates become intertwined with governmental organizations (Nellemann et al. 2016). Because of the extensive international pathways that allow people to commit wildlife crime and the complexity to prosecute, punishment for wildlife crimes is rare and primarily implemented through low fines and minimal jail time. This has led to rapid growth of wildlife crime, as these criminals operate in low-risk environments compared to other illegal activities (Warchol 2018). Further complicating matters, wildlife crime syndicates have diverse and adaptable networks, with replaceable participants and the ability to change trade routes and destinations in response to enforcement and new markets, allowing wildlife criminal operations to maintain profits and survival (Warchol 2018).

Negative Impacts of Wildlife Crime

It is estimated that since 1970, global populations of living vertebrates—fish, amphibians, reptiles, birds, and mammals—declined by as much as 52 percent as a result of habitat loss and degradation (McLellan 2014). Overexploitation, which includes illegal wildlife trade and poaching, is now a predominant cause of global wildlife decline, just behind habitat loss and degradation, and is considered one of the greatest threats to the long-term survival of wildlife populations around the globe (McLellan 2014; Maxwell et al. 2016). Many charismatic species of wildlife are imperiled due to poaching for illegal trade of products (e.g., ivory from African elephants [*Loxodonta africana*]) or for overharvesting of species for bushmeat or unsustainable harvest practices (e.g., African lions [*Panthera leo*]) (Wyatt 2013a) (Fig. 2). The extinction of native species caused by wildlife crime leads to degraded environments and loss of ecosystem function, which may bring about societal collapse by displacing people from homes and threatening their security while also negatively impacting income gathered from the legal use of plants, wildlife, and their associated products (Wyatt 2013a). Most importantly, these negative impacts result in loss of government revenues and threaten the strength and economic aspiration of most developing nations (Nellemann et al. 2014, 2016).

The global expansion of wildlife crime also threatens the security and prosperity of local communities. For example, loss of native biodiversity leads to loss of pharmaceuticals, loss of pollination services, increased malnutrition, and increased spread of disease (Pimentel et al. 1997). Illegal wildlife trade has also led to the spread of invasive species and zoonotic disease across borders, threatens public health, and negatively impacts legal businesses involved in agriculture and forestry (Wyatt 2013a). Because wildlife trade is heavily associated with the trade of illegal drugs, weapons, and human trafficking, it involves not only traditional organized criminal organizations but also political insurgents, rebel militias, and terrorist organizations. These criminal organizations threaten the lives of not only law enforcement personnel but also local citizens (Wyatt 2013b). As a result, illegal wildlife trade causes both biological and political instability while hindering economic progress and trade for many nations.

The Need for Conservation Forensics

Trade data on the harvest and shipment of illegally obtained wild fauna and flora from Africa and Asia indicates that nearly 1000 species involved in wildlife trade are listed as endangered species, as defined by the Convention on International Trade in Endangered Species (CITES) (Outhwaite and Brown 2018). Other reports

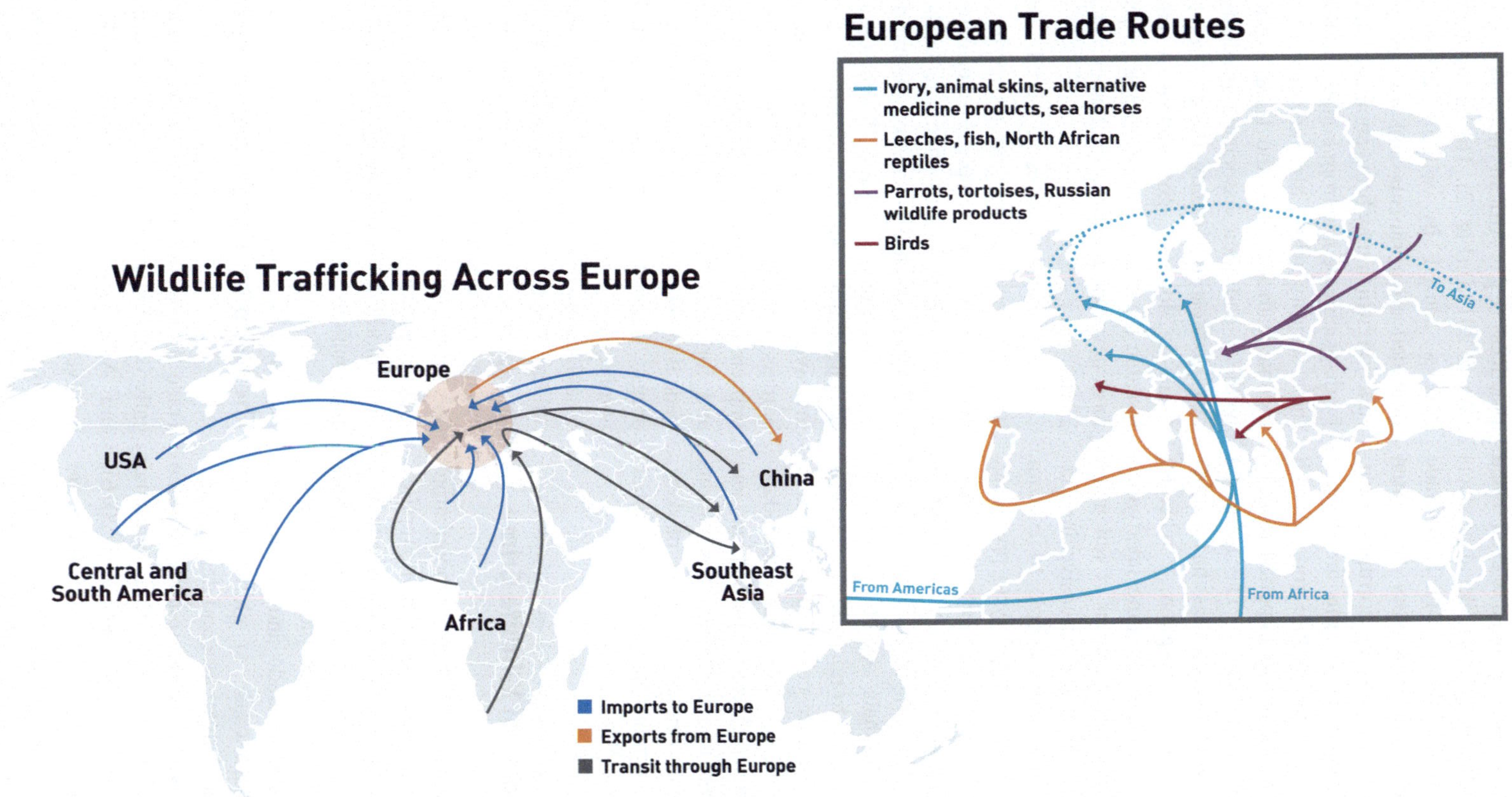

Fig. 1 A map illustrating major trade routes of illegal wildlife trade from North and South America and Africa to Europe and Asia. Inset illustrates trade routes within and out of Europe for specific species and products (Modified from Sina et al. 2016; Ruiz 2017) (Illustration Credit: Lindsay Matter)

Fig. 2 Charismatic wildlife species becoming imperiled due to (**a**) illegal wildlife trade (e.g., ivory), (**b**) bushmeat hunting, and (**c**) unsustainable harvest (e.g., African lion [*Panthera leo*]). Photo credits (**a**) J. Petula (**b**) A. Haines and (**c**) A. Haines

suggest that the volume of international trade of plants and animals not categorized by CITES may be 10X greater (Nuwer 2018). It has been reported that millions of CITES-listed wild animals are traded annually, but data currently does not exist on what level of harvest or extraction is biologically sustainable (Nijman 2010).

Currently, policymakers struggle to develop informed solutions to address the growing issue of wildlife crime, as little research exists on poaching detection and the effectiveness of punishments for wildlife crime (Haines et al. 2015; John 2018). There is a need to combine biological and law enforcement data, as well as to form collaborative governments, in order to help combat organized crime, and to establish large-scale policy solutions that will promote the pursual and prosecution of illegal wildlife criminals (Sundström and Wyatt 2017). The aforementioned needs have led to increased interest in wildlife crime and the establishment of two complementary fields of study—green criminology and conservation criminology.

The concepts of green and conservation criminology have a broad focus when addressing the negative impacts of wildlife crime on the environment and society. The aim of green criminology is to shape public environmental policy by combining political and practical action. It explores what it means to "harm" wildlife by extending victimhood to animals and plants, so they have rights that help to prevent cruel treatment. Thus, green crime involves criminal acts against nonhuman species, and green criminology views wildlife crime against nonhuman animals akin to crimes against powerless human individuals (Wyatt 2013a; Brisman and South 2018). However, many of the perpetrators who are on the ground committing acts of poaching come from impoverished backgrounds; thus a potential drawback of this philosophy is that there is an increase in incarceration and fine rates to already powerless human communities (Warchol 2018; Brisman and South 2018).

Conservation criminology is considered an interdisciplinary approach for research on wildlife crimes, focusing more on poaching and trafficking, as well as the effectiveness of methods used to investigate environmental and wildlife crimes. It is an applied paradigm for understanding programs and policies associated with global conservation risks involving natural resource management, criminal justice, risk, and decision-making (Gore 2017). Conservation criminology infuses theoretical and methodological rigor into the research of wildlife crimes to understand motivations, dispositions, and why such environmental crimes occur (Kahler and Gore 2017). Conservation criminology needs to be a multi-stage approach where deterrent strategies should be coupled with efforts to engender local populations to help law enforcement, by developing a holistic response to wildlife crime through collaborative research efforts (Gore 2017).

The concepts of green and conservation criminology emphasize the importance of addressing the social, political, and biological complexity of wildlife crime. However, there is a need to better focus field and laboratory techniques to capture and prosecute wildlife criminals to prevent wildlife crimes. This calls for a more applied field of conservation crime science that fits within the broader frameworks of green and conservation criminology. Such a field of study, entitled *conservation forensics*, would directly apply hard science techniques used to gather wildlife crime data, such as genetics, chemical analysis, geographical spatial analysis, statistics, artificial intelligence, and computational modeling, toward techniques that can directly benefit the efforts of law enforcement personnel involved in protecting imperiled wildlife. The focus of conservation forensics would address the concerns of Haines et al. (2015), who stated that more wildlife law enforcement research needs to be preventative and that the scientific community needs to revitalize research dealing with wildlife crimes, forensics, and enforcement. Examples of research questions that may be answered with conservation forensics include the following:

- Does consulting with ex-poachers improve enforcement effectiveness?
- Do conservation drones improve poacher apprehension rates or increase wildlife crime prevention?
- Do spatial models of predictive poaching patterns reduce wildlife crime?

Developing the field of conservation forensics to answer these questions, apply new ideas, and validate techniques for wildlife law enforcement will be vital in reducing wildlife crime in an effort to improve wildlife conservation.

Conservation forensics incorporates the tools and techniques needed to capture and prosecute wildlife criminals, as well as to prevent illegal activity, with the ultimate goal of protecting and preserving biodiversity and its evolutionary potential. While some crimes against wildlife are perpetrated on endangered and protected taxa, much of this illegal activity is not. Conservation forensics would therefore focus on the protection of biodiversity and imperiled species to prevent extirpation and extinction. Imperiled species would include those identified as vulnerable, endangered, and critically endangered by the International Union for the Protection of Nature (IUCN, www.iucnredlist.org). In the broad sense, conservation forensics would include those crimes against imperiled animals or plants themselves, as well as against the habitat in which these organisms live. For this chapter, the term conservation forensics is used to address crimes perpetrated on imperiled organisms and to describe how these crimes can be prevented or mitigated.

Application of Forensic Tools and Techniques

In the past five decades, the field of wildlife forensics has grown in response to the increased volume of crimes against wildlife, including both plants and animals. Within the constructs of natural and cultural sciences, this aspect of forensic science applies the knowledge of biological, chemical, and anthropological sciences in the court of law (Wallace and Ross 2012). Compared to forensic science sensu lato, wildlife forensics is a relatively young field that has adopted technologies most useful in investigations of crimes perpetrated against wildlife. Therefore, by definition, these same tools and techniques may be used to fight crime against taxa which are protected under our definition of conservation forensics. Specifically, methodologies useful in preventing and/or prosecuting crimes against protected wildlife are listed in Table 2.

The methodologies used in wildlife forensics and their application to conservation forensic issues evolved from (1) the scientific advances made over hundreds of years and their application in human forensics and (2) the national and international attempts to legislate protection via numerous treaties, acts, and laws that provide protection for wildlife, especially threatened or rare taxa in the last 120 years (Wallace and Ross 2012). Pathology, microscopy, entomology, and conservation genetic techniques within the biological sciences are often coupled with chemical science approaches to address critical questions asked in wildlife crimes.

The Crime Scene

Every death has an associated death scene (i.e., location where an illegal act has occurred) and contains a wealth of physical evidence that can be recovered (Horswell 2004). The crime scene of a wildlife crime is similar to human crime, allowing wildlife law enforcement to proceed using established protocols similar to those used for crimes against humans (Fox and Cunningham 1973; Adrian 1996). Preserving the scene with minimal disturbance is essential for discovery and proper collection of wildlife-related evidence. After evidential collection, methodologies such as those listed in Table 2 can be processed and pieced together. The proper documentation, collection, and preservation of material evidence such as cuts of meat (e.g., illegal harvest of deer), specific organs, targeted body parts (e.g., gallbladders from black bears [*Ursus americanus*]), or weapons used in committing such crimes, can facilitate the prosecution of individuals committing wildlife crimes (Hamilton and Erhart 2012).

Pathology and Toxicology

The use of a post-mortem pathological investigation on human remains (known as an autopsy) can be traced back to 367–282 BCE during ancient Greek times (Choo and Choi 2012). The use of such examinations on wildlife remains, termed necropsy, is a more recent practice. A major component in the investigation process is to provide

Table 2 Established methodologies used in the prevention and prosecution of crimes against humans and wildlife and applications for imperiled taxa

Methodology	Forensic use	Conservation application
Crime scene processing	Location/delineation of crime scene	Evidence collection
Chemistry/toxicology	How/when the crime occurred	Pesticide poisonings, dating of harvest via isotope analysis
Pathology	Cause of death	Wound analysis
Microscopy	Taxonomic identification	Hair, skin, feathers, and plants
Entomology	When/where the crime occurred	Insects associated with remains
Conservation genetics	Taxonomic identification	Any tissue analysis and isotope analysis

information on species identification through gross, microscopic, and molecular methods, including cause of death (Cooper 2013). Forensic steps such as wound analysis on live versus dead animals, shape/size of wounds (e.g., circular holes from gunshot or lacerations from knives), or patterns and paths of firearm projectiles (indicating the distance from animal to shooter or determining the orientation of the animal relative to the shooter) have elucidated a great deal on the cause of wildlife death (Roscoe and Stansley 2012). Toxicological analyses are also often revealing with respect to wildlife deaths. Since the advent of toxic insecticides and rodenticides, there have been numerous inadvertent poisonings of protected avian species such as migratory species and birds of prey (Best and Fisher 1992), as well as deliberate poisonings of nuisance birds and mammals. Toxicological screenings of avian gastrointestinal and other systems during necropsies have shown the extent of direct ingestion of such pesticides, as well as indirect or secondary poisoning via the consumption of poisoned animals (Roscoe and Stansley 2012). Poisoning of wildlife is often targeted at particular species, especially those which provide high-value products; however poisoning can also impact rare non-target species. For example, the IUCN Vulture Specialist Group of the Species Survival Commission has established a database to collect data on current and future incidents of inadvertent wildlife poisonings from chemicals or lead ammunitions in order to better understand the impact of such threats to vultures and other scavenging wildlife (African Wildlife Poison Database 2018).

Microscopy and Morphology

Since the advent of the microscope in the 1600s and the subsequent evolution of this tool for observing specimens, light and scanning microscopic techniques have been applied for nearly a century to the hair, skin, and bones in animals, as well as to morphological differences in plants (Housman 1920; Hardy and Wallace 2012). Microscopy on the hair, skin, horns, feathers, and plants has been critical in poaching investigations of game species, including protected taxa (Knecht 2012; Linacre 2009). In regard to enforcing international and national wildlife laws on illegally collected protected species, light microscopy, either with compound or dissecting microscopes, has been used in identifying plant and pollen samples from the stomachs of illegally killed endangered grizzly bears (*Ursus arctos*), as well as cyanobacteria from the livers of federally protected sea otters (*Enhydra lutris*). Endangered and internationally protected reptile skins and scales (e.g., Amazonian tree boas [*Corallus hortulanus*]) have been identified based on morphological differences determined through various forms of microscopy and spectroscopy (Hainschawang and Leggio 2006; Berthé et al. 2009; Klein et al. 2010). A major limitation in this area of methodology is the ability to identify skins and other reptilian products to species level.

Forensic Entomology

The utilization of insect evidence in human death investigations dates back to the thirteenth century (Schoenly et al. 2007). The use of insects in wildlife crimes is a more recent application that holds great potential (Anderson 1999; Tomberlin and Sanford 2012). The primary questions this type of evidence can address are temporal and spatial in nature; in other words, forensic entomologists can determine the time since colonization that can match up to the approximate time of death for an animal. Because some forensically important insects have narrow geographic distributions, entomologists can determine geolocation of remains to some extent and whether remains have been moved. Terrestrial insect-based evidence, as well as other types of invertebrates in aquatic systems, shows potential in both human and wildlife death investigations. This field is well-established in the criminal and civil courts and can bridge easily with investigations focused on protected wildlife, such as with harbor seals (*Phoca vitulina*) along the New England coast and impala (*Aepyceros melampus*) in Africa (Lord and Burger 1984, El-Kady 1999). Future developments in this area suggest that insect-microbe interactions may provide more sophisticated approaches to establish a time of colonization and perhaps even a time of death interval on the animal in question (Tomberlin et al. 2011).

Conservation Genetics and Isotope Analysis

Where morphological or microscopic methodologies have fallen short in the identification of illegally collected rare plants and animals, the use of molecular tools involving DNA analysis has picked up the pace. While many studies published to date have suggested that wildlife DNA analyses for individual, taxonomic/species, and geographic origin identification can be done using similar guidelines as those used for human DNA analyses, there is no consensus among authors in terms of best practices (Linacre et al. 2011; Moore and Kornfield 2012). However, the Organization of Scientific Area Committees for Forensic Science (OSAC), as administered by the National Institute of Standards and Technology (NIST), is writing standard operating procedures for wildlife forensics addressing wildlife forensic general standards, morphology, report writing, validation standards, and DNA standard procedures. Since the advent of forensic DNA analyses in the mid-1980s, genetic tools have been used in the prosecution of illegal trade in ivory, horns, olive oil, rice, timber, and many other illegally harvested species of conservation concern (McGraw et al. 2012). The use of minisatellites (VNTRs), mitochondrial markers (mtDNA), cytochrome b, cytochrome c oxidase, and pyrosequencing techniques has contributed to the prosecution of many cases involving protected species such as Chinese sika deer (*Cervus nippon*), rare Amazonian parrots (*Amazona* spp.), tigers (*Panthera tigris*), elephants (*Elaphus* spp.), orangutans (*Pongo pygmaeus*), and banteng (*Bos javanicus*, a species of wild cattle). DNA profiling has also been used on carrion feeding insects such as blowflies for biodiversity surveys of mammals in Malaya (Lee et al. 2015). Of special note is the work done by Wasser et al. (2015) in which their genetic analyses provided valuable information on elephant poaching hotspots in Africa, thereby directing future law enforcement efforts. This field has grown exponentially in the last decade with real-time PCR and qPCR techniques and will continue to expand, as scientists begin to explore microbial evidence using advanced molecular methodologies such as necro- and microbiome analyses.

Ivory-driven poaching of elephants over the past 50 years has led to a significant decline in elephant populations and skewed sex ratios, impacting long-term survival (Lamieux and Clarke 2009). Sampling ivory to understand the species involved, its geographic origin, and age can be facilitated either by DNA or more recently via isotope map analyses (West et al. 2006; UNODC 2014). As defined, isotopes are different forms of earth's elements that, due to the differing numbers of neutrons, have dissimilar massing (UNODC 2014). Many of these measured isotopes represent the assimilated nutrients in an animal's diet. Though most of the isotopes on earth are stable, some are radioactive and can be aged based on the decay rate characteristic of their half-life cycle (UNODC 2014). Various forms and ratios of isotopes preserved in elephant ivory such as $^{14}C/^{12}C$, $^{15}N/^{14}N$, $^{18}O/^{16}O$, and $^{87}Sr/^{86}Sr$ have been used to track ivory trade, and this form of mapping has allowed researchers to understand not only the historic trade of ivory and time it had been harvested but also the major hotspots of poaching (Coutu et al. 2016; Cerling et al. 2016). Moreover, these studies indicate the power of radiocarbon dating to reveal lag times between date of death and seizure, a technique that can be utilized for other wildlife products such as rhinoceros horn, pangolin scales, pelts, furs, and even timber, to provide valuable information to international and national law enforcement, conservation, government, and non-government agencies in fighting wildlife crimes (Cerling et al. 2016).

Despite the cumulative wildlife forensic experience and knowledge gained in the last 50 years, there remain numerous technological challenges facing law enforcement, research laboratories, and governments charged to protect endangered plants and animals (Espinoza et al. 2012). However, scientific technologies continue to develop, advancing needs such as taxonomic identification, geolocation of where crimes occur, and identification of captive-bred versus wild-caught animals/plants. The next section on emerging technologies addresses such advances in an attempt to elucidate future directions of conservation forensics.

Emerging Technologies

Advances in technology have enabled criminals to develop networks, communicate more effectively, find rare species more easily, and create supply chains that remain undetected. This leads to more organized, complicated, and successful covert criminal operations that have infiltrated large economic markets for illegal wildlife products. This results in wildlife criminals obtaining more funds for technological purchases as compared to governmental agencies that typically are tied to flat or decreasing annual budgets (Kretser et al. 2015). The greater economic backing of crime allows criminals to assess, test, and try new technologies at an ever-increasing rate. Those that provide criminals with an advantage are adopted, even if the technology is relatively expensive, because the return on investment is greater than the cost of the technology. Alternatively, Haines et al. (2016) found that advanced technologies are also actively used by wildlife law enforcement and are important to help curtail criminal activity and slow the decline of biodiversity loss (Fig. 3).

It would be nearly impossible to review all technologies that contribute to the field of conservation forensics. We will instead provide

Fig. 3 Examples of technologies useful for conservation forensics: (**a**) radio frequency identification (RFID) tags, (**b**) global positioning system devices, (**c**) unmanned aerial vehicles, and (**d**) autonomous recording units (Photographs provided by Noble Research Institute, LLC, Ardmore, Oklahoma)

Table 3 Overview of selected technologies that can be used in conservation forensics to aid in detection/surveillance, deterrence, data collection, and analysis

Technology/tool	Detection/surveillance	Deterrent	Data collection/analysis
UAV, drone	✓	✓[a]	✓
Thermal camera	✓		✓
Night vision	✓		
GPS			✓
Smart device			✓
Cameras	✓		✓
Acoustics	✓		✓
RFID tags	✓	✓[a]	✓
GPS collars	✓	✓[a]	✓
Accelerometers	✓		✓
Temperature sensor	✓		e
Pressure Pad[b]	✓		e
Break-beam sensor	✓		e
Satellite imagery	✓[c]		✓
Prediction maps		✓[d]	
Software		✓[d]	✓
Cloud-based apps			✓
AI, machine learning			✓

[a]Pressure pads can be designed for use to detect human or vehicle traffic
[b]Satellite imagery is used to learn an area and to develop patrol routes in the absence of prediction/planning maps
[c]Presence of a drone, visual tag, or collar on an animal can act as a deterrent
[d]Considered a deterrent when used to plan physical presence such as patrols on landscape
[e]Could be designed to collect and log data for future analysis

an overview of some of the more common or well-developed technologies, with applications in three primary areas. The first is technologies associated with detection and surveillance. Second is technologies that act as a deterrent which may help to prevent a crime from occurring. Third is advanced analytical software that can analyze data to identify patterns and develop models that may further aid in the detection or deterrence of wildlife crime.

Technologies for Surveillance and Detection

Much technology can be applied to conservation forensics when drawing from areas such as precision agriculture and precision livestock farming. Precision livestock farming relies on animal technology and the use of real-time automated processes to collect, analyze, and interpret a wide range of metrics on individual animals for making management decisions, reducing economic losses, and increasing overall animal health and productivity (Webb 2019). Similar to the field of precision livestock farming, conservation forensic technologies will draw on the technology itself, as well as to real-time analytics about animal well-being to make "smart" decisions that aid in detection/surveillance, deterrence, and data collection/analysis (Table 3). See also Tables 2 and 3 in Kamminga et al. (2018) for a comparison of sensor technologies and an overview of poaching detection technologies.

Animal Devices

Much development of tracking technologies has stemmed from the research needs for studying wild animals, while many of the "smart" technologies are being developed on domestic animals (e.g., pets and livestock). Wearable technology is the use of sensors and devices placed directly onto an individual and provides data in real time, such as movement, geographic position, health, and disturbance of well-being (Webb 2019). Another term in the literature often associated with anti-poaching systems (APS; Kamminga et al. 2018) is mobile biological sensors (MBS) that utilize free-ranging wildlife by attaching sensors and transmitters that alert responders to changes in animal behavior, possibly indicating mortality or long-distance movement outside of the normal capabilities of the animal (Banzi 2014).

Radio Frequency Identification Tags

One of the oldest wearable animal technologies is a radio frequency identification (RFID) tag (Fig. 3a), which is now a standard in the livestock industry. RFID tags are most common when used in conjunction with a visual ear tag. These tags come in many shapes and sizes and can be worn by many animal species (Bonter and Bridge 2011). RFID tags use low-frequency radio waves, which only allow the tags to be read from a short distance. Providing a longer read range, typically within line of sight, involves ultrahigh-frequency (UHF) tags, which operate the same way as RFID tags. RFID tags are economically priced, and due to their wide range of sizes and designs, they can be hidden easily on the animal if necessary, or made visible, allowing detection at reasonably close range with the naked eye or longer distances with the aid of optics or other technologies. Despite their broad application for animal identification and coarse animal tracking, from a conservation forensics perspective, there are several drawbacks to this technology for use on wildlife species. First, these are physical tags that must be attached to the animal, meaning that animals must be captured to receive the tag (UNEP 2014). The capture of animals typically requires permits, expenses associated with capture, and significant personnel time. RFID tags also are passive, which means they do not transmit data. Instead, data loggers or readers (e.g., a wand) must be deployed to capture the radio frequency of each tag at close range, which limits the ability to track animals across the landscape. However, RFID tags offer much potential for highly prized species or parts (e.g., rhino horn or elephant tusks). In these instances, RFID tags can be embedded with horn or tusk during animal capture, making them nearly invisible to poachers (Intel 2015; Kamminga et al. 2018). When an animal part is en route to market, strategic scanning locations (e.g., at ports of entry or customs) use scanning devices to detect hidden tags in illegal products. However, in many cases it is likely that either the poacher or another individual involved with marketing the wildlife product will come across the RFID tag. Although the tags are passive, there are opportunities to deploy reading devices on aircraft such as planes or helicopters during patrols or surveys or on unmanned aerial vehicles (UAV, also known as drones). This may allow detection of the tag earlier either to identify a crime or to identify whether the animal is still alive. Due to the passive nature of these tags, the primary application of RFID tags is for detection of an animal or animal part (Table 3).

Spatial Technologies

Global positioning system (GPS) collars are a well-tested and reliable technology, mostly applicable to wildlife research, but increasing in popularity when coupled with anti-poaching systems that incorporate other sensors and technologies (Banzi 2014). GPS receivers have been developed that weigh as little as 2 g (Fig. 3b), allowing them to be placed on small species such as lizards, birds, or small mammals (UNEP 2014). While GPS data is invaluable to researchers, it may provide benefits to on-the-ground rangers. GPS collars receive signals from satellites, but GPS units can also communicate back through satellites to base stations or a user's computer or smart device. One of the most frequently used satellite constellations is known as the Iridium satellite constellation, originally designed to have 77 satellites, thus giving rise to the name iridium (the element with the atomic number 77). These satellites are low earth orbiting satellites that provide visibility and coverage. Data onboard the GPS receiver can be sent via Iridium communication so that users can receive near real-time information on animal location or mortality (i.e., if equipment with activity sensors signal a mortality event). More importantly, users can also communicate with the GPS receiver to change settings or to set virtual boundaries (known as geofencing) that can send alerts (Wall et al., 2014). For example, if an animal leaves a national park, increasing its chance of being poached, a signal is sent to the user indicating such movement. Similar to RFID tags, drawbacks of GPS include the necessity of capture for device placement; limitations on the number of individuals tracked; and expense. However, much information can be gained in near real time (visualization of animal locations on an interactive web-based map such as Google Earth), as well as the information and planning tools that can be developed in the hands of a researcher. Therefore, GPS receivers allow for detection of a poaching event, collect large volumes of spatially explicit data for analysis, and may act as a visual deterrent (Table 3).

Biologgers

Biologgers are electronic data logging devices or sensors on animals for biological purposes. Biologgers are popular devices for humans interested in tracking their daily movements, for the purpose of fitness or health. One can think of biologgers as animal Fitbit activity trackers. Activity tracking devices for animals use a sensor known as an accelerometer, which measures acceleration or vibrations (e.g., such as earthquakes). Further data is provided when a magnetometer is coupled with an accelerometer. Magnetometers measure magnetism, or magnetic field strength, which allows a user to determine direction or change of the magnetic field relative to a particular orientation (Dewhirst et al. 2016). When these two sensors are combined, they can be used to estimate the position of the object or animal wearing the two sensors, making the data spatially explicit (Dewhirst et al. 2016). Even though the data can be spatially explicit in the end, these sensors on their own are unable to communicate the data or information to a user, requiring another technology such as Iridium GPS collars to equip them as smart devices.

Much work is being conducted to develop biologging technologies and applications, but there are many challenges to overcome before broad-scale adoption for conservation forensics (see Table 1 in O'Donoghue and Rutz 2016). Thus, a number of necessary adaptations are suggested for use in free-ranging wildlife. First, research is needed to identify how accurately accelerometer signals reflect true animal behavior (Wilson et al. 2014; Diosdado et al. 2015; Fehlmann et al. 2017). Once this process is complete, triggers or alerts can be set that will notify a user as to a change in behavior (e.g., lesser or greater activity or movement compared to normal) or a mortality event. Additionally, sensors will need to be linked to a communication device (e.g., cellular, satellite, radio) in order to transmit the data or to provide an alert. Linking biologgers to a communication device is referred to as bio-telemetry (Diosdado et al. 2015). Assuming communication is possible, a notification can be sent if onboard processing of data is available; otherwise there will be a delay between transmitting the data to a base station or to the cloud for processing before the alert is signaled to the user. Having an alert system based on defined parameters or thresholds is very useful, but only when linked to a communication system that is spatially explicit, which will be needed to find the animal (O'Donoghue and Rutz 2016). Lastly, animals will need to be captured to deploy these sensors along with other spatially explicit communication systems. Sensors are relatively inexpensive, consume little power, and are very small, which will allow them to be integrated into existing systems.

Non-animal Devices

Basic technologies for passive detection and surveillance of animals may include a GPS unit, smart device (such as a GPS-enabled smartphone), high-resolution digital camera (either with geotagging capabilities or as part of a GPS-enabled smartphone), and high-resolution optics. These devices are crucial for recording basic information about spatial location and the crime scene before more traditional wildlife forensic methods can be collected (see Application of Forensic Tools and Techniques). When a smart device is available with satellite or cellular communication, it can provide immediate information on GPS location, species identification, and crime scene description to rangers or forensics teams.

Night Vision and Thermal Optics

Other less common technologies that may provide tremendous aid during surveillance are night vision and thermal optics. Night vision optics provides greater ability to see in low-light conditions—the times when many poaching events may take place (e.g., see Haines et al. 2012). Thermal imaging optics are preferred, as they do not require ambient light. Thermal imaging relies on infrared energy, or heat, to detect differences in temperature. The infrared radiation collected by the thermal device creates an "electronic" image based on temperature differences. A human being, such as a poacher, will give off heat (known as a heat signature) that can be detected and identified by the user (Tan et al. 2016). The same process can be used to find and identify animal species (Christiansen et al. 2014), as their heat signature will differ from the surrounding environment and usually to a much greater degree. Even after death, heat from the animal will remain for some time, allowing rangers to identify illegally harvested animals after the poaching event takes place.

Remote Cameras and Break Beams

Rangers can use other advanced equipment as part of their surveillance program to help detect or identify a crime or criminals. Two very common technologies used are remote camera traps and infrared break-beam sensors (Williams 1995; UNEP 2014; Hossain et al. 2016; Kamminga et al. 2018). Currently, there are many remotely triggered game cameras on the market that send notifications and pictures directly to a person's smartphone, email inbox, or cloud-based application. If pictures are not transmitted, whether due to logistical issues related to image size or expense associated with transmitting data (with cellular technology), a notification can be sent, allowing an opportunity to view the photographs. As an example, the BoarBuster trap system notifies a user by email or text when the associated camera has detected motion. Next, the user logs into a website or mobile application (either using cellular or Internet connectivity) to view the photographic evidence. If necessary, the user may send a command to the camera to stream live video or change camera settings. Break-beam sensors alone do little for surveillance, but when combined with communication systems, patrol may be deployed, enhancing overall effectiveness (Williams 1995). Utilizing an array of break-beam sensors may help to indicate directional movement. One challenge to overcome will be to determine whether detected movement originated from human or animal.

Unmanned Aerial Vehicles

The "eyes in the skies"—unmanned aerial vehicles (UAV)—more commonly referred to as drones, are aircraft without a human pilot on board (Fig. 3c). UAVs can provide a novel tool in the arsenal against wildlife crime (Olivares-Mendez et al. 2015; Wich 2015; Bondi et al. 2018), but their usage comes with several limitations. UAVs utilized by rangers or wildlife personnel (researchers, biologists, managers) are primarily limited to models of moderate price, which can equate to a limit in capability or range of sensors. For example, rotary-wing UAVs are the most common type of UAV utilized; however they have a limited battery life due to payload capacity (Wich 2015). For this reason, applications such as wildlife surveys and large-scale surveillance will be limited to smaller areas or shorter time periods in the air (Wich 2015). However, UAVs can be strategically deployed when coupled with the alert systems previously discussed. Without much effort or disturbance, a UAV can be programmed to fly to a location identified in an alert system. To scale up surveillance, one would need to invest in a fixed-winged UAV, which can cover a much larger area (Wich 2015). UAVs could primarily be considered as a transport tool for other technologies, which can provide more of the information needed to fight crime. But, as with any new technology, especially with respect to national regulatory controls such as the FAA in the United States, users must work within the laws and restrictions of the country. Below are descriptions of some of the most common sensors that can be combined with a UAV to create an UAS (unmanned aerial system).

The type of sensor or camera attached to the UAV will be dependent upon the specific needs of the user (Wich 2015). To begin with, most users select a camera that uses red, green, and blue (RGB) light to produce a wide range of colors. This type of camera will be most common for obtaining high-resolution images (pictures) of the landscape (Wich 2015). Other wavebands such as ultraviolet or infrared have many applications in agriculture, including the detection of poachers or humans (Bondi et al. 2018). Next is thermal imaging cameras, which are similar to the thermal optics previously discussed. Combining a thermal camera with a UAV will offer a new perspective to surveillance—a top-down view of a larger spatial extent, which has applications to poaching where animals or humans can be identified (Mulero-Pázmány et al. 2014). This allows for more robust surveillance and detection, although there still are inherent challenges to the use of drones, namely, battery life and coverage. In such cases, the ranger and their vehicle can serve as a mobile base station carrying battery packs and multiple flight plans for full-blown reconnaissance flights that normally would require the utilization of multiple rangers on the ground. Lastly, light detection and ranging (LiDAR) may have some application to conservation forensics. LiDAR measures the distance to a target (or to the ground) using laser technology. The reflected light is sent back to the sensor for processing of distance, which is then converted into height or elevation. LiDAR is useful for building digital elevation models (DEM) and vegetation height, which may factor into surveillance, detection, and deterrent strategies. Subsequently, these digital products (DEM and/or vegetation height) can be incorporated into models of animal resource selection or for prediction of potential escape routes by perpetrators. In addition to the expense of the UAV, users need to factor in the costs of each sensor (many are standalone sensors), which can greatly increase the overall cost of hardware. Besides the hardware, many of these sensors require processing with software to maximize the benefits of imagery.

Autonomous Recording Devices

The last class of non-animal devices is referred to as acoustic or bioacoustic monitors, hereafter referred to as autonomous recording units (ARUs; Fig. 3d). ARUs are gaining in popularity for their use in biodiversity research as well as for auditory wildlife surveys on birds, bats, and elephants (Blumstein et al. 2011; UNEP 2014; Wrege et al. 2017; Kamminga et al. 2018). Similar to remotely triggered cameras, ARUs can be deployed for auditory surveillance across a landscape. Most commercially available ARUs can be programmed to begin at specified dates and times, or they can run 24 h a day. At present, most ARUs are used for research purposes, due to the large amounts of data generated (terabytes), which requires a substantial amount of computational power (Wrege et al. 2017). Commercial software is inefficient at processing large amounts of data, which presents an additional challenge (Wrege et al. 2017). The complex array of sound signatures also poses a challenge, as the sound signature of interest is embedded among other sounds and background noises. Artificial intelligence and deep learning techniques (e.g., artificial or convolutional neural networks) analyze the data to learn and recognize the "sounds" of interest, which are often analyzed as spectrograms (also known as sonograms, which visually represent a spectrum of frequencies) (Aide et al. 2013; Knight et al. 2019). However, for application in a conservation forensics framework, processing could be expedited because the number of signatures (i.e., unique sounds) would be less than trying to "match" all sounds. For instance, the sounds of interest may include gunshots, vehicle sounds (engine noise or doors closing), a human talking, or the sound of an animal such as an elephant, whose calls could help to enumerate the number of elephants or the area of inhabitation (Temple-Raston 2019).

This research is in its early stages, but preliminary findings indicate that elephants will not enter certain parts of the forest during specific times of the year. This means that anti-poaching parties could streamline their efforts and ignore certain areas of the forest based on acoustic monitoring of highly prized species (Temple-Raston 2019). Kalmár et al. (2019) are developing and researching ways to use acoustic technologies with GPS tracking collars to develop a smart system to identify gunshots near groups of animals, which then creates a real-time alert, relaying the spatial location of the incident so that rangers could be dispatched to the area. With acoustic monitors, there is a detection zone around the animal, or animals, so that a larger scale of surveillance or monitoring can take place. Research will need to be conducted to determine the detection function (i.e., distance) that acoustic monitors can detect each type of signature (e.g., the detection distance of a gunshot will cover a large radius around each monitored animal). The detection distance will influence the spatial accuracy of identifying where the crime took place, but with a larger detection zone, the real-time notification and assimilation of rangers to the general proximity will enhance effectiveness as compared to current notification systems.

Analytical and Outcome-Oriented Products

Predictive Mapping

Most of the aforementioned technologies can be leveraged to their fullest potential when combined with real-time data processing (i.e., analytics) onboard or through the cloud, as well as with real-time alerts. Technologies such as biologgers, acoustics, remotely triggered cameras, and remotely sensed imagery need to be analyzed before they provide tangible value. Data collected from such technologies must first pass through research and validation phases. In addition to active alert systems, data may be used to generate other tools such as maps that can facilitate patrol routes. Some of the more common uses of data include satellite imagery, in order to visualize landscape features and to develop patrol routes that are traditionally based on access roads (Critchlow et al. 2017; Krester et al. 2017). Other predictive maps can be generated to identify and prioritize habitats and areas used (e.g., resource selection studies) by species most likely at risk. There is a dire need to understand habitat use of wildlife species, as often non-patrolled areas are ideal wildlife habitats, which is where most illegal activity occurs (Shaffer and Bishop 2016). Similar resource selection methodologies can be used to study and predict human use or risk across the landscape. There are many approaches to develop useful maps to help with planning. For example, Haines et al. (2012) classified the major land uses and vegetation classes relative to how they would function for white-tailed deer (*Odocoileus virginianus*) as well as how they would function for a poacher. Studying where deer poaching events took place allowed a hotspot or risk map to be developed that displayed high priority areas where deer were likely to be poached, given available deer habitat and features that would allow poachers to go undetected or to escape quickly. Another example models the data in a slightly different way—in a study by Dzialak et al. (2011), researchers developed a resource selection function to identify non-random habitat selection by elk (*Cervus canadensis*). A risk map was developed to spatially depict locations across the landscape where elk were likely to survive. When the two maps were combined within a geographical information system (GIS), the resulting map depicted demographic productivity (survival) based on preferred habitats and areas that were less risky. This same framework can be used to overlay animal distribution onto human, hunter, or poacher risk maps to help prioritize areas where animals will frequent and where they may be at risk for poaching.

Data Analysis and Applications

These types of maps or tools provide a wealth of information but are still considered reactive, rather than proactive in their approach. Technology can be used to collect data, and then tools like software and databases can be used to digest and analyze the data (Table 3) to make informed decisions on where crimes are likely to occur—a strong step in crime deterrence. Although most tools are reactive—using past data to predict future events—the use of software, maps, and web applications has been successful in gaining greater knowledge about the spatio-temporal nature of wildlife crimes. As an example, the Spatial Monitoring and Reporting Tool (SMART) is an open source, non-proprietary, freely available software application that allows the collection, storage, communication, and evaluation of ranger-based data (e.g., patrol efforts, patrol results, threat levels) (Krester et al. 2017). SMART is a tool that can leverage technologies through the collection of data from multiple platforms (e.g., GPS devices) (Krester et al. 2017). Much effort is put into understanding criminals and their actions, but when dealing with wildlife crimes, knowledge of the species at risk is also needed to inform and adjust surveillance and deterrent strategies on a species-specific level. Linking spatially explicit animal distribution and abundance models to risk models of potential poacher hotspots, travel routes, and other spatial patterns can further prioritize areas in greatest need of protection (Dzialak et al. 2011; Haines et al. 2012).

On a global scale, patrols are the most widespread method to combat and prevent wildlife poaching (Fang et al. 2017). Despite the advantage of real-time processing and communication, troop mobilization and travel to the crime site are time-consuming (Bergenas et al. 2013). Unfortunately, rangers often arrive at a crime scene too late. Software and applications such as SMART are helping to reduce wildlife crimes through more strategic patrols or by deterring criminals. Knowledge of patrols in an area may be effective in deterring poaching or other criminal activity. This is referred to as situational crime prevention (SCP)—manipulation of environments to disrupt opportunities for crime to take place (Krester et al. 2017). Deployment of rangers to high priority areas helps to limit wasted resources and personnel (Haines et al. 2012), which may open up opportunities to hire additional rangers or to invest in other valuable technologies.

Technology holds much promise for protecting wildlife species, but there may be inherent technical barriers that hinder its use (Arts et al. 2015), not to mention the associated financial costs. If relevant technology and tools are underutilized, it is difficult to improve conservation outcomes (Sintov et al. 2018). Further testing and research using different technologies will be needed to meet specific needs of law enforcement personnel involved in conservation forensic efforts.

Conservation Forensics to Prevent Wildlife Crime

Wildlife forensics research is an emerging discipline that has in recent years enhanced the success of wildlife law enforcement. The main purpose of wildlife forensics is to identify violations and potential violators after a crime has occurred. However, due to the rapidly declining population sizes of many rare wildlife species (e.g., pangolin, tiger, different species of rhinoceros), advanced technologies and enforcement need to be preventative to mitigate loss of natural resources, rather than merely being reactive through forensic investigation. The goal of the field of conservation forensics is to provide more preventative solutions to wildlife crime enforcement, for the benefit of biodiversity preservation (Fig. 4).

Wildlife law enforcement officers reported in a US survey that technology plays a large role in wildlife crime, with poachers using a wide array of technology for their illegal activities (Haines et al. 2016). This technology included night-vision, real-time, or remote field cameras, smart devices, GPS tracking devices, and social media. Haines et al. (2016) also found that wildlife law enforcement officers require utilization of sophisticated devices and data collection techniques such as surveillance cameras, including body, trail, and pole cameras, as well as GPS, GIS, smartphones, and social media to aid in the apprehension of poachers. The role of technology and forensics in wildlife crime enforcement has produced a technological "arms race" between perpetrators and law enforcement officers (i.e., conservation officers, wardens, rangers, and their governments). Therefore, considerable conservation investment coupled with supportive

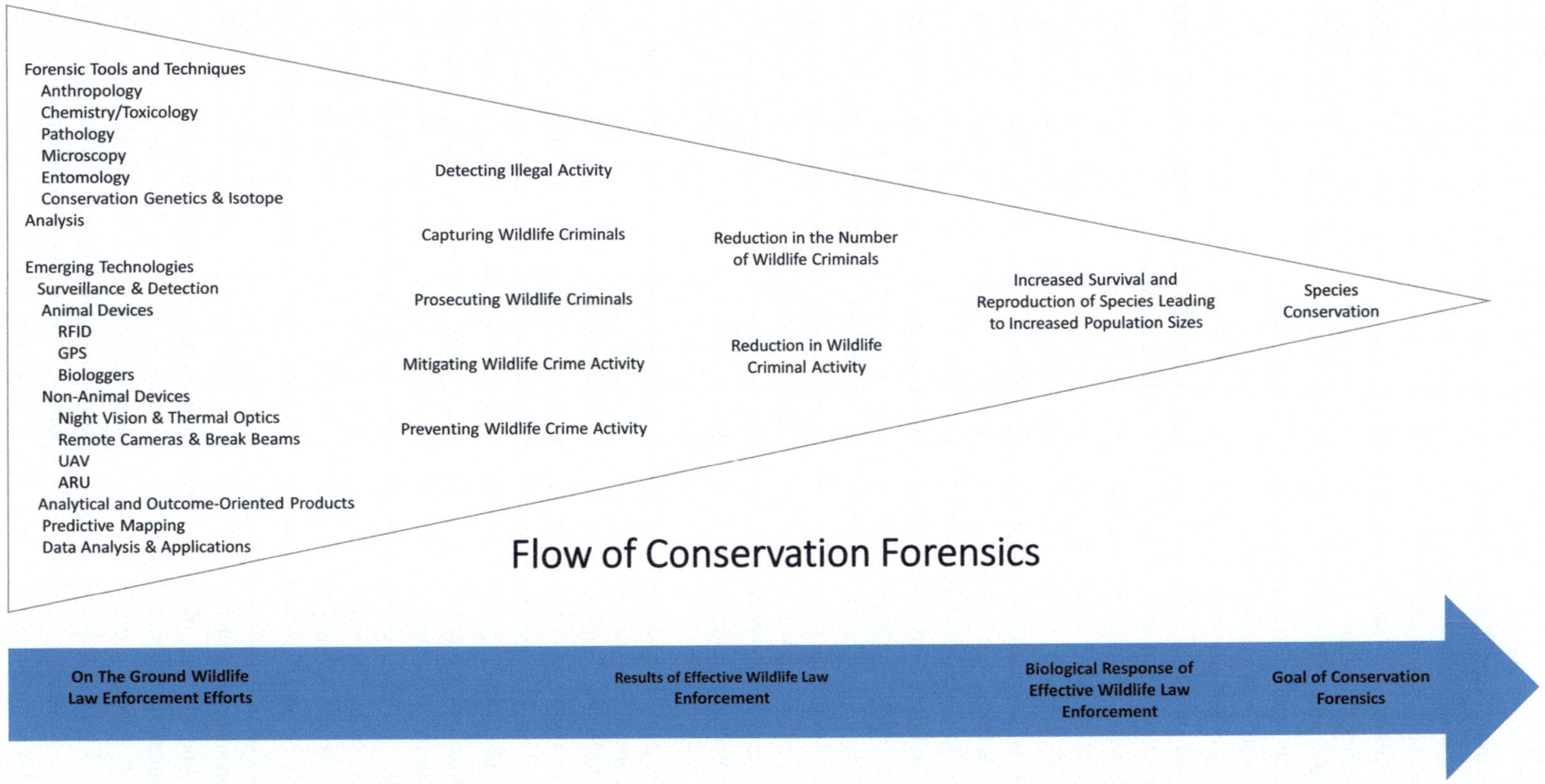

Fig. 4 Flow diagram outlining the focus and goal of conservation forensics, from developing on-the-ground wildlife conservation efforts to reducing wildlife crime and monitoring the biological response to ensuring species conservation. Acronyms: RFID (Radio Frequency Identification), GPS (global positioning system), UAV (unmanned aerial vehicle), and ARU (autonomous recording unit)

governments and communities is required for on-the-ground law enforcement conservation success (Gray et al. 2016).

In the future, smart devices and web applications will be critical tools for wildlife law enforcement. To power these applications, cloud-based computing will play a more critical role in the process (Wall et al. 2014). Emerging technological advances in conservation science may also be used in conservation forensics, such as DNA tracking, GPS collars, chips for spatial and temporal analyses, alarm fences, hidden cameras, conservation drones for high-tech surveillance systems, smartphone apps, as well as reward programs for reporting illegal activity. However, many of these technologies and techniques have lacked adequate scientific testing to determine whether they reduce wildlife crime or establish a level of accuracy necessary to convict wildlife criminals in a court of law. Data is a critical component of the decision-making process and is key to developing applied tools. With more and more data, a strong focus on data science and analytics will need to emerge into the realm of conservation forensics. Artificial intelligence and machine learning will play critical roles in finding relationships and developing predictive models. For example, machine learning is currently being used to automatically identify individuals, species, and/or their respective behaviors from photographs (Norouzzadeh et al. 2018).

Conclusion

In this chapter, the development and use of conservation forensics are proposed to capture and prosecute wildlife criminals, as well as to prevent their activities, as part of the greater effort to preserve biodiversity (Fig. 4). With the focus on conservation forensics, we have provided a brief review of the latest research and current technologies used in wildlife forensics to assist with wildlife law enforcement. Also outlined are the tools that can be used to achieve the goals of conservation forensics—to prosecute wildlife criminals and prevent or mitigate wildlife crime in an effort to conserve biodiversity. However, as with law enforcement efforts in conservation criminology, in order for conservation forensics to be successful, it must be part of a larger deterrent strategy involving collaborative research efforts. This approach is accomplished through the expansion law enforcement efforts to improve local support by empowering local populations to work with law enforcement, coupled with financially supportive functioning governments, in order to achieve conservation success (Gray et al. 2016).

References

Adrian WJ (1996) Wildlife forensic field manual. Association of Midwest Fish and Game law Enforcement Officers, Colorado, USA

African Wildlife Poison Database (2018) Vulture Specialist Group of the IUCN Species Survival Commission. https://www.africanwildlifepoisoning.org/. Accessed 15 Sep 2019

Aide TM, Corrada-Bravo C, Campos-Cerqueira M, Milan C, Vega G, Alvarez R (2013) Real-time bioacoustics monitoring and automated species identification. PeerJ 1:e103

Anderson G (1999) Wildlife forensic entomology: determining time of death in two illegally killed Black bear cubs. J Forensic Sci 44:856–859

Arts K, van der Wal R, Adams WM (2015) Digital technology and the conservation of nature. Ambio 44: S661–S673

Banzi JF (2014) A sensor based anti-poaching system in Tanzania national parks. Int J Sci Res Publ 4:105–111

Bergenas J, Stohl R, Georgieff A (2013) The other side of drones: saving wildlife in Africa and managing global crime. Conflict Trends 3:3–9

Berthé RA, Westhoff G, Bleckmann H, Gorb SN (2009) Surface structure and frictional properties of the skin of the Amazon Tree Boa (*Corallus hortulanus*; Squamata: Boidae). J Comp Physiol 195(3):311–318

Best LB, Fisher DL (1992) Granular insecticides and birds: factors to be considered in understanding exposure and reducing risk. Environ Toxicol Chem 11:1495–1508

Blumstein DT, Mennill DJ, Clemins P, Girod L, Yao K, Patricelli G, Deppe JL, Krakauer AH, Clark C, Cortopassi KA, Hanser SF, McCowan B, Ali AM, Kirschel ANG (2011) Acoustic monitoring in terrestrial environments using microphone arrays: applications, technological considerations and prospectus. J Appl Ecol 48:758–767

Bondi E, Fang F, Hamilton M, Kar D, Dmello D, Choi J, Hannaford R, Iyer A, Joppa L, Tambe M, Nevatia R (2018) SPOT poachers in action: augmenting conservation drones with automatic detection in near real time. In: The 32nd AAAI Conference on Innovative Applications of Artificial Intelligence, New Orleans, 2–7 February 2018

Bonter DN, Bridge ES (2011) Applications for radio frequency identification (RFID) in ornithological research: a review. J Field Ornithol 82:1–10

Brashares JS, Abrahms B, Fiorella KJ, Golden CD, Hojnowski CE, Marsh RA, McCauley DJ, Nuñez TA, Seto K, Withey L (2014) Wildlife decline and social conflict. Science 345(6195)

Brisman A, South N (2018) Perspectives on wildlife crime: the convergence of "green" and "conservation criminology". In: Moreto WD (ed) Wildlife crime: from theory to practice. Temple University Press, Philadelphia, PA, pp 135–149

Cerling TE, Barnette JE, Chesson LA, Douglas-Hamilton-I, Gobush KS, Uno KT, Wasser KT, Xu X (2016) Radiocarbon dating of seized ivory confirms rapid decline in African elephant populations and provides insight into illegal trade. PNAS 113(47):13330–13335

Choo TM, Choi YS (2012) Historical development of forensic pathology in the United States. Korean J Leg Med 36:15–21

Christiansen P, SteenKA JRN, Karstoft H (2014) Automated detection and recognition of wildlife using thermal cameras. Sensors 14:13778–13793

Cooper JE (2013) Working with dead animals. In: Cooper JE, Cooper ME (eds) Wildlife forensic investigation: principles and practice. CRC Press, Boca Raton, FL, p 238

Coutu AN, Lee-Thorp J, Collins MJ, Lane PJ (2016) Mapping the elephants of the 19th century east African ivory trade with a multi-isotope approach. PLoS One 11(10):e0163606

Critchlow R, Plumptre AJ, Alidria B, Nsubuga M, Driciru M, Rwetsiba A, Wanyama F, Beale CM (2017) Improving law-enforcement effectiveness and efficiency in protected areas using ranger-collected monitoring data. Conserv Lett 10:572–580

Dewhirst OP, Evans HK, Roskilly K, Harvey RJ, Hubel TY, Wilson AM (2016) Improving the accuracy of estimates of animal path and travel distance using GPS drift-corrected dead reckoning. Ecol Evol 6:6210–6222

Diosdado JA, Vázquez ZE, Barker HR, Hodges JR, Amory DP, Croft NJ, Bell E, Codling A (2015) Classification of behaviour in housed dairy cows using an accelerometer-based activity monitoring system. Anim Biotelemetry 3:15

Dzialak MR, Webb SL, Harju SM, Winstead JB, Wondzell J, Mudd JP, Hayden-Wing LD (2011) The spatial pattern of demographic performance as a component of sustainable landscape management and planning. Landsc Ecol 26:775–790

El-Kady EM (1999) Problems facing the application of forensic entomology. Pak J Biol Sci 2:280–289

Eriksson HB, Osterblom H, Crona B, Troell M, Andrew NL (2015) Contagious exploitation of marine resources. Front Ecol Environ 13(8):435–440

Espinoza EO, Espinoza JL, Trail PW, Baker BW (2012) The future of wildlife forensic. In: Huffman JR, Wallace JR (eds) Wildlife forensics: methods and applications. Wiley Blackwell, West Sussex, UK, p 343

Fang F, Ford B, Yang R, Tambe M, Lemieux AM (2017) PAWS: game theory based protection assistant for wildlife security. In: Core ML (ed) Conservation criminology. Wiley Blackwell, West Sussex, UK, pp 179–195

Fehlmann G, O'Riain MJ, Hopkins PW, O'Sullivan J, Holton MD, Shephard ELC, King AJ (2017) Identification of behaviours from accelerometers data in a wild social primate. Anim Biotelemetry 5:6

Forsyth YA, Forsyth CJ (2018) Ordinary folk transformed: Poacher's accounts of cultural contests and history. In: Moreto WD (ed) Wildlife crime: from theory to practice. Temple University Press, Philadelphia, PA, pp 135–149

Fox RH, Cunningham CL (1973) Crime scene search and physical evidence handbook. U.S. Department of Justice, National Institute of Justice, Washington, DC

Gore GL (2017) Global risks, conservation and criminology. In: Gore ML (ed) Conservation criminology. Wiley Blackwell Publishers, West Sussex, UK, pp 1–20

Gray TN, Hort S, Lefter E, Grosu R, Kong K, Keo O, Gauntlett S (2016) A decade of zero elephant poaching in the cardamom rainforest landscape, Cambodia. Gajah 45:35–38

Haines AM, Elledge D, Wilsing LK, Grabe M, Barske MD, Burke N, Webb SL (2012) Spatially explicit analysis of poaching activity as a conservation management tool. Wildl Soc Bull 36:685–692

Haines AM, Webb SL, Wallace JR (2015) Cracking down on wildlife crime: how wildlife law enforcement research helps conservation. Wildl Prof 9(1):35–37

Haines AM, Webb SL, Meshe F (2016) Forty years in the making: a survey of wildlife law enforcement needs. Wildl Prof 10(5):34–36

Hainschawang T, Leggio L (2006) The characterization of tortoise shell and its limitations. Gems Gemol 42 (1):36–52

Hamilton MD, Erhart EM (2012) Forensic evidence collection and cultural motives for animal harvesting. In: Huffman JE, Wallace JR (eds) Wildlife forensics: methods and applications. Wiley Blackwell, West Sussex, UK, pp 65–79

Hardy CR, Wallace JR (2012) Algae in forensic investigations. In: Hall DW, Byrd JH (eds) Forensic botany: a practical guide. Wiley, London, pp 145–173

Horswell J (2004) The practice of crime scene investigation. In: International forensic science and investigation service. CRC Press, Boca Raton, FL, p 450

Hossain ANM, Barlow A, Barlow CG, Lynam AJ, Chakma S, Savini T (2016) Assessing the efficacy of camera trapping as a tool for increasing detection rates of wildlife crime in tropical protected areas. Biol Conserv 201:314–319

Housman LA (1920) Structural characteristics of the hair of mammals. Am Nat 5:496–523

Intel (2015) Tiny Intel chip helping save huge rhinos. https://blogs.intel.com/csr/2014/08/rhino/#gs.aeo3ze. Accessed 15 Oct 2019

John J (2018) From poaching to smuggling. Wildl Prof 12 (4):18–24

Kahler JS, Gore GL (2017) Conservation crime science. In: Gore ML (ed) Conservation criminology. Wiley Blackwell, West Sussex, UK, pp 27–44

Kalmár G, Wittemyer G, Völgyesi P, Rasmussen HB, Maróti M, Lédeczi A (2019) Animal-borne anti-poaching system. In: The 17th annual international conference on Mobile systems, applications, and services, Seoul, Republic of Korea, 17–21 June 2019

Kamminga J, Ayele E, Meratnia N, Havinga P (2018) Poaching detection technologies – a survey. Sensors 18:1474

Klein MC, Deuschle JK, Gorb SN (2010) Material properties of the skin of the Kenyan sand boa *Gongylophis colubrinus* (Squamata: Boidae). J Comp Physiol A 196(9):659–668

Knecht L (2012) The use of hair morphology in the identification of mammals. In: Huffman JE, Wallace JR (eds) Wildlife forensics: methods and applications. Wiley Blackwell, West Sussex, UK, pp 129–143

Knight EC, Hernandez SP, Bayne EM, Bulitko V, Tucker BV (2019) Pre-processing spectrogram parameters improve the accuracy of bioacoustic classification using convolutional neural networks. Bioacoustics. https://doi.org/10.1080/09524622.2019.1606734

Krester H, Stokes E, Wich S, Foran D, Montefiore A (2017) Technological innovations supporting wildlife crime detection, deterrence and enforcement. In: Gore ML (ed) Conservation criminology. Wiley Blackwell, West Sussex, UK, pp 157–177

Kretser H, Wong R, Roberton S, Pershyn C, Huang J, Sun F, Kang A, Zahler P (2015) Mobile decision-tree tool technology as a means to detect wildlife crimes and build enforcement networks. Biol Conserv 189:33–38

Lamieux AM, Clarke RV (2009) The international ban on ivory sales and its effects on elephant poaching in Africa. Br J Criminol 49(4):451–471

Lee PS, Sing KW, Wilson JJ (2015) Reading mammal diversity from flies: the persistence period of amplifiable mammal mtDNA in blowfly guts (Chrysomya megacephala) and a new DNA mini barcode target. PLoS One 10(4):eo123871

Linacre A (2009) Forensic science in wildlife investigations. CRC Press, London

Linacre A, Gusmao L, Hecht W, Hellmann AP, Mayr WR, Parson W, Prinz M, Schneider PM, Morling N (2011) ISFG: recommendations regarding the use of non-human (animal) DNA in forensic genetic investigations. Forensic Sci Int Genet 5(5):501–505

Lord WD, Burger JF (1984) Arthropods associated with harbor seal (*Phoca vitulina*) carcasses stranded on islands along the New England coast. Int. J Entomol 26:282–285

Maxwell SL, Fuller RA, Brooks TM, Watson JE (2016) Biodiversity: the ravages of guns, nets and bulldozers. Nat News 536(7615):143–145

McGraw SN, Keeler SP, Huffman JE (2012) Forensic DNA analysis of wildlife evidence. In: Huffman JE, Wallace JR (eds) Wildlife forensics: methods and applications. Wiley Blackwell, West Sussex, UK, pp 253–269

McLellan R (2014) World Wildlife Fund 2014 living planet report. World Wildlife Fund, Gland, Switzerland

Moore MK, Kornfield IL (2012) Best practices in wildlife forensic DNA. In: Huffman JE, Wallace JR (eds) Wildlife forensics: methods and applications. Wiley Blackwell, West Sussex, UK, pp 201–236

Mulero-Pázmány M, Stolper R, van Essen LD, Negro JJ, Sassen T (2014) Remotely piloted aircraft systems as a rhinoceros anti-poaching tool in Africa. PLoS One 9 (1):e83873

Muth RM, Bowe JF (1998) Illegal harvest of renewable natural resources in North America: toward a typology of the motivations for poaching. Soc Nat Resour 11:9–24

Nellemann C, Henriksen R, Raxter P, Ash N, Mrema E (2014) The environmental crime crisis – threats to sustainable development from illegal exploitation and trade in wildlife and Forest resources. A UNEP Rapid Response Assessment. United Nations Environment Programme and GRID-Arendal, Nairobi and Arendal. www.grida.no

Nellemann C, Henriksen R, Kreilhuber A, Stewart D, Kotsovou M, Raxter P, Mrema E, Barrat S (2016) The Rise of Environmental Crime – A Growing Threat To Natural Resources Peace, Development And Security. A UNEPINTERPOL Rapid Response Assessment, United Nations Environment Programme and RHIPTO Rapid Response–Norwegian Center for Global Analyses. www.rhipto.org

Nijman V (2010) An overview of international wildlife trade from Southeast Asia. Biodivers Conserv 19:1101–1114

Norouzzadeh MS, Nguyen A, Kosmala M, Swanson A, Palmer MS, Packer C, Clune J (2018) Automatically identifying, counting, and describing wild animals in camera-trap images with deep learning. Proc Natl Acad Sci 115:E5716–E5725

Nuwer RL (2018) Poached: inside the dark world of wildlife trafficking. Merloyd Lawrence Publishing, Hachette, UK, p 384

O'Donoghue P, Rutz C (2016) Real-time anti-poaching tags could help prevent imminent species extinctions. J Appl Ecol 53:5–10

Olivares-Mendez MA, Fu C, Ludivig P, Bissyandé TF, Kannan S, Zurad M, Annaiyan A, Voos H, Campoy P (2015) Towards an autonomous vision-based unmanned aerial system against wildlife poachers. Sensors 15:31362–31391

Outhwaite W, Brown L (2018) Eastward bound: analysis of CITES-listed flora and fauna exports from Africa to East and Southeast Asia. TRAFFIC International, Cambridge, UK, p 150

Passas N (1999) Informal value transfer systems and criminal organizations; a study into so-called underground banking networks. A Study into So-Called Underground Banking Networks, 14 Dec 1999

Pimentel D, Wilson C, McCullum C, Huang R, Dwen P, Flack J, Tran Q, Saltman T, Cliff B (1997) Economic

and environmental benefits of biodiversity. Bioscience 47(11):747–757
Primack RB (2014) Essentials of conservation biology, 6th edn. Sinuar Associates Press, Sunderland, MA, pp 1–589
Roscoe DE, Stansley W (2012) Wildlife forensic pathology and in wound analysis and pesticide poisoning. In: Huffman JE, Wallace JR (eds) Wildlife forensics: methods and applications. Wiley Blackwell, West Sussex, UK, pp 109–143
Ruiz IB (2017) Europe, a silent hub of illegal wildlife trade. DW. https://p.dw.com/p/2W17D
Scheffers BR, Oliveira BF, Lamb I, Edwards DP (2019) Global wildlife trade across the tree of life. Science 366:71–76
Schoenly KG, Haskell NH, Hall RD, Gbur JR (2007) Comparative performance and complementarity of four sampling methods and arthropod preference tests from human and porcine remains at the forensic anthropology Center in Knoxville, Tennessee. J Med Entomol 44:881–894
Shaffer MJ, Bishop JA (2016) Predicting and preventing elephant poaching incidents through statistical analysis, GIS-based risk analysis, and aerial surveillance flight path modeling. Trop Conserv Sci 9:525–548
Shelley L, Kinnard K (2018) The convergence of trade in illicit rhino horn and elephant ivory with other forms of criminality. In: Moreto WD (ed) Wildlife crime: from theory to practice. Temple University Press, Philadelphia, PA, pp 109–134
Sina S, Gerstetter C, Porsch L, Roberts E, O' Smith L, Klaas K, Fajardo T (2016) Wildlife crime. Policy department A: Economic and Scientific Policy. Directorate General for Internal Policies
Sintov ND, Seyranian V, Tambe M (2018) Adoption of conservation technologies. In: Moreto WD (ed) Wildlife crime: from theory to practice. Temple University Press, Philadelphia, PA, pp p217–p238
Sundström A, Wyatt T (2017) Corruption and organized crime in conservation. In: Gore ML (ed) Conservation criminology. Wiley Blackwell, West Sussex, UK, pp 97–110
Tan RF, Teoh SS, Fow JE, Yen KS (2016) Embedded human detection system based on thermal and infrared sensors for anti-poaching application. IEEE Conference on Systems, Process and Control Conference on Systems, Process and Control, Melaka, Malaysia, pp 37–42
Temple-Raston D (2019) Solving the challenges to counting forest elephants. Weekend Edition, National Public Radio (NPR)
Tomberlin JK, Sanford MR (2012) Forensic entomology and wildlife. In: Huffman JM, Wallace JR (eds) Wildlife forensics: methods and applications. Wiley Blackwell, West Sussex, UK, pp 81–107
Tomberlin JK, Mohr R, Benbow ME, Tarone AM, VanLaerhoven S (2011) A roadmap for bridging basic and applied research in forensic entomology. Annu Rev Entomol 56:401–421
TRAFFIC Briefing paper: wildlife trade in the European Union, 2014. http://www.traffic.org/general-reports/traffic_pub_gen56.pdf. Accessed 15 Sep 2019
Treves A, Browne-Nunez C, Hogberg J, Frank JK, Naughton-Treves L, Rust N, Voyles Z (2017) Estimating poaching opportunity and potential. In: Gore ML (ed) Conservation criminology. Wiley Blackwell, West Sussex, UK, pp 197–212
UNEP (United Nations Environment Programme) (2014) Emerging technologies: smarter ways to Fight wildlife crime. https://na.unep.net/geas/archive/pdfs/GEAS_Jun2014_EmergingTechnologies_illegalwildlife.pdf. Accessed 15 Oct 2019
UNODC (2014) Guidelines on methods and procedures for ivory sampling and laboratory analyses United Nations, p 119
Wall J, Wittemyer G, Klinkenberg B, Douglas-Hamilton I (2014) Novel opportunities for wildlife conservation and research with real-time monitoring. Ecol Appl 24:593–601
Wallace JR, Ross JC (2012) The application of forensic science to wildlife evidence. In: Huffman JE, Wallace JR (eds) Wildlife forensics: methods and applications. Wiley Blackwell, West Sussex, UK, pp 35–50
Warchol G (2018) Wildlife crime and criminal organizations: can the theory of enterprise help explain the ivory and rhino horn trade. In: Moreto WD (ed) Wildlife crime: from theory to practice. Temple University Press, Philadelphia, PA, pp 81–105
Wasser SK, Brown L, Mailand C, Mondol S, Clark W, Laurie C, Weir BS (2015) Genetic assignment of large seizures of elephant ivory reveals Africa's major poaching hotspots. Science 349(6243):84–87
Webb SL (2019) Animal tech that could help you make decisions on the ranch. Noble News Views 37(8):16–19
West JB, Bowen GJ, Cerling TE, Ehleringer JR (2006) Stable isotopes as one of nature's ecological recorders. Trends Ecol Evol 21(7):408–414
Wich SA (2015) Drones and conservation. In: Kakaes K (ed) Drones and aerial observation: new technologies for property rights, human rights, and global development: a primer. New America, Washington DC, pp 64–70
Williams JD (1995) Advanced technologies for perimeter intrusion detection sensors. European Convention on Security and Detection, Brighton, pp 133–137
Wilson RP, Grundy E, Massy R, Soltis J, Tysse B, Holton M, Cai Y, Parrott A, Downey LA, Qasem L, Butt T (2014) Wild state secrets: ultra-sensitive measurement of micro-movement can reveal internal processes in animals. Front Ecol Environ 12:582–587
Wrege PH, Rowland ED, Keen S, Shiu Y (2017) Acoustic monitoring for conservation in tropical forests: examples from forest elephants. Methods Ecol Evol 8:1292–1301
Wyatt T (2013a) Wildlife trafficking: a deconstruction of the crime, the victims and the offenders. Palgrave Macmillan Press, Hampshire, UK, pp 1–198
Wyatt T (2013b) The security implications of the illegal wildlife trade. J Soc Criminol 2013:130–158

The Intersection of Forensic Techniques with Ecological Issues

Tiara Moore, Camille Gaynus, Phillip S. Levin, and Rachel Meyer

Abstract

Molecular forensic techniques to trace DNA from surfaces and sediments have played important roles in conservation and species protection for decades, but their application in the field of ecology is not as well established. In the 1990s, forensic sequencing was applied to non-human samples, illuminating how DNA shed in the environment can be used to track species, populations, and individuals, with implications for animal conservation and management cases. Concurrently, molecular markers such as DNA barcodes were increasingly being published for systematics and biodiversity surveys. Now, forensic techniques are being applied as metabarcoding and metagenomics for ecology and conservation, taking advantage of massive biodiversity DNA reference databases to trace and characterize everything in environmental samples from individuals to whole communities, revealing the provenance and timing of events. Sequencing environments across space and time provides key ecosystem biometrics that point to the causes—and often the culprits—of ecological change. Ecological forensics broadens translational research opportunities in conservation, policy, and environmental justice. This chapter specifically focuses on the increased use of environmental DNA (eDNA), which can be collected from soil, sediment, water, or air, in forensic ecological research from real-time to deep time. The rise of this new subdiscipline has the potential to shape the future of biodiversity management and discovery globally.

Keywords

Environmental DNA · eDNA · Ecological forensics · Ecology · Conservation · Biodiversity

An Introduction to Ecology, Biodiversity, and Forensics

Ecology is a broad scientific field, focused on describing organismal interactions with their associated environment (Owen and Owen 1974). Modern ecology focuses on understanding species presence, niches, and how species interact with the ecosystem across trophic levels (Futuyma and Moreno 1988). Documented increases in the severity and duration of

T. Moore (✉) · P. S. Levin
The Nature Conservancy, Seattle, WA, USA

School of Environmental and Forestry Sciences, University of Washington, Seattle, WA, USA

C. Gaynus
Department of Biology, University of Pennsylvania, Philadelphia, PA, USA

R. Meyer
Department of Ecology and Evolutionary Biology, University of California, Santa Cruz, CA, USA

S. C. Underkoffler, H. R. Adams (eds.), *Wildlife Biodiversity Conservation*,
https://doi.org/10.1007/978-3-030-64682-0_7

disturbances are making ecological-based studies more commonplace, forcing the integration of novel methods and techniques to assess how organisms influence ecosystem function, and vice versa.

Traditional ecological techniques such as visual surveys, video surveillance, and trapping are commonly employed to monitor species. These techniques, however, are time-consuming and involve personnel highly trained in a given taxonomy. The use of molecular methods may prove more reliable, or may complement traditional observation, and facilitate long-term monitoring. Monitoring multiple species at a time is important because environmental resilience and resistance to disturbance depend on species interactions in a system.

Biodiversity is integral to ecosystem function. Highly biodiverse ecosystems such as coral reefs (Hughes et al. 2007), rainforests (Hoorn and Wesselingh 2010), and kelp forests (Steneck et al. 2002) have complex food webs, biogeochemical cycles, and disturbance response strategies, potentially yielding natural systems more resilient to disturbances. Compromises to biodiversity, however, can result in lower functionality (e.g., coral reefs, kelp forest). Biodiversity measurements indirectly provide assessments of ecosystem function (Pauchard et al. 2018). Ecological forensics investigates changes to a system that affect biodiversity, which may require changes in human practices, such as modified fisheries policy (Ogden 2008) and criminal punishment for illegal or damaging actions such as oil spills (Mudge 2008). Ecological forensics also encompasses applications of forensics techniques to solve ecological mysteries such as trophic changes (e.g., Tolimieri et al. 2013; Jouta et al. 2017) that can be interpreted with regard to a legislative framework.

Biodiversity is measured at all organizational levels of the ecosystem, from the diversity of genes within a sample, genetic diversity of populations, or phylogenetic or taxonomic composition of communities to intersecting species occurrence dynamics in order to describe the macrosystem (Pauchard et al. 2018). Species in a community that provide similar functions, such as the ability to metabolize certain elements like nitrogen or sulfur, form functional guilds (Blondel 2003). Functional guilds are long-standing pillars of biodiversity organization for ecological study, for example, microbial functional guilds create the foundations upon which higher trophic levels can assemble (Borst et al. 2018). Strong richness and compositional diversity within a guild as well as between guilds create highly functional ecosystems, with a stronger resilience to disturbance, such as a storm, introduction of a pollutant, or altered nutrient availability. With these concepts in mind, ecological forensic research seeks to determine the root cause of ecological change that affects biodiversity on the genetic, phylogenetic, or functional guild level and to apply these findings to crime and policy.

Box 1 When a Hallmark Food Web Needs Ecological Forensics

A pillar of ecology is the recognition that species distributions and abundances are dynamically driven by trophic interactions. In some locations, kelp forest ecosystems undergo phase shifts (Estes et al. 2004) between kelp-dominated and crustose coralline algae-dominated, depending on the abundance of sea otters (*Enhydra lutris*). Sea otters maintain the ecosystem through a top-down trophic cascade, and their plight has earned them the place as a poster child of a keystone species that must be protected. Sea otters were nearly overhunted to extinction, both caused by humans since European arrival (Watson 2000). Once thriving with over 300,000 individuals, their numbers decreased to ~2000 in only 13 locations during the fur trade (Bodkin 2015). The reduction of grazing pressure on kelp resulting from the decline of sea otters broadly reshaped coastal kelp ecosystems. The recovery of sea otter populations has remarkably improved the functioning of kelp ecosystems and benefitted local human

(continued)

communities who can make use of increased fisheries resources associated with kelp forests (Gregr et al. 2020). Nonetheless, there is almost no genomic diversity in southern and northern sea otter populations (Beichman et al. 2019), which makes populations vulnerable.

A recent threat to sea otters has emerged: *Toxoplasma gondii*, which causes meningoencephalitis disease. In 2004, *Toxoplasma* caused 16% of sea otter mortality in recovered carcasses. Pathology and multilocus genotyping of the *Toxoplasma* parasite from otter carcasses and nearby cats revealed cats transmit the most lethal strains of the parasite (Shapiro et al. 2019). Snails, other mollusks, and crustaceans at the adjunct of freshwater runoff and coasts may bioaccumulate *Toxoplasma* and facilitate its ingestion by otters. University of California (UC) Davis researchers identified marine snow as another bioaccumulator that may enrich kelp beds with the parasite (Shapiro et al. 2012). While direct amplification of *Toxoplasma* from crustaceans and kelp has been proposed as biomonitoring strategies (Bigot-Clivot et al. 2016), community-based environmental DNA (eDNA) research has revealed that metabarcoding of the *18S* rRNA locus from soil, sediment, or water can all detect *Toxoplasma* without harvesting or harming any organisms (inset figure shows results of 13 positive detections from August 2020; Map from www.ucedna.com, no copyright, permission from author Rachel Meyer; inset by DPDx Parasite Image Library from https://en.wikipedia.org/wiki/Toxoplasma_gondii#/media/File:Toxoplasma_gondii_tachy.jpg). With this data, people can better protect the keystone otter-kelp forest ecosystem that our own species almost wiped out once and that activities such as allowing domestic cats to roam outside continue to threaten.

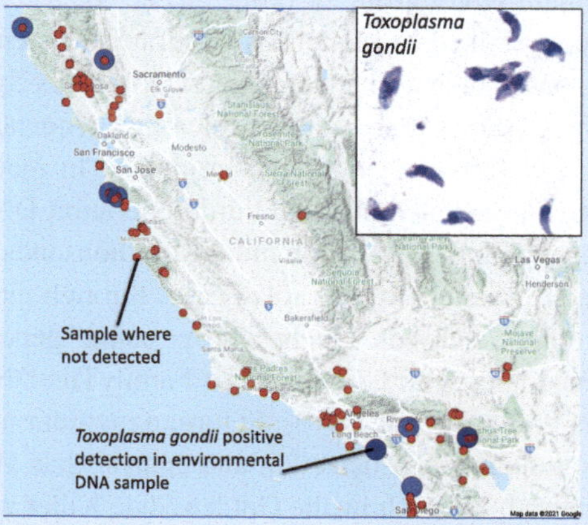

To explore dimensions of biodiversity and the ecosystem by surveying DNA molecules in the environment, such as using DNA sequences to discern guilds (Bouskill et al. 2012; Nguyen et al. 2016), researchers must consider the nature of deposition and the temporal footprint of molecules. Most organisms shed DNA throughout their development as they move through the environment. Certain events like spawning or release of pollen can increase DNA (Fahner et al. 2016) in the environment, as can death (Tillotson et al. 2018). DNA molecules can either rapidly degrade or remain preserved for many thousands of years (Taberlet et al. 2018), depending on the local environmental conditions, and thus can be useful to characterize systems and assess changes to biodiversity at many space-time scales (e.g., Petrou et al. 2019).

For the last 30 years, the field of forensics has heavily relied on DNA detection and sequencing to solve crimes (Roewer 2013). DNA fingerprinting, microsatellites, and whole genome resequencing have been used to track criminals from cells shed at the scene of a crime from the hair, skin, blood, or semen (Jeffreys et al. 1985). The quick touch of a mirror or door handle can leave behind sufficient genetic signatures for police and prosecutors to capture the criminal and solve the crime (Roewer 2013) by direct match or by associative evidence such as a location-specific microbiome (Pechal et al. 2014). Human DNA from a crime scene is usually directly linked to sequences in curated reference databases such as the Combined DNA Index System (CODIS) that contains reference sequences from victims and registered criminals. In recent years, next-generation and third-generation DNA sequencing has enabled hundreds of thousands of genetic markers to help solve cases through more complex analyses, such as by using genetic genealogies with GEDmatch or FamilyTreeDNA data (Greytak et al. 2019). In turn, a revolution in non-human DNA detection and analysis has also occurred. This is how the Golden State Killer was caught by scientist Barbara Rae-Venter.

The employment of forensics to match a cat hair to a specific cat helped solve a murder case (Menotti-Raymond et al. 1997) in the early 1990s and was quickly followed by use of the same techniques in conservation genetics to monitor poaching and trafficking (e.g., tigers; Kitpipit et al. 2012) as well as crime involving domestic species (Budowle et al. 2005). Programs like RhODIS were designed to mirror the CODIS databases, wherein select genomic loci were matched to solve confiscated rhino horn and poaching cases (Harper 2011; Harper et al. 2018). Concurrently, the establishment of DNA barcoding (Nanney 1982; Hebert et al. 2003) became important in basic biology (e.g., in systematics and ecology), and this also became valuable reference molecular information for environmental and conservation forensics, leading to the growth of DNA databases that represent the tree of life (e.g., National Center for Biotechnology Information, NCBI; Barcode of Life Database, BOLD). NCBI's GenBank exceeds 700,000 taxa in the year 2020.

Biodiversity sequencing aids in solving criminal cases, answering the questions of "where" and "when" an incident occurred. *Metagenomic* analysis of genetic sequence data reveals a microbiome that can be exploited to elucidate events, processes, and provenance. For example, the human skin microbiome is an indicator of the postmortem interval (Metcalf 2019). Trace environmental DNA at crime scenes became recognized as providing dozens to thousands of taxa that could resolve provenance robustly through community similarity, and the technique *DNA metabarcoding* was found to be useful for tracing DNA signatures of many species with substantially lower cost than metagenomics. Quickly, bioinformatic tools to aid metabarcoding analysis have emerged to encourage deployment of this technology in many kinds of forensics (Fløjgaard et al. 2019; Allwood et al. 2020).

Within environmental forensics, the subdiscipline of ecological forensics encompasses multiple fields of molecular investigation, but the field of environmental DNA (eDNA) encompasses most investigations. eDNA analysis commonly uses metagenomics and metabarcoding to probe genetic material found in environmental samples and identify species presence (Thomsen and Willerslev 2015) or characterize community composition (e.g., Stoeck et al. 2010). Researchers can then use this information to reveal missing information needed to explain ecological phenomena. With this in mind, environmental scientists become forensic detectives as they collect samples, extract the DNA, and analyze sequences to determine the genetic, phylogenetic/taxonomic, or functional composition of the ecosystem. This genetic information provides data that can help scientists solve fundamental questions as well as crimes that occur in the environment. These standard community ecology measurements of species richness and compositional diversity help to assess the current status of ecosystem health and aid in future conservation efforts (Deiner et al. 2016).

Methods Used to Study Environmental DNA

Environmental DNA is a mixed sample of DNA from many species that may exist in different levels of degradation and abundance. Relevant to ecological forensics are sample types ranging from bone to soil, water, surfaces, and even air. Shedding of genetic material does not happen on a universal level for organisms, and how organisms shed this material is important in its ability to transport through a variety of mediums, and its "shelf-life" in different environments impacts the types of organisms detected by eDNA sequencing and the duration for which they can be detected (for a review see Barnes and Turner 2016). Researchers mine environmental samples representing different slices of time and space primarily using four techniques: DNA metagenomics; DNA metabarcoding; quantitative PCR (qPCR); and hybridization capture.

DNA metagenomics includes a suite of methods to mine shotgun sequence data or sequences generated from fragments of sequenced DNA that are used to recreate an entire chromosome or genome and match the unknown sample to referenced sequences. Because most of the DNA in environmental samples is microbial, the depth of sequencing required is scaled somewhat with the body size and spatial density of the species of interest—microbial biomes can be characterized with a few hundred thousand reads (Dinsdale et al. 2008), but finding signatures of a specific plant or mammal DNA in a sediment sample may require 10–100 million reads. Even bone, left to decay exposed or buried, becomes overrun with microbial DNA over time, with diminishing endogenous DNA. As perhaps an extreme case, to obtain the entire genome from the bone of an extinct horse relative that roamed over 500,000 years ago, 12.2 billion sequences were required (Orlando et al. 2013). Fortunately, smaller and more prevalent organisms can serve as proxies for larger organisms. For example, dung fungus *Sporormiella* is an indicator of large mammal presence that can be found in sediments (see Perrotti and van Asperen 2019). The investment in metagenomic sequencing environmental samples can offer incredible returns for ecology, such as by offering multiple lines of evidence to determine whether humans or climate change led to past megafaunal extinctions (Wang et al. 2017) and paleoecological transitions (Hofman et al. 2015). Additionally, metagenomics offers the only means to obtain whole genomes from environmental samples that are needed to potentially de-extinct species (Shapiro 2017). Metagenomics also allows for determination of ancient versus contemporary DNA, because chemical damage to DNA after the organism's death causes deamination of cytosine (Hofreiter et al. 2001). Thus, metagenomics is conducive to historical forensic cases.

Metagenomes from contemporary samples also provide functional genetic information for microbial strains. Identifying metabolic pathway genes and their links on the same genome is critical to accurately modeling functional guilds and biotic interactions in ecosystems (Barnum et al. 2018). Protocols are continuously being improved to optimize DNA extraction (Rohland et al. 2018) and shotgun library preparation (e.g., SRSLY Kits from Claret Bioscience) for metagenome analysis from degraded samples that allow more genomic information to be obtained from smaller amounts of starting DNA. These same protocols are being used to examine historic human forensic samples (see Astrea Forensics; astreaforensics.com).

DNA metabarcoding is the amplification of target barcode loci to characterize specific groups of organisms such as plants, microbes, or animals. From a single sample, researchers can separately amplify multiple barcode loci and then combine amplicons into a single indexing reaction that is sequenced on Illumina or PacBio instruments to produce multiplexed metabarcoding data and analyzed by matching the DNA reads to reference DNA databases. Some of the most frequently metabarcoded loci include the *16S* ribosomal RNA (rRNA) locus, which is diagnostic for bacteria and for mammals; the *18S* rRNA locus, which is diagnostic for broad eukaryotes; the *internal transcribed spacer* (*ITS*) rRNA gene, often used for diagnosing plants and fungi; the *cytochrome C oxidase subunit 1* (*CO1*) mitochondrial gene, which is diagnostic of animals and protozoa; and the *12S* mitochondrial gene to diagnose vertebrates (Creer et al. 2016).

eDNA metabarcoding has proven to be a useful tool to assess community composition, where targeted amplification enhances the representation of specific taxa, even if their sequences are relatively rare compared to bacterial sequences. However, metabarcoding is not without limitations. Rare organisms may be missed (Barnes and Turner 2014), biases inherent in

PCR polymerase (Nichols et al. 2018) may skew results, and the inability to discern species from a short barcoding fragment alone may all hamper the abilities of this method to fully elucidate true communities. Additionally, ancient DNA damage patterns cannot be detected with metabarcoding. Nonetheless, with very few reads, an environmental sample usually has sufficient community information to match samples to a candidate provenance (Fløjgaard et al. 2019), or to detect changes in the community (Fahner et al. 2016), which are major objectives. For example, bacterial *16S* metabarcoding of chimpanzee feces revealed the "humanization" of the gut microbiome under captivity, which, when combined with deep shotgun sequencing metagenomic analysis, demonstrated a reduction in plants being consumed in captivity as the cause of the microbiome perturbation (Clayton et al. 2016). This signal of captivity and a propensity to disease have policy implications for the ecological structuring of parks, reserves, and zoos.

Quantitative PCR (qPCR) is a powerful tool to rapidly detect a species or group by amplification rather than sequencing. For qPCR, custom primers are typically designed to specifically and unequivocally amplify a short (~50 bp–200 bp) region of DNA from a target in a mixed sample. Either an endogenous control gene fragment is included for semi-quantitative PCR that is useful for comparison of cycle thresholds among samples, or a serial dilution of known target DNA is used as a scaling metric to estimate the absolute number of target DNA molecules in samples. Throughout this chapter we highlight many case studies that use qPCR in these ways. Species detection by qPCR is so heavily relied on for rapid quantification that some workflows have joined field and laboratory steps to actually succeed in obtaining qPCR results in the field within a few hours with portable thermocyclers that connect to mobile phones (Thomas et al. 2019).

Quantitative PCR is sometimes used on eDNA sample extracts to determine the presence of inhibitors such as humic acids, proteins, or salts (Lloyd et al. 2010). Inhibitors are common in the geochemistry of soils and certain plants. Failing to remove inhibitors during DNA extraction can skew results derived from standard methods such as metabarcoding because target DNA will not have an equal chance of being amplified across samples. Many forensic sample types contain inhibitors which can be removed by a variety of purification steps (see Hedman and Rådström 2013). Obtaining qPCR cycle thresholds for a range of sample dilutions can signal the appropriate working dilution where inhibitors have no perceived effect (see Murray et al. 2015).

Hybridization capture using enrichment techniques such as bead-bound "baits" in solution or microarrays with baits on a solid surface combines the benefits of targeted sequencing and metagenomics. Baits are synthesized DNA that probes for and can capture thousands of non-repetitive polymorphic DNA fragments for organisms that are rare in environmental samples. Capture probes allow even old, degraded fragments of DNA as short as 70 bases to be sequenced. Examples of baits used include population-level diagnostics to sort ancient from contemporary DNA to detect Neanderthal relatives of humans (Gansauge and Meyer 2014), to obtain entire mitochondria from degraded human skeletal samples (Marshall et al. 2017), to reconstruct mammalian communities from sediments (e.g., PalaeoChip; Murchie et al. 2019), and to analyze historical plant specimens in herbaria (e.g., Angiosperm353; Johnson et al. 2019) and arthropod collections in museums (Knyshov et al. 2019) at thousands of homologous loci. Like metagenomes, baits-captured DNA molecules retain ancient damage signatures. Functional gene families involved in stress response can be probed in the environment using hybridization to a gene chip such as the GeoChip (He et al. 2007). The GeoChip has been a key to characterize microbial functioning affected by contaminants, pollutants, and naturally occurring elements, from aquifers to groundwater and wastewater (Jiang et al. 2019; Li et al. 2019). Hybridization capture methods have also been successfully used in tracing infectious disease in humans (Gaudin and Desnues 2018), invertebrates (Carpi et al. 2015), and vertebrates such as koalas (Tsangaras et al. 2014).

Tracking Species of Interest with eDNA

Conservation ecology evaluates human impacts on biodiversity and develops practical approaches to simultaneously maximize the preservation of biodiversity and the improvement of human well-being (Kareiva and Marvier 2012). A main objective of conservation science is to establish pragmatic methods for preserving or restoring species and associated ecological communities while reconciling concerns and supporting needs of people. As a consequence, conservation research often focuses on monitoring biodiversity and identifying threats to species or ecosystems. eDNA has the potential to contribute to both biodiversity and threat assessments, and we discuss these below.

Monitoring species of conservation concern is an appealing use for eDNA because the approaches can be sensitive enough to detect rare species and because obtaining eDNA is usually non-destructive. To date, the bulk of studies have focused simply on the presence/absence of species (Shelton et al. 2019), and in many situations such information is sufficient to inform conservation efforts. For instance, Spear et al. (2015) used eDNA and quantitative PCR (qPCR) across 61 sites to assess the presence of the Eastern hellbender (*Cryptobranchus alleganiensis alleganiensis*), a declining, secretive salamander (Spear et al. 2015). The cryptic nature of hellbenders makes traditional, non-genetic sampling methods extremely difficult, and knowing sites where the salamander resides is a first step in prioritizing conservation efforts. Other studies have used eDNA metabarcoding to assess the occurrence of assemblages of species. For example, sampling natural saltlicks in tropical forests revealed the presence of diverse mammals, including orangutan and banteng (Ishige et al. 2017). In another example, Deiner and colleagues (2016) detected 296 families of eukaryotes, spanning 19 phyla across the catchment of a river. These examples highlight how such spatially integrated snapshots of biodiversity provide useful means for assessing conservation and restoration efforts of rare, imperiled species, as well as the total biodiversity of whole landscapes.

While presence/absence of data is valuable, knowing the relative abundance of individuals can be essential for many conservation applications. Thus, a critical step toward making eDNA methods even more useful in conservation settings is developing the ability to use eDNA quantitatively. Recent evidence suggests that in many situations using eDNA to estimate relative abundance is quite feasible. For example, Tillotson and colleagues (2018) report a strong quantitative relationship between eDNA concentration and the abundance of spawning sockeye salmon in a small stream in Alaska, USA. Similarly, in threatened Chinook salmon (*Oncorhynchus tshawytscha*) from an estuary in Washington, USA, abundance indices derived from eDNA qPCR reflect the seasonal migration of salmon and provide virtually identical quantitative information as conventional net-based methods (Shelton et al. 2019). Water samples were used to monitor sea otter and fish abundance in California kelp forest (Port et al. 2016), and these abundances were also found to align closely with visual dive survey data.

In addition to assessing the occurrence or abundance of species, eDNA has been used to monitor or evaluate threats to biodiversity. For example, because the structure of communities varies with the primary pollutants in an environment as well as the magnitude of contaminants (Wang et al. 2019b), eDNA is a powerful tool for assessing impacts of pollution. In the case of oil pollution, hydrocarbon-degrading bacterial families will be present, varying in a manner that can be related to the time since an oil spill (Xie et al. 2018). Similarly, the presence of specific bacterial families can indicate a wider sensitivity to oil pollution (Xie et al. 2018). Other researchers have used eDNA focused on eukaryotes to diagnose environmental impacts. For instance, using eDNA technology focused on zooplankton, Yang and colleagues developed an assay for ammonia nitrogen pollution based on the relative abundances of rotifers, copepods, and Cladocera (Yang et al. 2017).

Detection of the presence and dispersal of invasive species has been a mainstay of eDNA conservation research. Early detection of an incipient invasion by a harmful species increases the feasibility of rapid responses to eradicate the species or contain its spread, and eDNA has proven useful as an early detection tool. Jerde et al. (2011) used eDNA to delimit the invasion fronts of two species of Asian carps in area canals and waterways feeding into the Laurentian Great Lakes, leading to better management of these invaders. Assays of Burmese python eDNA in concert with occupancy modeling were able to detect the leading edge of the python invasion in Florida, USA, improving conservation efforts in this region (Hunter et al. 2015). eDNA surveillance of an invasive fish (ruffe, *Gymnocephalus cernua*) revealed a much more rapid spread of the invader than predicted by conventional sampling techniques (Tucker et al. 2016). As a final example, the non-native American bullfrog, an invasive species in French wetlands (Ficetola et al. 2008), was successfully detected from water samples. Successful detection can give researchers inferences regarding the range dynamics of non-native species, helping to classify the non-native species as invasive. While this technology is generally employed after invasion, there is interest in employing eDNA as a preventative measure. For example, ballast water, a common transport of marine invasive species, could be used to detect non-native species before they invade an ecosystem (Mahon et al. 2014; Egan et al. 2015). The introduction of non-native species can cause major problems for an ecosystem, especially if a population rises to invasive status, making the use of eDNA to track these species a viable and appealing management strategy.

Conservation projects utilizing eDNA have occurred in many different terrestrial and aquatic systems from the tropics to the desert and the Arctic and from wild to captive systems. For example, eDNA was successfully acquired from water holes and used to identify inhabitants of a Japanese zoo (Ushio et al. 2017), suggesting a promising future for forest mammal biodiversity monitoring. eDNA was shown to be a useful tool to inventory mammals from artificial wallows in Colorado, USA (Williams et al. 2018), and from spring and stock tank water in Utah, USA (Rodgers and Mock 2015). In African parks, water hole mammal DNA hybrid capture has become a new tool for biomonitoring (Seeber et al. 2019). These recent findings are leading researchers to ask how deep down in lake and pond sediments can plant and animal DNA be detected from these hot climate areas (Bremond et al. 2017). It may be possible to also use sediments that date back in time to reconstruct the landscape composition of areas before and during civil wars so we know what landscape reconstruction should resemble.

Case Study on eDNA from Sediments: Species Networks to Diagnose Processes

Since the first eDNA study focused on sediment (Ogram et al. 1987), the patterns of occurrence in both microbial and macrobial taxa have shown their value to diagnose ecosystem processes. The co-occurrence of species provides hypotheses about species interactions, which in turn can be used to propose hypotheses regarding the ecological role of species (e.g., keystone species) to disturbance of ecosystems (e.g., invasive species) (Ficetola et al. 2008). Patterns of species can be found in sample sets that vary across both space and time using software such as SPIEC-EASI (Kurtz et al. 2015). In this example case study with preliminary results from Moore (2019), we describe the study in a forensics framework and then show how network analysis of eDNA leads us to posit the mechanism of disturbance processes.

Lagoons

Moore (2019) sought to identify the effects of large macroalgal blooms on the community composition of lagoon systems, specifically during decomposition of the blooms (Moore 2019). Algal blooms are thought to be caused by the increased input of nutrients from human agricultural and wastewater systems. Standard ecological methods were used to identify a cascade of biogeochemical changes caused by these blooms. They then used the ecological forensic method of metabarcoding to characterize biodiversity changes to the lagoon microbiome as microbes are involved in decomposition processes. In between lagoon sampling events, the largest wildfire in California history (the Thomas fire, December 2017) occurred near the lagoons. Because the fire was human-caused, this case study provided evidence of the interaction between two human-caused hazards (suspected culprits), wildfire and nutrient enrichment, on fragile lagoon systems, from the lens of the microbiome. Which culprit has a larger effect, and how does the system respond?

The crime report: Large macroalgal blooms of *Ulva* sp. in nutrient-enriched lagoon, Carpinteria Salt Marsh in California, lead to large amounts of decomposing material in summers causing eutrophied and acidic water columns. A nearby lagoon in Newport, which is perceived to have more nutrient input from urbanization, does not experience the extreme bloom and eutrophication as Carpinteria, but does experience summer acidification. In addition, the Thomas Fire (December 2017) occurred near the Carpinteria lagoon, which caused movement of biological materials and is expected to alter nutrient pools. The Thomas Fire is also suspected to be exacerbating nutrient deposition and erosion.

The evidence: Algal biomass surveys and acidification measurements (decreased pH) from November 2017 and June 2018.

The victims: Water quality, which is now acidified, and carbonate organisms, which are directly impacted by acidification.

The suspects: Human nutrient pollution from upstream and nutrient-rich sediment runoff following a human induced wildfire.

The weapon: Nutrient alterations interacting with the decomposer microbial species in the system. The way this weapon harmed the victims specifically is unsolved.

The forensic eDNA investigation (Fig. 1a–c): Metabarcoding of the *16S* locus targeting bacteria (Caporaso et al. 2011) from multiple samples in November 2017 (winter) and in June 2018 (summer) showed both the mouth and the middle of the Carpinteria lagoon (where the bloom was located) sampled in the summer after the fire were very different from the lagoon head. Because the head did not compositionally shift in the beta diversity (species diversity between samples) ordination plot (Fig. 1a), and because Carpinteria has a much larger bloom than Newport lagoon, wildfire was ruled out as a suspect in causing the microbiome change, and human nutrient pollution was determined to be the culprit.

To characterize the new-found culprit further, the taxon presence and abundance patterns were used to calculate a species co-occurrence network, and of 2046 taxa detected with the *16S* metabarcode, a robust network of 32 taxa was discovered (Fig. 2). While many of these microbes are yet uncharacterized or poorly characterized, some, such as the Deltaproteobacteria and Myxococcales, are known decomposers found in lagoon sediments, and gammaproteobacteria, Chromatiales, are sulfide reducers known to increase with increasing decomposition due to elevated nutrient loads (Aires et al. 2019). Myxococcales act as an endophyte in halophytic plants and select green algae (Aires et al. 2015). The presence of Myxococcales in this network suggests it is associated with *Ulva*. In addition, another member of the network, Micrococcales, has been associated previously linked with *Ulva*, but not linked to its decomposition (FitzGerald et al. 2015), while other studies have noted Micrococcales performing cellulolysis and have observed a correlation with lower pH on farms in China (Wang et al. 2019a). These results link these members of the lagoon microbiome with *Ulva* and with eutrophication via their activity as an ecological community.

These findings are typical of classical ecological studies that assess how disturbances affect ecosystems, but by employing ecological forensics methods to characterize the culprit, an additional layer of evidence was discovered: the assessment of the microbial community in the sediment found particular decomposing bacteria that could be linked as co-conspirators in the crime. Due to the increase of these decomposing bacteria, other healthy bacteria typically found in

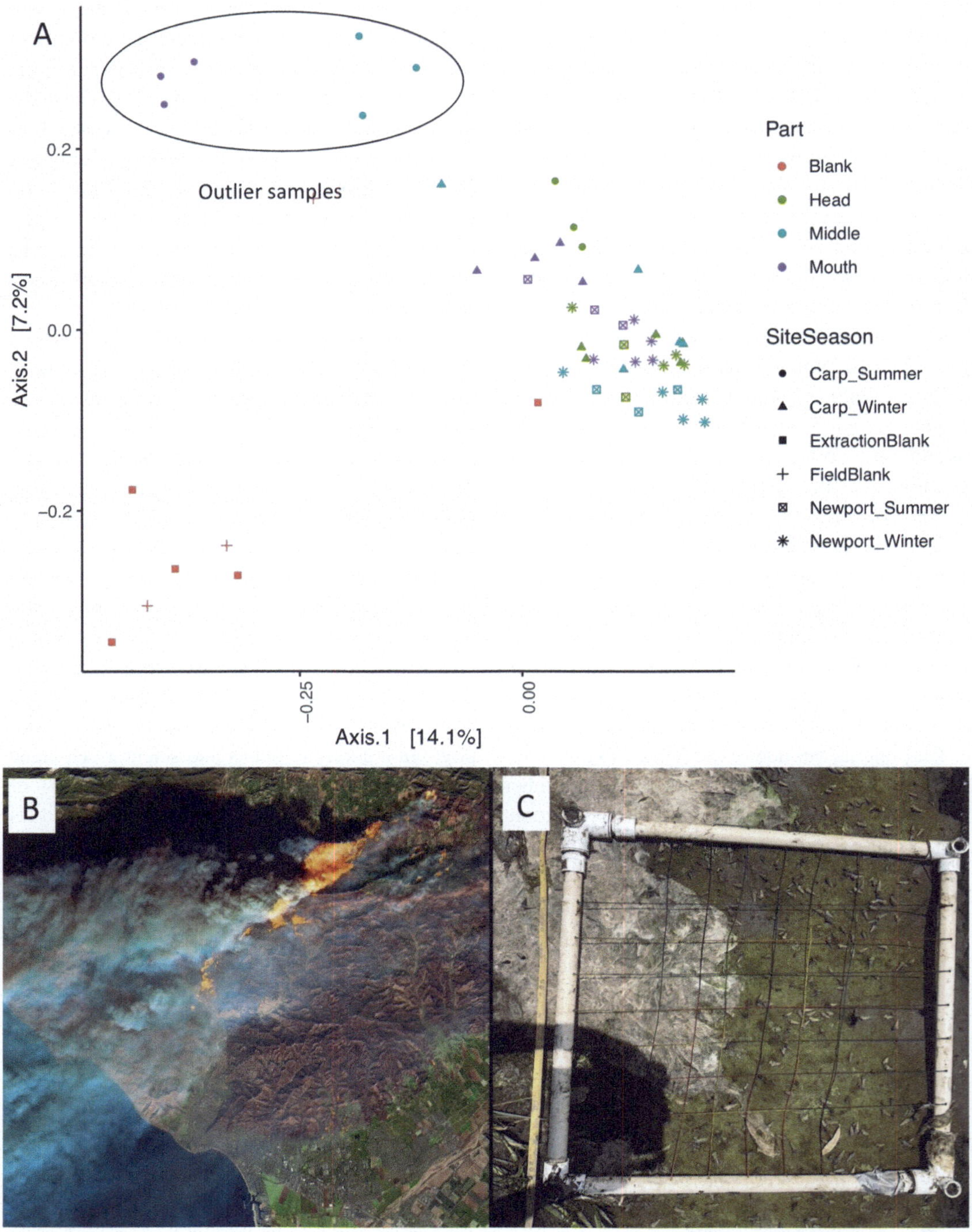

Fig. 1 (**a**) Beta diversity ordination plot. (**b**) Aerial photo of Carpinteria Salt Marsh during the Thomas Fire. (NASA MODIS data from LANCE/EOSDIS Rapid Response and Sentinel-2 from https://en.wikipedia.org/wiki/Thomas_Fire#/media/File:Thomas_Fire_burn_scar_on_Dec_7.jpg) (c) Conventional ecological algal biomass survey with both healthy and decomposing *Ulva* sp. and carbonate snails present. (Photo by Tiara N. Moore)

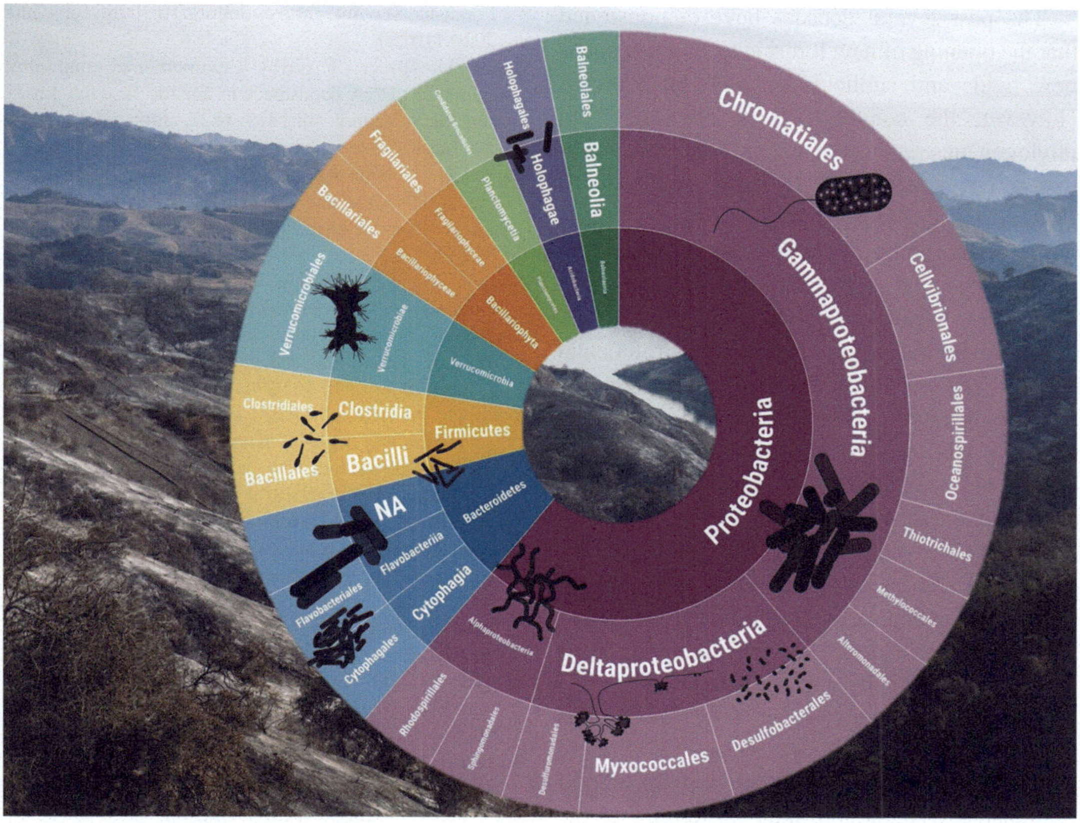

Fig. 2 Taxa present in the ecological co-occurrence network of microbial communities identified across sites and samples. Finding these taxa in a network suggests their cooperation to perform the metabolic function of decomposition. (Background Forest Service photo by Stuart Palley from https://commons.wikimedia.org/wiki/File:Thomas_Fire_(24469612857).jpg)

lagoon sediment were suppressed, potentially catalyzing changes in ecosystem services and making other species unexpected victims. The wildfire was also eliminated as a suspect for causing the community turnover by considering it should have effects on the entire system. Ultimately, the paired use of ecological forensics and conventional surveys provided enough evidence to solve our crime and close the investigation.

The Future

Biodiversity is vanishing at an astounding rate and facing the sixth greatest mass extinction (Ceballos et al. 2015; Shivanna 2020). Molecular forensic techniques offer a means to non-invasively and rapidly monitor biodiversity and develop a deeper understanding of how it is maintained on a systems level. The ecological forensics community, however, will need to confront the nature of how DNA molecules distribute and preserve. Perhaps every animal and plant has associated microbes and parasites that can serve as surrogate evidence for their presence. Small organisms are easier to detect than larger ones, and they may be highly diagnostic of patterns and processes, as revealed by the success of microbiome use in forensics and molecular ecology discussed above. The next years will require tremendous effort to better align the signals from eDNA and the signals from conventional ecological methods.

The past several decades have demonstrated that the opening of new forensic techniques opens new fields and subfields of study in ecology. However, the convergence of forensics with phylogenetics and systematics may have produced the greatest gain. Like the usefulness of CODIS and GEDMatch, biodiversity references have revolutionized our precision. Questions must be pursued at the intersection of multiple disciplines, ultimately enriching science. Now that there are many ways in which environmental samples and eDNA can technically be integrated into biodiversity, conservation, and ecological research, can another revolutionary advance for understanding all of the components in an ecosystem simultaneously be produced? Can the acquisition of big data encompassing biodiversity baselines, reference communities, and known patterns of change from disturbances and environmental gradients across space and time allow researchers to finally mechanistically solve the outstanding questions about how ecosystems work and who the key species are? Can these data cause legislation to be revised to support sustainable systems and undo the crimes of the Anthropocene to the environment? In order to focus forensics and policy on systems rather than species, a concerted effort will be required to bring molecular biological interactions in ecosystems and ecological forensics together with social science and public education to become genomics literate and appreciative of microbial biodiversity.

References

Aires T, Moalic Y, Serrao EA, Arnaud-Haond S (2015) Hologenome theory supported by cooccurrence networks of species-specific bacterial communities in siphonous algae (Caulerpa). FEMS Microbiol Ecol. https://doi.org/10.1093/femsec/fiv067

Aires T, Muyzer G, Serrão EA, Engelen AH (2019) Seaweed loads cause stronger bacterial community shifts in coastal lagoon sediments than nutrient loads. Front Microbiol. https://doi.org/10.3389/fmicb.2018.03283

Allwood JS, Fierer N, Dunn RR et al (2020) Use of standardized bioinformatics for the analysis of fungal DNA signatures applied to sample provenance. Forensic Sci Int. https://doi.org/10.1016/j.forsciint.2020.110250

Barnes M, Turner CR (2014) Environmental conditions influence eDNA persistence in aquatic systems. Environ Sci Technol 48:1819–1827. https://doi.org/10.1021/es404734p

Barnes MA, Turner CR (2016) The ecology of environmental DNA and implications for conservation genetics. Conserv Genet. https://doi.org/10.1007/s12686-012-9843-y

Barnum TP, Figueroa IA, Carlström CI et al (2018) Genome-resolved metagenomics identifies genetic mobility, metabolic interactions, and unexpected diversity in perchlorate-reducing communities. ISME J. https://doi.org/10.1038/s41396-018-0081-5

Beichman AC, Koepfli KP, Li G et al (2019) Aquatic adaptation and depleted diversity: a deep dive into the genomes of the Sea Otter and Giant Otter. Mol Biol Evol. https://doi.org/10.1093/molbev/msz101

Bigot-Clivot A, Palos Ladeiro M, Lepoutre A et al (2016) Bioaccumulation of toxoplasma and Cryptosporidium by the freshwater crustacean Gammarus fossarum: involvement in biomonitoring surveys and trophic transfer. Ecotoxicol Environ Saf. https://doi.org/10.1016/j.ecoenv.2016.07.006

Blondel J (2003) Guilds or functional groups: does it matter? Oikos 100:223–231. https://doi.org/10.1034/j.1600-0706.2003.12152.x

Bodkin JL (2015) Historic and contemporary status of sea otters in the North Pacific. Sea Otter Conservation

Borst ACW, Verberk WCEP, Angelini C et al (2018) Foundation species enhance food web complexity through non-trophic facilitation. PLoS One. https://doi.org/10.1371/journal.pone.0199152

Bouskill NJ, Tang J, Riley WJ, Brodie EL (2012) Trait-based representation of biological nitrification: model development, testing, and predicted community composition. Front Microbiol. https://doi.org/10.3389/fmicb.2012.00364

Bremond L, Favier C, Ficetola GF et al (2017) Five thousand years of tropical lake sediment DNA records from Benin. Quat Sci Rev. https://doi.org/10.1016/j.quascirev.2017.06.025

Budowle B, Garofano P, Hellman A et al (2005) Recommendations for animal DNA forensic and identity testing. Int J Legal Med. https://doi.org/10.1007/s00414-005-0545-9

Caporaso JG, Lauber CL, Walters WA et al (2011) Global patterns of 16S rRNA diversity at a depth of millions of sequences per sample. Proc Natl Acad Sci USA. https://doi.org/10.1073/pnas.1000080107

Carpi G, Walter KS, Bent SJ et al (2015) Whole genome capture of vector-borne pathogens from mixed DNA samples: a case study of Borrelia burgdorferi. BMC Genomics. https://doi.org/10.1186/s12864-015-1634-x

Ceballos G, Ehrlich PR, Barnosky AD et al (2015) Accelerated modern human-induced species losses: entering the sixth mass extinction. Sci Adv. https://doi.org/10.1126/sciadv.1400253

Clayton JB, Vangay P, Huang H et al (2016) Captivity humanizes the primate microbiome. Proc Natl Acad Sci U S A. https://doi.org/10.1073/pnas.1521835113
Creer S, Deiner K, Frey S, et al (2016) The ecologist's field guide to sequence-based identification of biodiversity. Methods Ecol Evol
Deiner K, Fronhofer EA, Mächler E et al (2016) Environmental DNA reveals that rivers are conveyer belts of biodiversity information. Nat Commun. https://doi.org/10.1038/ncomms12544
Dinsdale EA, Edwards RA, Hall D et al (2008) Functional metagenomic profiling of nine biomes. Nature. https://doi.org/10.1038/nature06810
Egan SP, Grey E, Olds B et al (2015) Rapid molecular detection of invasive species in ballast and harbor water by integrating environmental DNA and light transmission spectroscopy. Environ Sci Technol 49:4113–4121. https://doi.org/10.1021/es5058659
Estes JA, Danner EM, Doak DF et al (2004) Complex trophic interactions in kelp forest ecosystems. Bull Mar Sci
Fahner NA, Shokralla S, Baird DJ, Hajibabaei M (2016) Large-scale monitoring of plants through environmental DNA metabarcoding of soil: recovery, resolution, and annotation of four DNA markers. PLoS One. https://doi.org/10.1371/journal.pone.0157505
Ficetola GF, Miaud C, Pompanon F, Taberlet P (2008) Species detection using environmental DNA from water samples. Biol Lett 4:423–425. https://doi.org/10.1098/rsbl.2008.0118
FitzGerald JA, Allen E, Wall DM et al (2015) Methanosarcina play an important role in anaerobic co-digestion of the seaweed *Ulva lactuca*: taxonomy and predicted metabolism of functional microbial communities. PLoS One. https://doi.org/10.1371/journal.pone.0142603
Fløjgaard C, Frøslev TG, Brunbjerg AK et al (2019) Predicting provenance of forensic soil samples: linking soil to ecological habitats by metabarcoding and supervised classification. PLoS One. https://doi.org/10.1371/journal.pone.0202844
Futuyma DJ, Moreno G (1988) The evolution of ecological specialization. Annu Rev Ecol Syst Vol. https://doi.org/10.1146/annurev.es.19.110188.001231
Gansauge MT, Meyer M (2014) Selective enrichment of damaged DNA molecules for ancient genome sequencing. Genome Res. https://doi.org/10.1101/gr.174201.114
Gaudin M, Desnues C (2018) Hybrid capture-based next generation sequencing and its application to human infectious diseases. Front Microbiol
Gregr EJ, Christensen V, Nichol L et al (2020) Cascading social-ecological costs and benefits triggered by a recovering keystone predator. Science. https://doi.org/10.1126/science.aay5342
Greytak EM, Moore CC, Armentrout SL (2019) Genetic genealogy for cold case and active investigations. Forensic Sci Int. https://doi.org/10.1016/j.forsciint.2019.03.039
Harper C (2011) RhODIS–DNA profiling and a DNA database as a tool to protect the rhino. In: Proceedings of the tenth meeting of the IUCN African Rhino Specialist Group
Harper C, Ludwig A, Clarke A et al (2018) Robust forensic matching of confiscated horns to individual poached African rhinoceros. Curr Biol
He Z, Gentry TJ, Schadt CW et al (2007) GeoChip: a comprehensive microarray for investigating biogeochemical, ecological and environmental processes. ISME J 1:67–77. https://doi.org/10.1038/ismej.2007.2
Hebert PDN, Cywinska A, Ball SL, DeWaard JR (2003) Biological identifications through DNA barcodes. Proc R Soc B Biol Sci. https://doi.org/10.1098/rspb.2002.2218
Hedman J, Rådström P (2013) Overcoming inhibition in real-time diagnostic PCR. Methods Mol Biol. https://doi.org/10.1007/978-1-60327-353-4_2
Hofman CA, Rick TC, Fleischer RC, Maldonado JE (2015) Conservation archaeogenomics: ancient DNA and biodiversity in the Anthropocene. Trends Ecol Evol
Hofreiter M, Jaenicke V, Serre D et al (2001) DNA sequences from multiple amplifications reveal artifacts induced by cytosine deamination in ancient DNA. Nucleic Acids Res. https://doi.org/10.1093/nar/29.23.4793
Hoorn C, Wesselingh FP (2010) Amazonia, landscape and species evolution: a look into the past
Hughes TP, Bellwood DR, Folke CS et al (2007) No-take areas, herbivory and coral reef resilience. Trends Ecol Evol
Hunter ME, Oyler-McCance SJ, Dorazio RM et al (2015) Environmental DNA (eDNA) sampling improves occurrence and detection estimates of invasive Burmese pythons. PLoS One. https://doi.org/10.1371/journal.pone.0121655
Ishige T, Miya M, Ushio M et al (2017) Tropical-forest mammals as detected by environmental DNA at natural saltlicks in Borneo. Biol Conserv. https://doi.org/10.1016/j.biocon.2017.04.023
Jeffreys AJ, Wilson V, Thein SL (1985) Hypervariable 'minisatellite' regions in human DNA. Nature 314:67–73. https://doi.org/10.1038/314067a0
Jerde CL, Mahon AR, Chadderton WL, Lodge DM (2011) "Sight-unseen" detection of rare aquatic species using environmental DNA. Conserv Lett. https://doi.org/10.1111/j.1755-263X.2010.00158.x
Jiang Z, Li P, Wang Y et al (2019) Arsenic mobilization in a high arsenic groundwater revealed by metagenomic and geochip analyses. Sci Rep. https://doi.org/10.1038/s41598-019-49365-w
Johnson MG, Pokorny L, Dodsworth S et al (2019) A universal probe set for targeted sequencing of 353 nuclear genes from any flowering plant designed using k-Medoids clustering. Syst Biol. https://doi.org/10.1093/sysbio/syy086
Jouta J, Dietz MW, Reneerkens J et al (2017) Ecological forensics: using single point stable isotope values to infer seasonal schedules of animals after two diet

switches. Methods Ecol Evol. https://doi.org/10.1111/2041-210X.12695

Kareiva P, Marvier M (2012) What is conservation science? Bioscience. https://doi.org/10.1525/bio.2012.62.11.5

Kitpipit T, Tobe SS, Kitchener AC et al (2012) The development and validation of a single SNaPshot multiplex for tiger species and subspecies identification - implications for forensic purposes. Forensic Sci Int Genet. https://doi.org/10.1016/j.fsigen.2011.06.001

Knyshov A, Gordon ERL, Weirauch C (2019) Cost-efficient high throughput capture of museum arthropod specimen DNA using PCR-generated baits. Methods Ecol Evol. https://doi.org/10.1111/2041-210X.13169

Kurtz ZD, Müller CL, Miraldi ER et al (2015) Sparse and compositionally robust inference of microbial ecological networks. PLoS Comput Biol. https://doi.org/10.1371/journal.pcbi.1004226

Li Q, You H, Xie W, et al (2019) Review in recent researches and applications of technology of environmental microbiology metagenomics in water treatment engineering. In: IOP conference series: earth and environmental science

Lloyd KG, MacGregor BJ, Teske A (2010) Quantitative PCR methods for RNA and DNA in marine sediments: maximizing yield while overcoming inhibition. FEMS Microbiol Ecol. https://doi.org/10.1111/j.1574-6941.2009.00827.x

Mahon AR, Nathan LR, Jerde CL (2014) Meta-genomic surveillance of invasive species in the bait trade. Conserv Genet Resour 6:563–567. https://doi.org/10.1007/s12686-014-0213-9

Marshall C, Sturk-Andreaggi K, Daniels-Higginbotham J et al (2017) Performance evaluation of a mitogenome capture and Illumina sequencing protocol using non-probative, case-type skeletal samples: implications for the use of a positive control in a next-generation sequencing procedure. Forensic Sci Int Genet. https://doi.org/10.1016/j.fsigen.2017.09.001

Menotti-Raymond MA, David VA, O'Brien SJ (1997) Pet cat hair implicates murder suspect. Nature 386 (6627):774

Metcalf JL (2019) Estimating the postmortem interval using microbes: knowledge gaps and a path to technology adoption. Forensic Sci Int Genet

Moore TN (2019) Nutrient enrichment promotes eutrophication in the form of macroalgal blooms causing cascading effects in two Anthropogenically disturbed coastal ecosystems. University of California, Dissertation

Mudge SM (2008) Environmental forensics and the importance of source identification

Murchie TJ, Kuch M, Duggan A et al (2019) PalaeoChip Arctic1.0: an optimised eDNA targeted enrichment approach to reconstructing past environments. bioRxiv 730440. https://doi.org/10.1101/730440

Murray DC, Coghlan ML, Bunce M (2015) From benchtop to desktop: important considerations when designing amplicon sequencing workflows. PLoS One. https://doi.org/10.1371/journal.pone.0124671

Nanney DL (1982) Genes and phenes in tetrahymena. Bioscience. https://doi.org/10.2307/1308971

Nguyen NH, Song Z, Bates ST et al (2016) FUNGuild: an open annotation tool for parsing fungal community datasets by ecological guild. Fungal Ecol. https://doi.org/10.1016/j.funeco.2015.06.006

Nichols RV, Vollmers C, Newsom LA et al (2018) Minimizing polymerase biases in metabarcoding. Mol Ecol Resour. https://doi.org/10.1111/1755-0998.12895

Ogden R (2008) Fisheries forensics: the use of DNA tools for improving compliance, traceability and enforcement in the fishing industry. Fish Fish. https://doi.org/10.1111/j.1467-2979.2008.00305.x

Ogram A, Sayler GS, Barkay T (1987) The extraction and purification of microbial DNA from sediments. J Microbiol Methods. https://doi.org/10.1016/0167-7012(87)90025-X

Orlando L, Ginolhac A, Zhang G et al (2013) Recalibrating equus evolution using the genome sequence of an early middle Pleistocene horse. Nature. https://doi.org/10.1038/nature12323

Owen DF, Owen J (1974) Species diversity in temperate and tropical Ichneumonidae. Nature. https://doi.org/10.1038/249583a0

Pauchard A, Meyerson LA, Bacher S et al (2018) Biodiversity assessments: origin matters. PLoS Biol 16: e2006686

Pechal JL, Crippen TL, Benbow ME et al (2014) The potential use of bacterial community succession in forensics as described by high throughput metagenomic sequencing. Int J Legal Med. https://doi.org/10.1007/s00414-013-0872-1

Perrotti AG, van Asperen E (2019) Dung fungi as a proxy for megaherbivores: opportunities and limitations for archaeological applications. Veg Hist Archaeobot

Petrou EL, Drinan DP, Kopperl R et al (2019) Intraspecific DNA contamination distorts subtle population structure in a marine fish: decontamination of herring samples before restriction-site associated sequencing and its effects on population genetic statistics. Mol Ecol Resour. https://doi.org/10.1111/1755-0998.12978

Port JA, O'Donnell JL, Romero-Maraccini OC et al (2016) Assessing vertebrate biodiversity in a kelp forest ecosystem using environmental DNA. Mol Ecol. https://doi.org/10.1111/mec.13481

Rodgers TW, Mock KE (2015) Drinking water as a source of environmental DNA for the detection of terrestrial wildlife species. Conserv Genet Resour. https://doi.org/10.1007/s12686-015-0478-7

Roewer L (2013) DNA fingerprinting in forensics: past, present, future. Investig Genet

Rohland N, Glocke I, Aximu-Petri A, Meyer M (2018) Extraction of highly degraded DNA from ancient bones, teeth and sediments for high-throughput sequencing. Nat Protoc. https://doi.org/10.1038/s41596-018-0050-5

Seeber PA, McEwen GK, Löber U et al (2019) Terrestrial mammal surveillance using hybridization capture of environmental DNA from African waterholes. Mol Ecol Resour 19:1486–1496. https://doi.org/10.1111/1755-0998.13069

Shapiro B (2017) Pathways to de-extinction: how close can we get to resurrection of an extinct species? Funct Ecol 31:996–1002. https://doi.org/10.1111/1365-2435.12705

Shapiro K, Silver MW, Largier JL et al (2012) Association of Toxoplasma gondii oocysts with fresh, estuarine, and marine macroaggregates. Limnol Oceanogr. https://doi.org/10.4319/lo.2012.57.2.0449

Shapiro K, VanWormer E, Packham A et al (2019) Type X strains of Toxoplasma gondii are virulent for southern sea otters (*Enhydra lutris nereis*) and present in felids from nearby watersheds. Proc R Soc B Biol Sci. https://doi.org/10.1098/rspb.2019.1334

Shelton AO, Kelly RP, O'Donnell JL et al (2019) Environmental DNA provides quantitative estimates of a threatened salmon species. Biol Conserv. https://doi.org/10.1016/j.biocon.2019.07.003

Shivanna KR (2020) The sixth mass extinction crisis and its impact on biodiversity and human welfare. Resonance. https://doi.org/10.1007/s12045-019-0924-z

Spear SF, Groves JD, Williams LA, Waits LP (2015) Using environmental DNA methods to improve detectability in a hellbender (*Cryptobranchus alleganiensis*) monitoring program. Biol Conserv. https://doi.org/10.1016/j.biocon.2014.11.016

Steneck RS, Graham MH, Bourque BJ et al (2002) Kelp forest ecosystems: biodiversity, stability, resilience and future. Environ Conserv

Stoeck T, Bass D, Nebel M et al (2010) Multiple marker parallel tag environmental DNA sequencing reveals a highly complex eukaryotic community in marine anoxic water. Mol Ecol. https://doi.org/10.1111/j.1365-294X.2009.04480.x

Taberlet P, Bonin A, Zinger L, Coissac E (2018) Environmental DNA: for biodiversity research and monitoring. Oxford University Press, Oxford

Thomas AC, Tank S, Nguyen PL et al (2019) A system for rapid eDNA detection of aquatic invasive species. Environ DNA. https://doi.org/10.1002/edn3.25

Thomsen PF, Willerslev E (2015) Environmental DNA - an emerging tool in conservation for monitoring past and present biodiversity. Biol Conserv 183:4–18. https://doi.org/10.1016/j.biocon.2014.11.019

Tillotson MD, Kelly RP, Duda JJ et al (2018) Concentrations of environmental DNA (eDNA) reflect spawning salmon abundance at fine spatial and temporal scales. Biol Conserv. https://doi.org/10.1016/j.biocon.2018.01.030

Tolimieri N, Samhouri JF, Simon V et al (2013) Linking the trophic fingerprint of groundfishes to ecosystem structure and function in the California current. Ecosystems. https://doi.org/10.1007/s10021-013-9680-1

Tsangaras K, Siracusa MC, Nikolaidis N et al (2014) Hybridization capture reveals evolution and conservation across the entire koala retrovirus genome. PLoS One. https://doi.org/10.1371/journal.pone.0095633

Tucker AJ, Chadderton WL, Jerde CL et al (2016) A sensitive environmental DNA (eDNA) assay leads to new insights on Ruffe (Gymnocephalus cernua) spread in North America. Biol Invasions. https://doi.org/10.1007/s10530-016-1209-z

Ushio M, Fukuda H, Inoue T et al (2017) Environmental DNA enables detection of terrestrial mammals from forest pond water. Mol Ecol Resour 17:e63–e75. https://doi.org/10.1111/1755-0998.12690

Wang Y, Heintzman PD, Newsom L et al (2017) The southern coastal Beringian land bridge: cryptic refugium or pseudorefugium for woody plants during the last glacial maximum? J Biogeogr. https://doi.org/10.1111/jbi.13010

Wang CY, Zhou X, Guo D et al (2019a) Soil pH is the primary factor driving the distribution and function of microorganisms in farmland soils in northeastern China. Ann Microbiol. https://doi.org/10.1007/s13213-019-01529-9

Wang P, Yan Z, Yang S, et al (2019b) Environmental DNA: an emerging tool in ecological assessment. Bull Environ Contam Toxicol

Watson J (2000) The effects of sea otters (Enhydra lutris) on abalone (Haliotis spp.) populations. In: Eorkshop on rebuilding abalone stocks in British Columbia

Williams KE, Huyvaert KP, Vercauteren KC et al (2018) Detection and persistence of environmental DNA from an invasive, terrestrial mammal. Ecol Evol. https://doi.org/10.1002/ece3.3698

Xie Y, Zhang X, Yang J et al (2018) eDNA-based bioassessment of coastal sediments impacted by an oil spill. Environ Pollut. https://doi.org/10.1016/j.envpol.2018.02.081

Yang J, Zhang X, Xie Y et al (2017) Ecogenomics of zooplankton community reveals ecological threshold of Ammonia nitrogen. Environ Sci Technol. https://doi.org/10.1021/acs.est.6b05606

Wildlife Forensic Genetics and Biodiversity Conservation: The Intersection of Science, Species Management, and the Law

Mary K. Burnham-Curtis, Dyan J. Straughan, Brian C. Hamlin, Hope M. Draheim, Teagen K. Gray Partin, and Darren J. Wostenberg

Abstract

The essential purpose of biodiversity conservation is to protect individual species and to preserve habitats and ecosystems which have declined in quantity and quality over time as a result of human activities. In some instances, laws have been passed to protect vulnerable resources from overexploitation and to support efforts that address biodiversity conservation. Enforcement of laws intended to protect natural resources can only be accomplished if it can be demonstrated that violations have occurred. With respect to biodiversity conservation, this means being able to protect both the species impacted and their habitat. The protection of individual species requires the ability to characterize species, as well as how those species are structured in animal communities. Forensic science is deeply rooted in international legal systems, dating back to the first use of fingerprints to identify deceased humans in China in the 700s. The field of wildlife forensic genetics is currently used to support wildlife law enforcement investigations. The evolution of wildlife forensic genetics closely follows the use of genetic technologies in the field of human forensic identification, with some significant differences in its application for wildlife law enforcement. In this chapter, we will provide an overview of how wildlife forensic genetics intersects with biodiversity conservation efforts and provide case studies which illustrate the diverse application of forensic genetics to the enforcement of wildlife conservation laws.

Keywords

Wildlife trafficking · Wildlife forensics · Conservation genetics · DNA · Species identification

Introduction

The vast majority of wildlife importation violations I find are the result of ignorance. People simply don't know they can't bring some wildlife items into the U.S. Some people often don't make the connection that what they have is wildlife; they claim jewelry, clothing, or medicine on their Customs forms and are genuinely surprised when I inform them that they actually are importing wildlife pieces. USFWS OLE Wildlife Inspector

A Brief History of Wildlife Forensics

Forensic science, the application of science to the rule of law, has deep roots in international legal

M. K. Burnham-Curtis (✉) · D. J. Straughan · B. C. Hamlin · H. M. Draheim · T. K. Gray Partin · D. J. Wostenberg
U.S. Fish and Wildlife Service, OLE National Fish and Wildlife Forensic Laboratory, Ashland, OR, USA
e-mail: mary_curtis@fws.gov

S. C. Underkoffler, H. R. Adams (eds.), *Wildlife Biodiversity Conservation*,
https://doi.org/10.1007/978-3-030-64682-0_8

systems. The first use of fingerprints to identify humans reportedly occurred in China in the 700s, but documented use of fingerprints for identification dates back to the late 1800s (Galton 1892; Ashabaugh 1999; Caplan 1990). Forensic science disciplines have both spurred and followed the development of new techniques and technologies, progressing from simple comparisons of visible morphological character states to the use of state-of-the-art instrumentation. Some disciplines of forensic science grew directly out of the needs of law enforcement and are now working to provide foundational science to support their conclusion (e.g. Ulery et al. 2011, 2012), whereas others, like forensic genetics (both wildlife and human), chemistry, wildlife forensic morphology, and anthropology, are applications of technology from existing research fields with deep scientific and theoretical underpinnings (Bell et al. 2018). Despite the genesis of each forensic discipline, the aim of examiners is to uncover links between the suspect, the victim, and the scene of the crime. This is commonly referred to as "the forensic triangle." Forensic science has more recently come into use to support wildlife law enforcement investigations in the wake of high profile human crimes that brought science and statistics into the courtroom (Breyer 2000). The passage of laws and treaties on a national and international scale, including the Lacey Act in 1900, the Migratory Bird Treaty Act (MBTA) in 1918, the Bald and Golden Eagle Protection Act (BGEPA) in 1940, the Marine Mammal Protection Act (MMPA) in 1972, the Convention on International Trade in Endangered Species of Wild Fauna and Flora (CITES) in 1973, and the Endangered Species Act (ESA) in 1973, as well as the expansion of wildlife crimes into multinational trafficking operations, has highlighted the need for the use of science in the courtroom to combat wildlife crimes (Anderson 1995; Burnham-Curtis et al. 2015). In particular, wildlife forensic genetics has emerged as an indispensable tool for these efforts.

The first well-recognized wildlife forensics case occurred in nineteenth-century France, for which the methodology was published by Robin and Salmon (1857). In that case, an individual was thought to have been murdered, and the primary evidence was a blue smock suspected to be stained with blood and mud. The suspect's alibi was that the blood was from a butchered duck. At that time early blood type databases only included human samples, and there were no databases available for non-human animal blood types. The scientist investigated antibody reactions with known human blood, known waterfowl blood, mud, and clothing dye, in an effort to support a forensic analysis which concluded that the blood on the smock was of human origin. This published account represents one of the first instances in which scientific investigation was used to validate an analytical technique for use in the identification of an animal involved in a crime. Since then, wildlife forensics has evolved to answer ever more complex questions in legal investigations and to intersect more often with conservation activities.

The purpose of this chapter is not to present the reader with a comprehensive tutorial on how to conduct wildlife forensic genetic analyses, but rather to present a discussion of the ways that wildlife forensic genetics intersects with biodiversity conservation. Conservation of species often requires regulatory intervention, which ultimately leads to legal considerations (Breyer 2000). We will briefly present the foundational information about wildlife forensic genetic methods and procedures, followed by a series of case studies that demonstrate the diverse ways in which wildlife forensics and biodiversity conservation intersect.

The Link Between Wildlife Forensics and Biodiversity

"Wildlife crimes" are defined as actions that threaten wild animals, plants, or their habitats and which constitute a breach of regional, national, or international law. Wildlife crime broadly falls into four categories (Huffman and Wallace 2012): illegal take or poaching; illegal possession; illegal trade, import, or export; and harm or harassment of protected species. Offenses can range from minor issues such as subsistence hunting for animals without a permit to major issues involving criminally organized international trafficking of endangered species. The global reach of the Internet has exacerbated these problems by creating a novel illicit marketplace that is easily accessible by buyers and sellers alike (TRAFFIC 2019). Wildlife is one of the most profitable subjects of illegal trafficking, alongside drugs, arms, timber, and human trafficking, and is often facilitated by organized crime syndicates (World Wildlife Fund 2018) (Fig. 1).

Illegal trade in wildlife and their products is a constant threat to biodiversity, particularly for those species which are endemic, rare, or already threatened or endangered—each of these conditions makes a particular species much more desirable in the target marketplace. Illegal wildlife products support a variety of consumer desires, both practical and ornamental (TRAFFIC 2019), including: (1) remedies provided by traditional Eastern Asian medicine (e.g., traditional Asian medicines or TAMs); (2) food for human consumption (e.g., bushmeat, native cultural dishes, caviar); (3) symbols of wealth (e.g., ivory, rhinoceros horns, hunting trophies); (4) tourist souvenirs (e.g., framed insects, tortoise shell jewelry, coral jewelry); and (5) the live pet trade (e.g., reptiles, tropical fish, wild cats). In the USA, wildlife forensic science is conducted by a variety of different entities at the federal, state, private, or local (universities) levels, each of which is focused on a different aspect of wildlife crime (see Box 1).

Many countries have laws and regulations to monitor legal wildlife trade as a direct consequence of human-mediated activities that commercialize wildlife resources. Wildlife forensics has evolved to aid in the enforcement of laws and regulations (including exonerations), and wildlife forensic genetics in particular has emerged as a critical tool for the identification of animal and plant species, populations, and individuals in wildlife law enforcement investigations (Linacre and Tobe 2013; Burnham-Curtis et al. 2015; Dormontt et al. 2015; CITES 2019). The existence of such laws and regulations is a direct result of the recognition that the earth's natural resources are not an unlimited commodity, and protecting those resources requires efforts to conserve the diversity of plants and animals that in some cases are endemic to specific areas of the earth. This is the foundational link between biological conservation and wildlife forensic genetics.

The intersection of wildlife forensic genetics and biodiversity conservation is anchored in a network of connections to research, wildlife management, and law enforcement communities (Fig. 2).

The research community offers access to state-of-the-art techniques that can be applied to forensic analyses. These techniques are used, often with some modification, to address scientific validation required by the judicial system in the USA, in line with Federal Rules of Evidence (U.S. Pub. L. 93–595, §1 1975). Collaboration with the wildlife management community provides access to applied information about species biology and ecology (e.g., population abundance, migration patterns, species distribution) which may impact statistical analysis and data interpretation, as well as estimation of error for forensic analyses. Wildlife forensic analysts, in turn, inform the research and management communities about species they encounter that are underrepresented in biological research or species trafficking trends that may be undetected by wildlife management agencies.

For example, the trade in exotic pets is dominated by trade in rare reptiles, but the taxonomic status of many reptile species is commonly disputed by experts, or in flux due to a lack of knowledge of cryptic or rare species in the wild (e.g., Ogden et al. 2009). Some of these species are provided as evidence in wildlife trafficking investigations, and wildlife forensic scientists may be unable to provide a conclusive species identification due to lack of knowledge in the field at large. If the purported species is of conservation concern, additional specimens obtained through wildlife trafficking seizures provide crucial information to the biodiversity conservation community about trends in exploitation and population abundance that can be investigated further. Important to note is the critical link between

Fig. 1 Collection of confiscated illegal wildlife products. Photo credit: USFWS

taxonomic experts who characterize species with morphological characteristics and population geneticists who identify diagnostic genetic characteristics. If taxonomic classifications are modified in one field, they must also be recognized and modified in the other, particularly if the species is of conservation or legal concern.

Another example involves species succession and substitution in the wildlife trade marketplace. Caviar has been a prized delicacy on the international market for centuries—supported primarily by US production in the nineteenth century and more recently by Russian production, after the decline in North American sturgeon populations (Harris and Shiraishi 2018). Continued extraction

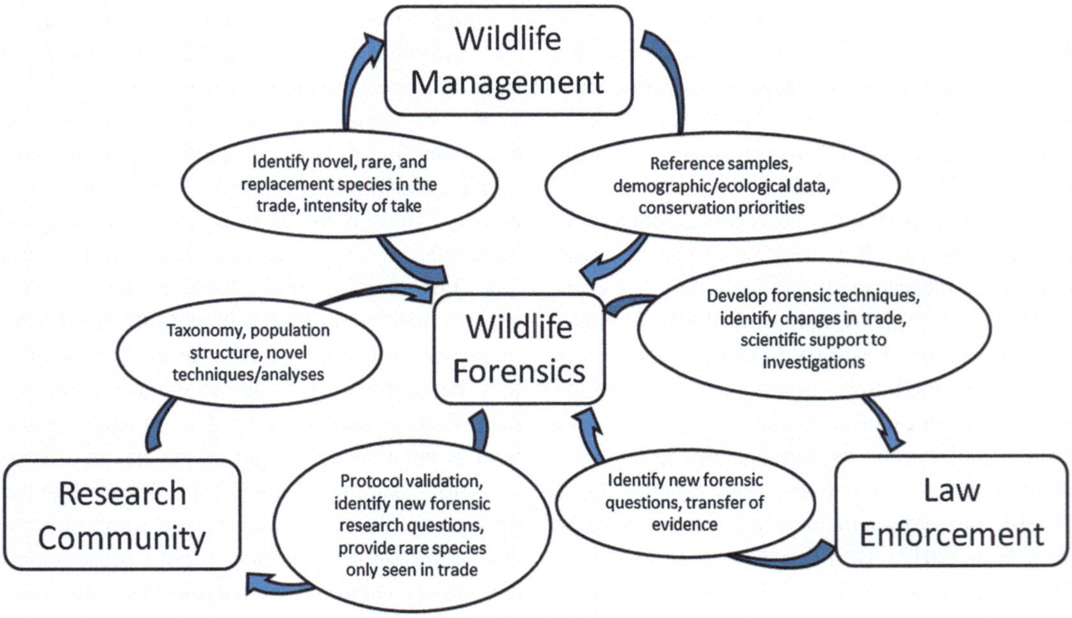

Fig. 2 Wildlife forensics works at the intersection of science, wildlife management, and the law

of sturgeon for the caviar market led to precipitous declines in populations of virtually every species of sturgeon in the twentieth century, to the point where the trade of all extant sturgeon species was regulated under CITES in 1998 (Williamson 2003). Since then, at least four species of sturgeon have gone extinct, and several others are at risk of extinction, prompting international efforts for sturgeon species conservation (Williamson 2003). However, the illegal market did not go along with conservation efforts and instead shifted their focus to newly recovered sturgeon and paddlefish populations in North America. In the early 2000s, international organized crime syndicates shifted to North American sturgeon and paddlefish caviar as substitutes for Eurasian caviar in several high-profile poaching cases, which were resolved with the support of wildlife forensic genetic analyses (Profita 2015; Knight 2017). Species substitution and species succession in the commercial markets are not a novel occurrence (Smith 1968), and the information provided by the wildlife forensic community related to exploitation trends is invaluable to the biodiversity conservation community (Williamson 2003; Bailey 2003; Profita 2015).

Biodiversity Conservation Research and Wildlife Forensic Genetics: Similarities and Differences

The identification of species and their abundance and distribution are the intersection of wildlife forensics and biodiversity conservation. Conservation efforts seek to protect and preserve species for posterity, and their success depends upon the ability to accurately characterize the species of interest, as well as to determine how that species is unique in regard to its biological and ecological characteristics. In a legal framework, wildlife forensics provides scientific data that can be used in the courts when conservation efforts encounter legal conflicts (Anderson 1995; Burnham-Curtis et al. 2015). Criminal and civil prosecutions rely on DNA data with increasing frequency for both human and wildlife crimes. Unlike human forensics, evidence in wildlife forensics comes from a multitude of species; therefore, wildlife forensic geneticists need to have an understanding of animal biodiversity, taxonomy, phylogenetics, and population genetics.

The definition of some terms should be clarified within the context of how genetic technology is applied to wildlife forensic science, primarily to distinguish the intent of wildlife forensic genetic analyses from human forensic genetic analyses. These include the terms "identification" and "individualization." In wildlife forensic analyses, identification is used to refer to the taxonomic classification of an organism (plant or animal), ideally to the lowest level possible, e.g., species or subspecies. In wildlife forensic genetics, this is most often accomplished using DNA sequence analysis and sometimes serological analyses (AAFS Standards Board 2019a, b). The term individualization is used to describe the classification of a specific organism as unique among other conspecific organisms, i.e., when comparing a gut pile from a kill site to meat in a freezer, to evaluate the probability that the two samples originate from the same animal or from different, unrelated individuals. Individualization in wildlife forensics is only conducted when it is possible to assign a measurement of error to the comparison (e.g., likelihood ratios or match probabilities) and is most often conducted using lineage markers such as simple tandem repeats (STRs) or single nucleotide polymorphisms (SNPs) (AAFS Standards Board 2019b, c). Under those criteria, individualization in wildlife forensics is currently only conducted with DNA-based techniques. STR and SNP analyses are also used for population genetic applications, such as determining the population origin or geographic provenance of a specific item, or the presence of admixtures, as with anadromous fisheries (Kalinowski 2004; Anderson 2010).

Population Genetics

Wildlife forensics requires knowledge of the fundamental principles of population genetics such as the structure and function of DNA, methods for nucleic acid detection, Mendelian inheritance, the roles of mutation and selection, and estimation of population genetic metrics. Population genetics is the study of genetic variation within and among populations and the examination of the evolutionary influences that drive the observed variation (Hartl 2000). With the advent of molecular technologies, the field of conservation genetics has developed in which the biologist uses population genetic analysis to address species conservation and management questions (reviewed in Hedrick 2001; Palsboll et al. 2007; Allendorf et al. 2013). Wildlife forensic genetics uses standard population genetic techniques in the process of analyzing unknown evidence items to determine the species or population origin, individual identity, or relationship between samples. The bench and analytical techniques are the same for both conservation population genetics and wildlife forensic genetics, as is the fundamental importance of peer review for novel methods and research results. The difference between research population genetics and wildlife forensic genetics lies in the ultimate goal of the analysis. Whereas research studies characterize species abundance, biogeography, and evolutionary history, forensic analyses use those characterizations to answer legal questions about the identity of evidence items, where the underlying scientific analysis fulfills the criteria required for acceptance of scientific evidence in the courtroom (Title 28, U.S. Pub. L. 93-595, §1 1975). In order to satisfy these criteria, wildlife forensic genetics goes a step further, through protocol validation and standardization of methods and interpretation guidelines (e.g., Title 28, U.S. Pub. L. 93-595, §1 1975; Daubert v Merrill Dow Pharmaceuticals 509 U.S. 579, 113 S. 1993; AAFS Standards Board 2019a, b, c, d, e).

Mitochondrial DNA (mtDNA) is most often used in wildlife forensics as a genetic marker for species identification. The mtDNA genome contains roughly 37 genes across approximately 17,000 base pairs. It is maternally inherited and present in thousands of copies within a single cell. Taxonomic and phylogenetic studies of mtDNA often reveal a minimal level of intraspecific variation and sufficient interspecific variation to allow for assignment of species identity with high certainty (Moritz et al. 1987). The loci commonly used in wildlife forensic studies for such classifications are the cytochrome *b* (Cyt *b*) and the cytochrome *c* oxidase 1 (*COI*) genes. Other mitochondrial genes (e.g., 12S and 16S rRNA loci, the NADH family of genes, or the D-loop of the control region) have been employed when the intraspecific sequence variation of a common target locus has insufficient resolution at the taxonomic level of interest with Cyt *b* or *COI*. While species identification methodologies are generally standardized in the discipline (Linacre and Tobe 2013; AAFS Standards Board 2019a, c), the specific genetic approach for species identification is dependent on the taxon, the variability of the genetic marker, ecological factors, and evolutionary influences that impact how genetic variation is distributed within and among species (Fig. 3).

Nuclear DNA (e.g., autosomal and sex chromosomes or Y-STRs) is the source for genetic markers (lineage markers) used to answer investigative questions about parentage, sex, relatedness, population or geographic origin, and hybrid origin (Karl and Avise 1993; Zhang and Hewitt 2003; Nagy et al. 2012). Simple tandem repeat (STR) loci are fragments of DNA sequence that vary in length among conspecific individuals based on the number of repeated 2–6 base segments of nucleotides. A single nucleotide polymorphism (SNP) is a variation at a single nucleotide position on a targeted DNA sequence. The resolution for STR and SNP markers is high and enables identification at the level of the individual. Suites of STR loci and SNP loci are commonly used lineage markers for the purpose of determining relatedness between and among individuals and populations, including parentage analyses. In conservation genetic research, these types of genetic markers are used to characterize species populations, determine sex of the individual, investigate processes such as migration and introgression, and determine population origin of admixed individuals (Zhang and Hewitt 2003; Courtios et al. 2003). The high level of resolution characteristic of nuclear DNA markers requires appropriate reference databases, often comprised

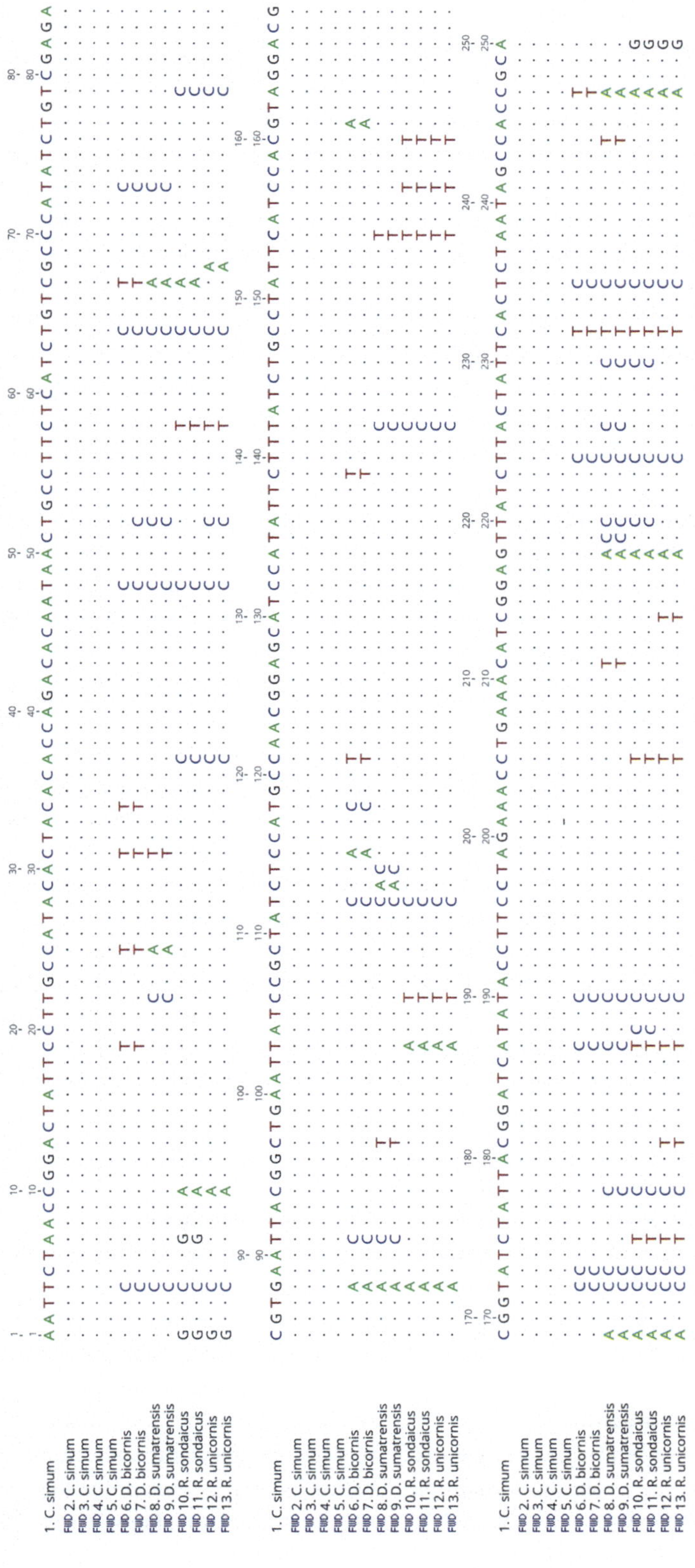

Fig. 3 Sequence alignment of a portion of the mtDNA cytochrome *b* locus used to discriminate among extant species of Rhinocerotidae. Interspecific sequence variation allows discrimination among the different rhinoceros species, while low intraspecific sequence variation can be used in comparison with an unknown item for species identification

of single species or limited groups of very closely related species (Banks et al. 1999; Waits et al. 2001; Smith and Seeb 2007; Jobin et al. 2008; Cronin et al. 2009). The range of species for which STR analysis is conducted in wildlife forensic applications is limited to those that appear most often in wildlife investigations, the most common being recreationally hunted animals (e.g., deer, elk, bear) and other charismatic macrofauna (e.g., eagles, wolves, cougar). The ability of STR and SNP markers to provide resolution for population-level identification makes them ideal candidates for future development in determining geographic provenance of animals involved in international wildlife trafficking operations.

High-throughput sequencing platforms such as next-generation sequencing (NGS) and massively parallel sequencing (MPS) are emerging as novel tools in both conservation and wildlife forensic genetics. These platforms are capable of producing large amounts of sequence data that can be used to identify low copy number DNA present in the environment (e.g., environmental DNA or eDNA) or in wildlife products. Complex bioinformatic data analysis is required to produce appropriate depth and coverage of substrate genomes. Criteria for data integrity that would fulfill the requirements for acceptance of this scientific evidence in court have been compiled for use in human forensic identification (SWGDAM, U.S. Dept. Justice 2019). However, the complexities of dealing with multiple species mean it has not been routinely applied in the wildlife forensic field as of the date of this chapter publication. NGS is increasingly being applied as a more efficient and less expensive method of genetic marker development that may have forensic applications. Discovery of new SNP markers could improve and/or resolve species identification questions that have been historically challenging (Ogden 2011).

Phylogenetics and Evolution

Wildlife forensics requires knowledge of the taxonomy, phylogenetics, and evolution of multiple species, genera, and families. In contrast to human DNA forensics, the identity of evidence in a wildlife forensic investigation could involve multiple species, rather than just one (e.g., *Homo sapiens*). Knowledge of how animal species are related at different taxonomic levels informs the choice of the appropriate test to answer the investigative question, as well as proper and accurate interpretation and reporting of the analytical results. While many species identification questions can be addressed with routine sequence comparison, more challenging questions involving poorly resolved or poorly characterized taxa, as hybrid origin, or subspecies classification cannot be answered without a basic knowledge of phylogenetic principles (Wright 1978; Watrous and Wheeler 1981; Felsenstein 2001; Wiley et al. 2013). Wildlife forensic geneticists are often confronted with unknown items for species identification that do not have representation in reference databases (laboratory or public databases). Knowledge of phylogenetic principles is crucial in order to make educated choices when comparing unknown DNA sequences to those from closely related species. Similarly, knowledge of phylogenetic principles is necessary to appropriately address the potential existence of cryptic species, subspecies, and hybrids, to avoid misidentifications that could have legal ramifications (Jacobs and Baker 2018; CITES 2019; Endangered Species Conservation Act 1969). Knowledge of appropriate taxonomic authorities is also crucial for accurate analysis (e.g., Uetz and Etzold 1996; Froese and Pauly 2000; Wilson and Reeder 2005; Dickinson and Remsen 2014, ITIS http://www.itis.gov), as is maintaining familiarity with taxonomic revisions as they occur (Simkins et al. 2019).

Statistical Analysis

Statistical analysis is a cornerstone of scientific investigation and provides a measure of confidence in the interpretation of analytical results. The judicial system relies on estimates of error to gauge the robustness and reliability of a particular scientific test. Therefore, it is crucial for wildlife forensic geneticists to understand the role of statistical analyses in examination of evidence and its proper implementation (Weir 1996; Evett and Weir 1998).

Forensic analyses at the level of populations and individuals (e.g., geographic assignment or individual identification) are conducted primarily with lineage markers such as STRs. STR analyses are easily set up in a hypothesis testing framework, in which prior and post probabilities can be estimated and compared to acquire likelihood ratios (LRs) and match probabilities (Paetkau and Strobeck 1994; Weir 1996; Evett and Weir 1998). In contrast to strictly academic studies of population genetics, forensic genetic analysis of evidence that takes population structure into account must incorporate not only biologically relevant geographic boundaries into the statistical models but also boundaries that may be relevant to law enforcement (Ogden and Linacre 2015). While statistical analysis of lineage markers for wildlife populations follows the general lead of human forensics with the use of LRs and associated estimates of error, there is no consensus approach to data analysis for wildlife forensic geographic assignment. This is primarily due to the fact that analyses and databases for different species, or closely related species groups, must be developed separately and uniquely. Population characteristics that influence calculations of population subdivision vary across different taxonomic groups (Ogden et al. 2009; Ogden and Linacre 2015). Factors such as demographics, inbreeding, migration, translocation, historical introgression, and admixture can confound efforts to determine the relationship between population genetic boundaries and corresponding geographic boundaries (Waits et al. 2001; Kalinowski 2004; Ogden et al. 2009). To address these issues, markers are often chosen to maximize polymorphism for the purpose of obtaining greater power for geographic or individual identification (Anderson 2010). Highly polymorphic markers also provide a buffer in the event that an incomplete suite of loci can be amplified from compromised or low copy number substrates commonly provided as wildlife evidence.

Population genetic databases intended for use in forensic applications should have loci that are well characterized for standard population genetic metrics, including heterozygosity, linkage disequilibrium, Hardy-Weinberg equilibrium, allelic richness, allelic diversity, population subdivision (e.g., F_{ST}), polymorphic information content, and statistical power (Wright 1978; Weir 1996; Hartl 2000). For wild populations for which subdivision is measurable, inbreeding coefficients should also be estimated, as the level of inbreeding will influence estimation of individual assignment metrics. In such cases, a theta value calculated from the inbreeding coefficient is added to probability of identification estimates (Weir 1996). All of this information is used to produce a match probability or LR for presentation in court, if needed. At the most basic level, these estimates provide the court with information about the likelihood that two different, unrelated individuals could have contributed evidence samples that share the same composite genotype. If the legal question is whether the suspect, crime scene, and animal victim can be connected, evidence gathered from a crime scene (e.g., illegal hunting site) and evidence collected from a suspect (e.g., blood on a vehicle or meat in a freezer) are analyzed to determine if they share the same multilocus genotype. The LR then provides an estimate of the probability that items which share the same multilocus genotype may have originated from two unrelated individuals. For species which have high genetic diversity, the LR will likely be exponentially low, in the range of one in a billion, trillion, or more. When compared to the estimated population size of the species of interest, the conclusion drawn is that the evidence items had the same contributor.

Sequence comparison is one of the most common methods used in wildlife forensic genetics for taxonomic identification. Since taxonomic identification is a form of classification and not individualization, and there are a number of biological and legal factors that inform identification, it is not appropriate to assign likelihood ratios or other statistical metrics to describe the surety of the classification (Moore and Frazier 2019; AAFS Standards Board 2019e). Irrespective of the locus used for taxon identification, sequences obtained from PCR and direct sequencing of unknown evidence items will be compared to those of one or more reference sequences from known species to look for similarities. For example, if the comparison of the unknown sequence is identical at all base positions (100% homologous) to a known sequence, and 95–98% similar to the next closest species (a difference commonly referred to as the "barcode gap"), the taxonomic identity of the unknown sample is usually assigned as that of the species with the greatest similarity, e.g., the one with 100% homology in this example. Alternatively, if the sequence comparison is conducted in the framework of a phylogenetic analysis, bootstrap estimates from that analysis can be used to inform the identification (Hillis and Bull 1993). However, confidence in species determination is dependent on a few critical assumptions for which published taxonomic studies provide crucial supporting data. First, the database used for comparison to the unknown sequence should represent all possible species in the taxon of interest (e.g., all species in a genus). Second, the sequence chosen for comparison is a locus for which the intraspecific variation of the species of interest is much less than the interspecific variation among conspecifics. However, despite the rapid rate of sequence accumulation in public databases such as GenBank (Benson et al. 2013) and the Barcode of Life (Ratnasingham and Hebert 2007), these assumptions are violated if there are too few sequences available for comparison with the taxon of interest. Hybrids in a taxonomic group present an additional challenge for species identification with mitochondrial DNA sequencing because only the maternal DNA is characterized. For items of suspected hybrid origin, nuclear DNA targets need to be developed if sequence comparisons are being made. The wildlife forensic science professional community has produced standards and guidelines for the conduct of DNA sequence analyses and interpretation of results (AAFS Standards Boards 2019a, b, c, d, e).

Cautionary Points

While the application of genetic techniques is well suited to wildlife forensic analysis, there are some aspects of genetic analyses that should be treated with caution. These include (but are not limited to) low copy number DNA samples, contamination issues, presence of exogenous DNA, stochastic versus genetic effects, nuclear mitochondrial DNA (NuMTs), hybrids and subspecies, gene order changes among animal taxa, database issues such as insufficient representation, use of public databases, examination of rare species for which no data is available, and cognitive bias.

Contamination is a major concern when working with low copy number DNA, as well as compromised samples, which are often exposed to the environment and contain exogenous DNA. For these instances, the use of positive and negative controls in all stages of analysis is critical and is foundational across all forensic DNA protocols (Federal Bureau of Investigation DNA Advisory Board 2000). Positive controls are known DNA samples that have been previously characterized with the marker set used for analysis. A successful result for the positive control provides evidence that the analytical step (PCR, sequence analysis, STR analysis) was conducted accurately and correctly. If the positive control does not work in any stage of the reaction, or produces an unexpected result, the analyst knows that an error occurred in the process, and the error should be investigated. A negative control is a sample included in an analysis that by intent does not contain DNA template. Negative controls provide important information about the integrity of the stages of analysis, from DNA extraction through data interpretation. When DNA product is detected in a negative control at the post-extraction quantitation step, the PCR amplicon detection step, or the sequencing or STR analysis step, the control fails. This failure may indicate contamination of the original DNA eluate, reagents used during the analytical process, or other sources of exogenous DNA. Clean laboratory techniques, such as washing lab surfaces with a bleach solution or DNA elimination product after use, using laminar flow hoods to prevent cross-contamination among samples, and separating pre- and post-PCR activities, are essential parts of standard laboratory practice.

Artifacts of genetic origin during PCR may confound wildlife forensic analysis. Such artifacts include genetic mutations in primer binding sites

(Hillis et al. 1996), nuclear mitochondrial DNA (NuMT, Richly and Leister 2004), and gene order changes among taxa (Mindell et al. 1998). Little can be done to avoid such artifacts, except to be aware of their potential. This sort of knowledge informs modifications to analytical protocols in order to minimize the influence of these artifacts.

The absence of data from local or public databases is often the result of the lack of samples of rare or endangered species due to low population abundance or the inability to obtain sampling, export, or import permits. In these instances, museum collections may provide sources of physical material for genetic analysis (McCormack et al. 2017). When all else fails, the wildlife forensic geneticist's knowledge of phylogenetic principles enables *them* to piece together information from closely related species for which data is available, and often an identification can at least be determined at the genus level.

Finally, cognitive bias is emerging as an important factor for forensic scientists to consider in the conduct of their work, e.g., confirmation bias, situational bias, expectation bias, or personal bias (Jeanguenat et al. 2017). Objectivity is crucial in all scientific investigations, particularly in forensic science, and acknowledging the potential for bias in an investigation, whether it is pure research or a forensic analysis, is the first step in preventing bias from exerting undue influence on scientific decision-making. Training and rigorous technical review processes should also be implemented to help catch bias-driven mistakes.

Standards and Guidelines

The ultimate goal of forensic science is the presentation of scientific evidence in a court of law. What differentiates a wildlife forensic geneticist from a conservation geneticist is that the wildlife forensic geneticist is qualified as an expert witness in a courtroom. This can be a tricky issue, as the presiding judge in a trial is the gatekeeper for determining what evidence is or is not admitted in the proceedings, as well as which experts are admitted to provide testimony. Federal Rules of Evidence provide guidance for the acceptance of scientific analyses and qualification of expert witnesses (Title 28, U.S. Pub. L. 93–595, §1 1975). In general, forensic science data is accepted if it meets the Daubert criteria (Frye v United States, 293 F. 1013, 1923; Daubert v Merrill Dow Pharmaceuticals Inc., 509 U.S. 579, 1993; Fenner 1996; Kumho Tire Co. v Patrick Carmichael, 119 S. Ct. 1167, 1999): (1) the technique has been tested; (2) the technique has been subjected to peer review; (3) the potential error rate of the technique is known or can be tested; (4) the technique is standardized with validated protocols; and (5) the technique is widely accepted in the relevant scientific community. A foundation of the scientific method is the peer review process which provides other scientists with the necessary information to repeat an experiment for confirmation or improvement. Forensic scientists must go one step further and create validated and standardized protocols for the analyses they intend to provide for legal investigations. Since the judge is the gatekeeper who determines the admissibility of evidence in a trial, and the judge is not necessarily an experienced scientist, validation and standardization provide the court with a way to evaluate the reliability and accuracy of a scientific method as they weigh its suitability for use in legal judgments.

To address the need for standardized methods, the forensic science community, including the wildlife forensic science community, has participated in efforts to construct a series of standards and guidelines that address best practices in both science and the application of that science to the rule of law (Society for Wildlife Forensic Sciences [SWFS] Technical Working Group 2018; AAFS Standards Board 2019a, b, c, d, e). Laboratory accrediting bodies, such as the Association of Crime Laboratory Directors-Association of National Accrediting Bodies (ASCLD/ANAB), use these standards, along with ISO 17025 criteria for testing laboratories (International Standards Organization ISO/IEC 2017) to evaluate forensic science providers, including those doing wildlife forensic analyses. Wildlife forensic geneticists also participate in proficiency testing and certification programs (SWFS, http://www.wildlifeforensicscience.org/programs), all of which provide additional confidence to the courts that analyses are being conducted with integrity, following standard procedures to minimize error.

Box 1 Examples of Forensic Testing Laboratories That Conduct Wildlife Forensic Genetics Analyses

Side Box A: Examples of wildlife forensic laboratories

Department of Interior, US Fish and Wildlife Service at the OLE National Fish and Wildlife Forensic Laboratory in Ashland, Oregon (USFWS OLE-NFWFL)—service to federal, state, and international wildlife law enforcement; ISO 17025 accredited (ANSI/ANAB); multi-disciplinary—including genetics, veterinary pathology, morphology, and chemistry.

Department of Commerce, National Oceanic and Atmospheric Administration, National Marine Fisheries Service, Northwest Fisheries Science Center (NOAA-NFSC) in Charleston, South Carolina and Seattle, Washington—service to federal and international wildlife law enforcement; primary discipline is genetics of marine taxa, with a focus on protected species and species of commercial or recreational interest.

Wyoming Game and Fish Department, Game and Fish Laboratory in Laramie, Wyoming—service to Wyoming and other western US states; primarily fish disease analyses and genetics of state recreational game and fish species.

University of California at Davis Veterinary Forensic Genetics Laboratory—private, university-sponsored, ISO 17025 accredited (ANSI/ANAB) laboratory that provides fee-based wildlife forensic genetic analyses for civil and criminal investigations including animals involved in human crimes and animal abuse cases.

DNA Solutions, Norman, Oklahoma—private, for-profit, ISO 17025 accredited (ANSI/ANAB) genetic analysis laboratory that offers services for animal identification for wildlife law enforcement investigations as well as animal husbandry and private civil investigations.

Thailand Department of National Parks, Wildlife Conservation Office, Wildlife and Plant Conservation—Wildlife Forensic Science Unit in Bangkok, Thailand, and Department of Wildlife and National Parks (PERHILITAN)—Wildlife Forensic Laboratory in Kuala Lumpur, Malaysia are wildlife forensic laboratories at the forefront of the international response to wildlife trafficking and conduct morphological and genetic identification of wildlife species.

How Wildlife Forensic Genetics Intersects with Biodiversity Conservation: Case Studies

Wildlife forensic genetics and biodiversity conservation are intimately linked through the national and international laws and treaties that protect natural resources, particularly for species at risk. The remainder of this chapter will provide some real-world examples of how wildlife forensic genetics has been used to provide scientific support to investigations of crimes that violate laws created to protect our biological resources (Fig. 4).

The Endangered Species Act (ESA)

Since the turn of the twentieth century, the USA has passed dozens of major wildlife laws and officially entered into three international treaties dealing exclusively with the protection of endangered species (MBTA 1918; CITES 1973; United Nations 1982). The ESA is perhaps the most widely known wildlife protection law in the USA, and it protects close to 2000 different species of plants and animals by making it illegal to export, import, extract, take, possess, or transport any threatened or endangered species. The ESA is

Fig. 4 Photos of different evidence types processed at the USFWS OLE National Fish and Wildlife Forensic Laboratory (left to right: assorted gall bladders, confiscated eagle parts, analyst sampling the scales of an Asian arowana *Scleropages formosus*). Photo credit: USFWS OLE-NFWFL

unique in several aspects, including language that allows for the protections of the ESA to be applied to critical habitat necessary for protected species survival. The ESA also allows for implementation of provisions from the CITES treaty that have been approved through periodic review by the Conference of Parties (COP or signatory countries). Other similarities between the ESA and CITES (discussed below) help enhance the ability of US and international conservation groups and law enforcement organizations to protect both US and international species of concern. Violations of the ESA are prosecuted through civil suits, or as criminal felony violations, which could lead to imprisonment and forfeiture of possessions. Criminal violations are prosecuted by attorneys from the US Department of Justice Environment and Natural Resources Division (USDOJ-ENRD) as well as US Attorneys' offices. Wildlife forensic genetics plays a role in enforcement of the ESA in the interests of biological conservation by providing analyses to determine species and individual identity, as well as geographic provenance determination for protected species, subspecies, and distinct population segments (DPS) (Waples 2006).

ESA Case Study: Wildlife Forensic Identification of North American Gray Wolf Populations

One of the more widely recognized endangered species is the North American gray wolf (*Canis lupus*) (Fig. 5).

Fig. 5 North American gray wolf, *Canis lupus*. Photo credit: USFWS

Gray wolves were initially listed on the ESA (Endangered Species Preservation Act of 1966) in 1967 after they were extirpated from the contiguous USA, with the exception of a small number of animals in Minnesota. Following the passage of the ESA in 1973, North American gray wolves were classified as endangered, with the Minnesota population classified as threatened. In 1994, when experimental populations were introduced into the Yellowstone National Park (YNP) and other locations in Idaho, biologists and

(continued)

managers recognized the need for population assignment capabilities (Wayne et al. 1992; Chambers et al. 2012). Due to the varying degrees of ESA classifications, the DPS from which a deceased wolf originated would determine whether or not violations of the ESA had occurred. For example, a dead wolf found in North or South Dakota could plausibly originate from any of three populations (Minnesota, YNP, or Canada). If the death was not of natural causes, different penalties could be applied depending on the source population. In this instance, wildlife forensic science could provide answers to multiple questions: (1) how did the animal die; (2) what species is it; and (3) from what population did it originate? When a wolf arrives at the USFWS OLE-NFWFL, the animal is sent to the veterinary pathology lab to determine cause of death (COD). COD can include natural causes (e.g., disease, injury, starvation, killed by other predators) or human-mediated causes (e.g., collision with a car, poisoned, shot). After a necropsy is performed, the carcass may be sent to the morphology lab where examinations may provide insight as to whether the animal was captive (e.g., wear on foot pads, tartar buildup on the teeth) or free-ranging. Ultimately, tissues from the animal are sent to the genetics lab, where a combination of mitochondrial DNA sequencing, nuclear STR, and Y-STR analyses is used to first identify the species of the individual (wolf, dog, coyote, or hybrid). If the animal is determined to be a wolf, statistical assignment methods are used to determine the DPS origin of the individual, which may have bearing on what charges are ultimately filed, if any (USDOJ Press release 2019). These same techniques are also used for other species of conservation concern, particularly those managed for recreational hunting and fishing, e.g., salmon, bear, deer, and elk (Banks et al. 1999; Waits et al. 2001; Cronin et al. 2009).

CITES: The Convention on International Trade in Endangered Species of Wild Fauna and Flora

The CITES treaty is "an international agreement between governments" with the intent "to ensure that international trade in specimens of wild animals and plants does not threaten their survival…" (CITES 1973). International governments recognize that trade in wild animals and plants crosses international borders and the effort to regulate it requires international cooperation to safeguard certain species from overexploitation. CITES was conceived in the spirit of international cooperation among governments and wildlife conservation groups, and today it accords varying degrees of protection to more than 35,000 species of animals and plants, whether they are traded as "…live specimens, fur coats, or dried herbs…" (https://www.cites.org/eng/disc/what.php). Both the ESA and CITES afford different levels of protection based on a species' likelihood of extinction. Similar to the ESA status of endangered, CITES species listed as Appendix I are species considered to be on the brink of extinction and for which commercial trade is prohibited. Exceptions to this may be allowed when the purpose for importation is not commercial, such as for scientific research or law enforcement purposes. Species listed on CITES Appendix II as well as species labeled as threatened under the ESA are species which may not be on the brink of extinction now, but could be in the foreseeable future. CITES also has a third category, Appendix III, which regulates trade in certain species at the request of a country which already regulates their trade, but needs the cooperation of other countries in order to prevent illegal exploitation. Wildlife forensic genetics plays a role in the enforcement of the CITES treaty in multiple ways. Analytical work helps to identify species involved in violations of international wildlife laws when only pieces and parts are available for analysis (e.g., TAMs, processed foods, crafted items). Species identifications inform conservation biologists and managers as well as law enforcement entities about trends and changes in commercial and illegal wildlife trade. Wildlife forensic genetics plays a more prominent role in decision-making among the COPs regarding petitions to list species in CITES appendices (CITES 2019). Wildlife forensic genetic laboratories also work together on an international scale to provide technical assistance and capacity building in both source and range countries.

CITES Case Study: Pa Lor

The USFWS OLE is responsible for enforcement of the CITES treaty through its Wildlife Inspection Program, Special Agents, and the USFWS OLE-NFWFL. Cases involving CITES species are commonplace, but often start in surprising ways. In October of 2005, a small elderly woman was returning to Minneapolis, MN, USA from a visit to Laos in Southeast Asia. The woman, Pa Lor, like all international passengers, was required to fill out a US Customs form to declare certain items before being allowed re-entry into the USA. Lor did not declare any wildlife products upon arrival. However, during the routine interview, the Customs agent became suspicious, and Lor's luggage was given a secondary screen. A USFWS Wildlife Inspector was called in to inspect Lor's luggage and noticed several items suspected to be wildlife products. Two items in particular caught the inspector's eye—primate and elephant parts. "A monkey hand or foot looks very much like a human, so when you see it lying on a counter, it attracts your attention ... and it leaves an impression," the Inspector stated (USFWS Wildlife Inspector Susan Benson, pers. comm.) (Fig. 6).

The Inspector seized the suspected wildlife products from Pa Lor and subsequently coordinated additional investigation with the USFWS Special Agent in the region. The investigation uncovered that Lor and her daughter ran a booth in an ethnic market in Minnesota, USA, where they sold wildlife products for use in TAMs, as well as items used to ward off evil. The USFWS Special Agent sent in a Laotian-speaking agent to make covert purchases. Over the course of a 2-year investigation, agents seized 1388 animal products that were sent in to the USFWS OLE-NFWFL for identification. Using morphological, chemical, and genetic analyses, the forensic scientists at the laboratory identified over 30 different species, including pangolin (*Manis* spp.), Asian elephant (*Elephas maximus*), Asian otters (*Aonyx cinereus*), Asian golden cat (*Catopuma temminckii*), clouded leopard (*Neofelis nebulosa*), sperm whale (*Physeter macrocephalus*), and multiple species of primates. In particular, Pa Lor had in her possession over 300 individuals of the Laos warty newt (*Paramesotriton laoensis*). This species had only been described a few years prior from approximately a dozen specimens. Lor and her daughter had been running the booth for many years prior to getting caught, a fact evident during the investigation when Pa Lor gave the undercover agent instructions on how to avoid detection when shipping illegal items. In January of 2009, Lor was sentenced to 2 years' probation and a US $9000 fine and ordered to perform community service by offering public seminars on prevention of wildlife trafficking (United States Attorney's Office-District of Minnesota 2009). This case is only one of many examples of wildlife trafficking events that have alerted conservation agencies to species in illegal trade that are unknown or not well described in the biological literature.

Fig. 6 Bushmeat evidence waiting to be processed at USFWS OLE-NFWFL. Top, hands of one or more douc langur seized from Pa Lor; bottom, flying fox, bat, barracuda, and the nose of an elephant. Photo credit: USFWS OLE-NFWFL

The Lacey Act

When the Lacey Act was passed in 1900, it became the first law in US history to specifically address wildlife conservation. The Lacey Act is the oldest wildlife protection law that is still effective today (Anderson 1995). By the late 1800s, many game species such as American bison (*Bison bison*), passenger pigeon (*Ectopistes migratorius*), and heath hen (*Tympanuchus cupido cupido*) were rapidly disappearing from the North American landscape. The purpose of the act was to target illegal trade of native species by penalizing individuals who killed protected game in one state with the intent to sell it in another. Since its passage, the Lacey Act has been amended several times but remains focused on the transport of protected animals and plants across state and federal borders. Amendments to the Lacey Act in 1981 increased coverage to internationally protected species in response to increased illegal trade in fish and wildlife both in the USA and abroad (Wisch 2003). Wildlife forensic genetics plays a role in the enforcement of the Lacey Act by providing analyses that help to determine the identity of and relationships among protected species involved in violations of the law, as well as information about geographic provenance that can inform agents who are trying to determine if transport across state or international borders has occurred.

Lacey Act Case Study: House Mountain Elk

Although North American elk (*Cervus canadensis*) are a species of least conservation concern (IUCN 2019) in the USA, they are an important component of both the ecology and economy of many western US states. Therefore, management and protection of their populations are a conservation and law enforcement priority. In the fall of 1999, wildlife law enforcement officers with Idaho Fish & Game located a suspicious hunting camp in the mountains of Idaho, USA. Scattered around the camp were various pieces of hide, hair, bone, and biological material that appeared to be from multiple elk. Six separate kill sites were located by the officers, and seven

(continued)

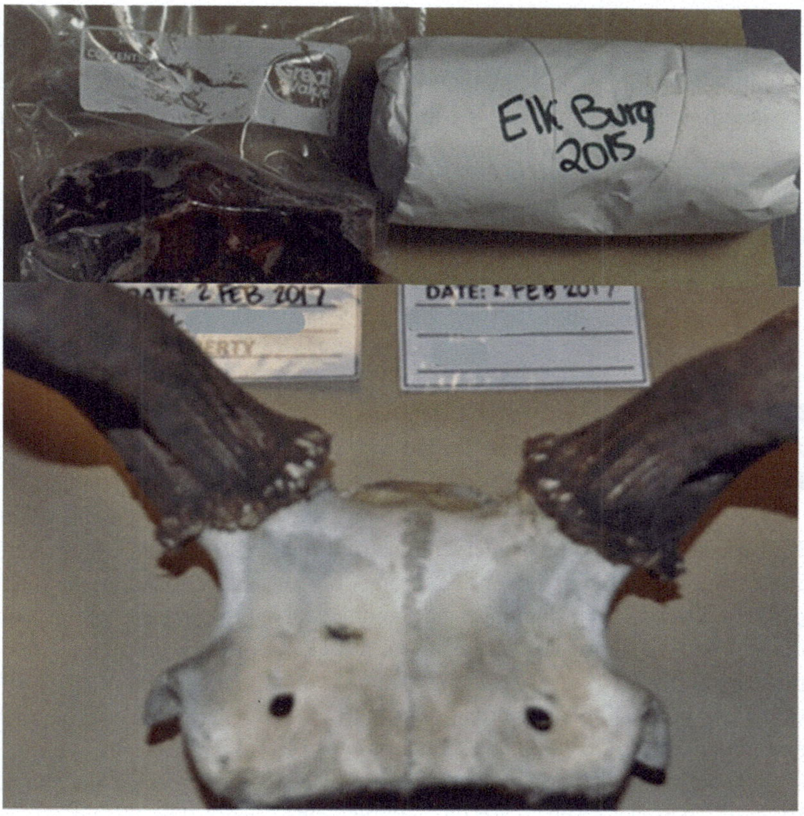

Fig. 7 Elk antlers and meat submitted as evidence for processing at the USFWS OLE-NFWFL. Photo credit: USFWS OLE-NFWFL

individuals from Minnesota, Michigan, and Idaho, USA were suspected of illegal take and waste (Fig. 7).

Officers obtained search warrants for the suspects' residences, including freezers, and items were subsequently seized. Items seized from the campsite, kill sites, and suspects' residences were submitted to the USFWS OLE-NFWFL for genetic analysis to determine species source, sex, total number of individual animals, and individual identification. The central question of the submitting officers was whether any of the items collected from the six separate kill sites originated from the same individuals as those evidence items collected from either the campsite or freezer searches at the suspect's residences. Serological techniques (isoelectric focusing and protein immunodiffusion), comparative electrophoresis of PCR amplicons, and STR analyses were used to determine the species source, sex, and individual identity of the evidence. Multiple items analyzed from the camp and freezer searches shared the same STR genotypes as items from each of the six kill sites. LR and match probability calculations were made utilizing data from known elk reference materials collected from across the state of Idaho, USA, including management unit 39 where this case originated. The LRs calculated for the evidence items were conservatively estimated to be in the range of 1 in 145,000. At the time of analysis, there were estimated to be approximately 120,000 total elk in Idaho, USA and approximately 7000 elk in managment unit 39 (pers. comm. Idaho Fish & Game).

Under threat of federal prosecution, six defendants from Minnesota and Michigan,

(continued)

USA returned to Idaho, USA and plead guilty to various state hunting violations, including four charges of waste and two Lacey Act violations. In addition to court fines and miscellaneous civil penalties of about US$750 each, the six individuals had their hunting licenses revoked for 2 years in Idaho, USA and their state of residence (U.S. Fish and Wildlife Service 2001).

Migratory Bird Treaty Act (MBTA) and the Bald and Golden Eagle Protection Act (BGEPA)

Bird species have required intervention from the conservation community over the years due to overexploitation from hunting and commercial trade. The MBTA makes it illegal to "...take, possess, import, export, transport, sell, purchase, or barter, or offer for sale, purchase, or barter..." any migratory bird, or the parts, nests, or eggs of such a bird, except under the terms of a valid Federal permit. The MBTA implements conventions among the USA and four countries (Canada, Mexico, Japan, and Russia) to protect migratory birds from overexploitation. The impetus for this treaty emerged from demand for the feathers of waterfowl and songbirds in the garment and hat industries in the late nineteenth century. Excessive demand for bird feathers, used to adorn the hats of wealthy women, led to the extinction of the heath hen, passenger pigeon, Labrador duck (*Camptorhynchus labradorius*), and great auk (*Pinguinus impennis*) and the near extinction of numerous other migratory waterfowl including the sandhill crane (*Grus canadensis*), great egret (*Aldea alba*), and bald eagle (*Haliaeetus leucocephalus*). It was outrage at the slaughter of so many birds that led Harriet Hemenway and Minna B. Hall to found the Audubon Society (Greenspan 2015). Through the efforts of Audubon members and other early conservationists, the MBTA was brought to fruition.

The BGEPA was initially implemented to protect populations of the bald eagle, the USA's most recognized symbol, from the effects of excessive hunting and habitat encroachment. The act was expanded from the original Eagle Act in 1962 to include the golden eagle (*Aquila chrysaetos*) due to population declines and similarities in juvenile appearance between the two species. This act also provided allowances for Native American possession of eagle feathers for religious purposes with appropriate permits. Bald and golden eagles were listed for protection under the ESA in the lower 48 US states in 1978. The ESA and the BGEPA were instrumental in conservation efforts during the mid-twentieth century when the use of pesticides, in particular DDT, was implicated in the near extinction of bald eagles and dramatic decline in abundance of golden eagles due to multigenerational reproductive failure (Carson 1962). Bald eagles were removed from the ESA in 2007 and are held as a premier example of an ESA conservation success story. However, protections remain for bald and golden eagles under MBTA and BGEPA. Wildlife forensic genetics plays a role in the conservation of migratory birds when birds are killed due to illegal hunting, nest poaching, or illegal trade in bird parts. The MBTA also plays a role in establishing restitution and remediation for deaths of migratory birds due to electrocution and impact mortalities from solar and wind energy structures.

MBTA/BGEPA Case Study: Operation Dakota Flyer

Late in 2014, US Federal wildlife officers were alerted to an extensive black market trading network for bird feathers and parts, including those of bald and golden eagles. In 2015, an informant established a relationship with four of the traffickers, some of whom were tribal members from several midwestern Native American tribes. After purchasing an eagle feather fan for US $350, the informant continued to work with the traffickers, who told the informant that not only could they supply feathers and parts from eagles, hawks, raptors, and other North American migratory birds, but also macaws, caracara, woodpeckers, hornbills,

(continued)

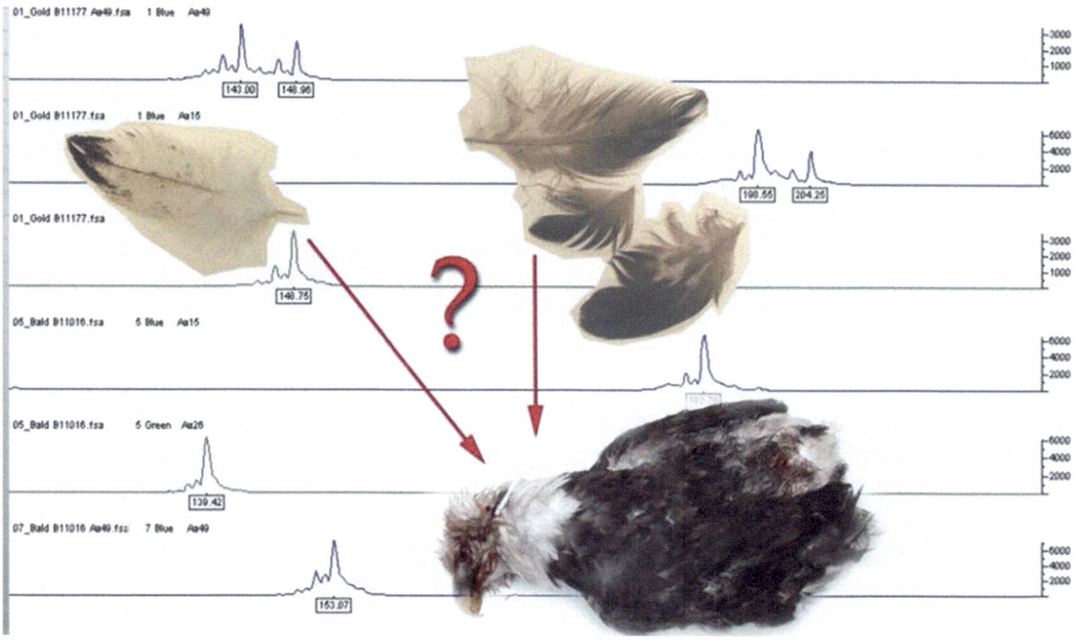

Fig. 8 Bald eagle *Haliaeetus leucocephalus* carcass and feathers suspected to have originated from it were tested with STR analysis to determine if they are from the same eagle. Photo credit: USFWS OLE-NFWFL

and even a Philippine serpent eagle (*Spilornis holospilus*). One of the lead traffickers also told the informant that he had a contact in Wyoming, USA who could get him whole eagle carcasses and set him up with an eagle hunt. The USFWS set up numerous undercover buys from the traffickers, which eventually turned out to be a network of over 51 individuals from 5 US states selling bird feathers and parts, primarily those of bald and golden eagles, from their homes and two pawn shops.

As undercover purchases were being made, the items became some of more than 250 pieces of evidence submitted for forensic analysis at the USFWS OLE-NFWFL. Forensic scientists at the USFWS OLE-NFWFL used both morphological identification and genetic analysis to determine the species origin, individual identification, and minimum number of individuals that comprised the submitted evidence. Crafted items for sale to the Native American community included beaded prayer sticks, bustles, and headdresses containing eagle and hawk feathers, staffs and prayer sticks with eagle heads and talons, and assorted bags and boxes that contained hundreds of loose feathers from multiple species. Morphological analysis of eagle feathers and parts is the easiest method for species identification, using visual identification of characters that define bird feathers and body parts to species. Genetic analysis was used to confirm smaller feathers and bird parts that no longer had species defining physical characteristics. Most importantly, genetic analysis of DNA markers (STRs) was used to determine the minimum number of individuals included in the collection of bald and golden eagle feathers and parts, as well as determining which

(continued)

items originated from the same individual bald or golden eagle (Fig. 8).

In the end, over 240 individual birds from 35 species, primarily bald and golden eagles, and other hawks, raptors, and migratory birds, were identified among the evidence, with some items from the same eagle found in different suspects' homes. Fifty-one individual suspects were indicted, and 31 of those individuals were convicted in US Federal Court for conspiracy to commit wildlife trafficking, violations of the MBTA, the BGEPA, and the Lacey Act. The most egregious violations netted the suspects prison terms of up to 5 years and fines up to US$250,000. Most significantly, the investigation was conducted without infringing on the Native American use of eagle parts for spiritual purposes (Actman 2020; United States Department of Justice 2018).

US Animal Conservation Laws for International Species of Interest

The USA is considered to be an international leader in conservation of animal species and is also one of the primary destinations for animal products, many of them through illicit markets. Laws such as the African Elephant Protection Act of 1988 (16 USCS § 4201), the Rhinoceros and Tiger Conservation Act of 1994 (16 U.S.C. 5306), and the Marine Mammal Protection Act of 1972 (MMPA, 16 U.S.C. §§1361-1383b, 1401-1406, 1411-1421h) were created to afford extra levels of protection to species deemed as critically endangered and important ecologically to the USA and to the world.

The African Elephant Protection Act of 1988 provides additional protection for African elephants and authorizes financial assistance for African elephant conservation programs, establishes moratoria on ivory import if specific criteria are not met, and creates criminal and civil penalties for illegal ivory import or export. Several US states (e.g., Washington, Oregon, California, Illinois, New York, New Jersey, Hawaii, Nevada, New Hampshire) have enacted their own state-level bans on ivory imports in recent years (Bale 2017). Identification of ivory tusks and crafted products obtained from wildlife inspections or special investigations is made using both morphological and genetic analyses. Morphological analysis is used to determine first if the item is made of real ivory or if it is fake (plastic or another polymer). Once an item is determined to be from a real elephant, observations of the Schreger lines (visual artifacts in the cross-sectional matrix of ivory) are used to determine if the item is from an extant (African or Asian elephant) or extinct (mammoth or mastodon) proboscidean (Schreger 1800; Espinoza and Mann 1993). If insufficient information is provided from morphological observation, genetic analysis utilizes a partial sequence of a variable region of the mtDNA cytochrome *b* to identify the species source of the item. In some cases, if the item is determined to be from an African elephant, the sequence can be compared to a database of other African elephants to infer putative region of population origin (e.g., *Loxodonta Localizer*, Ishida et al. 2013). Efforts to identify illegal trade routes for ivory in Africa demonstrated that DNA methods could also be used as an investigative tool to identify elephant tusks in the current trade that have originated from recently killed animals, as well as ivory stockpiles (Wasser et al. 2007, 2008, 2015).

Case Study: Ivory Crush to Combat Trade of Ivory

The international market for ivory is extensive. Ivory comprises antiquities that are important religious and cultural artifacts (Wylie 1980). Trade in ivory has increased during the last decades of the twentieth century for various reasons (Underwood et al. 2013), but regardless of the reason, the trade in ivory has resulted in the loss of millions of wild elephants from the continent of Africa in the last three decades (Chase et al. 2016). However, not all of

(continued)

Fig. 9 Elephant ivory carvings submitted for an ivory crush. Top, Ivory carvings to be crushed; bottom, crushed ivory is collected for disposal. Photo credit: Gavin Shire USFWS

the ivory in trade has its origin from recently killed animals—stockpiles of ivory and crafted ivory items remain in trade, some legally and some not. In 2013, the USFWS presided over an event now known as an ivory crush, at which over 6 tons of ivory were demolished by a rock-crushing machine (Christy 2013). The intent of ivory crushes was to deter the entry of ivory from held stockpiles into the illegal international market. In 2015, the USFWS conducted their second ivory crush (https://www.fws.gov/le/elephant-ivory-crush.html). Since that crush, several governments throughout Europe, Africa, the Middle East, and Asia have also destroyed ivory to highlight that only a worldwide solution will stop poaching of

(continued)

elephants (Denyer 2014; Earth League International 2016; International Fund for Animal Welfare 2016; Pincott 2018) (Fig. 9).

Case Study: Operation Crash

The Rhinoceros and Tiger Conservation Act of 1994 created a fund to assist the conservation of rhinoceros and tigers by supporting the conservation programs of nations whose activities directly or indirectly affect rhinoceros and tiger populations. The act also prohibits the sale, import, export, or the attempted sale, import, or export, of any product, item, or substance intended for human consumption, or application containing, or advertised as containing, any substance derived from any species of rhinoceros or tiger. The importance of this act lies in the wording that includes the labeling of items purported to contain rhino- or tiger-derived substances—an effort to combat fraudulent labeling of consumer products.

Operation Crash is, as of 2021, an ongoing nationwide criminal investigation led by the USFWS OLE. The investigation has focused on the illegal trade in rhinoceros horn and elephant ivory conducted by international poaching and smuggling syndicates to supply demand for hunting trophies, TAMs, and carvings viewed as good luck pieces and signs of wealth. In 2012, law enforcement officers from multiple US regions and supporting federal agencies executed search warrants in 13 US states. To date, over 30 individuals have been convicted for violations of the ESA and Lacey Act, conspiracy, smuggling, money laundering, international money laundering, mail fraud, tax evasion, bribery, and false documentation (Eisele 2017; U.S. Fish and Wildlife Service 2018). While most of these infractions involved financial crimes, it was up to wildlife forensic geneticists at the USFWS OLE-NFWFL to determine the species source of many of the evidence items that could not be conclusively identified to species by morphological examination. These items included partial horns, skin, and toenails, as well as numerous carved and crafted items that originated from rhino horns. In addition to identifying at least two species of rhinoceros in the collection of evidence (*Diceros bicornis*, *Ceratotherium simum*), wildlife forensic geneticists identified several species of domestic cattle (*Bos* sp.) as well as horse (*Equus caballus*). The identification of substitutions in the trade is not unexpected, and without the ability to use genetic tools, these substitutions would go undocumented. Identification of rhinoceros horns in source countries is also making use of lineage markers to determine individual identification of horns and illegally killed rhinos in the wild (Harper et al. 2013a, b). These efforts are helping to identify both horns poached in the wild and stockpiled horns that have reentered the illegal trade.

Aquatic resources are not immune to exploitation pressures that have led to widespread declines in abundance of marine fish, mammals, and reptiles. Marine mammal and fish populations, including sharks, are constantly at risk of accidental capture in commercial fishing operations and fishing line entanglement, particularly from illegal, underreported, and unregulated (IUU) fisheries worldwide (Fig. 10).

The MMPA (16 U.S.C. §§1361-1383b, 1401-1406, 1411-1421 h) protects all cetaceans (e.g., whales, dolphins, porpoises), sirenians (e.g., manatees and dugongs), and several marine carnivores (e.g., otters, seals, sea lions, walrus, and polar bear) from illegal exploitation. The ESA provides similar protection for marine reptiles, including all species of sea turtle (Chelonidae). Forensic genetic analyses are used to determine species source and minimum number of individuals of dismembered parts of animals when intercepted by law enforcement agents. Evidence can include shark fins and sea turtle meat that shows up in the food trade, sea

Fig. 10 Walrus tusks seized by USFWS law enforcement agents. Photo credit: USFWS

turtle leather and seal skins in the clothing trade, and walrus tusks, whale teeth, baleen, and whale bones in the craft trade. A comprehensive background in genetics, evolution, phylogenetics, and statistical analysis is a crucial suite of skills for a wildlife forensic geneticist, because sometimes the universe of possible species identifications is vast.

Case Study: Operation Northwest Caviar

Twenty-three years after the international trade in sturgeon and sturgeon caviar was regulated, the illegal trade in sturgeon caviar continues to threaten the survival of the 27 extant sturgeon species, 16 of which are listed by IUCN as critically endangered (Williamson 2003; Harris and Shiraishi 2018; IUCN 2019). The major markets for caviar in the nineteenth and twentieth centuries were primarily the USA and Canada, and the target species was lake sturgeon (*Acipenser fulvescens*) from the North American Great Lakes. By the end of the 1800s, lake sturgeon populations were decimated by the combination of overharvest, habitat loss to dams, and pollution. A shift to Eurasian sturgeon species for the caviar trade began to supplement the increasing demand for caviar from Europe and eastern Asia, and beluga (*Huso huso*), sterlet (*A. ruthenus*), Siberian (*A. baerii*), and Russian (*A. gueldenstaedtii*) sturgeons were the target of the caviar market (Ludwig 2008; Fain et al. 2013). By the late twentieth century, populations of sturgeon species worldwide were in serious decline, resulting in the intervention of CITES to regulate the trade. By this time, aquaculture facilities had been introduced in both open water and enclosed formats (Bronzi et al. 2011), which both helped and aggravated the pressure on wild sturgeon populations. As the pressure on Eurasian species increased, markets began to explore substitution species, leading them back to North America. By the early 2000s, exploitation pressure on North American sturgeon species (*Scaphirhynchus platorynchus*, *S. albus*, *S. suttkusi*) was increasing and expanding to include American paddlefish (*Polyodon spathula*) and white sturgeon (*A. transmontanus*), both of which were desirable substitutes for the ever-diminishing supply of Siberian sturgeon caviar (Ludwig 2008; Bronzi et al. 2011; U.S. Department of Justice 2015) (Fig. 11).

(continued)

Fig. 11 Sturgeon carcasses submitted for species identification at USFWS OLE-NFWFL. Photo credit; USFWS OLE-NFWFL

In 2002, wildlife forensic geneticists at USFWS OLE-NFWFL were asked to identify the species source of caviar obtained from black market suppliers in Vancouver, WA, USA. Individuals known to law enforcement were buying gravid female sturgeon caught in the sport fishing industry in the Sacramento and Columbia rivers and processing the caviar for sale in Oregon and Washington, USA (Bailey 2003). A suspected link to this caviar trafficking network was individuals who were harvesting oversized white sturgeon (>50 inches) from below the Bonneville Dam on the lower Columbia River (Profita 2015). Wildlife forensic scientists from USFWS OLE-NFWFL were asked to identify fingerprints from hundreds of mason jars of home-preserved product. Agents also requested the species source of the caviar, as well as any information related to the river of origin (Beamesderfer 1993; Anders and Powell 2002; Hildebrand et al. 2016), as charges against the suspects carried different consequences in US state and federal courts.

In addition to the caviar evidence, wildlife agents also obtained partial sturgeon carcass pieces being sold for meat from suspects thought to have dragged a tethered sturgeon from the Columbia River under cover of night (Profita 2015). Wildlife forensic geneticists at the USFWS OLE-NFWFL analyzed four pieces of meat that weighed between 15 and 19 lbs. each and a white sturgeon head (rostrum to just past the operculum) that weighed over 19 lbs. The species source was confirmed to be white sturgeon, and STR markers designed for genetic monitoring of white sturgeon populations (Anders et al. 2011) were used to demonstrate that the four flesh pieces and the head originated from the same fish. This data, in combination with the estimated weight of the confiscated pieces, the head length of the sturgeon, and morphometric data (head length-total length and head length-total weight relationships; Anders 1991; North et al. 1996), determined that the total length of the white sturgeon represented by these five pieces was over 11 feet long. This was 6 feet in excess of the 5-foot slot limit for

(continued)

white sturgeon from the Columbia River. The offending suspects received varied sentences in court, from monetary fines of thousands of US dollars to house arrest and minimal time in prison (Bailey 2003).

Summary

The use of wildlife forensic genetics for species and individual identification, as well as determination of geographic provenance, has been identified as a critical tool in international judicial systems. A number of organizations are active in international conservation efforts, including US Federal and State wildlife forensic laboratories; NGOs including the International Union for Conservation of Nature (IUCN), World Wildlife Fund, National Geographic Society, Natural Resources Defense Council, National Wildlife Federation, International Rhino Foundation, African Wildlife Foundation, Wildlife Alliance, Earthwatch Institute, and TRAFFIC; and others. In recent years, these organizations have joined forces with CITES signatory countries and international law enforcement organizations to combat the epidemic of illegal wildlife trafficking by forming a number of Wildlife Enforcement Networks. These coalitions, such as the Association of Southeast Asian Nations' Wildlife Enforcement Network (ASEAN-WEN) and the Central American Wildlife Enforcement Network (CAWEN) are designed to integrate the skills of law enforcement, field biologists, research scientists, and government agencies to protect indigenous natural resources in the face of the pressure from wildlife traffickers. As technology progresses, more useful tools are emerging in the effort to stem the illegal wildlife trade by identifying species that are being exploited. While technology may be our best tool to detect illegal species exploitation, public awareness and stewardship of environmental resources will be the most effective tool for protecting at-risk species of interest from overexploitation.

Acknowledgments The authors would like to extend their thanks to our colleagues in the field of wildlife forensic science who have contributed to this paper through many different conversations about science, casework, standards, and guidelines for the wildlife forensic science discipline. We thank Barbara A. Shields and M. Katherine Moore, whose comments greatly improved earlier versions of this manuscript. Finally, we dedicate this work to the memory of Robert Schulyer Anderson (1954–2016), whose tireless efforts with the US Department of Justice Environment and Natural Resources Division, on behalf of the animal resources of the world, continue to serve as an inspiration to all wildlife forensic scientists. The findings and conclusions in this chapter are those of the authors and do not necessarily represent the views of the U.S. Fish and Wildlife Service.

References

AAFS Standards Board (2019a) ANSI/ASB standard 047: wildlife forensic validation standards – validating new primers for sequencing, 1st edn. AAFS Standards Board, LLC

AAFS Standards Board (2019b) ANSI/ASB standard 048: wildlife forensic DNA standard procedures, 1st edn. AAFS Standards Board, LLC

AAFS Standards Board (2019c) ANSI/ASB Standard 046: Wildlife forensic validation standards – STR analysis, 1st edn. AAFS Standards Board, LLC

AAFS Standards Board (2019d) ANSI/ASB standard 019, wildlife forensics general standards, 1st edn. AAFS Standards Board, LLC

AAFS Standards Board (2019e) ANSI/ASB standard 029, report writing in wildlife forensics: morphology and genetics, 1st edn. AAFS Standards Board, LLC

Actman J National Geographic Photo Essay https://www.nationalgeographic.com/photography/proof/2017/08/eagle-trafficking/. Accessed 02 Feb 2020

African Elephant Conservation Act of 1988. 16 USCS § 4201. U.S.G.P.O. 1988

Allendorf F, Luikart G, Aiken SN (2013) Conservation and the genetics of populations, 2nd edn. Wiley, West Sussex, UK

Anders PJ (1991) White sturgeon (*Acipenser transmontanus*) movement patterns and habitat use in the Kootenai River system, Idaho, Montana, and British Columbia. MS Thesis, Eastern Washington University, 153 pp

Anders PJ, Powell MS (2002) Geographic and frequency distributions of control region length variation in the mtDNA genome of white sturgeon (*Acipenser transmontanus*) from the Columbia River basin chapter 2. In: Anders PJ Conservation biology of white sturgeon. Ph.D. dissertation, University of Idaho, Aquaculture Research Institute, Center for Salmonid and Freshwater Species at Risk, 221 pp

Anders PJ, Drauch-Shreier A, May B, Rodzen J, Powell M, Narum S, Crossman J (2011) A review of genetic evaluation tools for conservation and management of sturgeons: roles, benefits, and limitations. J Appl Ichthyol 27(Suppl 2):3–11

Anderson RS (1995) The Lacey Act: America's Premier weapon in the fight against unlawful wildlife trafficking. 16 Public Land Law Review (Public Land and Resources Law Review) 27

Anderson EC (2010) Assessing the power of informative subset of loci for population assignment: standard methods are upwardly biased. Mol Ecol Res 20:701–710

Ashabaugh DR (1999) Quantitative-qualitative friction ridge analysis: an introduction to basic and advanced Ridgeology. Taylor & Francis, Boca Raton, FL

Bailey E (2003) 8 held in probe of alleged sturgeon poaching. Los Angeles Times, May 10, 2003 https://www.latimes.com/archives/la-xpm-2003-may-10-me-caviar10-story.html

Bald and Golden Eagle Protection Act (1940 et seq) 16 U.S.C. 668a – d. U.S. GPO, Washington

Bale R (2017) Why D.C. Is the New Hub for U.S. Ivory Sales. National Geographic News. https://www.nationalgeographic.com/news/2017/07/wildlife-watch-washington-elephant-ivory-for-sale-united-states/ Accessed 19 December 2019

Banks MA, Blouin MS, Baldwin BA, Rashbrook VK, Fitzgerald HA, Blankenship SM, Hedgecock D (1999) Isolation and inheritance of novel microsatellites in Chinook salmon (*Oncorhynchus tshawytscha*). J Hered 90(2):281–288

Beamesderfer RC (1993) A standard weight (Ws) equation for white sturgeon. California Fish and Game

Bell S, Sah S, Albright GSJ Jr, Denton MB, Casadevall A (2018) TDA call for more science in forensic science. PNAS 115(18):4541–4544. https://doi.org/10.1073/pnas.1712161115

Benson DA, Cavanaugh M, Clark K, Karsch-Mizrachi I, Lipman DJ, Ostell J, Sayers EW (2013) GenBank. Nuc Acids Res 41:36–42

Breyer S (2000) Science in the courtroom. Issues Sci Technol 16(4):1

Bronzi P, Rosenthal H, Gessner J (2011) Global sturgeon aquaculture production: an overview. J Appl Ichthyol 27:169–175

Burnham-Curtis MK, Trail PW, Kagan R, Moore MK (2015) Wildlife forensics: an overview and update for the prosecutor. United States Attorney's Bull 2015:53–68

Caplan RM (1990) How fingerprints came into use for personal identification. J Am Acad Dermatol 23 (1):109–114

Carson R (1962) Silent Spring. Houghton Mifflin, Boston

Chambers SM, Fain SR, Fazio B, Amaral M (2012) An account of the taxonomy of North American wolves from morphological and genetic analyses. N Am Fauna 77:1–67

Chase MJ, Schlossberg S, Griffin CR, Bouché PJC, Djene SW, Elkan PW, Ferreira S, Grossman F, Kohi EM, Landen K, Omondi P, Peltier A, Selier SAJ, Sutcliffe R (2016) Continent-wide survey reveals massive decline in African savannah elephants. PeerJ 4:e2354

Christy B (12 November 2013) Opinion: U.S. ivory crush should be just a first step. Natl Geogr Accessed 22 Feb 2020

CITES (1973) Convention on International Trade in Endangered Species of Wild Fauna and Flora, 3 March 1973, 993 U.N.T.S. 243

CITES (2019) Notification to the parties no. 2019/018. Geneva, March 2019. https://www.cites.org/sites/default/files/notif/E-Notif-2019-018.pdf

Courtios R, Bernatxchez L, Oullet J-P, Breton L (2003) Significance of caribou (*Rangifer tarandus*) ecotypes from a molecular genetics viewpoint. Cons Gen 4:393–404

Cronin MA, Renecker LA, Patton JC (2009) Genetic variation in domestic and wild elk (*Cervus elaphus*). J Anim Sci 87(3):829–834

Daubert v. Merrell Dow pharmaceuticals, Inc, 509 US. 579 (1993)

Denyer S (2014) China's first 'ivory crush' signals it may hoin global push to protect African elephants. The Washington Post, January 06, 2014. https://www.washingtonpost.com/world/chinas-first-ivory-crush-signals-it-may-join-global-push-to-protect-african-elephants/2014/01/06/267e432c-76d6-11e3-b1c5-739e63e9c9a7_story.html Accessed 22 February 2020

Dickinson E, Remsen J (2014) The Howard & Moore complete checklist of the birds of the world: passerines. Aves Press

Dormontt EE, Boner M, Braun B, Breulmann G, Degen B, Espinoza E, Gardner S, Guillery P, Hermanson JC, Koch G, Lee SL (2015) Forensic timber identification: it's time to integrate disciplines to combat illegal logging. Biol Conserv 191:790–798

Earth League International (2016). https://earthleagueinternational.org/italy-first-ivory-crush/. Accessed 22 February 2020

Eisele T (2017) Operation Crash: Stopping a Network of Illegal Wildlife Traffickers in United States. Newsletter of the National Association of Conservation Law Enforcement Chiefs. October 11, 2017. https://www.naclec.org/press-pages/2017/10/11/operation-crash-stopping-a-network-of-illegal-wildlife-traffickers-in-united-states Accessed 26 February 2020

Endangered Species Act of 1973 (1973 et seq). U.S. GPO, Washington, DC

Endangered Species Conservation Act of 1969 (1969) Public Law 91–135. U.S. G.P.O. Washington

Endangered Species Preservation Act of 1966 (Public Law 89–669). U.S. G.P.O. 1966

Espinoza EO, Mann M-J (1993) The history and significance of the Schreger pattern in Proboscidean ivory characterization. J Am Inst Cons 32(3):241–248

Evett IW, Weir BS (1998) Interpreting DNA evidence. Sinauer Associates, Sunderland, MA

Fain SR, Straughan DJ, Hamlin BC, Hoesch RM, Le May JP (2013) Forensic genetic identification of sturgeon caviars traveling in world trade. Conserv Genet 14:855–874

Federal Bureau of Investigation (2000) DNA Advisory Board for Federal Bureau of Investigation, Quality assurance standards for forensic DNA testing laboratories. Forensic Sci Commun 2(3). http://www.fbi.gov/about-us/lab/forensic-science-communications/fsc/july2000/codispre.htm/codis2a.htm. Accessed 24 Feb 2020

Felsenstein J (2001) Taking variation of evolutionary rates between sites into account in inferring phylogenies. J Mol Evol 53(4–5):447–455

Fenner GM (1996) The Daubert handbook: the case, its essential dilemma, and its progeny. Creighton Law Rev 29:939–1089

Froese R, Pauly D (eds) (2000) FishBase 2000: concepts, design and data sources. ICLARM, Los Baños, Laguna, Philippines. 344 p

Frye v. United States, 293 F. 1013 (D.C. Cir. 1923)

Galton F (1892). Finger prints. Macmillan and Company

Greenspan J (2015) The History and Evolution of the Migratory Bird treaty Act. Audubon, May 22, 2015. https://www.audubon.org/news/the-history-and-evolution-migratory-bird-treaty-act Accessed 03 March 2020

Harper CK, Ludwig A, Clarke AB, Makgopela K, Yurchenko A, Guthrie A, Dobrynin P, Tamazian G, Emslie R, Heerden M, Hofmeyr M, Potter R, Roets J, Beytell P, Otiende M, Kariuki L, Toit R, Anderson N, O'Brien DJ (2013a) Robust forensic matching of confiscated horns to individual poached African rhinoceros. Curr Biol 28(1):R13–R14

Harper CK, Vermeulen GJ, Clarke AB, de Wetc JI, Guthried AJ (2013b) Extraction of nuclear DNA from rhinoceros horn and characterization of DNA profiling systems for white (*Ceratotherium simum*) and black (*Diceros bicornis*) rhinoceros. For Sci Int Genetics 7 (4):428–433

Harris L, Shiraishi H (2018) Understanding the global caviar market. Results of a rapid assessment of trade in sturgeon caviar. TRAFFIC and WWF joint report. https://www.traffic.org/publications/reports/understanding-the-global-caviar-market/ Accessed 28 February 2020

Hartl DL (2000) A primer of population genetics, 3rd edn. Sinauer Associates, Sunderland, MA

Hedrick PW (2001) Conservation genetics: where are we now? Trends Ecol Evol 16(11):629–636

Hildebrand LR, Drauch Schreier A, Lepla K, McAdam SO, McLellan J, Parsley MJ, Paragamian VL, Young SP (2016) Status of white sturgeon (*Acipenser transmontanus* Richardson, 1863) throughout the species range, threats to survival, and prognosis for the future. J Appl Ichthyol 32:261–312

Hillis DM, Bull JJ (1993) An empirical test of bootstrapping as a method for assessing confidence in phylogenetic analysis. Syst Biol 42(2):182–192

Hillis DM, Moritz C, Mable BK (eds) (1996) Molecular systematics, vol 23. Sinauer Associates, Sunderland, MA

Integrated Taxonomic Information System (ITIS). http://www.itis.gov

Huffman JE, Wallace JR (eds) (2012) Wildlife forensics – methods and applications. Wiley, Sussex

International Fund for Animal Welfare (2016). https://www.prnewswire.com/news-releases/ifaw-largest-ivory-crush-ever-in-kenya-sends-message-to-poachers-300260399.html. Accessed 22 Feb 2020

International Standards Organization (2017) ISO/IEC 17025: General requirements for the competence of testing and calibration laboratories, (1/12/17). https://www.iso.org/publication/PUB100424.html. Accessed 01 Mar 2020

International Union for the Conservation of Nature (2019) The IUCN Red list of threatened species. Version 2019-3. http://www.iucnredlist.org. Accessed 10 Dec 2019

Ishida Y, Georgiadis NJ, Hondo T, Roca AL (2013) Triangulating the provenance of African elephants using mitochondrial DNA. Evol Appl 6:253–265

Jacobs RL, Baker BW (2018) The species dilemma and its potential impact on enforcing wildlife trade laws. Evol Anthropol Issues News Rev 27(6):261–266

Jeanguenat AM, Budowle B, Dror IE (2017) Strengthening forensic DNA decision making through a better understanding of the influence of cognitive Bias. Sci Justice 57(6):415–420

Jobin RM, Patterson D, Zhang Y (2008) DNA typing in populations of mule deer for forensic use in the province of Alberta. For Sci Int 2(3):190–197

Kalinowski ST (2004) Genetic polymorphism and mixed-stock fisheries analysis. Can J Fish Aquat Sci 61:1075–1082

Karl S, Avise JC (1993) PCR-based assays of Mendelian polymorphisms from anonymous single-copy nuclear DNA: techniques and applications for population genetics. Mol Biol Evol 10(2):342–361

Knight J (2017) Operation Roadhouse: how inter-agency collaboration stopped illegal paddlefish depredation. National Association of Conservation Law Enforcement Chiefs News. https://www.naclec.org/press-pages/2017/9/17/operation-roadhouse-how-inter-agency-collaboration-stopped-illegal-paddlefish-depredation Accessed 28 February 2020

Kumho Tire Co. v. Patrick Carmichael, 119 S. Ct 1167 (1999)

Lacey Act (1900 et seq)16 U.S.C. §§ 3371–3378.U.S. GPO, Washington

Linacre AMT, Tobe SS (2013) Wildlife DNA analysis – applications in forensic science. Wiley-Blackwell, Hoboken

Ludwig A (2008) Identification of Acipenseriformes species in trade. J Appl Ichthyol 24:2–19

Marine Mammal Protection Act (1972) 16 U.S.C. §§1361-1383b, 1401–1406. U.S. GPO, Washington, pp 1411–1421h

McCormack JE, Rodríguez-Gómez F, Tsai WLE, Faircloth BC (2017) Transforming museum specimens into genomic resources. In: Webster MS (ed) The extended specimen: emerging frontiers in collections-based ornithological research. CRC Press, Boca Raton, FL, pp 143–156

Migratory Bird Treaty Act (1918 et seq) 16 U.S.C. 703 – 712. U.S. GPO, Washington

Mindell DP, Sorenson MD, Dimcheff DE (1998) Evolution of multiple independent origins of mitochondrial gene order in birds. Proc Natl Acad Sci U S A 95:10693–10697

Moore MK, Frazier K (2019) Humans are animals, too: critical commonalities and differences between human and wildlife forensic genetics. J Forensic Sci 64 (6):1603–1621. https://doi.org/10.1111/1556-4029.14066

Moritz C, Dowling TE, Brown WM (1987) Evolution of animal mitochondrial DNA: relevance for population biology and systematics. Ann Rev Ecol Syst 18 (1):269–292

Nagy S, Poczai P, Cernak I, Gorji AM, Hegedus G, Taller J (2012) PICcalc: an online program to calculate polymorphic information content for molecular genetic studies. Biochem Gen 50:670–672

North JA, Rien TA, Farr RA (1996) pp 5–36. In: Beiningen KT (ed) Effects of mitigative measures on productivity of white sturgeon populations in the Columbia River downstream from McNary Dam, and status and habitat requirements of white sturgeon populations in the Columbia and Snake rivers upstream from McNary Dam. Annual Progress Report to Bonneville Power Administration, Portland, OR

Ogden R (2011) Unlocking the potential of genomic technologies for wildlife forensics. Mol Ecol Res 11:109–116

Ogden R, Linacre A (2015) Wildlife forensic science: a review of genetic geographic origin assignment. For Sci Int Gen 18:152–159

Ogden R, Dawnay N, McEwing R (2009) Wildlife DNA forensics - bridging the gap between conservation genetics and law enforcement. Endanger Species Res 9(3):179–195

Paetkau D, Strobeck C (1994) Microsatellite analysis of genetic variation in black bear populations. Mol Ecol 3 (5):489–95; Erratum in: Mol Ecol 1995 4(1):133

Palsboll P, Berube M, Allendorf F (2007) Identification of management units using population genetic data. Trends Ecol Evol 22(1):11–16

Pincott S (2018) Australia's First Ivory Crush – For the love of Elephants. The Zimbabwean, March 07, 2018. https://www.thezimbabwean.co/2018/03/australias-first-ivory-crush-for-the-love-of-elephants/. Accessed 22 February 2020

Profita C (2015) World's appetite for Caviar Sends Poachers after Columbia River Sturgeon. OPB/EarthFix May 16, 2015 https://www.opb.org/news/series/wildlife-detectives/worlds-appetite-for-caviar-sends-poachers-after-columbia-river-sturgeon/ Accessed 28 February 2020

Ratnasingham S, Hebert PDN (2007) BOLD: the barcode of life data system (http://www.barcodinglife.org). Mol Ecol Notes 7(3):355–364

Rhinoceros and Tiger Conservation Act of 1994. 16 U.S.C. 5306. U.S.G.P.O. 1994

Richly E, Leister D (2004) NUMTs in sequenced eukaryotic genomes. Mol Biol Evol 21(6):1081–1084

Robin C, Salmon A (1857) Translation of Memoire Concernant I'Examen, a I'Aide du Microscope, de Taches de Sang sur une Blouse de Coton Bleu dans un Cas d'Assassinat. Annales de Hygiene Publique et de Medecine Legale 8 (2nd series): 368–397. English edition: Gaensslen RE, ed (1983). Sourcebook in Forensic Serology, Immunology, and Biochemistry: Unit IX: Translations of Selected Contributions to the Original Literature of Medicolegal Examinations of Blood and Body Fluids. U.S. Dept of Justice, National Institute of Justice, Washington, DC

Schreger BNG (1800) Beitrag zur geschichte der zähne. Beitrage für die Zergliederungskunst 1:1–7. German

Simkins AT, Buchanan GM, Davies RG, Donald PF (2019) The implications for conservation of a major taxonomic revision of the world's birds. Anim Conserv

Society for Wildlife Forensic Science Programs https://www.wildlifeforensicscience.org/programs/

Society for Wildlife Forensic Science Technical Working Group (2018) Standards and Guidelines for Wildlife Forensic Analyses Version 3.0. Ed. LJ Webster. Society for Wildlife Forensic Science. 19 Nov 2018. pp 21. https://www.wildlifeforensicscience.org/wp-content/uploads/2018/11/SWFS-Standards-and-Guidelines_Version-3_19-11-18.pdf. Accessed 28 Feb 2020

Smith SH (1968) Species succession and fishery exploitation in the Great Lakes. J Fish Res Bd Can 25:667–693

Smith C, Seeb L (2007) Number of alleles as a predictor of the relative assignment accuracy of short tandem repeat (STR) and single-nucleotide-polymorphism (SNP) baselines for Chum salmon. Trans Am Fish Soc 137 (3):751–762

SWGDAM (2019) Scientific working group on DNA analysis methods. Addendum to "SWGDAM interpretation guidelines for autosomal STR typing by forensic DNA testing laboratories" to address next generation sequencing. April 23, 2019. U.S. Dept justice. https://1ecb9588-ea6f-4feb-971a-73265dbf079c.filesusr.com/ugd/4344b0_91f2b89538844575a9f51867def7be85.pdf. Accessed 28 February 2020

Title 28 - Judiciary and Judicial Procedure. Federal Rules of Evidence (1975 et seq) Article VII. Opinions and Expert Testimony. Rule 702. Testimony by Expert Witnesses. Pub. L. 93–595, §1, Jan. 2, 1975, 88 Stat. 1937; April 17, 2000, eff. December 1, 2000; April 26, 2011, eff. December 1, 2011

TRAFFIC (2019) Combating wildlife crime linked to the internet: global trends and China's experiences. TRAFFIC, Cambridge, UK

U.S. Fish and Wildlife Service (2001) Division of Law Enforcement Annual Report 2001

U.S. Fish and Wildlife Service (2018). https://www.fws.gov/le/pdf/Operation%20Crash%20Fact%20Sheet%20January%2011,%202018.pdf. Accessed December 15, 2019

Uetz P, Etzold T (1996) The EMBL/EBI reptile database. Herpetol Rev 27(4):174–175

Ulery BT, Hicklin RA, Buscaglia J, Roberts MA (2011) Accuracy and reliability of forensic latent fingerprint decisions. PNAS 108(19):7733–7738. https://doi.org/10.1073/pnas.1018707108

Ulery BT, Hicklin RA, Buscaglia J, Roberts MA (2012) Repeatability and reproducibility of decisions by latent fingerprint examiners. PLoS One 7(3):e32800. https://doi.org/10.1371/journal.pone.0032800

Underwood FM, Burn RW, Milliken T (2013) Dissecting the illegal ivory trade: an analysis of ivory seizures data. PLoS One 8(10):e76539

United Nations (1982) United Nations Convention on the Law of the Sea, Part XI, Section 5, Dec. 10, 1982, 1833 U.N.T.S. 397. (http://www.un.org/depts/los/convention_agreements/texts/unclos/part2.htm) Accessed February 20, 2020

United States Attorney's Office, District of Minnesota (2009) Annu Rep 2009:30

United States Department of Justice (2015) Caviar Prosecutions https://www.justice.gov/enrd/caviar-prosecutions Accessed 28 February 2020

United States Department of Justice (2018) Press release. https://www.justice.gov/usao-sd/pr/project-dakota-flyer-sentences-announced-0 August 02, 2018. Accessed 20 February 2020

United States Department of Justice (2019) Press release. https://www.justice.gov/usao-or/pr/lane-county-man-pleads-guilty-shooting-endangered-gray-wolf. Accessed 20 February 2020

Waits L, Taberlet P, Swenson JE, Sangegren F, Franzen RT (2001) Nuclear DNA microsatellite analysis of genetic diversity and gene flow in the Scandinavian brown bear (*Ursus arctos*). Mol Ecol 9(4):321–341

Waples RS (2006) Distinct population segments, pp 127–149. In: Scott JM, Goble D, Davis FW (eds) The Endangered Species Act at Thirty, Vol 2: Conserving biodiversity in human dominated landscapes. Washington Island Press

Wasser SK, Mailand C, Booth R, Mutayoba B, Kisamo E, Clark WJ, Stephens M (2007) Using DNA to track the origin of the largest ivory seizure since the 1989 trade ban. Proc Nat Acad Sci 104(10):4228–4233

Wasser SK, Clark WJ, Drori O, Kisamo E, Mailand C, Mutayoba B, Stephens M (2008) Combating the illegal trade in African elephant ivory with DNA forensics. Cons Biol 22(4):1065–1071

Wasser SK, Brown L, Mailand C, Mondol S, Clark W, Laurie C, Weir BS (2015) Genetic assignment of large seizures of elephant ivory reveals Africa's major poaching hotspots. Science 349:84–87

Watrous L, Wheeler Q (1981) The out-group comparison method of character analysis. Syst Biol 30:1–11

Wayne R, Lehman N, Allard M, Honeycutt R (1992) Mitochondrial DNA variability of the gray wolf: genetic consequences of population decline and habitat fragmentation. Cons Biol 6(4):559–569

Weir BS (1996) Genetic Data Analysis II: Methods for discrete population genetic data, 2nd edn. Sinauer Associates, Sunderland, MA

Wiley E, Causey D, Brooks D, Funk V (2013). The Compleat Cladist: a primer of phylogenetic procedures. University of Kansas Museum of Natural History, vol 19, Lawrence, KS

Williamson DF (2003) Caviar and conservation: status, management and trade of North American sturgeon and paddlefish. TRAFFIC North America, Washington, DC

Wilson DE, Reeder DM (eds) (2005) Mammal species of the world. A taxonomic and geographic reference, 3rd edn. Johns Hopkins University Press, Baltimore, MD, p 2

Wisch RF (2003) Overview of the Lacey Act (16 U.S.C. §§ 3371–3378). Animal Legal and Historical Center. Michigan State University College of Law. https://www.animallaw.info/article/overview-lacey-act-16-usc-ss-3371-3378. Accessed 29 April 2020

World Wildlife Fund (2018) Living planet report - 2018: Aiming Higher. In: Grooten M, Almond REA (eds) WWF, Switzerland

Wright S (1978) Evolution and the genetics of populations. The University of Chicago Press, Chicago, pp 1–4

Wylie KC (1980) Ivory, elephants, and man: a survey. Elephant 1(5):3–18

Zhang D-X, Hewitt GM (2003) Nuclear DNA analyses in genetic studies of populations: practice, problems and prospects. Mol Ecol 12:563–584

Carrion Ecology

Adrienne Brundage

Abstract

Detritus is the term for any source of non-living organic matter and is the basal trophic stage of most food webs, believed to be produced from phototropic sources. While plants make up a good portion of the detrital pool, carrion, also considered detritus, is the decaying flesh of dead animals and has a significant impact on the ecosystem. Carcass decomposition changes the surrounding environment both in the short term and in the long term. Understanding this process is therefore important to understanding the way an environment works as a whole. Decomposition in any given ecosystem includes actions of vertebrate scavengers, invertebrates, plants, and microorganisms. From a physical standpoint, after an animal dies, its molecular structure begins to disaggregate. Scavengers feeding upon carrion help to recycle carcass nutrients and resources throughout the ecosystem. In addition to impacting landscape heterogeneity through decomposition pattern destruction, decomposers such as scavengers also impact the structure of other trophic levels through their encounters with predators, parasites, and other organisms. Spatial and temporal dynamics of detritus subsidies, along with competitive hierarchy of species using these resources, are important drivers of terrestrial system decomposition processes and are ultimately related to necrophages through evolutionary processes. Carrion is an inevitable and ecologically important component of the ecosystem. Understanding its position in the overarching ecological system is crucial to the overall understanding of ecology. Although the importance of carrion decomposition in the atmosphere has long been recognized, the processes behind it have yet to be elucidated.

Keywords

Decomposition · Carrion · Detritus · Nutrient cycling · Ecosystem · Decomposition ecology

Introduction

Ecology is a branch of biology which describes the complexity of interactions among organisms and their biophysical environment and includes both biotic and abiotic components. This chapter will explain some of the more well-established concepts of ecology. Ecology encompasses everything around us, how and why it works, and where humans fit into the equation.

Detritus is any source of non-living organic matter and is the basal trophic stage of most food webs (Lindeman 1942; Odum 2014). Detritus decomposition is an integral component of ecosystem maintenance and health. The most

A. Brundage (✉)
Texas A&M University, College Station, TX, USA
e-mail: brundage@forensicentomologist.com

S. C. Underkoffler, H. R. Adams (eds.), *Wildlife Biodiversity Conservation*,
https://doi.org/10.1007/978-3-030-64682-0_9

abundant portion of the detrital pool consists of phototropic sources such as dead plants, algae, and other photosynthetic materials. It is these sources that provide a fundamental understanding of role of the dead particulate organic matter present in the ecosystem, collectively described as the detrital pool (Odum 2014). While understanding the decomposition of phototrophic aspects of ecosystems is vital, it ignores the importance of the heterotrophic portion of the detrital pool: carrion.

By definition, carrion is a term used for the carcasses of deceased vertebrates. In an ecological context, however, the term can be expanded. The ecological definition of carrion, then, is the rotting carcass of any once-living, heterotrophic organism. This includes not only vertebrate animals but also microscopic eukaryotes such as rotifers, nematodes, and macroinvertebrates. This expanded definition helps to integrate all invertebrate carrion research into available knowledge of the ecological impacts of carrion.

Most reviews lump carrion into the overarching pool of detritus, including it with all decaying organic matter. A select few give carrion more ecologically specific conceptual meanings, for example, defining carrion as a decomposition island or an ephemeral resource (Barton et al. 2019; Carter et al. 2007; Collins et al. 2020). To date, a complete understanding of the role of carrion in ecosystems has been largely under-researched. This chapter is a literature review of the current knowledge of carrion ecology, with emphasis on gaps in the research.

Biological Processes of Decomposition

The impact of carrion on the surrounding environment begins with the irreversible cessation of function of the cardiovascular system, respiratory system, and central nervous system of a heterotrophic organism; i.e., it begins at death. Taphonomy, or the study of decay, has long characterized the process by which organisms decompose while exploring their effect on the surrounding environment. Much of our current understanding of carrion ecology is simply expanded taphonomy.

It is essential to understand the processes of decomposition occurring in an animal carcass in order to understand how the biota changes during carcass decomposition. Hence, the deterioration process of carrion or corpses in terrestrial environments was historically defined by forensic literature. Researchers have attempted to classify decomposition into temporally discrete stages, despite decomposition being a continuum of soft tissue removal, with tissue degradation mediated by both chemical and biological processes.

Carrion is a valuable food source for a wide variety of species. Decomposition is dependent upon the presence of invertebrates and microorganisms but may also involve vertebrates. Vertebrate carrion use is referred to as scavenging (Deron et al. 2006), although some invertebrates such as aquatic crustaceans are also referred to as scavengers. These are all active "nutrient-aggressive" creatures or organisms that can quickly locate and dominate carrion resources. Microbes are most productive in warm humid environments and interact with both scavengers and insect colonizers (Anderson 1995). In concert with carrion-feeding arthropods, bacterial development contributes to rapid soft tissue decline by moving decomposing organisms to the carcass and feeding on the carcass itself. Under such conditions, within 7 days of the post-mortem period, insects and microbes account for 85% (50 kg) of tissue destruction in large carcasses (Spicka 2011). Nevertheless, most of the available carcasses are easily scooped up by vertebrates, if the remains are found before decomposition occurs (Putman 1977).

Arthropods and Carrion

Arthropods are particularly skilled at soft tissue removal and have a significant ecological role to play in the disposal of a large proportion of transport waste, making nutrients available to other organisms in the food web. Carrion becomes attractive to insects almost immediately after death, and remains are utilized as a nutrient-rich food supply during decay (Anderson 1995, 2010; Kreitlow 2010). Climate, inter- and intraspecies interactions, and other natural occurrences lead to a predictable ecological succession during decomposition. The carcass, its inhabitants, and the environment become an ephemeral ecological system.

Within minutes of death, arthropods are attracted to the volatile gases produced during all stages of decomposition. During autolysis, blow flies (Diptera: Calliphoridae) and flesh flies (Diptera: Sarcophagidae) are attracted to compounds released from the carcass. Fly oviposition is required during this stage of decomposition, before the tissue dries and becomes inhospitable for larval growth. Bacterial decay is the primary process of decay for invertebrates or vertebrates; decay occurs at a much slower rate than when assisted by invertebrate feeding (Payne 1965; Carter et al. 2007; Anderson 2010).

In moist areas of the carcass, early insects oviposit or larviposit young, which use the carcass as a food source during growth (Kreitlow 2010). Larval feeding digests the skin and liquefies soft tissue during the period of putrefaction (Carter et al. 2007). Diptera is otherwise attracted to the remains as the internal soft tissue is exposed due to bacterial deterioration. Predacious beetles (Coleoptera) may arrive and feed during rot, attracted by the mass of unprotected prey larvae and the decomposing carrion resource. There is rapid loss of biomass during putrefaction as the voracious larval mass feeds. Succession of insect larvae develops to pupae, while soft tissue rots and disintegrates due to bacterial putrefaction. As soft tissue becomes scarce, the hard tissue of the bone is left behind. A number of species of Coleoptera may remain during this phase, feeding on the dry, mummified tissue that sticks to the bones (Kreitlow 2010). Skin beetles (Coleoptera: Dermestidae) can perforate mummified and desiccated tissue and help decompose the skeleton by disarticulating the skeleton as a whole and allowing for physical degradation processes to reach the hard tissues. Many of these species are also responsible for the deposit of pupae, exuvia, and insect carcasses in the nutrient pool in the environment. This is crucial, as it contributes to the recalcitrant cycling of capital.

Once the skeletal system is exposed, it can be degraded through physical, chemical, or biological processes. Certain Lepidoptera species have the capability to feed directly on hard tissue. The moth species *Tinea deperdella* (Lepidoptera: Tineidae) larvae feed organic horn components, while *Ceratophaga vicinella* (Lepidoptera: Tineidae) can feed on tortoise shell plates (Behrensmeyer 1978; Deyrup et al. 2005). Termites (Blattodea: Termitoidae) are known to tunnel through the bone and are thought to derive nitrogen and phosphorus nutrients from the bone (Prado et al. 2012). The bulk of bone degradation, however, is accomplished by scavengers and physical weathering.

Carrion Use by Vertebrates

Carrion use by vertebrates has interesting implications for ecosystems. From a physical perspective, the molecules that make up an animal's body become progressively less aggregated after death. The speed and the extent to which these molecules are dispersed are explicitly related to the scavengers that feed on the carcass. Microbial decomposition creates well-defined and well-documented cadaver decomposition islands (CDIs), as nutrients released through microbial action are incorporated into the detrital pathway immediately below and near where the carcass decomposes (Johnson 1975; Melis 2007; Moore et al. 2004). Subsequently, those nutrients are mobilized by the adjacent flora, creating localized effects. For large carcasses, elevated nutrient levels coincident with a CDI have been detected in soil and foliar samples for years after carcass decomposition (Danell et al. 2002; Melis 2007). The additive effects of arthropod pupation and dispersal plus microbial action increase the final nutrient dispersion from the carcass further into the surrounding environment (Carter et al. 2007).

Vertebrate scavenging represents the widest dispersal of nutrients and energy from carcasses as vertebrate movement scales away to the broader landscape (Barton et al. 2012; Payne and Moore 2006). In this process, vertebrate scavengers diminish the formation of CDIs by scattering and incorporating portions of the carcass into the surrounding ecosystem. Thus, the spatial heterogeneity of resources that carcasses contribute to the landscape depends in part on the identity of the scavengers that feed on them. Carcasses, then, can contribute nutrients and energy to initiate resource hot-spots that add to landscape complexity. In other instances, carrion is effectively recycled among higher trophic levels by scavengers with limited direct inputs to the detrital pathway (Bump et al. 2009). We are just beginning to understand the ecological effects that carcasses have on the landscape, and further work is needed to determine how vertebrate scavengers influence those effects.

In addition to impacting landscape heterogeneity by the disruption of decomposition patterns, scavengers may also shape the structure of vertebrate communities through interactions with predators. In systems where predators are smaller in size than their prey, some portion of each fresh kill may not be immediately consumed by the predator (Houston 1979; Hunter et al. 2007). For example, the gleanings from partially consumed or usurped predator kills on the African savannah sustain an entire guild of vertebrate scavengers. Scavenging activity such as this is often grouped with kleptoparasitism in the literature, and it places a limit on the amount of resources the original predator can gain from each kill with potentially cascading effects (Vucetich and Peterson 2004). In Isle Royale National Park, USA, raven scavenging of wolf (*Canis lupus*)-killed moose (*Alces alces*) carcasses is apparently extensive enough to make increased pack sizes beneficial, despite the smaller share of each moose that an individual wolf receives when hunting in these larger packs. This is due to the larger pack's ability to share the energy necessary to chase ravens away from a carcass resource; smaller packs are unable to manage large raven flocks and will lose a substantial portion of uneaten carrion (Vucetich and Peterson 2004). Eurasian Lynx (*Lynx lynx*) in areas of Europe have been shown to increase their kill rates when the carcasses they cache are scavenged by grizzly bears (*Ursus arctos*) (Krofel et al. 2012). Further, a review of the literature revealed lower but still significant scavenging of felid-killed carcasses from Europe, North America, and Africa, indicating that the phenomenon is not restricted to any particular ecosystem (Krofel et al. 2012). Given the importance of top-down effects in many ecosystems, even a minor alteration to predation rates as driven by vertebrate scavengers may cause a significant flux in community composition.

Another possible effect of vertebrate scavengers on vertebrate communities is related to the fact that many facultative scavengers (e.g., those organisms that supplement their main diet with carrion; e.g., crows, raccoons, canines) are also predators. Researchers found that the abundance of prey species (i.e., hares, *Lepus* sp., and squirrels, *Sciurus* sp.) decreased in sectors containing a carcass based on evidence from tracks in snow. An interesting hypothesis emerged, in which the scavengers that are recruited to a carcass may have temporarily played the dual role of increasing predator abundance near each carcass (Krofel et al. 2012). It is unclear from this study whether tracks of the prey species might have declined because incidental predation by facultative scavengers reduced the abundance of prey species near carcasses or because of nonconsumptive effects such as altered behavior by prey species in the vicinity of carcasses due to the presence of predatory scavengers (Alvarez et al. 1976). However, the second possibility adds an interesting dimension to the "landscape of fear" hypothesis, in which movements and foraging decisions by prey species are influenced by a perceived risk of predation.

Some of the most visible and well-studied effects of vertebrate scavengers on communities can be seen in the interactions of scavengers on carrion. Dominance hierarchies and structural differences within and among species of vultures in both the Old and New World (Alvarez et al. 1976; Bellan et al. 2013; Houston 1986; Hunter et al. 2007; Shivik 2006) have established competitive frameworks that delineate resource acquisition at carcasses. Superficially, larger species tend to dominate smaller species for access to carcasses, but larger species also facilitate access to carcasses for a variety of less-specialized reasons, whereas smaller species facilitate access by breaking through the thick hides of larger carrion (Bellan et al. 2013; Green et al. 2004; Oaks et al. 2004; Ogada et al. 2012).

Ecosystem Services Provided by Vertebrate Scavengers

Vertebrate scavengers provide various ecosystem services, including cultural (e.g., spiritual value) and nutrient cycling (supporting services and carcass removal from the environment [regulating services]) (Markandya et al. 2008; Pain et al. 2003). Four of the most prominent services provided are critical linkages in food webs, nutrient distribution within and among ecosystems, economic benefits related to sanitary measures, and altered disease dynamics. Vertebrate scavengers, both obligate (e.g., those scavengers that rely entirely on carrion as a food source; e.g., vultures) and facultative, play under appreciated but pivotal roles in maintaining healthy ecosystem function.

Critical Linkages in Food Webs

Historically, the prevalence of scavenging activities has been greatly underestimated. However, this is based upon recognition that:

1. In most ecosystems, a large number of animals die from causes other than predation and thus become available to scavengers.
2. Most carcasses are scavenged by vertebrates before they are completely decomposed by arthropods and bacteria.
3. Almost all carnivorous animals are facultative scavengers; the importance of scavenging in food webs seems unsurprising (Payne 1965).

It has been estimated that links are underrepresented in food web research by 16-fold (Wilson and Wolkovich 2011). The omission of these connections in ecological models is striking, especially considering the role that the number and strength of trophic connections are known to play in promoting the persistence and stability of ecological communities and the ecosystem services they deliver (Costello et al. 2009; Gill-King 1997). The omission of scavenging activities from food web models has largely been the result of oversimplification, that is, the treatment of all types of detritus (from low-quality dead plant material to high-quality animal carrion) as a single resource pool (Wilson and Wolkovich 2011). However, food webs are increasingly recognized as complex and highly interconnected (e.g., Polis 1981; Polis and Strong 1996), and the importance of detritus, especially animal carrion as a distinct resource, is becoming widely accepted (Benninger et al. 2008; Moore et al. 2004; Payne 1965; Wilson and Wolkovich 2011).

Scavenging activities may be especially important for the resiliency of food webs. The stabilizing nature primarily results from scavengers feeding on carrion from a wide variety of species, resulting in a high number of interspecies links from scavenging, making webs more interconnected (Wilson and Wolkovich 2011). Also, the use of carrion as a supplemental food resource during prey shortages for species that are primarily predators might add to the stabilizing nature of scavenging (Gill-King 1997; Payne 1965). As ecosystems are increasingly subject to multiple stressors from human activities, it is important to gain a better understanding of the intrinsic properties of food webs that promote stability, and carrion use by vertebrates appears to be one such factor.

Distribution of Nutrients Within and Among Ecosystems

In addition to acute visual and/or olfactory abilities for detecting carcasses, one of the key attributes of the most successful scavengers is the ability to quickly and efficiently travel great distances in search of carrion, which in many cases is unpredictable and ephemeral (Kellerman et al. 1976; Metcalf et al. 2015; Micozzi 1986). As a result, obligate scavengers such as vultures generally have very large home ranges (Payne

1965; Shivik 2006). These scavengers, as well as some facultative scavengers such as certain bottom-dwelling marine species, often disperse carrion across large areas (Lyman and Lyman 1994; Sagara 1995). This dispersion of carrion biomass by vertebrates is especially evident when carrion is initially concentrated spatially. For example, single large carcasses like whales or elephants are often visited by multiple scavengers that range widely and therefore transport the nutrients from those carcasses over great distances (Tibbett and Carter 2003; Sagara et al. 2008). Periodic salmon die-offs represent a significant annual pulse of marine-derived nutrients that can be disseminated from aquatic to terrestrial ecosystems through vertebrate scavenging (Evans 1963; Menon et al. 2011). Movement of salmon carcasses to terrestrial habitats by feeding scavengers also links terrestrial invertebrate communities to marine-derived nutrients by providing substantial carrion subsidies to ovipositing flies and terrestrial invertebrate scavengers unable to access carrion in aquatic habitats (Hocking and Reimchen 2006). Cross-habitat nutrient transport can produce a variety of important outcomes in recipient systems, and scavengers can play a significant role in moving "ecological subsidies" between habitats. For example, the use of ocean-derived carrion by terrestrial mammals and birds is extensive and may strongly influence dynamics of coastal food webs (Putman 1978; Schlacher et al. 2013; Van Belle et al. 2009).

The Role of Scavengers in Disease Ecology

Large carcasses may serve as incubators for many types of infectious materials. Because many mammalian and avian species often visit single large carcasses, and scavengers often have large home ranges (discussed earlier), it has been suggested that vultures and other wide-ranging vertebrate scavengers might facilitate the spread of pathogens across large areas (Jennelle et al. 2009). For example, vultures might harbor infectious materials on their feet and feathers and introduce those materials across their habitat (Houston 1986). It has also been suggested that scavengers might exacerbate production of anthrax spores by opening carcasses and thus suppressing the ability of anaerobic bacteria residing inside carcasses to antagonize vegetative anthrax cells and, as a result, impede sporulation.

However, vultures are generally very resistant to diseases, a trait common among many successful scavengers (Shivik 2006). The digestive tract of griffon vultures is likely to kill most pathogenic bacteria, given that the pH of the stomach ranges from 1 to 2. Thus, as vultures forage, they likely reduce the proliferation of diseases, at least at the local scale, by removing infected carcasses from the landscape (Houston 1979; Ogada et al. 2012).

It is thus unclear the extent to which scavengers remove infectious materials from the landscape or, alternatively, spread pathogens across large areas. Even so, it seems likely that the identity of the vertebrate species scavenging at carcasses determines, at least in part, whether or not diseases are proliferated or impeded. For example, the near-extinction of several vulture species in south Asia due to the use of toxic livestock chemicals allowed cattle carcasses to remain in the landscape for longer time periods and thus were made available for consumption by feral dogs and rats, which apparently resulted in population increases in those species (Green et al. 2004; Oaks et al. 2004; Pain et al. 2003; Paine 1969). Markandya et al. (2008) estimated that the total costs to human health (including rabies cases from dog bites) that resulted from the severe vulture declines totaled over $34 billion from 1993 to 2006 (Markandya et al. 2008). This is in addition to the increased carcass decomposition times resulting from the exclusion of vultures from large animal carrion (Ogada et al. 2012). In the absence of vultures, the number of mammals scavenging carcasses, the average time spent at carcasses by mammals, and the number of contacts between mammals increased substantially. Such increases in inter- and intraspecific interactions due to increased persistence times of carcasses could increase the probability of disease spread within and among species (especially mammals), particularly for some important zoonotic diseases including rabies. Clearly, the role of vertebrate scavengers in disease proliferation is complex, and more research is needed to elucidate factors that influence disease dynamics with regard to scavenging ecology (Jennelle et al. 2009).

Microorganisms

Living organisms are home to a wide range of microorganisms. Most body parts house a substantial population of bacteria, including the skin, mouth, and intestines (Costello et al. 2009). These populations are integral for decomposition. In particular, the microbial biomass in the gut of the carcass drives decomposition, resulting in putrefaction, which significantly affects the appearance, odors, and chemistry of decomposing organisms (Benninger et al. 2008; Gill-King 1997). Microorganisms digest macromolecules (carbohydrates, proteins, lipids, and nucleic acids) into simpler compounds, which are then released into the surrounding environment (Benninger et al. 2008). Microorganisms in the genera *Bacteroides*, *Bifidobacterium*, *Clostridium*, *Enterobacter*, and *Streptococcus* are likely the main drivers behind the putrefaction of large carcasses (Kellerman et al. 1976; Micozzi 1986). Recent studies of carcass skin and mouth microbiology with respiratory aerobic Actinobacteria have seen a shift from a population dominated by aerobic bacteria to one dominated by anaerobic energy bacteria, especially *Lactobacillus*, *Staphylococcus*, and *Streptococcus* (Metcalf et al. 2015). This is likely because the decomposing carcass is a specialized environment that selects bacteria able to withstand an environment that is dry and acidic and has low oxygen availability. The skin of a carcass must degrade to allow oxygen to return to the internal carcass cavity and resume aerobic microbial activity (Carter et al. 2007). Bacterial proliferation requires moderate temperatures and humidity; dehydration and cooling temperatures inhibit bacterial growth (Lyman and Lyman 1994).

Carcass degradation conditions support many bacterial taxa, and there is often ample moisture and soft tissue to support other microorganisms as well. Fungus can proliferate as moisture and nutrient pools deplete, since hyphae cross gaps once these resources are depleted (Sagara 1995; Sagara et al. 2008; Tibbett and Carter 2003). Fungi are often found on the skin and other exposed carcass surfaces, and it is common for hyphae to spread throughout the carcass. Bacteria and fungi work together to actively decompose all tissues in a carcass, including hydrocarbons, the skin, hair, and even bone. Decomposer organisms possess several reduction enzymes, many of which are rare in other fungal classes or bacteria (Sagara 1995).

Fungi may develop carrion-related fruiting structures (Sagara 1995). Tibbett and Carter (2003) studied taphonomic fungi from Ascomycota and Basidiomycota, including various fungal taxa. Also known as putrefaction fungi, they can be used to process the more recalcitrant parts of the carcass: the hair, skin, and bone (Menon et al. 2011). Putrefaction fungi prevalence often directs the carcass decomposition effect on soil microbial communities. The presence of soil microorganisms due to increased microbial biomass and carbon evolution reacts quickly to newly available carcasses (Carter et al. 2008; Putman 1977). The result on soil microbial biomass is not unexpected: carcass breakdown leads to large amounts of chemical substances being released into soil. Considerably higher amounts of nitrogen, phosphorus, potassium, and sodium are found in carrion-related soils than in soils not affected by carrion (Benninger et al. 2008; Bump et al. 2009; Parmenter and Lamarra 1991; Van Belle et al. 2009). Parmenter and MacMahon (2009) have observed relatively higher loss levels of sulfur in carcasses, which could also be shown to increase soil sulfur.

It is through these mechanisms that the nutrients bound up in carrion exit into the terrestrial environment. There are notable differences, however, when dealing with carrion resource pulses in aquatic systems.

Carrion Ecology in the Aquatic Ecosystem

Decomposition ecology has focused primarily on the role of carrion ecology in terrestrial ecosystem functioning (Swift et al. 1979). Carrion decomposition in aqueous environments may follow similar patterns to what has been observed in terrestrial ecosystems; however there are major differences in processes constrained to aquatic ecosystems. Among the Earth's ecosystems, aquatic habitats are comprised of three types:

1. Freshwater including standing waters (lentic), ponds, and lakes and moving water systems (lotic)
2. Transitional communities such as estuaries or embayments and other wetlands
3. Marine systems that include shorelines, inland saline lakes, and open ocean

Decomposition of carrion in each of these environments follows a predictable pattern with unique elements according to the environment. What is learned about the differences can be gathered from comparisons of lentic and lotic freshwater systems, as well as marine systems. Thus, with water bodies of all size and type, it is a series of complex physical, chemical, and biological interactions that allow biogeochemical cycles and biotic communities to function, which in turn ultimately influences the role of decomposition in these systems (Bump et al. 2009).

Overall, different physical and chemical characteristics influence and sustain ecological systems (Barton et al. 2012; Payne and Moore 2006). Although decomposition in aquatic environments may be mediated by biological mechanisms such as microbial trophic interactions with macroinvertebrates, it is the collection of physicochemical parameters which determine the dominant direction of decomposition in aquatic habitats (Kjorlien et al. 2009).

Just like terrestrial ecosystems, the most significant parameters in aquatic carrion decomposition include temperature, hydrology, and nutrient availability. The thermal variability of aquatic species varies spatially and temporally across all types of aquatic environments, and this variation can be caused by both natural and anthropogenic changes (Reeves 2009). In lotic systems such as streams and rivers, both geographic and meteorological influences typically explain the thermal variability patterns, which are affected by the size, shape, and continuity of the aquatic system (Kjorlien et al. 2009).

Thermal stratification, or the vertical water temperature layering in lakes and oceans, has profound effects on the physical, chemical, and biological properties of these environments (Reeves 2009). Such disparate regions of these structures are (1) the epilimnion, the lowest and warmest; (2) the metalimnion or thermocline, the intermediate layer showing a rapid decline in depth temperature; and (3) the hypolimnion, the deepest and coldest, darkest regions below the thermocline. While epilimnion is a region where photosynthesis fixes nutrients and where allochronic organic material required for biological communities is produced, hypolimnion is the environment where decomposition products accumulate. Hypolimnion temperatures may be close to 4 °C—the average water and temperature density requiring special organism adaptations to survive and metabolize organic matter decomposition products (Schindler 1991).

Temperature is a major factor in the life history of aquatic organisms which are closely involved in the decomposition of organic material from aquatic bacteria invertebrates to freshwater and marine fish (Burkepile and Hay 2006; Pepin and Myers 1991; Pomeroy and Wiebe 2001; Sweeney and Vannote 1978; Ward and Stanford 1982; White et al. 1991). For example, when carrion half-life was plotted against temperature in a variety of freshwater systems from streams to lakes, there was a negative correlation between carrion half-life and rising temperatures; this was probably due to the positive relationship between temperature and most metabolic rates of bacteria and scavengers (Chidami and Amyot 2008).

Therefore, it should not be surprising that the decomposition rate of organic matter in lotic and lentic systems tends to rise depending on temperature, regardless of system or seasonal fluctuations in temperate water bodies or in more stable climates such as tropical seas. The effect of temperature on poikilothermic species involved in decomposition activities is considered to be a dominant factor in the processing of aquatic organic matter (Merritt and Wallace 2001; Webster and Negri 2006).

Gas Exchange

Dissolved oxygen and temperature are two of the essential factors influencing the ecology of lakes and ponds (Addy and Green 1997). Dissolved oxygen is the amount of oxygen in the solution and may fluctuate frequently depending on temperature, salinity, altitude, groundwater inflow, anthropogenic activities, or events (Addy and Green 1997). For example, more oxygen can be retained in cold water solution than warm water, varying from diel changes, i.e., 24 h, to seasonal fluctuations.

In solution, many aquatic species are naturally bound to breathe oxygen; their survival and subsequent role in decomposition require oxygen. Since oxygen solubility in water is negatively associated with temperature, certain classes of aquatic invertebrates that play a significant role in carrion tissue consumption may be excluded from such low-oxygen environments (Merritt and Wallace 2001). Therefore, depending on the oxygen concentration in a given habitat, the fauna available to colonize the carrion may be quite different (Hobischak 1997). During high rates of decomposition, such as after algal bloom, dissolved oxygen levels will crash and cause massive mortality of higher organisms involved in the lotic and lentic decomposition process (Osmond and Grace 1995). Although deep-sea benthic ecosystems are complex, dissolved oxygen concentrations are either very low or completely absent. This is a problem for most marine organisms, suggesting either non-existent or very slow decomposition at depths below 2000 m through bacteria (Heinrich 1974; Tunnicliffe et al. 2003).

Nutrient Limitations

In terms of nutrient availability, it is important to understand how nutrients and other chemicals in aquatic environments influence the basis of any food web by decomposing plant matter.

The fate of nutrients and compounds released from decomposing plant material in aquatic ecosystems may be transferred to soluble or particulate form (streams), absorbed into the soil (wetland) by microbial decomposers, or eventually transformed biotically and released into the atmosphere. Phosphorus supply in freshwater ecosystems such as streams and some wetlands is considered to be a resource that restricts plant growth (Marzolf et al. 1994). In these forms of aquatic ecosystems, specifically northern latitude arctic and subarctic waters with salmon subsidies as well as subtropical wetlands (e.g., Everglades, Florida, USA), a fish detritus pathway is likely to compensate for seasonal variations in the supply of phosphorus and, in turn, to facilitate the necessary nutrient cycling.

Other Chemical Factors

Some water chemistry factors, such as pH and salinity, may indirectly affect the decomposition of carrion by influencing the variety of cycle-initiating species, specifically primary and secondary decomposers such as bacteria, fungi, and invertebrates, thus affecting the functioning of the ecosystem in general. With the exception of anthropogenic sources, natural freshwater ecosystem pH ranges from 6.5 to 8.5, while normal freshwater pH ranges from 7.8 to 8.4 (Berezina et al. 1999; Paletta 1999). Nevertheless, anthropogenic activities have acidified streams and rivers on a global scale, and this process has been shown to have a negative impact on a number of aquatic micro- and macro-organisms important for the decomposition of aquatic organic material, including carrion (Schindler 1991).

A microorganism community of fungi and bacteria contributes to a breakdown of dead plant or animal matter in streams (Anderson and Sedell 1979). Because few terrestrial fungi can live in an aqueous environment, aquatic hyphomycete fungi are considered more important than bacteria when decomposing organic matter (Barlocher and Kendrick 1974). However, many of these fungi are unable to grow in water at a pH of 5.0. Interestingly, for the polysaccharide biofilm matrix that colonizes leaf matter, a pH above 5.0 is needed for bacterial growth and survival (Richards and Villet 2009).

Carrion falling into freshwater lakes or marine environments may cause long-term disruptions over large temporal and spatial scales through the release of carbon and other nutrients (Asaoka et al. 2008; Frost and Hicks 2012). Because phosphorous (P) can be a limiting nutrient in aquatic ecosystems, and nutrient budgets are closely linked to fish, a significant portion of total fish P is housed in the bones and scales. Parmenter and Lamarra (1991) proposed vertebrate wildlife carrion as a P sink in the P cycle rather than a recyclable nutrient storage pool. However, a dead fish or other wildlife carrion's trajectory typically begins with an initial sinking, followed by a surface rise and subsequent sediment fall. Eventually, carrion sources can undergo additional changes that may lead to altered decomposition results (Haefner et al. 2004). In a matter of days, littoral fish carcasses may be contaminated by terrestrial and aquatic invertebrates and sometimes by vertebrates (Chidami and Amyot 2008). Conversely, in deeper waters, fish carcasses falling into pelagic zones may show a slower rate of decomposition due to thermocline and may take months to re-enter the mineralized food chain. This half-life carcass gap dictates how quickly carcass-bound P will qualify for biomass production and reintegrate into aquatic food webs (Chidami and Amyot 2008).

Most of the world's benthic oceans are nutrient-limited, with very slow cycling from surface waters to deep depths. As in freshwater streams, organic detritus input significantly affects nutrient cycling and therefore the structure of the marine ecosystem. The death of large whales (30–160 tons of adult body weight) and their resulting deep-sea collapse also subsidize deep-sea benthic areas with massive bursts of labile organic matter (Smith and Baco 2003).

Because whale biomass contributes to about 5% of particulate organic carbon (POC), the flow of POC subsidies to the marine POC is estimated to be less than 0.5% (Dahlgren and Eggleston 2000). Whales, however, do not sink as a large object, but are scavenged and sink as large, organic-rich lumps that can be spread over areas of 50 m^2 or more. As a result, the sediments below these whale falls were estimated to be maintained in one pulse, equal to approximately 2000 years of abyssal history of POC flux (Smith 1975). Smith and Baco (2003) have consistently estimated that with the North Pacific Gray whales' (*Eschrichtius robustus*) range, annual whale-fall pulse events may cause carcass bits nearest <16 km. Scattered carbon pulses, as whales fall on sea beds, essentially create high-quality or organic-rich "islands" in a nutrient-depleted benthic system that could last long periods of time (Stockton and DeLaca 1982). Within these habitats, the whale carrion detritus pool may be directly linked to key ecological services such as nutrient cycling within and near these carrion-island ecosystems.

The structure and role of organic ecosystem ecology and the connections between land and water food webs are well researched (Moore et al. 2004). Nonetheless, some studies find detritus as a low-nutrient resource pool and neglect the existence of carrion (Wilson and Wolkovich 2011). For example, plant waste is considered low quality and slow to decline in aquatic and terrestrial ecosystems (Cebrian 2004; Odum 2014). Therefore, detritus can be seen as a multiple resource reserve with a price gradient from the lowest to the highest for animal tissue (Moore et al. 2004; Wilson and Wolkovich 2011).

Carrion and the Environment

A carcass has significant impact on the environment (Risch et al. 2020; Benninger et al. 2008; Barton et al. 2019; Bump et al. 2009; Coe 1978; Danell et al. 2002). The extent of this effect is related to carcass size, but even a mouse carcass will result in a significant increase of insect activity for several hours (Putman 1977). The flush of compounds released into the soil and atmosphere by carrion decomposition has an effect on the chemistry of the environment around the carcass. It also impacts the biology of the area, including vegetation and soil (Yang et al. 2010). Nitrogen, phosphorus, and potassium release into the soil causes a significant increase in microbial activity and a significant change in the area's chemistry over time (Anderson and Weigel 2003). This level of disruption usually results in the death of all associated plant life (Towne 2000; Anderson and Weigel 2003), a phenomenon found in several studies (Anderson 2004; Spicka 2011; Towne 2000; Yang et al. 2008). Eventually, carcasses cause secondary succession within the ecosystem through input of high levels of nutrients that first destroy organisms in the immediate vicinity and then support of new colonization of the area (Hawlena et al. 2012; Murphy et al. 2012).

Biological factors, like microorganisms, invertebrate decomposers, and vertebrate scavengers, are the main contributors to the decomposition of soft tissue. Density-independent environmental factors that influence the decomposition process include environmental effects such as temperature, precipitation, moisture, and solar radiation (Benbow et al. 2013).

Ambient temperature is the most powerful of environmental factors, as warm temperatures encourage microorganism development in both the carcass and soil. Enhanced enzymatic and microbial degradation will result in rapid accumulation and subsequent release of carcass gases, attracting invertebrate and vertebrate scavengers (Carter et al. 2007). Under warmer temperatures, Diptera on a carcass are more common, and fly eggs survive better. Increased larval insect and scavenging activity may help digest soft tissue quickly, speeding the decomposition process (Bass 1997; Mann 1990).

Cold temperatures are less conducive to decomposition due to limited microbial, invertebrate, and vertebrate development. Temperatures below freezing pause insect development, and eggs and larvae cannot survive (Mann 1990). The decomposition cycle is delayed, resulting in lack of putrefactive gases to attract vertebrate scavengers. The effect is soft tissue survival indefinitely or until the environment allows carrion to decompose again (e.g., seasonal temperature changes) (Piombino-Mascali et al. 2017).

Equally important for reproduction and proliferation of bacteria and insect larvae is moisture (Bass 1997). Humidity and precipitation are therefore important environmental factors that can mediate these biological activities. While insects may still be present in arid environments, lack of moisture may desiccate fly eggs and inhibit larval growth. Rapid loss of moisture is consistent with soft tissue desiccation, helping maintain mummification (Carter et al. 2007).

The degree of solar radiation to which remains are exposed impacts larval mass carcass location. In a shaded area, larval masses emerge on the carcass surface. However, when the carcass is in direct sunlight, larvae usually create a darker cavity environment (Bass 1997). Larvae feed on internal tissues, leaving the skin as a protective sunlight shield. The result will be a skin layer matching the carcass body shape with few, if any, tissues or organs remaining in the cavity (Benbow et al. 2015). Besides climatic conditions, the decomposition process will also affect soft tissue degradation. Most animals die in terrestrial (surface) weather. Nevertheless, many scavengers can bury carrion to ensure that other scavengers cannot use the food source. Burial effect decreases biological and chemical decomposition processes. Depending on burial depth, temperature may be lower than ambient temperature as the soil acts as a solar radiation foil (Rodriguez and Bass 1985). Soil is vital to the process, as it covers remains. Microbial biology, soil pH, oxygen content, and water-holding and ion-exchange ability will become important factors that can affect the degree of decomposition or preservation of a carcass. Typically, buried animals decompose more slowly than surface animals (Rodriguez and Bass 1985).

The process of plant death during carcass decomposition has not been extensively studied. Plants are often exposed to significant pH shifts and concentrations of unusually experienced nutrients, one of the primary reasons for carcass decay. Larger carcasses are associated with chlorotic and dead vegetation through maggot migration. The maggots will eventually die, ultimately fostering the robust growth of new plants. Pioneer species often outperform their native counterparts. Growth of these pioneering species may occur after 1 year for large carcasses and within 1 year for smaller carcasses (Anderson and VanLaerhoven 1996; Towne 2000). The smallest carcasses, such as mice, may not affect the surrounding vegetation. They may simply be too small, or ants and flies may absorb nutrients rather than release them into the soil. Whatever the exact mechanisms, carcasses may cause disturbances within a plant community which facilitate the creation of pioneering species that would otherwise be absent.

Both the atmosphere and the soil release volatile organic compounds (VOCs). It has long been established that even small carcasses emit excessive carbon dioxide into the atmosphere (Carter et al. 2007; Putman 1977). More recently, it has been discovered that a wide range of VOCs are actually released into the atmosphere through the decomposition process (Dent 2004; Paczkowski 2011). Short-chain alcohols (e.g., C3–C8), polysulfide compounds (e.g., dimethyl disulfide, dimethyl trisulfide, and dimethyl tetrasulfide), volatile fatty acids (e.g., C2–C6), and aromatics (e.g., phenol, indole, and skatole) are often released. These chemicals have a significant impact on soil-dwelling organisms and are used by cadaver detection dogs in locating remains (Forbes 1925).

Although many aspects of carcass decomposition have been clarified in recent years, it has long been understood that carcass decomposition leads to nutrient cycling in and between environments. Carrion serves as a nutrient island in the ecosystem and thus contributes to biodiversity by serving as a habitat for organisms using carcass material for food and/or shelter. Carcass-related successions are driven by changing environmental and soil conditions as resources are consumed (Bornemissza 1957; Hewadikaram and Goff 1991; Schoenly 1983).

Succession is a series of developments in which each stage, through its behavior, promotes the development of the next stage. For example, the initial oviposition of blowfly eggs leads to the development of the first instar blowfly larvae. The presence of these blowfly larvae promotes the growth of other larvae which feed insects such as rove beetles (Schoenly 1983). In contrast, inhibition occurs when the colonization of resources by one species inhibits the entry of other organisms into the system. This may occur by predating one species to another, by reducing resources to a degree that does not sustain subsequent species, or by competing with toxins (Brundage 2012). Tolerance is a developmental sequence in which some species colonize and create an environment, regardless of the presence or absence of other species (Anderson 2011).

Nutrient Cycling in a Food Web

Intra- and inter-kingdom competition for such a high-quality food resource is a general cause of detritus breakdown by bacteria, invertebrates, or vertebrates (Deron et al. 2006). Carrion quality may affect nutrient cycling, food web structure, and food web stability. Higher-quality, more nutrient-dense carrion will lead to efficient assimilation of nutrients bound in the detrital pool. To control market dynamics, further assimilation productivity could also regulate the prey species population and slowly stabilize food chains (Wilson and Wolkovich 2011). Consumption of vertebrates floating in water systems through terrestrial arthropod taxa may accelerate the carrion's ephemeral status at the surface and facilitates nutrient transport from the surface to the benthos.

The difference in spatial and temporal distribution of transport resources is twofold among terrestrial and marine environments (Baldock et al. 2004). First, the transportation mechanism is directly linked to the physical properties of water and air, such as narrow temperatures, relative water density, and the inherent three-dimensional nature of aquatic systems, which may affect movement on the later side of the water column through wave action (Beasley et al. 2012). For example, turbulent water due to current or waves causes downward and horizontal movement of carrion in water, while bacterial gas production leads to upward vertical movement of the carcass. This movement increases exposure to water-borne detritivores and an increased chance of consumption.

Second, the transit indicator species on terrestrial systems are defined in an ecological way by their coevolution to the transportation of feed and their trophic relationship with the decomposition of carcasses (Merritt and Wallace 2001). Terrestrial species utilizing carrion must locate the stationary carrion resource, necessitating coevolved mechanisms to allow for efficient search-and-find behaviors. Aquatic detritivores do not need such sophisticated behaviors, since the carrion is not stationary and can therefore move to the scavengers just as efficiently as the scavengers can move to it. For example, in freshwater systems, a taxonomic "normal" community does not occur (Bo et al. 2014). Instead, an approach from a practical feed group (PFG) may be better suited to the freshwater carrion arthropod.

The PFG approach classifies aquatic invertebrates on the basis of a set of morphological and behavioral adaptations that have developed into their basic nutritional resources—carcass or algal and fungal materials growing on the carcass. Transportation invertebrates in all aquatic environments, however, are more known to be ecologically oriented to scavenging: populated patterns of species using natural forms of transportation (e.g., whale falls) or artificial (e.g., pig carcasses used in research) (Anderson 1999, 2009; Bo et al. 2014; Smith 1986).

Carrion succession animals in freshwater systems often consist mainly of crayfish (Decapoda: Cambaridae) and other omnivores, while shredders and some aquatic predators also participate in scav-gathering by removing small bits of tissue (Merritt and Wallace 2001; Wallace et al. 2008). Historically, areas of fast turbulent water called "riffles" found throughout pools and streams were believed to account for the bulk of tissue removal in submerged carcasses. Observation has recently shown, however, that shredders play a more significant role than riffles in pool environments, and most shredders are present during all phases of decomposition. Collector filters are prevalent in riffle habitats, such as caddy-spinning and black larvae (Diptera: Simuliidae) (Merritt and Wallace 2001). Scraper mayflies were reported to have grazed on a microbial matrix containing a large algal portion during the fresh to floating degradation stages (Ephemeroptera: Heptageniidae). The main problem with the mixing of aquatic invertebrate taxa with a certain carrion decay stage is that aquatic invertebrates were not known to be strictly sarcophageal feeders. Several studies have therefore shown that a wide variety of taxa can occur spontaneously or in an unpredictable sequence in the decomposition of aquatic caries (Potapov et al. 2019; Barrios and Wolff 2011; Parmenter and Lamarra 1991; Wallace 2015; Sweeney and Vannote 1978).

Ecological Effect of Carrion

Spatial and temporal dynamics of detritus subsidies, together with a competitive hierarchy of organisms using these instruments, are important drivers of terrestrial system decomposition processes that are eventually related to necrophages through evolutionary processes (Beasley et al. 2012; Mondor et al. 2012). Transferring nutrients and carbon compounds from carrion through noninvasive means, such as leaching, or through active measures, such as detritivore feeding, can influence how effectively necrophagic populations may compete and exploit the ephemeral nature of carrion subsidies, thus affecting the rate of decomposition and ultimately the functioning of the ecosystem (Gessner et al. 2010) Thus, in systems such as freshwater and marine ecosystems, some species are essential to the transport of such nutrients. The impact of environmental factors such as climate is less researched, and thus it is less understood how they affect this rather poor balance between optional decomposers (e.g., invertebrates and vertebrates) and binding decomposers (e.g., food web stability bacteria). The reversal of nutrient subsidies affects not only the freshwater ecosystems that store salmon but also terrestrial insects and large mammalian fauna such as brown bears (*Ursus arctos* Linnaeus). If rising water temperature associated with climate change affects the pre-eminence of these invertebrate carrion feeders, processing and nutrient transfer of these subsidies can be changed in complex ways. Concomitantly, if water levels fall and water temperatures rise with impacts on salmon fry survival, potential subsidies to these programs are at risk of serious decline. Such a reduction in salmon carrion grants to these breeding waters may have a negative impact on riparian biodiversity by limiting the supply of carrion-dependent species (Wilmers et al. 2003). Conversely, climate change impacts in aquatic abyssal environments are not controlled as they are on land or, likely, freshwater systems (Beasley et al. 2012).

Chapter Summary

Carrion is an unavoidable and arguably critical aspect to ecology. Understanding its role in the overarching ecological system is vital to gaining a complete understanding of ecology in general. While the importance of carrion decomposition in the environment has long been understood, the mechanisms behind its importance have yet to be elucidated. This is an area ripe for investigation. Luckily, carrion is a model that is relatively easy to study and easier to replicate. There have also been groups working to synthesize research as it stands (see Benbow et al. (2015)). One hopes that future scientists will not ignore this critical area of ecological knowledge and will continue to investigate the role of carrion as it pertains to ecosystem health.

References

Addy K, Green L (1997) Dissolved oxygen and temperature. Nat Resour Facts 96(3):1–4

Alvarez F, de Reyna LA, Hiraldo F (1976) Interactions among avian scavengers in southern Spain. Ornis Scand 7:215–226

Anderson GS (1995) The use of insects in death investigations: an analysis of cases in British Columbia over a five year period. J Can Soc Forensic Sci 28 (4):277–292

Anderson GS (1999) Wildlife forensic entomology: determining time of death in two illegally killed black bear cubs. J Forensic Sci 44(4):856–859

Anderson GS (2004) Decomposition of carrion in the marine environment in British Columbia, Canada. Int J Legal Med 118(4):206–209

Anderson GS (2009) Factors that influence insect succession on carrion forensic entomology. CRC Press, Boca Raton, FL, pp 225–274

Anderson GS (2010) Decomposition and invertebrate colonization of cadavers in coastal marine environments current concepts in forensic entomology. Springer, Dordrecht, Netherlands, pp 223–272

Anderson GS (2011) Comparison of decomposition rates and faunal colonization of carrion in indoor and outdoor environments. J Forensic Sci 56(1):136–142

Anderson NH, Sedell JR (1979) Detritus processing by macroinvertebrates in stream ecosystems. Annu Rev Entomol 24(1):351–377

Anderson GS, VanLaerhoven SL (1996) Initial studies on insect succession on carrion in southwestern British Columbia. J Forensic Sci 41(4):617–625

Anderson TH, Weigel HJ (2003) On the current debate about soil biodiversity. Landbauforschung Volkenrode 53(4):223–233

Asaoka S, Yamamoto T, Yamamoto K (2008) A preliminary study of coastal sediment amendment with granulated coal ash-nutrient elution test and experiment on *Skeletonema costatum* growth. Mizu Kankyo Gakkaishi/J Jpn Soc WaterEnviron 31(8):455–462

Baldock J, Masiello C, Gelinas Y, Hedges J (2004) Cycling and composition of organic matter in terrestrial and marine ecosystems. Mar Chem 92(1–4):39–64

Barlocher F, Kendrick B (1974) Dynamics of the fungal population on leaves in a stream. J Ecol 62:761–791

Barrios M, Wolff M (2011) Initial study of arthropods succession and pig carrion decomposition in two freshwater ecosystems in the Colombian Andes. Forensic Sci Int 212(1–3):164–172

Barton PS, Cunningham SA, Lindenmayer DB, Manning AD (2012) The role of carrion in maintaining biodiversity and ecological processes in terrestrial ecosystems. Oecologia 2012:1–12

Barton PS et al (2019) Towards quantifying carrion biomass in ecosystems. Trends Ecol Evol 34(10):950–961

Bass WM (1997) Outdoor decomposition rates in Tennessee. Forensic taphonomy: the postmortem fate of human remains, 181–186

Beasley JC, Olson Z, DeVault T (2012) Carrion cycling in food webs: comparisons among terrestrial and marine ecosystems. Oikos 121(7):1021–1026

Behrensmeyer AK (1978) Taphonomic and ecologic information from bone weathering. Paleobiology 4 (2):150–162

Bellan SE, Turnbull PC, Beyer W, Getz WM (2013) Effects of experimental exclusion of scavengers from carcasses of anthrax-infected herbivores on *Bacillus*

anthracis sporulation, survival, and distribution. Appl Environ Microbiol 79(12):3756–3761
Benbow ME, Lewis AJ, Pechal JL, Tomberlin JK (2013) Seasonal necrophagous insect community assembly during vertebrate carrion decomposition. J Med Entomol 50(2):440–450. https://doi.org/10.1603/ME12194
Benbow ME, Tomberlin JK, Tarone AM (2015) Carrion ecology, evolution, and their applications. CRC Press, Boca Raton, FL
Benninger LA, Carter DO, Forbes SL (2008) The biochemical alteration of soil beneath a decomposing carcass. Forensic Sci Int 180:70–75
Berezina N, Kononenko N, Dvorkina G, Shel'deshov N (1999) Physicochemical Properties of Ion-Exchange Materials. Training Course (Kuban'State University, Krasnodar, 1999) [in Russian]
Bo T, Cammarata M, López-Rodríguez MJ, de Figueroa JMT, Baltieri M, Varese P, Fenoglio S (2014) The influence of water quality and macroinvertebrate colonization on the breakdown process of native and exotic leaf types in sub-alpine stream. J Freshw Ecol 29 (2):159–169
Bornemissza GF (1957) An analysis of arthropod succession in carrion and the effect of its decomposition on the soil fauna. Aust J Zool 5(1):12
Brundage AL (2012) Fitness effects of colonization time of Chrysomya rufifacies and Cochliomyia macellaria, and their response to intra-and inter-specific eggs and egg-associated microbes. Texas A&M University
Bump JK, Peterson RO, Vucetich JA (2009) Wolves modulate soil nutrient heterogeneity and foliar nitrogen by configuring the distribution of ungulate carcasses. Ecology 90(11):3159–3167
Burkepile DE, Hay ME (2006) Herbivore vs. nutrient control of marine primary producers: context-dependent effects. Ecology 87(12):3128–3139
Carter D, Yellowlees D, Tibbett M (2007) Cadaver decomposition in terrestrial ecosystems. Naturwissenschaften 94(1):12–24. https://doi.org/10.1007/s00114-006-0159-1
Carter DO, Yellowlees D, Tibbett M (2008) Temperature affects microbial decomposition of cadavers (*Rattus rattus*) in contrasting soils. Appl Soil Ecol 40:129–137
Cebrian J (2004) Role of first-order consumers in ecosystem carbon flow. Ecol Lett 7(3):232–240
Chidami S, Amyot M (2008) Fish decomposition in boreal lakes and biogeochemical implications. Limnol Oceanogr 53(5):1988–1996
Coe M (1978) The decomposition of elephant carcasses in the Tsavo (east) National Park, Kenya. J Arid Environ 1(1):71–86
Collins S, Stuart B, Ueland M (2020) Monitoring human decomposition products collected in clothing: an infrared spectroscopy study. Aust J Forensic Sci 52 (4):428–438
Costello EK, Lauber CL, Hamady M, Fierer N, Gordon JI, Knight R (2009) Bacterial community variation in human body habitats across space and time. Science 326(5960):1694–1697
Dahlgren CP, Eggleston DB (2000) Ecological processes underlying ontogenetic habitat shifts in a coral reef fish. Ecology 81(8):2227–2240
Danell K, Berteaux D, Bråthen KA (2002) Effect of muskox carcasses on nitrogen concentration in tundra vegetation. Arctic 55:389–392
Dent BB (2004) Review of human decomposition processes in soil. Environ Geol 45(4):576–585
Deron EB, Parker JD, Woodson CB, Mills HJ, Kubanek J, Sobecky PA, Hay ME (2006) Chemically mediated competition between microbes and animals: microbes as consumers in food webs. Ecology 87 (11):2821–2831
Deyrup M, Deyrup N, Eisner M, Eisner T (2005) A caterpillar that eats tortoise shells. Am Entomol 51 (4):245–248
Evans W (1963) The Microbiologícal Degradation of aromatic compounds. Microbiology 32(2):177–184
Forbes SA (1925) The lake as a microcosm. Ill Nat Hist Surv Bull 15(09):537
Frost PC, Hicks AL (2012) Human shoreline development and the nutrient stoichiometry of aquatic plant communities in Canadian shield lakes. Can J Fish Aquat Sci 69(10):1642–1650
Gessner MO, Swan CM, Dang CK, McKie BG, Bardgett RD, Wall DH, Hättenschwiler S (2010) Diversity meets decomposition. Trends Ecol Evol 25 (6):372–380
Gill-King H (1997) Chemical and ultrastructural aspects of decomposition. In: Forensic taphonomy: the postmortem fate of human remains. CRC Press, Boca Raton, FL, pp 93–108
Green RE, Newton I, Shultz S, Cunningham AA, Gilbert M, Pain DJ, Prakash V (2004) Diclofenac poisoning as a cause of vulture population declines across the Indian subcontinent. J Appl Ecol 41 (5):793–800
Haefner JN, Wallace JR, Merritt RW (2004) Pig decomposition in lotic aquatic systems: the potential use of algal growth in establishing a postmortem submersion interval (PMSI). J Forensic Sci 49(2):1–7
Hawlena D, Strickland MS, Bradford MA, Schmitz OJ (2012) Fear of predation slows plant-litter decomposition. Science 336(6087):1434–1438
Heinrich B (1974) Thermoregulation in endothermic insects. Science 185(4153):747–756
Hewadikaram KA, Goff ML (1991) Effect of carcass size on rate of decomposition and arthropod succession patterns. Am J Forensic Med Pathol 12(3):235–240
Hobischak NR (1997) Freshwater invertebrate succession and decompositional studies on carrion in British Columbia. Theses (Department of Biological Sciences)/Simon Fraser University
Hocking MD, Reimchen TE (2006) Consumption and distribution of salmon (Oncorhynchus spp.) nutrients and energy by terrestrial flies. Can J Fish Aquat Sci 63 (9):2076–2086

Houston DC (1979) The adaptations of scavengers. In: Serengeti: dynamics of an ecosystem. University of Chicago Press, Chicago, pp 263–286
Houston DC (1986) Scavenging efficiency of Turkey vultures in tropical forest. Condor 88(3):318–323
Hunter J, Durant S, Caro T (2007) Patterns of scavenger arrival at cheetah kills in Serengeti National Park Tanzania. Afr J Ecol 45(3):275–281
Jennelle CS, Samuel MD, Nolden CA, Berkley EA (2009) Deer carcass decomposition and potential scavenger exposure to chronic wasting disease. J Wildl Manag 73(5):655–662
Johnson MD (1975) Seasonal and microseral variations in the insect populations on carrion. Am Midl Nat 93:79–90
Kellerman T, Pienaar J, Anderson G, Naude T (1976) A highly fatal tremorgenic mycotoxicosis of cattle caused by Aspergillus clavatus. Onderstepoort J Vet Res 43 (3):147–154
Kjorlien YP, Beattie OB, Peterson AE (2009) Scavenging activity can produce predictable patterns in surface skeletal remains scattering: observations and comments from two experiments. Forensic Sci Int 188(1–3):103–106
Kreitlow K (2010) Insect succession in a natural environment. In: Byrd JH, Castner JL (eds) Forensic entomology: the utility of arthropods in legal investigations. CRC Press, Boca Raton, FL
Krofel M, Kos I, Jerina K (2012) The noble cats and the big bad scavengers: effects of dominant scavengers on solitary predators. Behav Ecol Sociobiol 66 (9):1297–1304
Lindeman RL (1942) The trophic-dynamic aspect of ecology. Ecology 23(4):399–417
Lyman RL, Lyman C (1994) Vertebrate taphonomy. Cambridge University Press, Cambridge
Mann RW (1990) Time since death and decomposition of the human body - variables and observations in case and experimental field studies. J Forensic Sci 35 (1):103–111
Markandya A, Taylor T, Longo A, Murty M, Murty S, Dhavala K (2008) Counting the cost of vulture decline—an appraisal of the human health and other benefits of vultures in India. Ecol Econ 67(2):194–204
Marzolf ER, Mulholland PJ, Steinman AD (1994) Improvements to the diurnal upstream–downstream dissolved oxygen change technique for determining whole-stream metabolism in small streams. Can J Fish Aquat Sci 51(7):1591–1599
Melis A (2007) Photosynthetic H 2 metabolism in *Chlamydomonas reinhardtii* (unicellular green algae). Planta 226(5):1075–1086
Menon LML, Prado KFB, Silva RHA d (2011) Histological evaluation of dentine and cementum after different periods of burial: an in vitro study. RSBO (Online) 8 (2):131–137
Merritt RW, Wallace JR (2001) The role of aquatic insects in forensic investigations. In: Forensic entomology: the utility of arthropods in legal investigations, pp 177–222
Metcalf JL, Carter DO, Knight R (2015) Microbiome studies of carrion decomposition. In: Carrion ecology, evolution, and their applications. CRC Press, Boca Raton, FL, pp 421–432
Micozzi MS (1986) Experimental study of postmortem change under field conditions: effects of freezing, thawing, and mechanical injury. J Forensic Sci 31 (3):953–961
Mondor E, Tremblay M, Tomberlin J, Benbow E, Tarone A, Crippen T (2012) The ecology of carrion decomposition. Nat Educ Knowledge 3:21
Moore JC, Berlow EL, Coleman DC, de Ruiter PC, Dong Q, Hastings A et al (2004) Detritus, trophic dynamics and biodiversity. Ecol Lett 7(7):584–600
Murphy SM, Wimp GM, Lewis D, Denno RF (2012) Nutrient presses and pulses differentially impact plants, herbivores, detritivores and their natural enemies. PLoS One 7(8):e43929
Oaks JL, Gilbert M, Virani MZ, Watson RT, Meteyer CU, Rideout BA et al (2004) Diclofenac residues as the cause of vulture population decline in Pakistan. Nature 427(6975):630–633
Odum EP (2014) The strategy of ecosystem development the ecological design and planning reader. Springer, New York, pp 203–216
Ogada DL, Keesing F, Virani MZ (2012) Dropping dead: causes and consequences of vulture population declines worldwide. Ann N Y Acad Sci 1249(1):57–71
Osmond C, Grace S (1995) Perspectives on photoinhibition and photorespiration in the field: quintessential inefficiencies of the light and dark reactions of photosynthesis? J Exp Bot 46:1351–1362
Paczkowski S (2011) Post-mortem volatiles of vertebrate tissue. Appl Microbiol Biotechnol 91(4):917–935
Pain DJ, Cunningham A, Donald P, Duckworth J, Houston D, Katzner T, Round P (2003) Causes and effects of temporospatial declines of Gyps vultures in Asia. Conserv Biol 17(3):661–671
Paine RT (1969) A note on trophic complexity and community stability. Am Nat 103(929):91–93
Paletta M (1999) Coral farming. Seascope, Spring, 1, 2
Parmenter RR, Lamarra VA (1991) Nutrient cycling in a freshwater marsh: the decomposition of fish and waterfowl carrion. Limnol Oceanogr 36(5):976–987
Parmenter RR, MacMahon JA (2009) Carrion decomposition and nutrient cycling in a semiarid shrub-steppe ecosystem. Ecol Monogr 79(4):637–662
Payne JA (1965) A summer carrion study of the baby pig *Sus scrofa* Linnaeus. Ecology 46(5):592–602
Payne LX, Moore JW (2006) Mobile scavengers create hotspots of freshwater productivity. Oikos 115 (1):69–80

Pepin P, Myers RA (1991) Significance of egg and larval size to recruitment variability of temperate marine fish. Can J Fish Aquat Sci 48(10):1820–1828

Piombino-Mascali D, Gill-Frerking H, Beckett RG (2017) The taphonomy of natural mummies. In: Taphonomy of human remains: forensic analysis of the dead and the depositional environment: forensic analysis of the dead and the depositional environment. Wiley, Hoboken, pp 101–119

Polis GA (1981) The evolution and dynamics of intraspecific predation. Annu Rev Ecol Syst 12(1):225–251

Polis GA, Strong DR (1996) Food web complexity and community dynamics. Am Nat 147(5):813–846

Pomeroy LR, Wiebe WJ (2001) Temperature and substrates as interactive limiting factors for marine heterotrophic bacteria. Aquat Microb Ecol 23 (2):187–204

Potapov AM et al (2019) Trophic position of consumers and size structure of food webs across aquatic and terrestrial ecosystems. Am Nat 194(6):823–839

Prado E, Castro C, Serrano A, Martins Da Silva P, GarcÍA MD (2012) Carrion flies of forensic interest: a study of seasonal community composition and succession in Lisbon, Portugal. Med Vet Entomol 26(4):417–431. https://doi.org/10.1111/j.1365-2915.2012.01031.x

Putman RJ (1977) Dynamics of blowfly, Calliphora erythrocephala, within carrion. J Anim Ecol 46 (3):853–866

Putman R (1978) Flow of energy and organic matter from a carcase during decomposition: decomposition of small mammal carrion in temperate systems 2. Oikos 31:58–68

Reeves NM (2009) Taphonomic effects of vulture scavenging. J Forensic Sci 54(3):523–528

Richards CS, Villet MH (2009) Data quality in thermal summation development models for forensically important blowflies. Med Vet Entomol 23:269–276

Risch AC et al (2020) Effects of elk and bison carcasses on soil microbial communities and ecosystem functions in Yellowstone, USA. Funct Ecol

Rodriguez WC, Bass WM (1985) Decomposition of buried bodies and methods that may aid in their location. J Forensic Sci 30(3):836–852

Sagara N (1995) Association of ectomycorrhizal fungi with decomposed animal wastes in forest habitats: a cleaning symbiosis? Can J Bot 73(S1):1423–1433

Sagara N, Yamanaka T, Tibbett M (2008) Soil fungi associated with graves and latrines: toward a forensic mycology. In: Soil analysis in forensic taphonomy. CRC Press, Boca Raton, FL, pp 79–120

Schindler D (1991) Aquatic ecosystems and global ecology. In: Fundamental of aquatic ecology, pp 108–122

Schlacher TA, Strydom S, Connolly RM (2013) Multiple scavengers respond rapidly to pulsed carrion resources at the land–ocean interface. Acta Oecol 48:7–12

Schoenly K (1983) Microclimate observations and Diel activities of certain carrion arthropods in the Chihuahuan Desert. J N Y Entomol Soc 91(4):342–347

Shivik JA (2006) Tools for the edge: what's new for conserving carnivores. Bioscience 56(3):253–259

Smith KG (1975) The faunal succession of insects and other invertebrates on a dead fox. Entomol Gazette 26:277

Smith KG (1986) A manual of forensic entomology

Smith CR, Baco AR (2003) Ecology of whale falls at the deep-sea floor. Oceanogr Mar Biol 41:311–354

Spicka A (2011) Carcass mass can influence rate of decomposition and release of ninhydrin-reactive nitrogen into gravesoil. Forensic Sci Int

Stockton WL, DeLaca TE (1982) Food falls in the deep sea: occurrence, quality, and significance. Deep Sea Res Part A Oceanogr Res Papers 29(2):157–169

Sweeney BW, Vannote RL (1978) Size variation and the distribution of hemimetabolous aquatic insects: two thermal equilibrium hypotheses. Science 200 (4340):444–446

Swift MJ, Heal OW, Anderson JM, Anderson J (1979) Decomposition in terrestrial ecosystems, vol 5. University of California Press, Berkeley, CA

Tibbett M, Carter DO (2003) Mushrooms and taphonomy: the fungi that mark woodland graves. Mycologist 17 (1):20–24

Towne EG (2000) Prairie vegetation and soil nutrient responses to ungulate carcasses. Oecologia 122 (2):232–239

Tunnicliffe V, Juniper SK, Sibuet M (2003) Reducing environments of the deep-sea floor. In: Ecosystems of the world. Elsevier, Amsterdam, pp 81–110

Van Belle LE, Carter DO, Forbes SL (2009) Measurement of ninhydrin reactive nitrogen influx into gravesoil during aboveground and belowground carcass (*Sus domesticus*) decomposition. Forensic Sci Int 193 (1–3):37–41

Vucetich JA, Peterson RO (2004) The influence of top–down, bottom–up and abiotic factors on the moose (*Alces alces*) population of Isle Royale. Proc R Soc Lond Ser B Biol Sci 271(1535):183–189

Wallace JR (2015) Aquatic vertebrate carrion decomposition. In: Benbow ME, Tomberlin JK, Tarone AM (eds) Carrion ecology, evolution, and their applications. CRC Press, Boca Raton, FL, pp 247–272

Wallace JR, Merritt RW, Kimbirauskas R, Benbow ME, McIntosh M (2008) Caddisflies assist with homicide case: determining a postmortem submersion interval using aquatic insects. J Forensic Sci 53(1):219–221

Ward JV, Stanford JA (1982) Thermal responses in the evolutionary ecology of aquatic insects. Annu Rev Entomol 27(1):97–117

Webster N, Negri A (2006) Site-specific variation in Antarctic marine biofilms established on artificial surfaces. Environ Microbiol 8:1177–1190

White PA, Kalff J, Rasmussen JB, Gasol JM (1991) The effect of temperature and algal biomass on bacterial production and specific growth rate in freshwater and marine habitats. Microb Ecol 21(1):99–118

Wilmers CC, Stahler DR, Crabtree RL, Smith DW, Getz WM (2003) Resource dispersion and consumer dominance: scavenging at wolf-and hunter-killed carcasses in greater Yellowstone, USA. Ecol Lett 6 (11):996–1003

Wilson EE, Wolkovich EM (2011) Scavenging: how carnivores and carrion structure communities. Trends Ecol Evol 26(3):129–135

Yang LH, Bastow JL, Spence KO, Wright AN (2008) What can we learn from resource pulses. Ecology 89 (3):621–634

Yang X, Quan Y, Vogt N, Looger LL, Lily Yeh J, Yuh Nung J (2010) Light-avoidance-mediating photoreceptors tile the Drosophila larval body wall. Nature 468(7326):921–926

Wildlife Forensic Pathology

John E. Cooper

"Kipya kinyemi ingawa kidonda"

(Swahili Proverb—a new thing gives pleasure, even if it is a wound).

Abstract

Global environmental threats present a challenge to scientists and the public alike. Both the disappearance of species and the accompanying decline in biodiversity urgently require a multidisciplinary and interdisciplinary scientific approach. This in turn warrants the application of techniques historically restricted to human and veterinary medical diagnosis or forensic investigations. In particular, there exists an unprecedented opportunity for input by pathologists and the appropriate use of their *modus operandi*, including necropsy and the laboratory examination of samples. The application of pathological methods to wildlife work is outlined, with particular reference to the investigation of crime and the emergence of conservation forensics as a subject in its own right.

Keywords

Pathology · Samples · Techniques · Wildlife · Forensic investigations · Conservation · Forensics

Introduction

Pathology was succinctly defined by the eminent Canadian veterinarian Reginald Thomson as "The science of the study of disease" (Thomson 1978). It is important to note that the discipline of pathology encompasses not only the examination of dead animals, which is a common association with the term, but also the investigation of samples obtained from the living, as well as the investigation of living patients.

In forensic parlance the term pathology is often employed in a rather different sense, implying the investigation of animals (including humans) and of samples in order to gain information for legal purposes.

Nevertheless, forensic pathology employs techniques most of which had originally been devised for the diagnosis and study of disease.

J. E. Cooper (✉)
Durrell Institute of Conservation and Ecology, The University of Kent, Canterbury, UK

S. C. Underkoffler, H. R. Adams (eds.), *Wildlife Biodiversity Conservation*,
https://doi.org/10.1007/978-3-030-64682-0_10

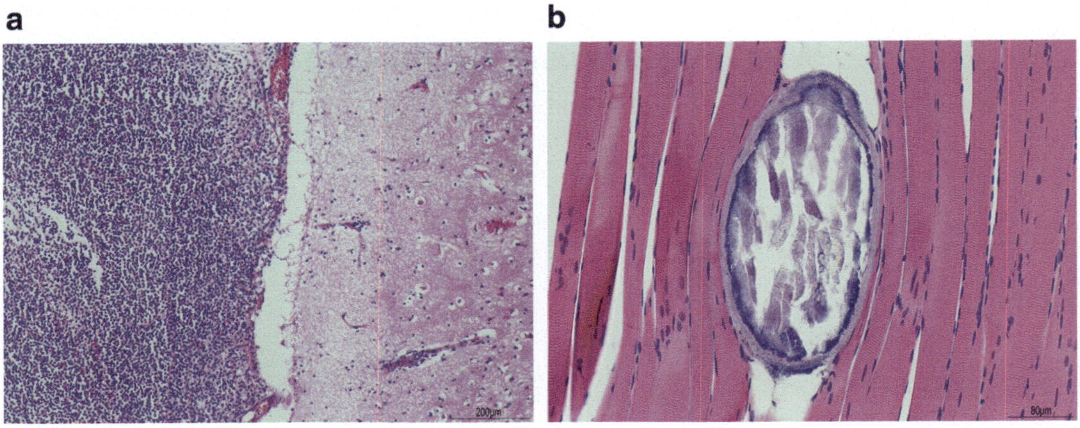

Fig. 1 Histological examination—inflammation and non-inflammation. HP (high power) of (**a**). Inflammation—section of brain showing meningitis—the right side of the picture is normal brain, the left a mass of purple (stained) inflammatory cells, and (**b**). A *Trichinella* parasite in skeletal muscle; there are no inflammatory cells around the parasite

Thus, in a forensic investigation, histological procedures may be used to identify structures, organisms and changes in tissues with a view to ascertaining whether they might throw light on a legal case. As an example, the presence or absence of an inflammatory reaction may significant (Fig. 1a, b).

Similarly, bacteriology and parasitology (both branches of pathology) may be helpful in determining the origin and provenance of a live or dead animal rather than throwing light on its health status (Fig. 2).

The development of wildlife forensic pathology as a discipline in its own right is linked inextricably with the evolution of the veterinary profession. The first veterinary college in Europe was founded in Lyon, France in 1762; the first in Britain in 1791. Veterinary schools were established in the United States of America (USA) in the early nineteenth century in Boston, New York, and Philadelphia.

The application of the science of pathology to animals was originally largely restricted to the study of disease mechanisms in domesticated animals. However, in the latter part of the last (20th) century, the veterinary profession became increasingly interested in and involved with non-domesticated species, including free-living and captive wildlife (see, for example, the books edited by Fowler and seminal papers by, for

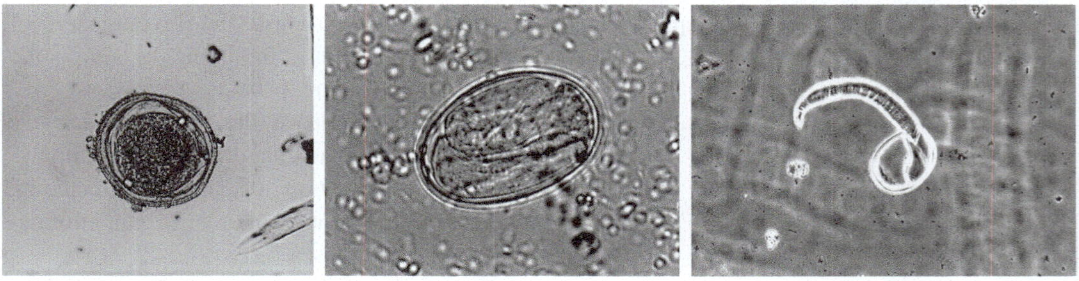

Fig. 2 Eggs of endoparasites can be significant, both in terms of an animal's health and as species' "markers" in forensic pathology studies. Some examples of what may be seen in unstained faecal samples: (left) an ascarid nematode egg; (centre) a strongle-type nematode egg containing a developing larva, and (right) a strongyle nematode egg from which a larval worm is emerging

instance, Jones (1982) and Woodford (1965)). Such work necessitated the use of gross *post-mortem* techniques as well as clinical procedures and, as a result, wildlife pathology began to emerge as a subject in its own right (Nielsen 1988; Munson 1990). In its wake, as society became more litigious, especially in Western countries (Parry and Stoll 2020), legislation relating to conservation grew (see Cooper ME, Wildlife Conservation Law, this volume), concern over global pollution and declining biodiversity burgeoned, and there was growing public insistence that action should be taken to challenge illegal and irresponsible damage to the planet. This all helped pave the way for the emergence of a discipline that could help investigate the health of animals and the environment and provide scientific information to those faced with a wildlife forensic conundrum.

Despite the above, wildlife forensic pathology is still in its infancy. Much of the information and apparent wisdom accumulated and used in legal cases is based on small datasets and personal experience rather than evidence. Munro and Munro (2013) reviewed and discussed the challenges in forensic veterinary pathology and the following points they made are germane to the context of wildlife forensic work:

> Firstly, extrapolation of results between species may jeopardise the reliability (and credibility) of the forensic opinion. Secondly, experimental studies may not truly reflect the spectrum of changes seen in actual cases (e.g. extent of injuries, infection, age and health of victim). The plethora of papers covering numerous aspects of wound healing in people and laboratory rodents may tempt the well-meaning, but unwary, to use this information during the compilation of forensic veterinary reports.

They concluded with the words of Lucius Accius (170–86 BC) "*vigilandum est semper; multae insidiae sunt bonis*" ("always be on your guard; there are many snares for the good").

The Role of Pathology in Wildlife Studies

Although this chapter is entitled "Wildlife Forensic Pathology," it will not be restricted to forensics in the narrow sense. Cooper and Cooper (2007) laid great stress on the forensic method, stressing the foundational value of such procedures as a strong chain of custody (evidence) and meticulous record-keeping in domains other than the collection of evidence for a court case. This includes the practice of biodiversity conservation, the subject of this book. Conservation biology is a rapidly growing subject that involves everybody—scientists from diverse disciplines, policymakers, and members of the public, including citizen scientists. Conservation biology needs a multidisciplinary and interdisciplinary approach if the many threats to our planet are to be tackled. Wildlife pathology has an important part to play in this venture.

Insofar as threatened species of animal are concerned, pathological expertise can be used to determine the cause and circumstances of death, to assist in the clinical investigation of ill health by examining biopsies, blood, faeces and other samples, to provide baseline data or normal values, to detect and interpret subtle changes in tissues and organs that may adversely affect function, productivity and survival, and to establish reference collections of material for retrospective studies (Cooper et al. 1998).

Both in forensic investigations and in conservation biology studies a number of pathology techniques may be used, each complementing the others. Thus, for example current efforts to save the swift parakeet (*Lathamus discolour*), which is threatened in the wild (Australia) by, amongst other things, logging of trees and predation by introduced sugar gliders, include captive breeding of the species. Birds brought into the programme are subject to tests that include genotyping, morphometrics and assessment of plumage coloration, and health monitoring. Each of these can be considered part of the forensic method.

In this chapter, the emphasis will be on *post-mortem* examinations (necropsies; singular, necropsy) and certain supporting tests. Pathologists and pathology techniques also play a pivotal part in the examination of live animals that are the subject of forensic investigation—see Cooper and Cooper (2007). Histology remains an important initial adjunct to *post-mortem* examination and enables changes in tissues to be studied in more detail so as (for example) to identify inflammatory changes (Fig. 1a, b). Live animals are not discussed in any detail here but some basic guidelines are given in Appendix 1 with reference (as an example only) to birds.

The Role of a Necropsy

Post-mortem examination of a dead animal or its parts can provide information on (1) causes of death, (2) causes of ill health, (3) underlying abnormalities, (4) the gross and histological appearance of tissues, (5) morphometrics, organ weights and organ-bodyweight ratios and (6) specific queries, such as the reproductive status of an animal (see, for example Wyllie and Newton 1999). All of these are important in investigations relating to biodiversity and conservation as well as in wildlife forensic studies.

Personnel

Wobeser (1996) wrote that ideally, the investigation of wildlife crime requires the services of a specialist veterinary pathologist. This was echoed by Munro (1998), who urged that trained pathologists should be considered the appropriate experts for the performance of *post-mortem* examinations and supporting laboratory tests on wildlife. Cooper and Hull (2017), writing in the context of gorillas, stated that ideally a necropsy should be carried out by a specialist veterinary pathologist—for example, a Diplomate of the ECVP (European College of Veterinary Pathologists) or the ACVP (American College of Veterinary Pathologists). Often, however, especially in the field, a specialist pathologist is not available, and the necropsy has to be carried out by a veterinary clinician, a biologist, a technician or a member of field staff, most of whom will not have had any detailed training in pathology. Students may be able to contribute usefully to a *post-mortem* examination, especially if the animal is large, but they must be carefully briefed beforehand, especially in respect of the importance of (a) chain of custody, and (b) health and safety (Cooper and Cooper 2007).

In this chapter, conscious that many readers will not be pathologists, the author adopts the same approach as he does in his teaching and in the field. This is that all those who might be involved in *post-mortem* examinations and sampling of wildlife, regardless of background, should have some understanding of the terminology and language used by pathologists, an appreciation as to how such specialists perform their skills—and then be able to participate and contribute to such work in a safe and productive way.

Familiarity with the biology and natural history of a wide range of species, vertebrate and invertebrate, coupled with a knowledge of comparative medicine, is an essential part of good wildlife pathology practices.

Post-mortem Methods

Standard forms should be used for the submission of material (See Appendix 2). A necropsy may be performed in a purpose-built *post-mortem* room (Fig. 3a) or in the field (see later and Fig. 3b).

There are numerous ways of carrying out the examination. Brownlie and Munro (2016) reviewed procedures and protocols for veterinary forensic necropsy but with a focus on domesticated animals. In wildlife forensic work, the wide range of species means that no one technique is likely to be applicable to all taxa. Recommended approaches have been promulgated for some species of wildlife—for example reptiles, amphibians and birds, including those that are the subject of forensic investigations (Cooper 2008, 2019; Cooper and Cooper 2016) and individual species such as gorillas (Cooper and Hull 2017). Some features relating to the necropsy of non-domesticated species are given in Table 1.

Very small specimens, such as fish, frogs, eggs, embryos and fetuses, will need a mini-necropsy. When several specimens are available, some can be embedded and serially sectioned for histological examination, a method pioneered by the late Edward Elkan (1981). Special

a b

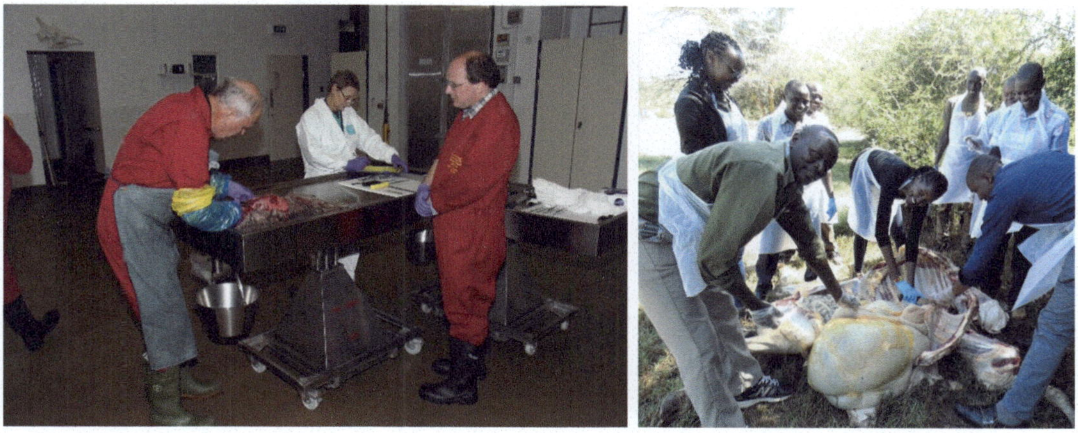

Fig. 3 (**a**) A necropsy being performed in a purpose-built *post-mortem* room. (**b**) A "field" necropsy—improvisation, using equipment that the pathologist and his team had with him or have borrowed locally

Table 1 Some features relevant to *post-mortem* examination of non-domesticated species (reproduced in a modified form from Cooper and Cooper 2007)

Group	Special features
Mammals (class Mammalia, phylum Chordata)	Many variations in external and internal anatomy, e.g. presence or absence of tail, structure of gastrointestinal (GI) tract, presence of a blow-hole in cetaceans. All mammals have hair to a greater or lesser extent.
Birds (class Aves, phylum Chordata)	Variations in external and internal anatomy, e.g. presence or absence of preen gland, crop, caeca. Certain specialised features, e.g. syrinx in male ducks, subcutaneous air sacs in some storks. All birds have feathers. All are oviparous.
Reptiles (class Reptilia, phylum Chordata)	Much anatomical variation between the main orders, e.g.: • Squamata—no functional limbs, elongated organs and only one functional lung in snakes. • Chelonia—presence of carapace and plastron and variation in their size and structure. • Crocodilia—tough (ectodermal) scales, right aortic arch. All reptiles have scaled skin. Some are oviparous. See Hanley and Hernandez-Divers (2003).
Amphibians (class Amphibia, phylum Chordata)	Marked distinction in most species between immature (larval—tadpole) stage with gills and other modifications to an aquatic lifestyle and mature (adult) with lungs and a terrestrial lifestyle. All amphibians have unscaled mucous skin. See Pessier (2003)
Fish (Pisces—Three classes, phylum Chordata)	Often variation in external and internal anatomy associated with life cycle and habitat (including whether freshwater or marine). Special features of most fish include mesodermal scales, fins, gills and swim bladder.
Invertebrates (Invertebrata—at least 30 phyla)	Much variation, associated with classification, life cycle and habitat. Features of relevance shown by some invertebrates include: • Arthropoda—exoskeleton, sometimes thick and strengthened by calcium salts, jointed body and limbs. • Mollusca—protective (calcareous) shell often present. Body soft, very prone to desiccation. • Metazoan invertebrates usually have haemolymph, containing haemocyanin. A few have haemoglobin.See Cooper and Cunningham (1991).

Fig. 4 Eggs of an endangered vulture submitted for forensic investigation, including toxicology

necropsy forms may be needed for the examination of eggs, embryos and fetus. Eggs require careful investigation (Fig. 4) and see Appendices 3 and 4.

If the remains of a carcase are required for display in a museum, a cosmetic *post-mortem* examination must be carried out in order to obtain the necessary information and material while inflicting minimum external damage.

There are several important stages of a *post-mortem* examination (see Box 1).

Box 1 Stages of a *Post-mortem* Examination

Stages of a *post-mortem* examination:

- Ascertaining background history and ascertaining the species to be examined
- Preparing facilities and equipment
- Performing the *necropsy*, including record-keeping and taking samples
- Retaining material and records
- Cleaning the facilities
- Collating records
- Analysing laboratory findings
- Producing a report
- Ensuring health and safety (see Cooper, ME, this volume)

Preparation of Equipment

Disparate species require different equipment. Much wildlife forensic work is performed initially in the field—see below—the features of which can be summarised as being away from one or more of the following: a home base, a normal environment, optimal equipment and back-up facilities.

Forensic Pathology in the Field

Fieldwork is a feature of both forensic and diagnostic wildlife pathology.

Sometimes forensic pathology in the field is very much opportunistic—for example if there is a report of an injured, sick or dead animal requiring immediate attention. Improvisation is then necessary, using any equipment that the pathologist has with him/her or can purchase or borrow locally. Some examples of simple improvisation are given in Table 2.

When fieldwork can be planned it will be performed more proficiently (Cork and Halliwell 2002; Cooper 2013a, b) and thereby will be more

Table 2 Field techniques—improvisation (reproduced in a modified form from Cooper and Cooper 2007)

Requirement	The ideal situation	Possible improvisation	Comments
Fixation of tissues for histological examination	Formalin or commercial alcohol	Alcoholic drinks, including spirits, wine and strong beer	See Cyrus (2005), Cooper and Cooper (2007) and Cooper (2013a, b)
Handling of contaminated or infected material	Rubber gloves Forceps	Plastic or paper bags Wooden spatulae	See Cooper and Samour (1997) and Cooper (2013a, b)
Sample collection	Laboratory containers	Re-used film pots	See Frye et al. (2001) and Cooper (2013a, b)
Clipboard for support when writing notes or preparing labels	Purpose-made—plastic or hardboard	A piece of cardboard and a large (bulldog) clip	Has the added advantage that the cardboard can, if contaminated, be incinerated after use

likely to produce reliable results that stand up to scrutiny in cross-examination. Appropriate equipment (such as hand-held microscopes) and techniques were discussed by Cooper (2013c). Kalema-Zikusoka and Rubanga (2013) described the establishment and use of field laboratories, with particular reference to gorilla research in Uganda. Haigh (2013) addressed the challenges in a cold climate. There are important legal, ethical and practical considerations to working in the field, as explained by Cooper (2013c). Table 3 and Appendix 5 list some of the equipment that may be needed under such circumstances.

Human medical texts can prove helpful; for example the book by Cheesbrough (2005) that describes district laboratory practice in tropical countries.

The design and composition of a field pathology kit depend upon (a) the circumstances; in particular, the type of case (live animal, dead animal, environmental issue) and (b) the location; for instance, whether the kit has to be carried by hand, on a bicycle or motorbike and the distance back to an appropriate laboratory where follow-up or more detailed investigations can be carried out. Microscopy in the field is becoming increasingly practicable, with the development of instruments that can be used in conjunction with a mobile phone. Both the handheld Newton Microscope and older instruments with mirrors still have an important role to play (Fig. 5).

Timing of the Necropsy

In routine diagnostic veterinary work, the general rule is that a carcase should be examined as soon as possible after death. It is also usually stressed that freezing of a carcase, in order to arrest autolysis, should be avoided because of the artefact that it can produce, in particular damage to tissues that might masquerade as pathological change. In forensic work, however, this drawback has to be balanced against the need to ensure that the necropsy is carried out as efficiently and meticulously as possible and helps answer the questions that have been posed. Therefore, although it is still advisable to perform the examination promptly, a delay that entails freezing may be acceptable if this means that a more methodical examination can be performed, and better results obtained.

Examination of Samples

Forensic pathology can encompass a whole range of investigative techniques on samples including (for example) bacteriology, histopathology, molecular biology and parasitology (Dolinak et al. 2005). A sample was defined succinctly by Blood and Studdert (1988) as "a specimen.... collected for analysis on the assumption that it represents the composition of the whole".

Table 3 Equipment for post-mortem examination

Species	Essential	Useful additions	Special precautions	Comments
All (vertebrate and invertebrate) species	*Post-mortem* table Instruments Protective clothing Incinerator, macerator or other means of disposal Refrigerator, freezer Balance/scales Disinfectants Steriliser or autoclave Bottles and fixatives	Protective hood or flow cabinet X-ray machine	Adequate drainage, disinfection and ventilation	Some requirements are likely to be a legal responsibility under health and safety legislation. See Cooper, ME, this volume. Disposal of very large carcases, such as cetaceans, may require specialised equipment
Mammals	As above Instruments should include bone forceps and saw for examination of the central nervous system (CNS)	Vacuum cleaner or similar to remove hair and dander	As above	As above Special care should be taken with certain species, e.g. primates, rodents, especially where infectious disease is suspected
Birds	As above	As above For feathers and feather dust	As above Dowsing carcase in disinfectant may reduce risk of airborne infection but hood or flow cabinet is preferable	As above Special care with psittacine birds and where infectious disease is suspected
Reptiles	As above Bone forceps and saw facilitate	Container for head of venomous species	As above (all)	As above (all) Care must be taken with venomous species
Amphibians	As above (all) Carcase should be kept damp during examination	As above (all)	As above (all)	As above (all) Care must be taken to avoid contact with parotoid and cutaneous secretions of some species
Fish	Instruments should include long forceps and scissors for dissecting species that have toxic spines	As above (all)	As above (all)	As above (all) Care must be taken with toxic species
Invertebrates	As above (all) Micro instruments (e.g. ophthalmic forceps, dissecting microscope or magnifying loop) are often needed—a mini necropsy Some aquatic species need to be kept damp or even immersed in saline or water (to prevent desiccation or collapse) during examination	As above (all)	As above (all)	As above (all) Care must be taken with toxic species. The small size of many invertebrates can make detailed examination difficult. Fixation in to may be advisable, followed by histological examination. See Cooper and Cunningham (1991)

Forensic samples may be taken from live animals, from dead animals or from the crime scene (Cooper et al. 2009). Any of these may constitute evidence. Mistakes can occur at various stages of sampling; these are summarised in Appendix 6, along with more detailed information regarding types of samples that may be obtained.

Fig. 5 Microscopy in the field. A range of microscopes can be employed successfully in locations where there is no electricity, including older styles that are equipped with mirrors. Here a battery-operated handheld Newton Microscope is used by a local member of the community on the Coast of East Africa

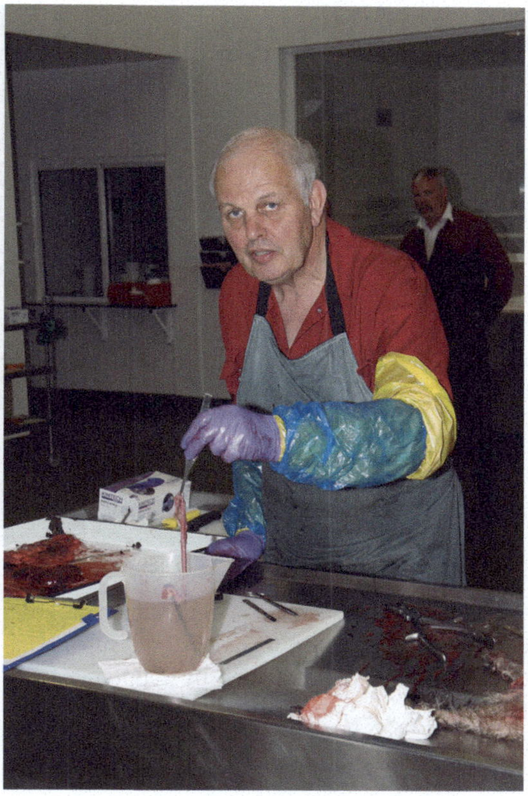

Fig. 6 Taking and testing of samples. Here tissues are being placed in saline for washing and closer examination. The same saline can be used to see whether lung and liver float or sink in fluid

Any changes in the appearance of a sample following collection should be noted, as should its behaviour when placed in fluid (water, saline or fixative). Does the sample float or sink? (Normal inflated lung usually floats; liver is only likely to do so if it is fatty or contains gas). Is there a change in colour following immersion? (Fig. 6)

Maintenance of a chain of custody (evidence) is a key part of forensic necropsy. Careful labelling, in particular, helps ensure that samples are traceable to the animal from which they came.

Health and Safety

Health and safety precautions, with appropriate risk assessments, are essential in *post-mortem* work and in the taking and processing of samples; see Cooper, ME, this volume. A "clean and dirty" division—that is, a demarcation between what is considered to be clean, possibly sterile and what is dirty, or used/contaminated—is essential, especially in the field. A colour code can be used to differentiate the two.

Performance of the Necropsy; *Post-mortem* Technique

It is particularly important in forensic examinations to have (and to employ) the appropriate equipment (Fig. 7). Plastic forceps should be used to handle shot, pellets or bullets (see later) because metal instruments may damage the surface or complicate analysis for certain metals.

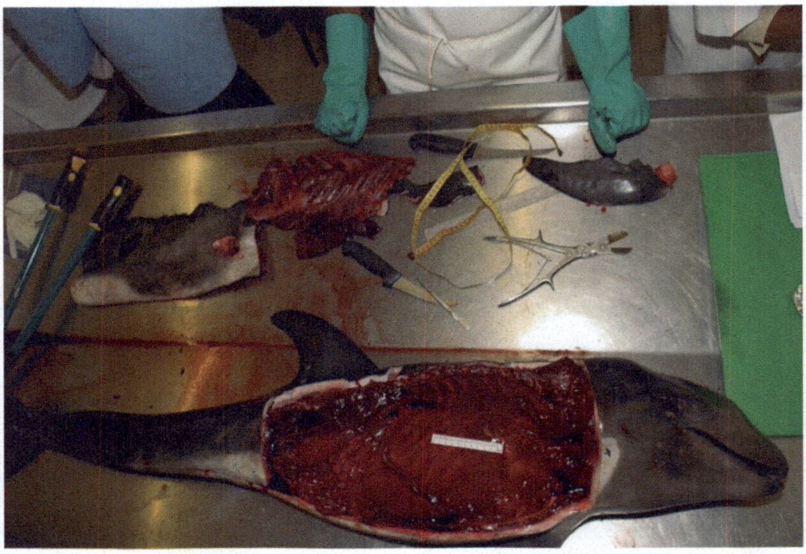

Fig. 7 Necropsy of a Risso's dolphin (*Grampus griseus*) illustrating some of the equipment being used

Skinning of the carcase is important in order to detect injury to the integument and superficial tissues. Removal of hair, by clipping, and feathers, by plucking, will help in the detection of subcutaneous bruising.

Hair, feathers and scales can provide much valuable information. Lesions suggestive or characteristic of damage by a bullet or shot or airgun pellet may be detected by light or scanning electron microscopy. Methods of investigation of such keratinous structures were detailed by Cooper and Cooper (2013).

Incisions must always be carried out with care, especially in cases where there is already traumatic injury or over-exuberant dissection may damage important evidence.

Assessing the Time of Death

There are various specialised methods for attempting to calculate the time of death, many of them developed for human work (DiMaio and DiMaio, 2001), others described for certain species of animal (Huffman and Wallace 2012; Cooper and Cooper 2013). The wildlife pathologist faces particular challenges if dealing with a novel or unusual species. Often an approximate estimate—or at least data that might be helpful in court—depends upon recording and analysis of basic criteria; see Box 2.

Box 2 Assessing the Time of Death: Indicators of Importance

Assessing the time of death: indicators of importance

- Cooling of the carcase
- *Rigor mortis*
- Desiccation
- Discoloration of skin
- Discoloration of internal organs
- Maggot and other invertebrate activity
- Degree of *post-mortem* change
- Scavenging patterns
- Circumstances of death and storage thereafter

Over two decades ago Munro (1998) suggested that, in animals, general estimates are usually all that can be given. He advocated the use of five broad categories—(1) recent death—less than 48 h; (2) several days; (3) weeks; (4) months and (5) years.

Determining How Death Occurred

This is an integral part of forensic investigation. In human forensic medicine, three aspects have to be considered (DiMaio and DiMaio 2001):

(a) Cause of death—for example, a gunshot wound or pneumonia.
(b) Mechanism of death—the physiological changes produced by the cause of death that made it fatal, such as haemorrhage or a blood-borne infection.
(c) Manner of death—how the cause of death came about, such as "accident".

A similar approach can be applied to animals but in the author's experience, it is often more straightforward and helpful in court to try to answer six questions (see Box 3).

Box 3 Questions to Ask When Ascertaining How Death Occurred
Ascertaining how death occurred:

- How did this animal die?
- Why did this animal die?
- How long did it take this animal to die?
- When did this animal die?
- Where did this animal die?
- Who was involved, intentionally or inadvertently, in its death?

Making a Diagnosis

Not all *post-mortem* examinations provide a diagnosis, even in standard medical or veterinary work. In a forensic case, however, the cause of death is not necessarily the important finding. Sometimes the cause of death is already known (e.g. the hawk was euthanised or the gazelle was killed by a vehicle on the road) in which case the role of the pathologist is to provide information on background pathology or other parameters. The forensic pathologist must always be resigned to the fact that some cases will be inconclusive and must be prepared to state this and not be swayed in any way to alter his/her opinion.

Descriptive pathology is usually the basis of a morphological diagnosis, such as "consolidation of the lung", as opposed to an aetiological diagnosis—which often consists of two words, e.g. "fungal dermatitis".

Definition and Description of Lesions

Abnormalities in organs or tissues are called "lesions" (Latin: "*laesio*"—injury). The word "change" can also be used. Knowing the NORMAL is important if one is to recognise abnormality (Fig. 8).

Before embarking on the necropsy of a less-familiar species it is sensible to examine reference material, for example the skeletons of cetaceans (Fig. 9).

Lesions in animals should be accurately described; see Box 4.

Box 4 Describing Lesions
Describing lesions.

- Numbers and distribution
- Location/orientation

(continued)

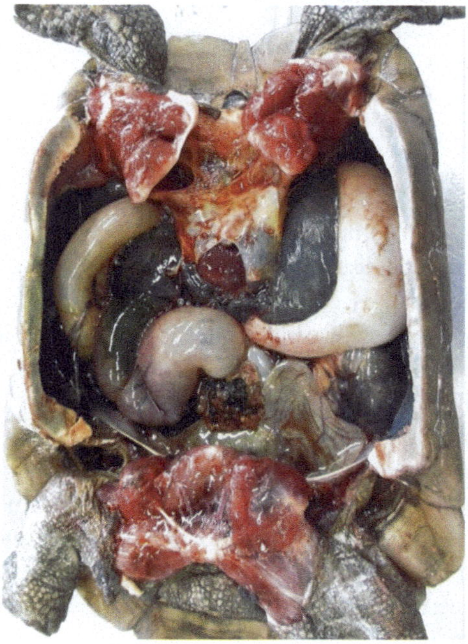

Fig. 8 Normal internal organs and tissues of a Mediterranean tortoise (*Testudo graeca*)

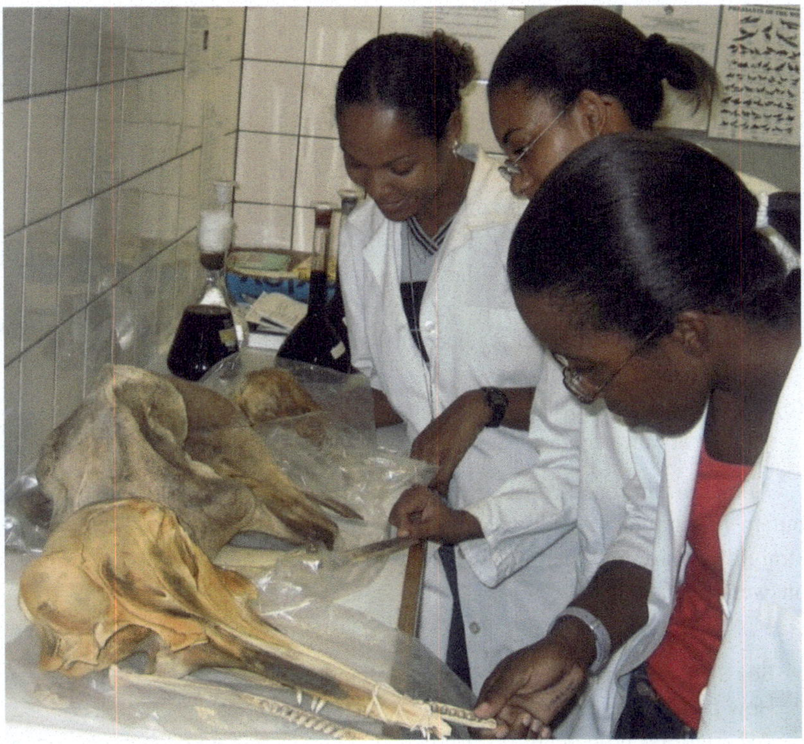

Fig. 9 Examining skulls of cetaceans prior to performing a necropsy on a dolphin in Trinidad, West Indies

- Raised, flat or depressed
- On surface only, or in deeper tissue
- Shape
- Size (cm and mm)
- Colour
- Consistency
- Odour

As in clinical work, all senses except taste should be used when examining a carcase or tissues—eyes, ears (crepitation etc.), nose (odours associated with certain infections, parasites and poisons), and touch. The Canadian physician, Sir William Osler (1849–1919) stated in his *Aequanimitas* (1914) 332, in the context of human medicine, "Observe, record, tabulate, communicate. Use your five senses. Learn to see, learn to hear, learn to feel, learn to smell, and know that by practice alone you can become 'expert'".

The colour of lesions can be important. Changes in colour have long been used as an indicator of pathological change; they can play a significant part in making a diagnosis. It is wise to standardise descriptions using a colour key.

Hypostasis (the gravitational pooling of blood in certain areas during or after the animal's demise), also termed livor mortis or "lividity", will produce colour changes and can occasionally be mistaken for bruising. It may be of relevance in ascertaining the circumstances of death.

Bruises also produce colour changes. They are important clues in any *post-mortem* examination but their detection in animals can be difficult because of the presence of hair, feathers, scales or pigmentation.

Trans-illumination (shining light through a tissue or organ) is very useful in *post-mortem* work. Strong light can often help detect lesions. In the field, especially in the tropics (where many high-profile wildlife crime investigations take place), the sun provides an excellent source of illumination through which (e.g.) the wall of the intestine can be viewed (Fig. 10).

Fig. 10 Transillumination being used to examine the wall of a piece of intestine. The red bowl, into which the sample will be returned, illustrates the use of a colour code (red, to indicate "dirty")

Morphometrics

In 1883 William Thomson, Lord Kelvin of Glasgow, in his "Electrical Units of Measurement" (PLA = Popular Lectures and Addresses), said: "*...when you can measure what you are speaking about, and express it in numbers, you know something about it; but when you cannot measure it, when you cannot express it in numbers, your knowledge is of a meagre and unsatisfactory kind...*".

Measuring of animals, their organs and their lesions must be accurate and consistent. It is preferable that one person performs all the measurements and uses the same equipment each time to avoid individual variations. Asymmetry must be confirmed by measurement, making allowance for observer bias. Fluctuating asymmetry (FA), which reflects small random deviations in otherwise bilaterally symmetrical characters, can be an indicator of stress.

Species Differentiation

The pathologist will want to know with which species s/he is dealing, and this may be a very pertinent question from a legal point of view—if, for example presumed bushmeat is being investigated (Fig. 11).

Difficulties often arise when parts of an animal are presented. Morphological methods for distinguishing carcases (or parts thereof) of some species are well-documented in books covering meat inspection and pathology, and DNA techniques (see Curtis, M., this volume) are now regularly used to distinguish meats, even those that are canned.

Age Determination of Animals and Their Derivatives

The ageing of animal remains can also be necessary, examples being when such tissues may provide evidence relevant to possible contravention of conservation legislation (e.g. Was a mounted crocodile prepared before or after the enactment of CITES legislation? Who might have been the taxidermist?). See Fig. 12.

Ageing of Lesions

The ageing (dating) of skin and soft tissue wounds, contusions and fractures can be of great importance in forensic cases, in both live and

Fig. 11 Bushmeat being sold from a roadside stall in the West Indies

dead animals—for example (1) in a live animal, to be able to estimate when a skin wound was inflicted, in a possible cruelty case, and (2) in a dead animal, to be able to age fractures or internal haemorrhage, in an alleged road traffic accident. However, the topic is fraught with difficulties.

Relatively few studies have been carried out on the progression and the ageing of lesions of animals, in contrast to the situation in humans—bruises, for example (DiMaio and DiMaio 2001). The subject becomes even more complicated once one moves into the realms of comparative forensic medicine because there is then an added dimension relating to body size and whether the species in question is endothermic ("warm-blooded") or ectothermic ("cold-blooded"). The metabolic rate of endotherms (mammals and birds) is inversely related to body mass. Therefore, a skin wound is likely to heal more rapidly in a mouse than in a mountain gorilla. Repair and/or regeneration takes place more rapidly at a high than at a low ambient temperature. In the case of ectothermic animals, the ambient temperature is the key factor; snakes with skin wounds will heal more rapidly at 24 °C than at 20 °C (Smith et al. 1988).

Approximate ageing of traumatic and inflammatory lesions is possible in some species, but the wildlife pathologist should be aware that there can be variation relating to such factors as species, age, sex of the animal (some of this a result of differences in metabolic rate—see above) and whether a lesion is infected or subjected to movement/trauma, which can delay "normal" healing. Criteria that can be used are listed in Box 5.

Box 5 Criteria Useful When Assessing Lesions

Criteria when assessing lesions

- Organisation of blood clots.
- Inflammatory infiltrates—acute versus chronic, numbers of pyknotic/karyorrhectic cells, presence or absence

(continued)

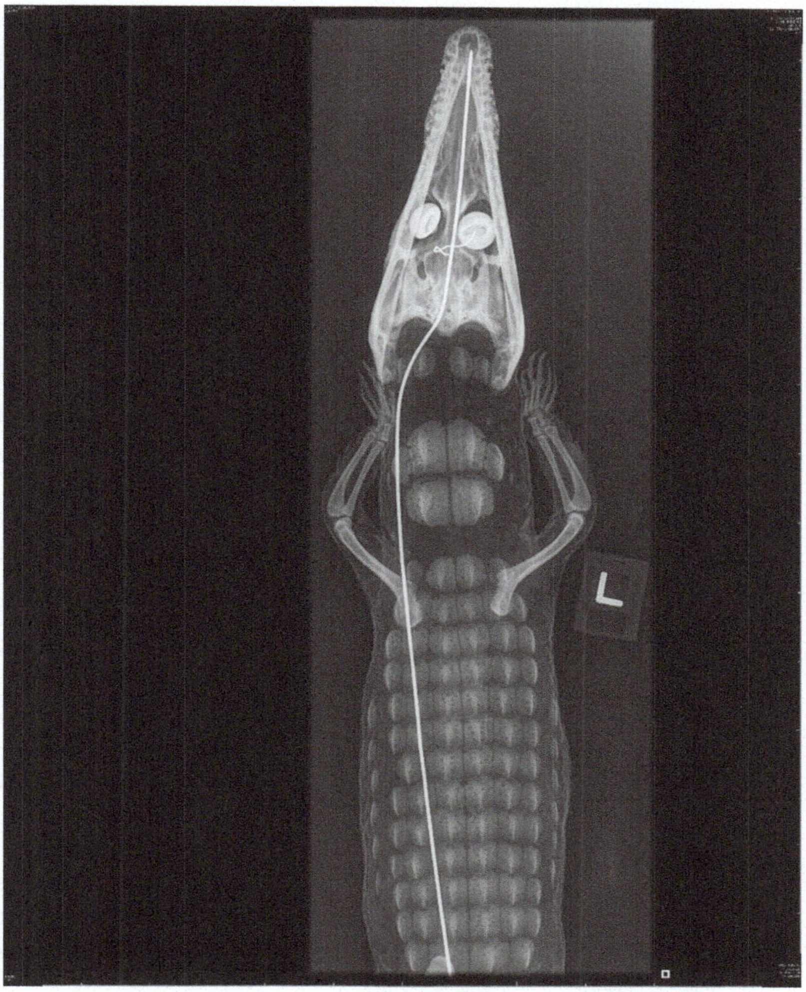

Fig. 12 A radiograph of a mounted Nile crocodile (*Crocodylus niloticus*), illustrating some of the materials used by the taxidermist

of phagocytosed organisms or other material.

- Fibroplasia.
- The healing of abrasions (fractures) presence of callus, its degree of ossification etc.

Post-mortem Imaging

Radiography and other imaging techniques are of great value in a forensic necropsy (Fig. 13) but not always feasible under field conditions. Whole-body radiography is recommended whenever possible.

Recognition of *Post-mortem* Change

Accurate differentiation of *ante-mortem* and *post-mortem* changes is important in all forensic cases. However, it is not always easy! Often one must include analysis of the history and circumstances since death, coupled with use of laboratory tests, e.g. histology (Cummings et al. 2011).

An ability to interpret *post-mortem* change is important because it may help to determine the time of death. Progression of *post-mortem* changes can depend upon many factors. See Box 6.

Box 6 Factors Affecting the Progression of Post-mortem Changes

Factors affecting the progression of *post-mortem* changes:

- Species of animal
- Health status before death
- Presence or absence of ingesta in the gastrointestinal (GI) tract
- Manner of death
- Body temperature at the time of death
- Environmental temperature and humidity
- Location and position of the body
- Handling and storage following retrieval of the body

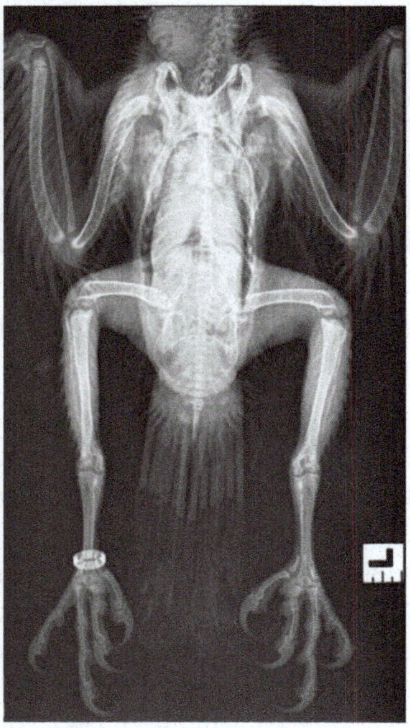

Fig. 13 Radiography and other imaging techniques are of great value in a forensic necropsy. A young peregrine falcon (*Falco peregrinus*) chick. There are no obvious lesions, but the radiograph confirms the immaturity of the bird—the bones are still growing, and the feathers have vascular sheaths

Useful clues to the animal's history, time since death (see earlier) and/or the environmental conditions to which the carcase has been exposed may be provided by the presence or absence of *rigor mortis*, *livor mortis* (hypostasis, pooling of blood etc.) and *algor mortis* (cooling); appearance of organs and tissues; evidence of predation/scavenging by other animals; infestation by maggots, carrion, beetles etc.

Organs and tissues undergo *post-mortem* autolysis at different rates, generally starting with the brain and GI tract and finishing with the skin and heart muscle and skeletal structures.

Early studies on *post-mortem* change were carried out on humans, rats and poultry (Cooper and Cooper 2007). Remarkably little has been published about *post-mortem* change in ectothermic animals other than fish (where the emphasis has been on spoilage); notable exceptions are the work by Frye (1999) and Cooper (2012). Frye emphasised that even when "clinically dead", ectothermic species can exhibit residual cardiac contractions. He listed *post-mortem* indicators in reptiles and amphibians, including bile-staining, chromatophore relaxation and expansion and the production of characteristic odour(s).

Post-mortem change is very relevant, both theoretically and practically, to wildlife forensic pathology. Wildlife cases are often presented as decomposed carcases or parts thereof. While it is not uncommon in standard diagnostic veterinary work for a laboratory to state that the carcase "was too decomposed for a *post-mortem* examination", in a forensic necropsy no amount of autolysis should be allowed to negate the value of examination. An animal's body may be only a liquefied "soup" but this is the evidence and must be properly examined. As Cooper and Cooper (2007) asserted "*Forensic pathology is not for the squeamish. A spirit of enthusiasm and curiosity, the hallmarks of a detective, rapidly numb the senses to foetid smells, maggot-infested viscera or an uninviting pile of arachnid-ridden crumbling vertebrae*".

Case Study 1

"How did the hippo die"?

Readers are invited to read and review the following case study and photographs (images)

Suggested answers are given at the end of the chapter.

The location of this incident was the edge of Lake Naivasha in Kenya. This freshwater area is home to many species of bird, amphibian and fish—and a large population of hippos.

A dead baby hippopotamus (*Hippopotamus amphibius*) was found at the foot of a steep but short drop (upper photo). Professional assistance was sought. However, the person asked to investigate ("first responder") had no forensic crime scene equipment with him. It was a Sunday and assistance from the Kenya Wildlife Service (KWS), which conserves and manages Kenya's wildlife, was not immediately available

Question 1: Describe what you see.

Question 2: Comment on how the investigator (first responder) is approaching and dealing with the case.

Question 3: How might samples from this animal throw light on the cause, mechanism and manner of its death?

Question 4: How might a knowledge of veterinary pathology contribute to the investigation of this case?

Fig. 14 Sifting and sieving of soil in the field in Kenya

Fig. 15 Direct examination of a faecal specimen in the field using a hand lens

Radiography and other imaging can be very helpful but must be considered an adjunct to the *post-mortem* examination, not an alternative to it.

Sometimes viscera and bones are removed by predators or scavengers and scattered widely. Detection of such remains together with teeth, hair, feathers, scales and exoskeletons of invertebrates (these may persist in large numbers if an animal has died with many invertebrates within its GI tract) is likely to require meticulous sifting of soil and vegetation (Fig. 14) followed by laboratory investigation (Cooper and Cooper 2007).

Forensic Entomology

The ability of maggots to transform and ultimately to destroy a carcase was recognised in the eighteenth century, notably by Linnaeus (1757), but the mechanisms and the genesis of the insects were not then understood. Forensic entomology is now a well-recognised and sophisticated tool in forensic work (Byrd and Castner 2011). The subject is discussed by Adrienne Brundage, this volume.

Laboratory Investigation

Additions to routine gross (macroscopic) *post-mortem* investigation that may be employed in forensic work are listed in Table 4. Laboratory tests can provide valuable supporting information.

Forensic Odontology

Bites made by a mammal, reptile or fish can injure or kill other animals or humans. Bite wounds may also be inflicted on an animal after it has died, as a result of scavenging.

The question that is often asked is, "What caused this bite?" This may refer to the species of animal (e.g. fox, seal and crow) or the individual animal (e.g. Mr. Smith's Jack Russell terrier). Sometimes the circumstances are contentious.

Table 4 Laboratory investigations (reproduced in a modified form from Cooper and Cooper 2007)

Technique	Comments
Direct examination with a hand lens, magnifying loupe or dissecting microscope	Recommended initially for any tissue or sample (Fig. 15). Use of ultraviolet light may reveal material that is not visible otherwise.
Radiography (part or whole-body)	Using digital, dental or low kv X-ray machines.
Histopathology	A selection of samples (lung, liver, kidney plus abnormalities) should be taken routinely, plus others as necessary. Buffered formol saline (BFS) is the standard fixative. See Cooper and Cooper (2007).
Bacteriology	If a bacterial infection is suspected. See Cooper and Cooper (2013).
Mycology	If a fungal infection is suspected. See Cooper and Cooper (2013).
Virology	If a viral infection is suspected—but virological examination is neither easy nor cheap. See Cooper and Cooper (2013).
Parasitology	A useful routine (ecto- and endoparasites) in all necropsies (Fig. 2). Combine faecal examination with cytology below. "Ecological tagging" uses helminths to identify the stock from which fish originate. The presence of parasites may help to pinpoint the origin of other species.
Cytology, specific organs, intestinal contents, exudates etc.	A whole range of simple and rapid tests that can provide useful information in legal cases. See Cooper and Cooper (2013).
Haematology	Small quantities of blood may be obtained from the heart or elsewhere of dying or freshly dead animals.
Biochemistry	As above.
Toxicology	Important in many forensic investigations. A wide array of substances can prove toxic to wild animals. Environmental toxins are discussed by Noel and Brown, this volume. Guidance on how to perform a necropsy if a toxin is suspected was provided by Rotstein (2008). Samples for toxicology often present logistical challenges and it is wise to consult the laboratory before taking material.
DNA studies (nuclear and mitochondrial)	Identification, sexing. Valuable. Appropriate material should always be stored.
Scanning electron microscopy (SEM)	Often of value in the investigation and differential diagnosis of lesions. Glutaraldehyde is the standard fixative. See Cooper and Cooper (2007).
Transmission electron microscopy (TEM)	Often of value in the investigation and differential diagnosis of lesions and studies on cellular changes. See Cooper and Cooper (2007).
Bone preparation, examination and analysis	Bone density and other studies.
Digestion studies	Studies on soft tissues.
Ashing and mineral analysis of organs or the animal's whole body.	Studies on hard tissues.
Radioisotope techniques	Ageing of remains. Studies on diet using stable isotope analysis of teeth (including ivory) and faeces.

No one method of investigation is likely to answer all the questions that may be asked in court, especially in cross-examination. For example, while the detection in a wound of DNA from a known animal provides evidence of contact between that animal and the victim, it does not necessarily mean that the former inflicted the wound on the latter. Other techniques may be needed to prove whether the suspect really did cause the wound *ante mortem* as opposed to its being a scavenger on the body *post mortem*. Such tests may also be helpful when more than one species or individual was involved in killing or injuring the victim.

It is therefore prudent, in all cases of presumed animal bites, to follow an established procedure that provides maximum information and that is likely to stand up to scrutiny in court. A recommended approach is given in Box 7.

Box 7 Recommended Approach: Presumed Animal Bites

Obtain as full a history of the alleged incident as possible.

If a live or a dead victim is available, investigate the presumed bite wound:

(a) Observe without touching. If available, use ultraviolet light.
(b) Photograph in black and white as well as colour/draw the wound(s) from different angles.
(c) Measure the external dimensions and depth of the visible wound using sterile techniques.
(d) Describe the lesion in medical (clinical/pathological) terms—see later.
(e) Take samples:
   - Multiple swabs, in appropriate transport medium, for microbiological examination and DNA studies, including the area surrounding the wound. Also, search for other materials that may harbour DNA from the predator (s) or from subsequent scavengers, e.g. saliva or scratch marks elsewhere on the victim's body, touch preparations for cytological examination and, if the victim is dead, a wedge of tissue for histological examination.
   - Hair from around and in the wound (these may be from the predator and/or from scavengers as well as from the victim).
   - Samples for a variety of tests from any scattered pieces of the victim's body; a search will need to be made for these, often some distance from the victim's carcase.

   Consider taking impressions of bite-marks, using a rubber or silicone-based material and/or plaster casts.

If the victim is dead, skin the carcase to help reveal hidden tooth punctures or imprints; use transillumination (Fig. 10), as well as reflected light, to examine the pelage.

If a specific animal is suspected of inflicting the injury, seek veterinary assistance so that the mouth of that animal can be properly examined, and, at the same time, appropriate samples taken (for laboratory investigations) and dental casts prepared for comparison with the wounds on the victim.

Bite-mark analysis is a specialised field (Dorion 2004). Its investigation can involve the use of scanning electron microscopy, digital imaging, histology, contrast medium-enhanced radiography and computerised axial tomography (Cooper and Cooper 2007). The seminal work by Colyer, revised by Miles and Grigson (1990), and the two books by Berkovitz and Shellis (2018a, b) are essential reading insofar as comparative aspects of dentition are concerned.

Recognition of the species or individual animal that caused a bite wound, whether or not it proved fatal, depends upon a combination of features (Box 8).

Box 8 Important Features of a Bite Wound

Important features of a bite wound

- Shape, including the dental arch.
- Type, for example whether canine teeth were involved, producing puncture wounds, or "cheek" (premolars and molars), causing crushing of tissues.
- Size of the bite marks (diameter and depth) and their distance apart.
- Site, for example domesticated cats tend to bite the face or hands whereas

(continued)

mustelids (e.g. badger and polecat) attack the lower limbs. Crows and other birds are often drawn to the eyes and orifices.

- Laboratory tests—for bacteria and fungi (different animal species harbour different species of organism in their oral cavities) and for DNA.

Road Traffic (Vehicular) Accidents/ Collisions

These are a common cause of injury or death in wild animals. The lesions are usually typical of blunt force trauma—tension ("pulling apart"), compression ("crushing") and friction/shearing ("application of differential force"). Diagnosis can be based on circumstantial evidence plus the detection on the body of plastic, paint or oil from the vehicle.

Much can be learnt from the study of "road kills". Some studies are linking patterns of road-killed animals with climate change (Capula et al. 2014).

Lesions Associated with Firearms, Other Weapons, Traps and Snares

Various weapons can be used to inflict damage on animals, ranging from hammers to garden forks. In poorer countries or more rural locations, a machette (termed a "*panga*" in many parts of Africa, a "cutlass" in the Caribbean) is often the offending item and produces a slash wound—which is longer than it is deep.

A stab wound is deeper than it is long. The wounds produced may be incisions or lacerations, sometimes a combination of the two. Williams et al. (1998) advocated that, when describing stab wounds, the following should be included: (a) position and angle of the wounds, (b) their dimensions, and c) the tissues or organs that are affected.

These criteria are applicable as well to wounds attributed to machettes or other cutting weapons but here the appearance of the wound will depend on the sharpness and structure of the implement, particularly its shape and the number of cutting edges (e.g. serrations) that are present.

Spears and arrows are still commonly used to hunt animals in poorer parts of the world, as are snares (Cooper and Hull 2017; Haggblade et al. 2019). Ageing of snare injuries can help in wildlife crime investigation but has apparently attracted little research to date.

Areas that need study include:

1. The types of tissue damage that different snares cause.
2. Which types of snare-induced lesions are most likely to become infected?
3. How does snare damage progress in terms of both local and systemic pathology?
4. The extent to which lesions due to snares differ depending upon the species, subspecies, age and sex of the trapped animal.
5. How are snare injuries affected by external factors, such as temperature, relative humidity, terrain (including soil type and vegetation)?

The author's suggested protocol for investigating snare injuries (lesions) is given in Box 9.

Box 9 Protocol for Investigating Snare Injuries

Protocol for investigating snare injuries

1. Observe the injury without touching. If available, use ultraviolet light.
2. Photograph in black and white as well as colour. If a camera is not available, draw the lesion.
3. Measure the dimensions and depth of the visible wound using sterile techniques.
4. Describe the lesion in medical (clinical/ pathological) terms—see later.
5. Take samples:
   (a) At least one swab, in appropriate transport medium, for microbiological examination (and possibly DNA studies).
   (b) At least one touch preparation, preferably two, for cytological

(continued)

examination from the edge of the lesion.

(c) If the victim is dead, take in addition a wedge of tissue from the edge of the lesion (representing a transition from normal to damaged tissue) for histological examination.

Whenever possible, snares that have been removed should also be examined in detail. They must be handled aseptically, wearing gloves. Recommended techniques for laboratory examination of snares are given in Box 10.

Box 10 Laboratory Examination of Snares

Laboratory examination of snares

- Photograph in detail the area(s) that were in contact with the animal's body.
- Use a hand lens (magnifying glass) to search the snare, especially the area (s) that were in contact with the animal's body, for evidence of blood or tissue (Fig. 16).
- If blood or tissue is found, take samples for cytology, bacteriology and histology.
- If no blood or tissue is found, take at least one sample apiece from the surface of the snare (the area(s) that were in contact with the animal's body) for cytology and bacteriology. The snare can be swabbed or irrigated in sterile saline.

Firearms contribute to injury and death in animals, both domesticated and wild. Gunshot wounds vary greatly in appearance and this is complicated in wildlife forensics because different morphology will influence the appearance of wounds—for example whether or not feathers or scales are present. So, also, will the type of firearm used; there are great differences between the damage done by a shotgun and a rifle. An airgun may not have sufficient energy to kill the animal; there may be a small entrance wound and the pellet is likely still to be in the body. Pathological change may be minimal.

The basic features of gunshot wounds were described by Williams et al. (1998), amongst others, and discussed in the context of animals by Cooper and Cooper (2007). Entry wounds can be summarised as in Box 11 (adapted from Williams et al. 1998).

Box 11 Entry Wounds

Entry wounds

- Contact wounds up to 2 cm diameter. A ragged, sometimes stellate, entrance wound is present. Soot or gunpowder may be obvious and carbon monoxide may be detectable in the tissues.
- Close range, 2–50 cm. The wound is likely to be circular and concave. Soot or gunpowder is unlikely to be detected macroscopically (cf. contact wound) but can be seen microscopically in histological sections. Feathers or hairs may appear burnt.
- Longer range. In the case of a bullet, the entrance wound is likely again to be circular and concave but in shotgun injuries there will be multiple peripheral separated holes ("satellite lesions"). Soot or gunpowder is not usually visible macroscopically or microscopically; ultraviolet light may assist.

Shotgun wounds show characteristic features relating to the distance and angle. A tangential shotgun injury may glance off the animal's body producing a wound that resembles a laceration. As indicated above, close shotgun wounds usually present as a rounded hole with a narrow rim of soot at close contact. Soot deposition and singeing of feathers or hair may be apparent. At a longer range, spread of the shot can result in a central entry hole with "satellite lesions". Measurements of the scatter pattern of the shot can be used to give an approximate range but, as in all ballistic work, it is best if this is correlated with the gun that fired the shot.

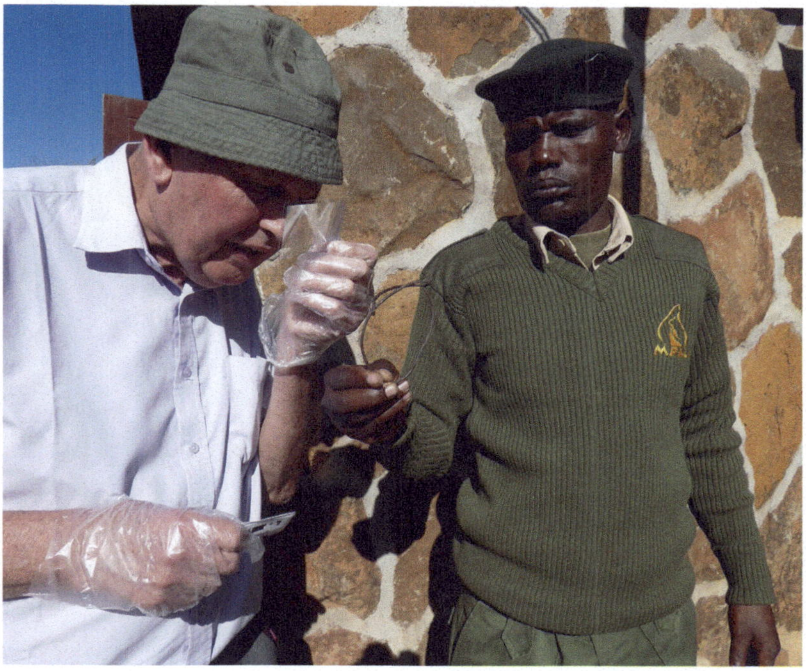

Fig. 16 Examining a snare, removed from a captured animal, for evidence of blood or tissue

There are important differences between the damage created by a low-velocity weapon, such as a handgun, and those from a high-velocity rifle. The latter cause severe injury and shock waves, created by the energy of the bullet, and can destroy tissues that are not in the pathway.

A very important part of the investigation of gunshot wounds is to ascertain the type of weapon that was used (Williams et al. 1998). Electronic databases are available that can help determine the weapon type that was most likely to have discharged a suspect bullet or cartridge case. This is a specialised field and an expert's advice is usually needed.

Whether a pathologist or not, the person who encounters a gunshot wound (or other lesions, such as a snare injury) must be prepared and able to describe the wounds. Differentiation of entry and exit gunshot wounds is important (see Table 5).

Radiography enables investigation of the bullet wound tract and will facilitate the finding and recovery of fragments from tissues. Wobeser (1996) advocated the use of both radiography and fluoroscopy in wildlife cases.

The microscopic examination of bullets and cartridge cases is an important part of firearm examination and is increasingly becoming automated. It is crucial that the pathologist removes, handles, and stores all ballistic material with care. For example, as mentioned previously, plastic forceps should be used to extract bullets because the damage inflicted by metal instruments may make subsequent identification of the weapon more difficult. Bullets and

Table 5 Differentiation of entry and exit gunshot wounds (reproduced in a modified form from Cooper and Cooper 2007)

Entry	Exit
Well circumscribed, round or oval	Often ragged edges, irregular shape
A "collar" of abrasion and discharge residue may be visible	Collar and discharge residues rarely present
The same size or smaller than the bullet	Variable size

fragments need to be gently rinsed and dried, to avoid oxidation, and then wrapped individually in soft material in a firm container.

Traumatic Injuries

Injuries may be due to assault by animals of the same species, by others of a different taxon and by human beings. Intraspecific injuries include fight wounds, some inflicted during courtship or competition for mates. Attacks by other species may involve "normal" predation (e.g. a hawk hunts a rabbit) or mass killings, as can be a feature when a red fox gains access to poultry or find itself amongst breeding seabirds (Cooper and Cooper 2007).

Case Study 2
"*Who killed the kukus*?"

Readers are invited to read and review the following case study and photographs (images).

Suggested answers are given at the end of the chapter.

These local chickens (*Gallus gallus domesticus*), known as "*kukus*" in Swahili, located near the coast of Kenya, were put out in the morning to forage. They were later found dead and largely dismembered. The photo below shows the first chicken that was collected. This incident occurred some distance from the nearest town, in the "bush", and so no personal protective equipment (PPE), such as a coat, gloves or eye protection, was available.

Question 1: Comment on the techniques being used to investigate and examine this dead chicken.

Question 2: What do you see in or on the chicken?

Question 3: How might a knowledge of veterinary pathology contribute to the investigation of this case?

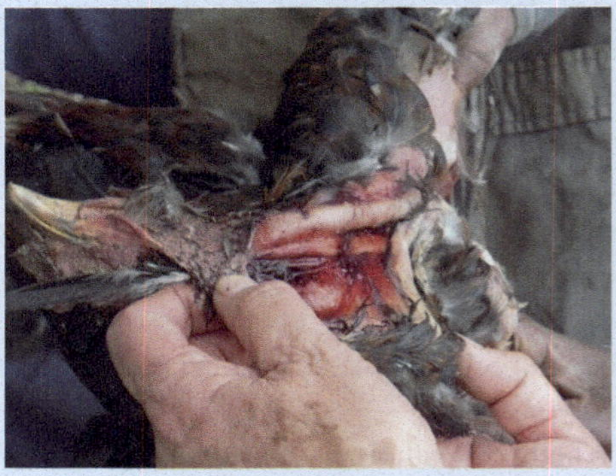

Both incised and lacerated wounds may be seen in attacks on humans and animals. These must be differentiated (see Table 6). Essentially, an incision is a break in the continuity of the skin caused by an object with a sharp edge, whereas a laceration is caused by blunt force.

Damage to the head can cause brain damage, often fatal, and necessitates careful dissection.

Table 6 Incisions and lacerations—some distinctions (reproduced in a modified form from Cooper and Cooper 2007)

Feature	Incision	Laceration
Appearance	Sharply outlined Linear, curved or angular Bruising is rare Strands of tissue bridging the wound are rarely present	Often irregularly outlined. Round, oval or stellate Bruising is common Tissue bridges are often present
Pelage	Little or no loss of hair, feathers or scales around the wound	Hair feathers or scales may be absent
Haemorrhage	Marked	Minimal or apparently absent

The spinal cord is also susceptible to trauma. A variety of *post-mortem* lesions may be shown, including haemorrhages (haematomyelia). Radiography and careful dissection may be needed. Gross *post-mortem* examination of the brain/spinal cord and surrounding tissues must be coupled with histological study.

Wild animals may suffer "non-accidental" injuries caused by humans (anthropogenic trauma). Great care must be exercised in interpretation, especially when large numbers of victims are involved. In individual animals, kicking or stomping injuries are sometimes a feature and are characterised by multiple lesions including external bruising, fractured bones and damage to internal viscera (Cooper and Cooper 2007).

Other Types of Pathology

These are covered very briefly below. Further information can be found in Cooper and Cooper (2007, 2013).

Strangulation and Hanging

Compression of the neck, causing death by occluding the trachea, by preventing venous return to the heart, by reducing arterial flow to the brain and causing asphyxia (see later) or by exerting specific effects on (e.g.) the vagus nerve or carotid sinuses (Shepherd and Simpson 2003). Animals may be strangled intentionally by an attack (human or animal) or as a result of self-inflicted damage, e.g. a bushbuck pulling on a snare around its neck.

Blast Injuries

Animals may be exposed accidentally (e.g. because they are trapped in a building that explodes) or intentionally (e.g. cattle used to detect land mines). Dogs trained to detect explosives or to locate cadavers are particularly at risk from injuries ranging from multiple fractures and soft tissue damage to extensive, often fatal, damage to the body.

Neonatal Deaths

Can be due to non-infectious factors (accidental or intentional) such as developmental abnormalities, trauma, suffocation, hypothermia, drowning or electrocution. Infectious disease can complicate these or cause neonatal disease in their own right.

Asphyxia

Refers to a reduction in oxygen (hypoxia) or a total absence of oxygen (anoxia). Commonly due to an obstruction to air flow. The ability of animals to withstand hypoxia varies greatly; reptiles, amphibians, fish and invertebrates are particularly resistant.

Electrocution

Wild animals may be exposed to electrical circuits, power lines and lightning. Shaving of the hair or removal of feathers is important prior to the examination of a carcase. Lesions characteristic of electrocution and/or burning (see below) may be detected by light or scanning electron microscopy (Cooper and Cooper 2013).

Electrocution damages tissues directly and by producing burns. Respiration can cease. Ventricular fibrillation is usually the cause of death. Burning or charring of the hair or feathers may be apparent. In severe cases, the carcase may be burned to the bone, giving a "fossilised" appearance. Lymph nodes are often haemorrhagic and there can be free blood in the respiratory tract, discoloration and rigor of muscles and black and unclotted blood. Lesions may be few, other than slight burns and petechial haemorrhages.

Fig. 17 Mediterranean tortoise (*Testudo hermanni*) with missing scutes and other lesions due to earlier burning by bushfire

Burns and Death in Fires

Fire causes death through inhalation of smoke and hot gases and burns. Fires may occur accidentally (in the home, in a zoo, in the bush) or may be started intentionally, either in order to harm humans or animals or to mask evidence of homicide, killing or other unlawful acts.

An increasing body of information about the effects of burning on wildlife has been amassed as a result of extensive bush fires in Australia. This affected seven states and territories and in an update on 8th January 2020 experts at Sydney University suggested that over 800 million mammals, birds and reptiles had died, either as a direct result of the fires, or due to excessive heat, dehydration, starvation and habitat loss. See: https://www.sydney.edu.au/news-opinion/news/2020/01/08/australian-bushfires-more-than-one-billion-animals-impacted.html

Live animals may be submitted with burn wounds and associated systemic changes. Old lesions may persist as scars and characteristic scute loss/damage in chelonians (Fig. 17).

Post-mortem diagnosis can be aided by detection of heat damage and soot deposition in the airways, toxicological analysis for carbon monoxide or other chemicals and histological examinations for evidence of inhalation of soot. Hair, feathers and scales may show characteristic microscopic changes. Artefacts associated with burning are a reduction in size and stiffening of the body. One must distinguish burnt dead bodies from those that were affected *ante mortem*.

Drowning

Defined as "the process of experiencing respiratory impairment from submersion or immersion in a liquid" (Bierens 2014). When water is inhaled, the surface area for gaseous exchange is reduced and bronchioles become blocked, causing anoxia. Osmotic effects are haemolysis in freshwater, or dehydration and pulmonary oedema in hypertonic seawater. Laryngospasm can cause asphyxia. The pathogenesis of drowning in humans is well understood, but there are no pathognomonic findings (DiMaio and DiMaio 2001) and diagnosis is sometimes one of the most difficult challenges in forensic pathology (Stephenson et al. 2019).

A valuable contribution to our understanding of drowning in animals was the description of the gross pathology in seabirds by Simpson and Fisher (2017) who explained that, when combined with contemporaneous observations, gross pathological lesions may permit a diagnosis of drowning, especially when a batch of freshly dead birds is examined. They cautioned that much of the recorded pathology relating to drowning (and other causes of mass mortality of seabirds) is derived from studies where specimens had been frozen prior to *post-mortem*

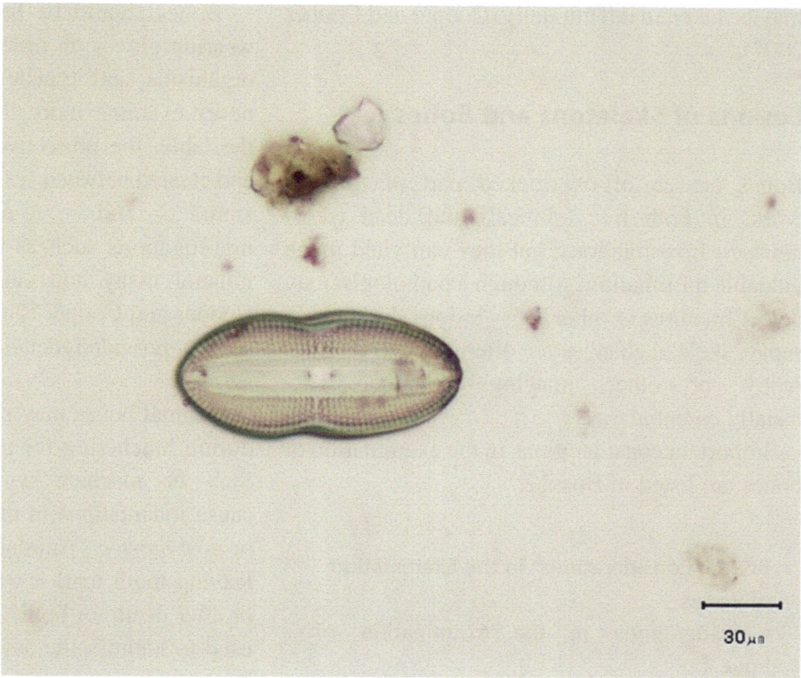

Fig. 18 A common and widespread diatom. There are very many species and genera, most requiring expert skills for identification

examination and some had shown degrees of autolysis. Certain lesions, especially those of the respiratory system, had thereby become obscured by artefactual changes.

In terrestrial mammals, water and frothy fluid (proteinaceous surfactant and macrophages) is found in the respiratory tract, including the air sacs of birds, and often in the oesophagus and stomach. The lungs may be distended and over-distended and sub-pleural petechiae may be present (Cooper and Cooper 2007).

Diatoms, micro-algae (Fig. 18), together with some other aquatic flora, may be found in the lungs of animals that are suspected of having drowned. Diatoms can be significant, but do not necessarily support a diagnosis of water inhalation *intra vitam* (while the animal was alive) or death by drowning. Diatoms can be inhaled at other times and not result in death. In marine mammals, in particular, they may be inhaled or ingested *post mortem* (Bortolotti et al. 2004). The book by Pollanen (1997) about "forensic diatomology and drowning" reviewed the medicolegal value of "the diatom test", including the examination of femoral bone marrow.

Notwithstanding the controversy over the relevance of diatoms in drowning, these organisms are increasingly being recognised as important evidence in other circumstances, including forensic geoscience. Diatoms tolerate a range of environmental conditions; they can provide important information (Scott et al. 2017).

Prolonged immersion of a carcase in water, whether following drowning or not, can produce swollen, corrugated skin—so-called "washerwoman's hands"—and distend soft tissues (DiMaio and DiMaio 2001).

Impaction of the Gastrointestinal Tract

Ingestion of debris causes morbidity or mortality in hundreds of thousands of animals each year. The problem is increasing as more and more non-degradable rubbish, especially plastic, builds up on land, in rivers and in seas. The presence of material in the gastrointestinal (GI) tract is not, however, always a sign of ill health. Some items are ingested accidentally, others (e.g. grit in certain birds, metal and other objects in ostriches) may be taken in intentionally (Cooper and Cooper 2007).

Lesions of Skeletons and Bones

Bones are an oft-overlooked part of forensic work, in both live (clinical) and dead (*post-mortem*) investigations, but they can yield much valuable information. Although a pathologist may detect fractures or other gross lesions during necropsy, skeletal changes are often missed. Radiography or other imaging techniques are usually essential.

Important considerations in the examination of bones are listed in Box 12.

Box 12 Considerations in the Examination of Bones

Considerations in the examination of bones:

- How was the bone prepared? Naturally (e.g. in the soil) or in the laboratory (boiling, chemical treatment, use of *Dermestes* beetles)? This can influence appearance.
- How has the bone been handled and stored since collection? Rough manipulation or packing bones together without protection can easily cause *post-mortem* artefact.
- Which species is it? There are differences in the vulnerability of bones to physical damage. Bones of flying birds, for instance, tend to be light in weight and less easily damaged in transit than are those of a mammal.
- Which bones are present? Some may be easy to identify, others may only be fragments or splinters.
- What was the age at death? This will require careful examination of epiphyses and teeth. Radiography may assist. For many species data are not available.

Bones should be handled gently, preferably wearing gloves, in order to reduce the spread of organisms and transfer of DNA. One should never examine more than one set of bones on the table: the protective sheet must be checked and cleared between sessions. Bones can be easily stored and re-examined. Laboratory investigations such as histological examination, mineral assay and carbon-dating are possible. Cooper and Cooper (2013) and Cooper and Hull (2017) provided detailed guidance on skeletal studies.

Animal bones may have been cut or damaged during butchering for meat or trophy collection. Stab or machete wounds—see earlier—may cause indentations in the periosteum, either *ante* or *post mortem*. Animal bites may damage bone, leaving tooth marks, when the victim was alive, or after death, or both. These tooth marks can be used to identify the assailant, whether it be dog, hyaena or crocodile. *Post-mortem* predation (scavenging) of bones is well recognised: both domesticated and wild animals may be attracted to animal remains, where they chew, break or swallow bones.

Although a pathologist may detect fractures in a carcase during necropsy, skeletal changes are often missed. Radiography is always important and in some cases, CT scanning or other techniques may be needed (Fig. 19).

A suggested format for recording skeletal findings, based on many years of study of gorilla bones, is given in Appendix 8.

Mummification

Mummification takes place when a carcase is exposed to environmental factors that permit breakdown of tissues and desiccation occurs but there is no associated microbial multiplication (Shepherd and Simpson 2003). This can occur, for example, in a cool, dry, place such as a cave. Such carcases are dry and may consist of only bones and dry tissues. Techniques for examination are poorly documented but useful guidance can be obtained from Cooper and Cooper (2007), and in publications by archaeologists,

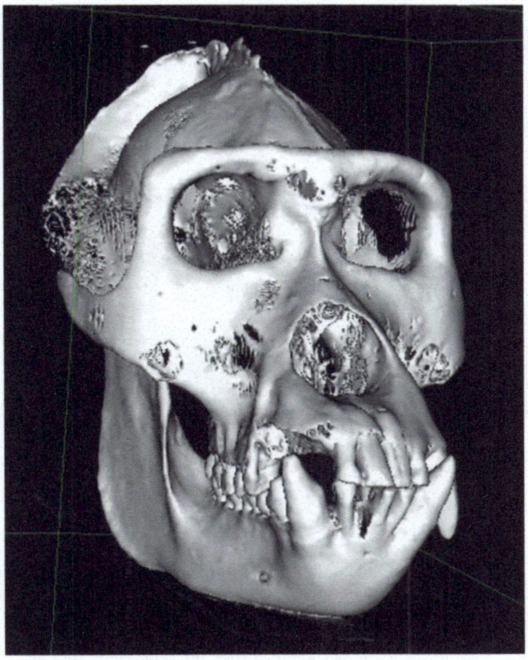

Fig. 19 The skull of a gorilla (*Gorilla gorilla*); external features demonstrated by CT (computed tomography) scanning

Egyptologists and others who examine mummies (see, for example, Taylor 2011).

Starvation

Starvation and dehydration may coexist. Gross *post-mortem* examination plays an important part in diagnosis of such cases. An emaciated appearance, with absence of subcutaneous fat and an empty GI tract, is suggestive of prolonged starvation. Organs may be reduced in size and weight. In ectothermic vertebrates, fat bodies may be absent or depleted.

Macroscopic examination will reveal the size of organs, their appearance (e.g. if a liver is pale, possibly fatty) and the amount of internal body lipid.

Microscopic (histological) examination enables the pathologist to assess more accurately the appearance of organs—to include whether they contain lipid or show other significant changes and will permit the measuring of hepatocytes or assessment of zymogen granules in pancreatic acini, which can sometimes be used as an indication of impaired nutrient intake.

Group, Herd or Flock Examinations

It may be impossible to necropsy all the animals that are presented—when, for example there is a "die-off" of fish on a lake. Under such circumstances, it is usually acceptable to take and examine a *sample* of the animals.

When many animals are involved, it is wise to save (chilled, frozen, fixed) carcases that cannot be examined immediately. If large numbers are to be examined immediately *post mortem*, there is merit in recording the main findings on a special "group or flock" form. This enables findings in different animals to be compared and contrasted rather more readily than if individual record sheets are used. Such an approach was used very successfully when investigating unexplained deaths in African flamingos under field conditions (Cooper et al. 2014).

Disposal of Carcases

Although, as a general rule, carcases and tissues relevant to a legal case should be retained, this is not always practicable.

Disposal of very large carcases, such as cetaceans, may require specialised equipment and scavengers are often a problem (Fig. 20a–c).

Reports and Reporting

Accurate and consistent reporting is vital in wildlife forensic work. Examples of report forms used by the author are given in Appendices 8 and 9. This and others are in a format that can be changed, as necessary, to suit the circumstances.

Sometimes there is a need to produce a preliminary or interim report before all laboratory tests are complete and/or a conclusion reached. As a general rule, one should be very wary of requests

Fig. 20 (**a–c**) Disposal of very large carcases, such as this large Bryde's whale (*Balaenoptera brydei*) will require specialised means of disposal and scavengers are often a problem. In b) note the vultures circling in the sky above

for a preliminary report. If obliged to provide some information ahead of a final report, this should be in the form of a short summary, listing the main findings and stressing that follow-up tests are in progress. See Appendix 9. Summary of *Post-mortem* Findings.

Verbal statements should be avoided: they may be misinterpreted.

Communities and Culture

Those working overseas, especially in rural settings, have community responsibilities (Cooper 1996)—see also chapters by, Lichtenfeld, L.; Kalema, G.; Owiti, G.: all in this volume). Animal tissues and bodies of animals are not usually as sensitive an issue as those of humans, but particular care should be taken where a species is of cultural, sometimes religious, importance (Cooper and Cooper 2007).

Conclusions

Global environmental change presents a challenge to all. The disappearance of species and the decline in biodiversity urgently require a multidisciplinary and interdisciplinary approach and the application of forensic principles—"conservation forensics". There are unprecedented opportunities for input by pathologists and the application of pathological methods. Sir William Osler, cited earlier, regularly told his students "*As is your pathology, so is your practice*". His medical aphorism is applicable also to very many wildlife forensic investigations.

Answers to Case Studies

The answers are only provided for guidance and are intentionally given in an abbreviated form in order to encourage readers to seek clues from this chapter and, where appropriate, other sections in the book.

Case Study 1
"*How did the hippo die*"?

Suggested Answers

Question 1: Describe what you see.

A small dead hippo with a lesion (a long, linear, skin wound) on its left flank. The lower middle photo shows large numbers of presumed maggots (dipterous larvae) and dark marks on the skin that might be lesions. The bottom picture confirms the presence of maggots, which were alive and wriggling at the time. Question 2: Comment on how the investigator (first responder) is approaching and dealing with the case.

The investigator and his assistant appear to have no personal protective equipment (PPE), such as gloves or eye protection. The investigator is using a plastic bag in lieu of a glove to investigate the wound and collect maggots.

*The area should be cordoned off using proper crime scene materials (tape, markers etc.) if available. In the absence of such items, every effort should be made to demarcate the location using natural materials such as vegetation and personal items such as scarves. Access must be denied to persons who are not part of the investigation.*Question 3: How might samples from this animal throw light on the cause, mechanism and manner of its death?

Pieces of skin could be examined histologically in order to study the main (open) wound and the dark marks on the skin that might be lesions. However, there is much post-mortem decomposition which is likely to hamper accurate interpretation of tissue changes.

*The maggots could help determine post-mortem interval (PMI). They would need to be identified and their age (which instar) determined. Both live and dead (fixed) maggots should be collected. The development and metamorphosis of insect larvae are temperature-dependent and therefore any calculations need to take into account the environmental and climatic conditions to which they and the dead hippo have been subjected.*Question 4: How might a knowledge of veterinary pathology contribute to the investigation of this case?

Pathology is "the science of the study of disease" and is not confined to the examination of dead animals. A knowledge of veterinary pathology would assist in describing the lesions and in the laboratory examination of pieces of skin.

A proper understanding and implementation of the "clean and dirty" principle would help protect both the humans involved in the investigation and the samples that they plan to process.

You may have other observations and suggestions; that is good. Forensic investigation requires an ability to "think outside the box" and to explore alternative scenarios and solutions.

Case Study 2

"Who killed the kukus?"

Suggested Answers

Question 1: Comment on the techniques being used to investigate and examine this dead chicken.

No personal protective equipment (PPE) is being used—as pointed out above. An attempt to protect those involved should have been made—for example, by adapting plastic bags to make simple gloves. Was it really necessary for all three people to handle the dead bird? Poultry can harbour pathogenic organisms and if, as was suspected here, baboons might have been responsible, they (like many other primates) can be a source of zoonoses. PPE also helps protect samples taken for forensic purposes.Question 2: What do you see in or on the chicken in the photos?

The chicken shows damage to the cervical (neck) region, with subcutaneous haemorrhage. Question 3: How might a knowledge of veterinary pathology contribute to the investigation of this case?

Pathology is "the science of the study of disease" and is not confined to the examination of dead animals. A knowledge of veterinary pathology can assist in detecting and describing lesions (in this case, the damage inflicted on the chicken's neck) and in determining the age and likely aetiology/pathogenesis of the lesion. Pathology also involves the taking and processing of samples that might throw more light on the episode.

A proper understanding and implementation of the "clean and dirty" principle would help protect both the humans involved in the investigation and the samples that they plan to process.

Some Additional Information Relevant to this Scenario

African baboons (Papio spp.*) are opportunistic feeders. They eat a variety of fruits and other plant materials but will also take meat in the form of invertebrates, reptiles, birds, rodents and certain larger mammals. They are considered pests in many African settings, where they raid farms and* "shambas", *taking crops and sometimes killing livestock. Protective measures include deterring them by using people (traditionally small boys armed with stones) and dogs, and building protective barriers around livestock pens.*

Baboons take eggs as well as the birds themselves.

You may have other observations and suggestions; that is good. Forensic investigation requires an ability to "think outside the box" and to explore alternative scenarios and solutions.

Acknowledgements John and Margaret Cooper would like to express their appreciation for the help and support of Susan Underkoffler, senior editor of this book, who encouraged them to contribute and who tolerates in such a good-humoured fashion their "olde worlde" approach to both professional and personal matters. They are also grateful to Dr. Dino Martins for opportunities to take photographs at Mpala Research Centre, Laikipia, Kenya, some of which are included in their chapters for this book. Blackwell-Wiley gave permission to reproduce in a modified form material from "Introduction to Veterinary and Comparative Forensic Medicine" (Cooper and Cooper 2007).

Appendices

Appendix 1: Forensic Examination of Live Animals: Birds

1. Remember that observation of the bird and its environment should precede restraint, handling or clinical examination. Note the bird's behaviour, including commenting on how docile, tame or habituated it appears to be, whether it might be imprinted, whether it is easily frightened by noises or other stimuli and whether it responds normally to routine

stimuli. Familiarise yourself with the natural history and biology of the species which is the subject of the investigation. Consider involving a colleague, bird-keeper or ornithologist, who may have more knowledge of such aspects than you do.

2. Carry out as full an investigation as possible. In addition to standard clinical examination record weight (mass) and standard measurements, give a condition score and report specifically on the following: plumage (including evidence of moult or pinioning), beak, claws, soles of feet, presence/absence of leg rings (bands), jesses, telemetry equipment or other attachments, presence of tattoo (s) or coloured dyes. Pay special attention to clinical signs or lesions that may be relevant to a history of alleged neglect, cruelty or persecution Throughout the clinical examination, practise strict hygiene, following a proper Risk Assessment.
3. Take photographs throughout the examination, even if only simple photographic equipment is to hand, especially if there are lesions that may change in appearance if there is any delay.
4. As a routine, take the following samples for laboratory tests: fresh droppings (faeces and urates), blood smears, blood for haematology/clinical chemistry, dropped feathers, pellets/castings (where available) and ectoparasites. Mark all specimens carefully and fully, with the label on the container, not on the lid.
5. Depending upon the circumstances, be prepared to carry out supporting investigations, e.g. radiography and ultrasonography, emesis and lavage. Imaging may reveal significant lesions that cannot be easily detected on palpation.

Some Further Reading

Cooper JE (2002) Birds of prey: health & disease. Blackwell, Oxford, UK

Cooper JE, Cooper ME (1986) Is this eagle legal? A veterinary approach to litigation involving birds. In: Proceedings of the Forensic Zoology Discussion Group. Zoo-technology, London, pp 27–30

Cooper JE, Cooper ME (1991) Legal cases involving birds: the role of the veterinary surgeon. Vet Rec 129:505–507

Cooper JE, Cooper ME (2013) Wildlife forensic investigation: principles and practice. Taylor & Francis/CRC Press, Boca Raton, Florida

Appendix 2: Submission and Report Forms

This is a suggested format that can be changed, as necessary, to suit the circumstances

Submitting Client/Agency.............................. Reference No ....

Lab Ref.. Date (in full)...................................

Address ....

....Post (Zip) Code.............................

Tel .... Fax................................ Email...

Veterinarian (where applicable)..

Address and contact details..

Other relevant persons, e.g. Police Officer, Wildlife Inspector......................................

Species of animal (English and scientific name)...

Local name.............................. Breed/variety..

Colour/markings...........................

Age....................... Sex .... Pet Name.......................................

Ownership of animal or sample(s)..

Number/Ring (Band)/Tattoo/microchip/Other methods of identification of material submitted...

Background to the case/history...

....

Signature:... ***Sheet Number 1/4***
Date: ....
Time:.......................................

Sample submitted………………………………………………………………………

Sent by hand/courier/post/other…………………….………..

Method and description of packing ………………………………………………………..

Tag No……….. Prior storage condition……………………………………………………

Signature of courier ……………………………………….. Date…………………………

Carcass…. …………………………………………………………………………………

Organs (state)………………………………………………………………………………

Other tissues…………………………………………………………………………………

Parasites……………………………………………………………………………………..

Blood (E/H/P/Ox/other)……………………………………………………………………

Swabs…………………………………………………………………………………………

Other………………………………………………………………………………………..

Comment on chain of custody (attach relevant paperwork …………………………………

…………………………………………………………………………………………………

Questions being asked (expand as necessary)

- Why did this animal die? ………………………………………………………
- When did this animal die? …………………………………………………………
- How did this animal die? ………………………………………………………..
- Did this animal suffer pain or distress? ……………………………………………

Signature:…………………………………………… ***Sheet Number 2/4***
Date: ………. …………………………….
Time:………………………………..

- What species is this? ………………………………………………………………
- What is its provenance/parentage, etc? ……………………………………………..
- What is this material? Species? Sex?………………………………………………
- Other questions ……………………………………………………………………..

Investigations required. …………………………………………………………………

Post-mortem examination ………………………………………………………………

Laboratory tests:
Toxicology ………… ……………………………………………………………..
Bacteriology………………………………………………………………………
Histology …………………………………………………………………………
Other ………………………………………………………………………………

Other tests as necessary to answer the questions above ………………………………..

……………………………………………………………………………………………...

……………………………………………………………………………………………...

Results to be sent to ……………………………………………………………………

……………………………………………………………………………………………..

Special instructions regarding the storage/transfer/disposal of samples/wrappings/carcass

……………………………………………………………………………………………..

……………………………………………………………………………………………..

……………………………………………………………………………………………..

……………………………………………………………………………………………..

Signature:……………………………………… ***Sheet Number 3/4***
Date: ………………………………………
Time:…………………………………..

Other comments, e.g. cruelty case, civil action, insurance claim, professional malpractice hearing, etc.

………………………………………………………………………………………………..

………………………………………………………………………………………………..

………………………………………………………………………………………………..

Signed by recipient at (location) ………………………………………………………………………..

Date ……. ……………………….. Time ………………………………….

Signature:…..…………………………………….. ***Sheet Number 4/4***
Date: ………………………………………..
Time:…………………………………..

SPECIMEN LABORATORY SUBMISSION AND REPORT FORMS

Species:...Reference No:..

Sample:...Collected by:...

Location:...Date:.............................Time:..............

Relevant History:..

....

....

Lab ID:..Stored in:...

Sample submitted by:...Date:........................Time:...................

Sample received by:...Date:........................Time:...................

Laboratory Investigation Report

Gross Examination:...

....

Microscopical Examination:...

....

Further Test(s)

Test	Submitted (date)	Results (Received date)	Comments
Microscopy			
Bacteriology			
Cytology			
Parasitology			
Haematology			
Biochemistry			
Other (specify)			

Summary of findings:..

Results/Comments/Interpretation:...

Reported by:..Date:......................Time:...................

Signature:.. ***Sheet Number 5/4***
Date: ....
Time:.....................................

Appendix 3: Examination of Eggs and Embryos

PROCEDURES

Egg received, given laboratory reference number and receipt acknowledged

Preliminary cleaning (record)

Candled (drawn and described)

Probably infertile

Weighed, measured, exterior drawn and described

Cleaned with ethanol/methanol

Opened - examined, drawn and described <u>in situ</u>; placed in Petri dish, samples taken as necessary

Contents frozen for toxicology, etc

Shell dried, weighed and retained

Report form(s) completed

Probably fertile

Weighed, measured, exterior drawn and described

Cleaned with ethanol/methanol

Opened - examined, drawn and described <u>in situ</u>; placed in Petri dish, samples taken for histology, bacteriology, etc

Contents fixed/frozen as necessary

Shell dried, weighed and retained

Report form(s) completed

REPORT FORM

Reference number:

Received: (date) .... (by) ....

Receipt acknowledged by: .... Date ....

Method of packing/wrappings:

History:

EGG / EMBRYO EXAMINATION
(to be completed for each specimen)

Species: (English name) .... (Scientific name) ....

Owner / Origin: ....

Weight of whole unopened egg: .... Length: .... Width: ....

External appearance (include photo or drawing - Fig. 1)

Appearance on candling (include photo or drawing - Fig. 2)

Embryo

Air cell

Blood vessels

Fluids

Appearance when opened (include photo or drawing - Fig. 3)

Contents:

Embryo: Length (crown-rump)

Amniotic cavity

Allantoic cavity

Yolk sac

Other comments:

Microbiology:

Histopathology:

Other tests:

Samples sent elsewhere:

Weight of dried eggshell: .... Thickness (measurement or index): ....

Samples stored

FINDINGS/DIAGNOSIS/ADVICE

Date: .... Signature: ....

JEC June 2020

Appendix 4: Turtle Egg/Embryo Examination Form

Reference Number:____________________ Date of examination: _____________
Necropsy Number: ____________________
Received (date): __________________________ by:___________________________
Location of collection: ____
Persons involved in examination:____
History/Environmental Conditions:

TURTLE EGG/EMBRYO EXAMINATION:
(to be completed for each specimen)

Species: (common name) _______________Scientific name:______________________
Owner/Origin: ____
Weight of unopened egg:______________g Lengh:________mm Width:________mm
Overall external appearance (including consistency of shell):

Appearance on candling:
Embryo:
Air cell:
Blood vessels:
Fluids:
Appearance when opened:

Contents
Embryo:
Measurements
Yolk sac:
Other comments on egg or embryo:

LABORATORY INVESTIGATIONS:
Microbiology:
Histopathology:
Other tests:
Samples sent elsewhere:
Weight of dried eggshell: ___________g Thickness (measurement or index) __________
Samples stored

COMMENTS:

FINDINGS/DIAGNOSIS/ADVICE:

Date: ____________________ Signature: ____________________________________

JEC June 2020

Appendix 5: Field Equipment Lists

Equipment—General Items

Recommended equipment is listed below:

- Protective clothing including boiler suits (coveralls), gloves (surgical, thick kitchen, long rubber etc.), masks, goggles, boots and overshoes.
- Barrier tape, flags, markers, cones and other crime scene security items.
- Protective gloves and clothing, barrier tape and tools, such as tongs and hooks for handling hazardous materials.
- Disinfectant and deodoriser to neutralise smells from carcases etc. (both used with caution at crime scenes as they may destroy trace evidence).
- Collection kits (prepacked) for taking samples for laboratory investigation, including toxicology, DNA analysis and haematology.
- Appropriate labels, tamper-proof tags, evidence seals etc.
- Marking pens.
- Scales/balances, callipers and micrometer for weighing animals, tissues and samples.
- Rules, tapes, centimetre scales etc. for measuring.
- Equipment for taking casts of dentition, imprints in bones, animal tracks etc., including plaster-of-Paris, other powders and silicone-based materials, waxes, plus retention frames or stiff cardboard.
- Trace-evidence collection equipment.
- Evidence packing, e.g. bags, boxes, tubes, envelopes and other supplies for packaging and storing evidence.
- Photographic kit, including (as appropriate) digital/still cameras, video cameras, magnifier, night vision equipment, aerial camera system and camcorder.
- Binoculars/field glasses.
- Torches (flashlights).
- Blue-Light kits and supplies.
- Egg candler.
- X-ray viewing screen.
- Magnifying glasses/hand lenses.
- Magnifying loupe or dissecting microscope.
- Clipboards and record sheets, plus pens and pencils.
- Chalk and crayons.
- Elastic bands and string.
- Tape-recorder (preferably voice-activated) and tapes.
- Evidence seals/tape.
- Kits and systems for collecting and preserving delicate crime scene evidence (see text).
- Computer and appropriate software, e.g. barcode tracking systems for property, evidence and crime scene reconstruction.
- Handheld computer.

Equipment: When Working in the Field

The detailed contents of field kits will depend upon the location and the type of investigations being carried out.

The following items are recommended in the field for (a) clinical, (b) *post-mortem* and (c) laboratory diagnostic work, respectively. In each case, the list should be supplemented with general items, as above.

(a) *Clinical equipment—when live animals have to be examined*
   - Stethoscope (lightweight)
   - Auriscope (otoscope) (lightweight)
   - Ophthalmoscope (lightweight)
   - Rigid endoscope (battery-operated)
   - Pen torch (flashlight)
   - Spare bulbs and batteries
   - Syringes and needles (disposable)
   - At least one boilable, reusable syringe and needle
   - Empty drinks cans, labelled "sharps boxes", for used needles, scalpel blades etc.
   - Disinfectant(s), including ethanol
   - Camping (gas cylinder-operated) stove—for sterilising, lighting and cooking
   - Pressure cooker for sterilising
   - Selected medicines, including local analgesics, sedatives and agents for euthanasia (plus gun if large animals may need to be killed)
   - Cotton wool
   - Dressings

- Suture materials
- Basic surgical ("cut-down") set and other instruments as necessary
- Disposable skin biopsy punch
- Cautery (battery operated)
- Clippers for claws, talons, beaks
- Ring (band) remover
- Spring balance(s) or battery-operated scales
- Cloth bags and other devices for restraining small animals
- Gloves—surgical and for handling
- Towel
- Oesophageal and other tubes
- Mouth gag/wooden spatulae
- Aluminium foil
- Sampling and other equipment for laboratory work (see later—list (c))

(b) *Post-mortem equipment—when dead animals have to be examined*

Standard necropsy items—portable/folding, lightweight/plastic where appropriate.

- Saw(s)
- Scalpels and blades
- Knives
- Forceps
- Probes—solid and flexible (rubber)
- Scoops for brain etc.
- Pen torch (flashlight)
- Spare bulbs and batteries
- Syringes and needles (disposable)
- Empty drinks cans, labelled "sharps boxes", for used needles, scalpel blades etc.
- Disinfectant(s), including ethanol/methanol/methylated spirits
- Camping (gas cylinder-operated) stove—for sterilising, lighting and cooking
- Pressure cooker for sterilising
- Cotton wool
- Spring balance(s) or battery-operated scales
- Scalpel handle and disposable sterile blades of several sizes and shapes. Dissecting scissors, curved haemostatic forceps, toothed and smooth-jawed fine-pattern thumb forceps and bone forceps
- Sampling and other equipment for laboratory work (see later—list (c))

(c) *Laboratory equipment*

- Microscope (solar or battery operated)
- Immersion oil (or methyl salicylate) with swabs and xylene for cleaning
- Pre-cleaned, frosted, ground-ended microscope slides and slide box or tray
- Pencils for marking glass slides
- Diamond-tipped pen for marking glass slides (if frosted not available)
- Worm-egg counting slide
- Coverslips
- Lens tissues
- Saline, saturated NaCl solution and other reagents and equipment for parasitology
- Transparent polythene strips and methylene blue/malachite green for the KATO method of cleaning faecal films for parasites, ova and cysts
- Fixatives—alcohol, formalin
- Selected stains for cytology
- Lightweight (plastic) staining jar or staining rack
- Urine and blood chemistry test strips
- Portable centrifuge
- Polypropylene capillary tubes, some coated with heparin or EDTA, plus commercial haemoglobin and PCV reader
- Handheld refractometer
- Transport medium for bacteria, viruses, mycoplasmas and *Trichomonas*
- Vacuum flask
- Buffer tablets for use—with local water
- Scalpel, scissors, forceps, artery forceps (haemostats)
- Wash bottles for alcohol, stains etc.
- Lightweight pots for specimens
- Disinfectant(s), including ethanol/methanol/methylated spirits
- Camping (gas cylinder-operated) stove—for sterilising, lighting and cooking
- Pressure cooker for sterilising

Recommended additional items when working overseas in the field

Sharp and other possibly dangerous items should not be placed in hand luggage when travelling by air, or through land or sea security checkpoints.

- Emergency pack containing business cards, letters of authorisation, protocols for snakebite (etc.), medicines and antidotes.
- Multipurpose Swiss Army-type pocket knife.
- Sewing kit with assorted needles, thread.
- Screwdrivers, pliers and an adjustable spanner.
- Elastic bands, string, dental floss, suture material, adhesive tape, insulating tape, duct tape, electrician's tape.
- Spare nylon cable ties for securing lid hasp of case during travel.
- Standard veterinary and other textbooks: where space is limited, the "Merck Veterinary Manual" is recommended (but caution must be exercised in quoting from this in court as it does not include references).
- Phrase books of appropriate languages.
- The "SAS Survival Guide" (Wiseman 1993), which contains much useful information that can be applied to difficult situations in the field.
- Appropriate clothing, e.g. the shoulders should be covered when working in a Moslem community, a tie is a courtesy in most countries when a man is meeting dignitaries.

Additional Equipment for Specific Investigations

Blood sampling from live animals for DNA and other investigative purposes

- Syringes (1 ml, 2 ml, 5 ml/10 ml) as appropriate
- Needles (20 g, 25 g, 22 g, 28 g) as appropriate
- Anticoagulant serum tubes
- Cotton wool
- Frosted glass slides and box or tray
- Coverslips
- Pencils for marking glass slides, plus sharpener
- Methanol
- Cards or solutions for blood collection
- Magnifying glass/hand lens
- Plastic bags and labels
- Electronic/spring balance
- Ruler, tape measure
- Clipboard, black pen and record sheets
- Basic clinical equipment
- Restraining equipment—nets, towels, gloves, hood, bags

Plus other items from earlier lists, as necessary

Entomological (invertebrate) collection

- Small ventilated plastic vials, lined with mesh, filled with filter paper (to reduce condensation and subsequent drowning) for holding live ticks or other invertebrates.

Relevant identification keys and texts

More specialised laboratory investigations in the field

- Vacuum flasks and portable, lightweight, cool box.
- Normal (isotonic) saline.
- Hypertonic NaCl or sugar (sucrose) solution for flotation/sedimentation examination.
- Tincture of merthiolate for staining faecal protozoa.
- Fixatives for blood and other body fluids, bone marrow and endo- and ectoparasites.
- Rapid acting stains for blood and other body fluids (sputum, urine, synovial and coelomic, cerebrospinal, bone marrow etc.) and touch/impression smear cytology.
- Gram, acid-fast and other special stains.
- Lactol-phenol cotton-blue for demonstrating fungi.
- Plastic pipettes.
- Slotted stain jar. Lightweight, unbreakable, plastic staining jars are preferable to heavy, fragile, glass Coplin jars.
- Mounting media for permanent preparations of blood and bone marrow films.
- Clearing and mounting media for small ectoparasites.
- Light-weight, slide-drying rack.
- Transport media for bacteria, viruses and protozoa (see earlier).

- Microbiological test strips.
- Urine and blood chemistry test strips.
- Rapid diagnostic test strips.
- Cardboard strips, which can be labelled in pencil or waterproof ink, and placed inside specimen containers.
- Safety matches, a small Bunsen burner or disposable butane cigarette lighter.
- Squeeze bottles for methanol etc.
- Specimen containers, filled with concentrated formaldehyde, for dilution with river or seawater.
- Tongue depressors, wooden applicator sticks and sterile cotton-tipped applicators. Plastic coffee spoons for use as spatulae.
- Non-lubricated condoms as finger covers.
- Plastic film canisters (pots) with labels attached for faecal collection, parasites etc. They usually can be obtained *gratis* from film dealers or photofinishing laboratories.
- Sterile disposable venous and urethral catheters; latex or plastic tubing.
- Plastic slide boxes, each prefilled with polished, frosted and glass microscope slides.

Checklist of battery-operated or solar-powered (direct sun or solar panels) equipment that can be used in isolated locations

- Miniaturised otoscope (auriscope)
- Ophthalmoscope
- Rigid endoscope
- Colorimeter
- Electrocautery equipment
- Blood pressure monitoring instrument
- Minicentrifuge
- Miniphotometer
- Respiratory monitor and pulse oximeter
- Refractometer

Further Reading

Cooper JE (2013a) What is fieldwork? J Exotic Pet Med 22(1):7–16

Cooper JE (2013b) Field techniques in exotic animal medicine. J Exotic Pet Med 22(1)

Cooper JE, Cooper ME (2007) Introduction to veterinary and comparative forensic medicine. Blackwell, Oxford

Cooper JE, Cooper ME (2013) Wildlife forensic investigation: principles and practice. Taylor & Francis/CRC Press, Boca Raton, Florida

Cooper JE, Cooper ME (2017) Appendix 3 Field pathology. In: Cooper JE, Hull G (eds) Gorilla pathology and health: with a catalogue of preserved materials. Elsevier, St Louis

Cooper JE, Cooper ME (2017) Appendix 4 Hazards, including Zoonoses. In: Cooper JE, Hull G (eds) Gorilla pathology and health: with a catalogue of preserved materials. Elsevier, St Louis

Appendix 6: Samples

Mistakes can occur at various stages of sampling:

Selection
↓
Taking
↓
Packing
↓
Transportation
↓
Reception
↓
Processing

An error or inconsistency introduced at any stage in the sequence above can adversely affect subsequent stages and easily prejudice results.

Selection of Samples

In forensic work, this depends upon the circumstances and the questions that are being asked. The choice of sample will be influenced by the purpose for which the test is to be performed. Some investigations and procedures may be routine, others more complex. Samples for toxicology often present particular challenges and it may be wise to consult the laboratory before taking such material.

Samples are not only taken for biomedical investigation. Some, such as explosives, ammunition and weapons will be examined by specialists in forensic and other laboratories.

Some samples can be taken from either live or dead animals but others may only be realistically obtained from one or the other.

Table 7 Samples from live and dead animals

Sample	Live animal	Dead animal
Faeces	Yes	Yes
Hair, feathers, scales	Yes	Yes
Touch preparations for cytology	Yes	Yes
Swabs, etc. for microbiology of external lesions	Yes	Yes
Swabs, etc. for microbiology of internal lesions	Sometimes, via an endoscope or a surgical incision	Yes
Tissue samples for histology, etc. of external lesions	Yes (biopsy)	Yes
Tissue samples for histology, etc. of internal lesions	Sometimes (biopsy)	Yes
Larger tissue samples for (e.g.) toxicology	Usually not feasible	Yes
Crop and stomach contents for toxicology	Sometimes by lavage or use of emetics	Yes
Blood	Yes	Occasionally (immediately after death)
Ectoparasites	Yes	Sometimes
Eggs and larvae of blowflies and some other insects	Sometimes (myiasis)	Yes

Table 8 Samples from animal remains

Remains	Comments
Bones and teeth	Often found, scattered or buried.
Hair, feathers, scales	Commonly present in small amounts (dried or caked with soil), even if the carcase has disappeared.
Gizzard stones	Commonly found where certain birds (e.g. galliforms) have died in numbers.
Gallstones	Very occasionally found. Usually readily recognisable, if sectioned, by laminated structure.
Bladder stones	As above.
Otoliths	A feature of some fish; may be present in large numbers where fish have been devoured.
Stomach or intestinal contents	May persist, especially if bone, fur, feather, scales or tough vegetation are present.
Pellets (castings)	As stomach contents, but usually more compact and drier.
Exoskeletons of invertebrates, including sea urchins	May remain in large numbers if an animal has died with large numbers of invertebrates within its GI tract.

Plastic forceps are important. They are used for a variety of purposes, including the handling of lead shot, pellets or bullets where metal instruments may damage the surface, making identification of the weapon less easy and the taking of samples for analysis of certain metals.

Samples for histological examination are usually fixed in buffered formol saline (BFS). Lung, liver, kidney and any abnormalities should be taken routinely.

Samples for toxicological assay (blood, urine, tissues) should be placed in separate containers; this is different from histological or electron microscopic examination where tissues may, if necessary, be placed in the same container of fixative.

Correct, secure, storage of material taken from a forensic case is vital—as is confidentiality. Samples must be handled carefully because of health and safety considerations and labelled or marked so that they cannot be misplaced or transposed.

Legal aspects

The collection, transportation and processing of samples are likely to be covered by legislation, including Post Office Regulations, health and safety, animal health (see Cooper ME, Wildlife Conservation Law, this volume).

Appendix 7: *Post-mortem* Examination Form

Species .... Reference No ....

Date of Submission .... Origin ....

Relevant history / circumstances of death:

Request - diagnosis (cause of death/ill-health), health monitoring, forensic investigation/research

Any special requirements re techniques to be followed, fate of body/samples

Submitted by .... Date ....

Received by .... Date ....

Measurements Carpus Tarsus Other Bodyweight (Mass)

Condition score: Obese or fat / good / fair or thin / poor

State of preservation: Good / fair / poor / marked autolysis

Storage since death: Refrigerator / ambient temperature / frozed / fixed

A number ("score") can be used for these

EXTERNAL OBSERVATIONS, including preen gland, state of moult, ectoparasites, skin condition, lesions etc

MACROSCOPIC EVALUATION ON OPENING THE BODY, including position and appearance of organs, lesions etc

ALIMENTARY SYSTEM

MUSCULOSKELETAL

CARDIOVASCULAR

RESPIRATORY

URINARY

necessary -

including

This section can be expanded, as

subheadings can be inserted,

1

REPRODUCTIVE checklist of organs and tissues

LYMPHOID (including bursa)

NERVOUS

OTHER
SAMPLES TAKEN

.... Bact Paras Hist DNA Cytology Other
(eg serology)

.... Bact Paras Hist DNA Cytology Other
(eg serology)

.... Bact Paras Hist DNA Cytology Other
(eg serology)

.... Bact Paras Hist DNA Cytology Other
(eg serology)

.... Bact Paras Hist DNA Cytology Other
(eg serology)

.... Bact Paras Hist DNA Cytology Other
(eg serology)

.... Bact Paras Hist DNA Cytology Other
(eg serology)

LABORATORY FINDINGS

Date: .... Initials: .... Reported to whom: ....

PRELIMINARY REPORT (based on gross findings and immediate laboratory results eg cytology)

Reported to .... Date .... Time ....

FINAL REPORT (based on all available information)

FATE OF CARCASS / TISSUES
destroyed / frozen / fixed in formalin(other) / retained for Reference Collection / sent elsewhere/submitted as evidence

PM examination performed by: .... Date .... Time ....

Reported by: .... Date ....

JEC June 2020

Appendix 8: Specimen Forms—Wildlife Forensic Cases

JOHN E. COOPER

Form I

SUBMISSION OF WILDLIFE FORENSIC EXHIBITS CHAIN OF CUSTODY RECORD

Laboratory case No: ....

Submitter's reference No: ....

Other relevant information ....

Special precautions necessary (e.g. hazards) ....

Exhibits submitted:

Item No.	Lab Exhibit No.	Brief Description

Chain of custody: received in sealed condition

Item No.	Received From	Received By	Date

Laboratory stamp received: ....

Laboratory stamp delivered: ....

....

The form given next is for the submission of a live animal, dead animal or sample for forensic investigation. The forms that follow are specific to certain types of examination and serve as report forms. All can be adapted, as needed, to suit requirements.

Form II

SUBMISSION FOR FORENSIC CLINICAL/ *POST-MORTEM*/LABORATORY EXAMINATION

Submitting Client/Agency .... Reference No ....

Lab Ref .... Date (in full) ....

Address ....

.... Post (Zip) code ....

Tel .... Fax .... E-mail...

Veterinary surgeon (veterinarian), where applicable...

Address and contact details...

Other relevant persons, e.g. police officer, wildlife inspector......................................

Species of animal (English and scientific name)..

Breed/variety .... Local name ....

Pet name (if appropriate).. Age .... Sex ....

Colour/markings ....

Ownership of animal or sample(s)...

Signature: ....

Date: ....

Time: .... *Sheet 1/5*

Number/Ring (Band)/Tattoo/Microchip/Other methods of identification

....

Other comments, e.g. cruelty case, civil action, insurance claim, professional malpractice hearing, etc.

....

....

....

Background to the case/history...

....

....

....

Sample submitted...

Sent by hand/courier/post/other ....

Method and description of packing...

Tag no (see chain of custody)..................... Prior storage condition ....

Signature of courier .... Date ....

Live animal .... Dead animal ....

Organs (state)...

Other tissues..

Parasites...

Blood (and details of how presented)..

Swabs..

Signature: ....

Date: ....

Time: ....

Sheet 2/5

Other...

Comment on chain of custody (attach relevant paperwork) ……………………………………..
……………………………………………………………………………………………….........
……………………………………………………………………………………………….........
……………………………………………………………………………………………….........

Questions being asked (expand as necessary)

- What is this material?

  ……………………………………………………………………………………

- Species/sub-species/breed?

  ……………………………………………………………………………………

- Age

  …. ………………………………………………………………………………

- What is its provenance/parentage, etc?

  ……………………………………………………………………………………

- Why did this animal die/how long has it been dead??

  ……………………………………………………………………………………

- When did this animal die?

  ……………………………………………………………………………………

- How did this animal die?

  ……………………………………………………………………………………

- Did this animal suffer pain or distress?

  ……………………………………………………………………………………

- Sex/entire/neutered

  ……………………………………………………………………………………

- Sexually mature/fertile/in oestrus

  ……………………………………………………………………………………

- Pregnant/gravid

  ……………………………………………………………………………………

- Lactating/incubating

  ……………………………………………………………………………………

- Origin – geographical

  ……………………………………………………………………………………

- Migrant or resident

  ……………………………………………………………………………………

- Origin – captive or free-living

  ………………………………………………………………………………

- Captive or free-living – for how long?

  ………………………………………………………………………………

Signature: ……………………………………..

Date: ………………………………………..

Time: ……………………………………….. *Sheet 3/5*

- Health status

  ………………………………………………………………………………

- Age of skin wound, fracture or other lesions

  ………………………………………………………………………………

- Other questions

  ………………………………………………………………………………

Investigations requested ………………………………………………………………………..

Clinical examination …………………………………………………………………………

Post-mortem examination ……………………………………………………………………

Laboratory tests:

Microscopy ………………………………………………………………………..

Cytology ……………………………………………………………………………..

Microbiology………………………………………………………………………..

Parasitology …………………………………………………………………………..

Histology ……………………………………………………………………………..

Toxicology ……………………………………………………………………………

DNA ……………………………………………………………………………………

Other (specify e.g. serology) ……………………………………………………….

Other tests as necessary to answer the questions given earlier ....

....

....

Results to be sent to ....

....

....

Signature: ....

Date: ....

Time: .... *Sheet 4/5*

Special instructions regarding follow-up action to be taken regarding a live animal or the storage/transfer/disposal of samples/wrappings/carcase or tissues

....

....

....

....

....

Signed by recipient at (location) ....

Date .... Time ....

Signature: ....

Date: ....

Time: .... *Sheet 4/4*

Form III

LABORATORY EXAMINATION FINDINGS (GENERAL) IN FORENSIC CASES

Reference Nos (delete if necessary):

Clinical .... Pathology .... Case No: ....

Species (English name): .... (Scientific name): ....

Age: .... Sex: .... Number/Ring (Band)/Tattoo/Microchip No: ....

Other methods of identification: ....

Relevant history/circumstances (including time) of death:

Accompanying dockets or other relevant papers: ....

Any special requirements regarding techniques to be followed, fate of body/samples/special precautions needed

Submitted by: ………………………………………..………….. Date: ……………………..

Received (date): …………………….. by: ………………………………………………….

Laboratory Report

Gross examination: …..…………………...……………………………………………..

……………………………………………………………………………..……………

Radiography, if performed.……………………………………………………………..

Microscopical examination:.……………...……………………………………………..

……………………………………………………………………...…………………..

Further Test(s)

Test	Submitted (date)	Results (Received date)	Comments
Microscopy			
Cytology			
Microbiology			
Parasitology			
Histology			
Toxicology			
DNA			
Other (Specify e.g. serology)			

Summary of findings:

…………………………………………………………………………………………………

…………………………………………………………………………………………...……

Results/Comments/Interpretation:

…………………..……………………………………....……………………………..……

………………………………………………………………………………………………......

Examination performed by: ……………………………….. Date: …………… Time: ………..

Reported by: ………………………………………………. Date: …………… Time: ………..

Method of reporting: electronic/fax/post/by hand

Form IV

GROUP OR FLOCK EXAMINATION FINDINGS IN FORENSIC CASES

Reference Nos (delete if necessary): Clinical: .... Pathology: .... Case No: ....

Species (English name): .... (Scientific name): .... Age: .... Sex: ....

Number/Ring (Band)/Tattoo/microchip No: .... Other methods of identification: ....

Relevant history/circumstances (including time) of death:

Accompanying dockets or other relevant papers: ....

Any special requirements regarding techniques to be followed, fate of body/samples/special precautions needed

Submitted by: .... Date: .... Received (date): ....by: ....

No.	Sex, Age, Size	External Findings	Internal Findings	Other Observations	Comments

Photographs taken (tick box) ....

Summary:

....

....

....

Examination performed by: .... Date: .... Time: ....

Reported by: .... Date: .... Time: ....

Method of reporting: electronic/fax/post/by hand

Form V

INITIAL RECORDING OF HISTOPATHOLOGICAL FINDINGS IN FORENSIC CASES

Slide No.	Comments on Preparation, Including Stain	Tissues on Slide	Observations	Scoring of Changes (Where Appropriate)	Comments	Interpretation

Some suggested abbreviations: AICI = acute inflammatory cell infiltration. CICI = chronic inflammatory cell infiltration; wnl = within normal limits. Scoring system: 1 = minimal; 2 = moderate; 3 = marked

Summary: ....

Further work needed: ....

Examination performed by: .... Date: .... Time: ....

Reported by: .... Date: .... Time: ....

Method of reporting: electronic/fax/post/by hand

Form VI

FORENSIC EXAMINATION OF LIVE ANIMALS

Reference Nos (delete if necessary):

Clinical: .... Pathology .... Case No: ....

Species (English name): .... (Scientific name): ....

Age: .... Sex: .... Number/Ring (Band)/Tattoo/Microchip No: ....

Other methods of identification: ....

Relevant history/circumstances (including time) of death:

Accompanying dockets or other relevant papers: ....
Any special requirements regarding techniques to be followed, fate of body/samples/special precautions needed

Submitted by: .... Date: ....

Received (date): ....by: ....

HISTORY AND RELEVANT DATA OBSERVATION – COMMENTS RESULTS OF CLINICAL EXAMINATION FURTHER TESTS (e.g. RADIOGRAPHY) SAMPLES TAKEN/LABORATORY RESULTS (use other form for details) SUBSEQUENT INVESTIGATIONS FOLLOW-UP	Sequential weights/measurements:-

COMMENTS (continue overleaf or attach additional papers as necessary)

Photographs taken (tick box) ....

Examination performed by: .... Date: .... Time: ....

Reported by: .... Date: .... Time: ....

Method of reporting: electronic/fax/post/by hand

Form VII

FORENSIC *POST-MORTEM* EXAMINATION FORM

Reference Nos (delete if necessary):

Clinical: .... Pathology .... Case No: ....

Species (English name): .... (Scientific name): No: ....

Age: .... Sex: .... Number/Ring (Band)/Tattoo/Microchip No: ....

Other methods of identification: ....

Relevant history/circumstances (including time) of death:

Accompanying dockets or other relevant papers: ....

Any special requirements regarding techniques to be followed, fate of body/samples/special precautions needed

Submitted by: .... Date: ....

Received (date): .... by: ....

Measurements: carpus tarsus other bodyweight (mass)

Condition: obese or fat/good/fair or thin/poor/numerical score if appropriate

Autolysis: fresh/minimal/moderate/marked autolysis/numerical score if appropriate

Storage since death: ambient temperature/refrigerator frozen/fixed/other

Radiography/other imaging, if performed: (findings)

EXTERNAL OBSERVATIONS, including mammary/preen gland, state of moult, ectoparasites, skin condition, lesions etc. (expand as necessary)

MACROSCOPIC EVALUATION ON OPENING THE BODY, including position and appearance of organs, lesions etc. (expand as necessary)

ORGAN SYSTEMS (expand each as necessary):

ALIMENTARY

MUSCULOSKELETAL

CARDIOVASCULAR

RESPIRATORY

URINARY

REPRODUCTIVE

LYMPHOID (including bursa of Fabricius)

NERVOUS

OTHER

SAMPLES TAKEN (expand as necessary)

.... Cyto Micro Paras Hist DNA Tox Other (e.g. serology)

.... Cyto Micro Paras Hist DNA Tox Other (e.g. serology)

.... Cyto Micro Paras Hist DNA Tox Other (e.g. serology)

.... Cyto Micro Paras Hist DNA Tox Other (e.g. serology)

.... Cyto Micro Paras Hist DNA Tox Other (e.g. serology)
.... Cyto Micro Paras Hist DNA Tox Other (e.g. serology)
.... Cyto Micro Paras Hist DNA Tox Other (e.g. serology)

LABORATORY FINDINGS (expand as necessary)

Date: .... Initials: ...... Reported by: .... To whom: ....

PRELIMINARY REPORT (based on gross findings and immediate laboratory results e.g. cytology)

Reported to: .... By whom: .... Date: .... Time: ....

FINAL REPORT (based on all available information)

FATE OF CARCASE/TISSUES
Destroyed/frozen/fixed in formalin (or other fixative)/retained as evidence/deposited in Reference Collection/sent elsewhere
Photographs taken (tick box) ....

Examination performed by: .... Date: .... Time: ....
Corroborating pathologist (if appropriate) .... Date: .... Time: ....
Reported by: .... Date: .... Time: ....
Method of reporting: electronic/fax/post/by hand *Sheet 1/2*

Form VIII

FORENSIC CYTOLOGY REPORT

Reference Nos (delete if necessary):
Clinical: .... Pathology .... Case No: ....
Species (English name): .... (Scientific name): No: ....
Age: .... Sex: .... Number/Ring (Band)/Tattoo/microchip No: ....
Other methods of identification: ....
Relevant history/circumstances (including time) of death:

Accompanying dockets or other relevant papers: ....
Any special requirements regarding techniques to be followed, fate of body/samples/special precautions needed

Submitted by: .... Date: ....
Received (date): .... (by): ....

Sample submitted: ....

Received (date): .... (by): ....

Stain(s) and number of preparation(s): ....

....

General comments on preparation(s): ....

....

....

Cell types	Numbers (+/++/+++)	Features	Comments

Photographs taken (tick box) ....

Other findings: ....

....

....

Summary or interpretation of findings: ....

....

....

Examination performed by: .... Date: .... Time: ....

Reported by: .... Date:.............. Time: ....

Method of reporting: electronic/fax/post/by hand

Form IX

FORENSIC HISTOPATHOLOGY REPORT

Reference Nos (delete if necessary):

Clinical: .... Pathology. .... Case No: ....

Species (English name): .... (Scientific name): No: ....

Age: .... Sex: .... Number/Ring (Band)/Tattoo/microchip No: ....

Other methods of identification: ....

Relevant history/circumstances (including time) of death:

Accompanying dockets or other relevant papers: ....

Any special requirements regarding techniques to be followed, fate of body/samples/special precautions needed

Submitted by: .... Date: ....

Received (date): .... by: ....

Findings: ....
....
....
....
Comments: ....
....
....
Further work needed: ....
....
....
Summary or interpretation of findings: ....
....
Advice or action if appropriate: ....
....
....
Photographs taken (tick box) ....
Examination performed by: .... Date: .... Time: ....
Reported by: .... Date:.. .... Time: ....
Method of reporting: electronic/fax/post/by hand

Form X

FORENSIC EXAMINATION OF EGGS/EMBRYOS

Reference Nos (delete if necessary):
Clinical: .... Pathology. .... Case No: ....
Species (English name): .... (Scientific name): No: ....
Age: .... Sex: .... Number/Ring (Band)/Tattoo/Microchip No: ....
Other methods of identification: ....
Relevant history/circumstances (including time) of death:

Accompanying dockets or other relevant papers: ....
Any special requirements regarding techniques to be followed, fate of samples/special precautions needed

Submitted by: .... Date: ....
Received (date): .... by: ....

Weight of whole unopened egg: ……..…….. Length: ………………… Width: ………………..

External appearance (to include drawing or photograph):

Appearance (to include drawing or photograph): Embryo
Air cell
Blood vessels
Fluids

Appearance when opened to (include drawing or photograph):

Contents:

Embryo: Length (crown-rump)
Amniotic cavity
Allantoic cavity
Yolk sac

Other comments:

Microbiology:

Histopathology:

Other tests:

Samples sent elsewhere:

Weight of dried eggshell: …………….. Thickness (measurement or index): …………………

Samples stored:

Summary or interpretation of findings

Photographs taken (tick box) ……………………………………………

Examination performed by: ………………………….. Date: …………… Time: ………..

Reported by: ……………………………………………… Date: …………… Time: ……….

Method of reporting: electronic/fax/post/by hand

Form XI

PELLET REPORT FORM

Reference Nos (delete if necessary):

Clinical: ……………………… Pathology ……..……………… Case No: ………………..…

Species (English name): ……………………… (Scientific name): No: ………………………

Age:……… Sex:……… Number/Ring (Band)/Tattoo/Microchip No:………………………

Other methods of identification: …………………………………………………………..

Relevant history/circumstances (including time) of death:

Accompanying dockets or other relevant papers: ....

Any special requirements regarding techniques to be followed, fate of samples/special precautions needed

Submitted by: .... Date: ....

Received (date): .... by: ....

1. External examination
   - Weight
   - Measurement (cm):
     i. Length
     ii. Width
     iii. Height
     iv. Circumference
   - Gross appearance/description, colour, consistency, apparent content, smell

     ....
     ....
     ....

Radiography, if performed

2. Internal examination – contents
   ....
   ....
   ....

3. Identification of pellet contents
   - Pellet material that floats, removed, placed on tissue paper and allowed to dry

     ....
     ....
     ....

   - Pellet material that sinks, preserved in dilute formalin to observe under low-power microscope

     ....
     ....
     ....

   - Laboratory results:
     i. Direct microscopy

     ....
     ....
     ....

ii. Culture

....
....
....

15. Material retained/submitted elsewhere:

....
....
....

16. Comments

....
....
....

17. Photographs taken/drawings

....
....
....

Examination performed by: .... Date: .... Time: ....
Reported by: .... Date: .... Time: ....
Method of reporting: electronic/fax/post/by hand
....

Form XII

FORENSIC EXAMINATION OF GORILLA SKELETON/BONES

SHEET 1/7

Study Ref:

Reference Nos (delete if necessary):

Clinical: .... Pathology .... Case No: ....
Species (English name): .... (Scientific name): ....
Age: infant/juvenile/young adult/old adult .... Sex: male/female/uncertain ....
Number/tattoo/microchip No: .... Other methods of identification e.g. noseprint.........
Relevant history/circumstances (including time) of death: ....
Origin of bones...
Method of preservation/preparation/storage of bones...
State of preservation...

Accompanying dockets or other relevant papers: ....
Any special requirements regarding techniques to be followed, fate of samples/special precautions needed.

....

Submitted by: ....Location.............................Date:....................

Received by: ....Date:..............

MATERIAL AVAILABLE/EXAMINED–GENERAL COMMENTS: ....

RADIOGRAPHY OR OTHER IMAGING PERFORMED...
ENDOSCOPY.............OTHER TESTS.......... PHOTOS: Y/N

Cranium (Skull)	Mandible (Jaw)	Post-Cranial Skeleton	Other (e.g. Skin)
SUMMARY OF PATHOLOGICAL FINDINGS:			
Cranium (Skull)			
Mandible			
Post-Cranial Skeleton			

SEE SUBSEQUENT SHEETS (2-7)

SHEET 2/7

Study Ref.

Inventory of bones present, with comments as appropriate

Cranium	
Mandible	
Sternebrae	

Vertebrae

Cervical

C1	C2	C3	C4	C5	C6	C7

Comment

Thoracic

T1	T2	T3	T4	T5	T6	T7	T8	T9	T10	T11

Thoracic/lumbar

T12	(T13)

Comment

L1	L2	L3	L4	L5

Comment

Comment

Comment

Sacrum

	Right			*Left*			Comment
	Present	Weight	Length	Present	Weight	Length	
Clavicles							
Scapulae							
Ribs							
Humerus							
Ulna							
Radius							

SHEET 3/7

Study Ref.

right hand

	sca	lun	tri	pis
	trm	trd	cap	ham
mc1	mc2	mc3	mc4	mc5
pp1	pp2	pp3	pp4	pp5
	mp2	mp3	mp4	mp5
dp1	dp2	dp3	dp4	dp5

left hand

pis	tri	lun	sca	
ham	cap	trd	trm	
mc5	mc4	mc3	mc2	mc1
pp5	pp4	pp3	pp2	pp1
mp5	mp4	mp3	mp3	
dp5	dp4	dp3	dp2	dp1

Comments

Weights and measurements of selected bones:			
Bone	Weight	Measurements(s)	Comments

SHEET 4/7

Study Ref.

	Right			*Left*			Comment
	Present	Weight	Length	Present	Weight	Length	
Innominate							
Femur							
Patella							
Tibia							
Fibula							
Pelvis							

Right foot

			calc	tal
			cub	nav
		cun3	cun2	cun1
mt1	mt2	mt3	mt4	mt5
pf1	pf2	pf3	pf4	pf5
	mf2	mf3	mf4	mf5
df1	df2	df3	df4	df5

Left foot

tal	calc			
nav	cub			
cun1	cun2	cun3		
mt5	mt4	mt3	mt2	mt1
pf5	pf4	pf3	pf2	pf1
mf5	mf4	mf3	mf3	
df5	df4	df3	df2	df1

Comments

Sexing

Sex		Determined from

Comments	

SHEET 5/7

Study Ref.

Ageing

(See Lovell, 1990)

Infant	Determined from
Juvenile	Determined from
Young adult	Determined from
Old adult	Determined from

SHEET 6/7

Study Ref.	

Dental record

UPPER (maxilla)

R	Other																	L
	Malocclusion																	
	Hypoplasia																	
	Abscesses																	
	Caries																	
	Recession																	
	Calculus																	
	Attrition																	
	Presence																	
	Permanent	8	7	6	5	4	3	2	1	2	3	4	5	5	6	7	8	
R	Deciduous				e	d	c	b	a	a	B	c	d	e				L

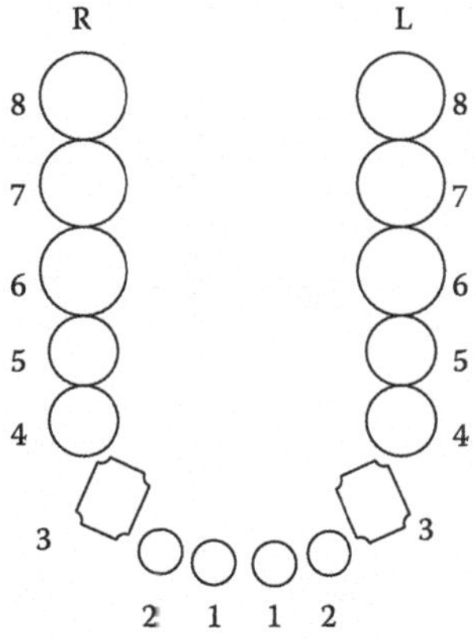

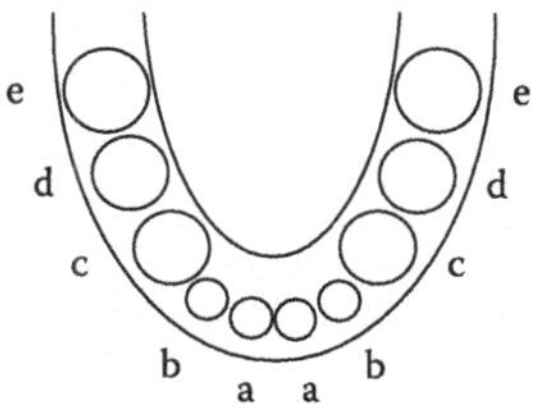

Lower deciduous

R	Deciduous				e	d	c	b	a	a	B	c	d	e				L
	Permanent	8	7	6	5	4	3	2	1	2	3	4	5	5	6	7	8	
	Presence																	
	Attrition																	
	Calculus																	
	Recession																	
	Caries																	
	Abscesses																	
	Hypoplasia																	
	Malocclusion																	
R	Other																	L

Key: N/A = not applicable * = severe ✓ = present χ = absent

COMMENTS ON DENTAL PATHOLOGY

SHEET 7/7

Study Ref.

Weight of skull:	Weight of mandible:

Selected measurements:

Symmetry:

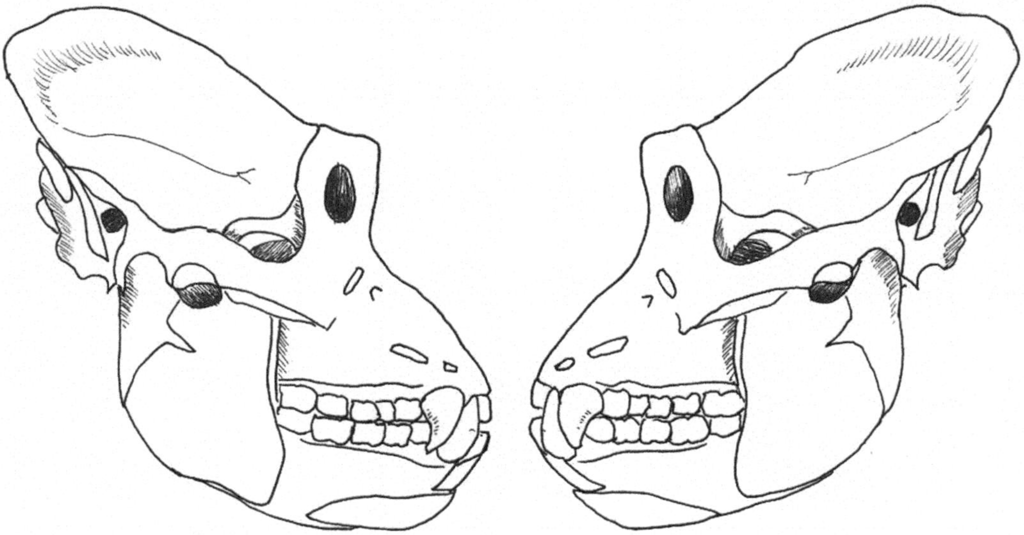

Key to lesions:
X = tooth missing ⇆ = deviation • = pathological hole in bone

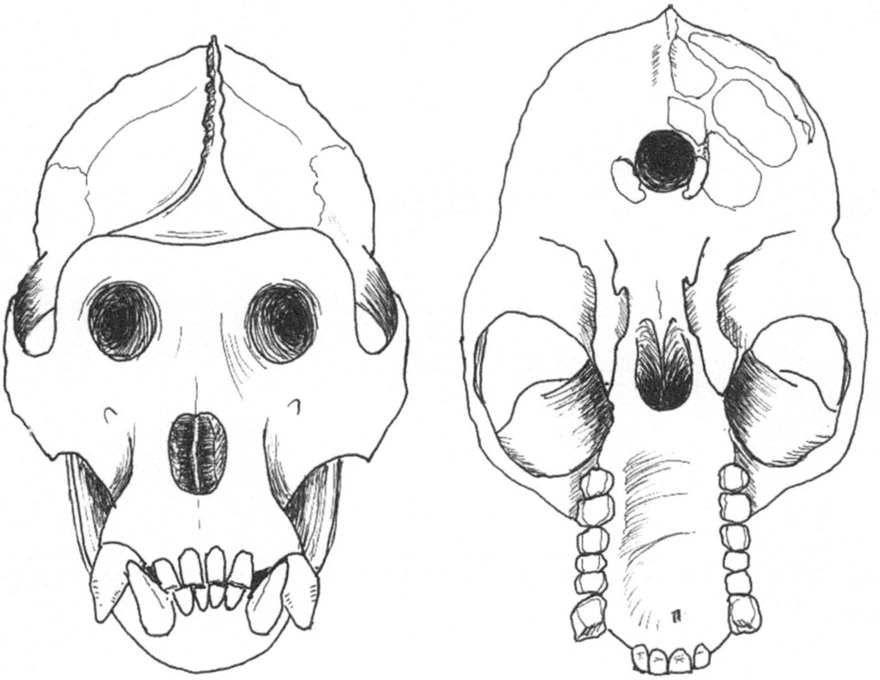

COMMENTS ON PATHOLOGY

Fate of material

Summary or interpretation of findings..

....

....

....

Examination performed by: .... Date: .... Time: ....

Reported by: .... Date: .... Time: ....

Method of reporting: electronic/fax/post/by hand

Appendix 9: Summary of *Post-mortem* Findings

SPECIES	NAME / ID	SEX
PM NUMBER		
EXTERNAL		
INTERNAL		
FURTHER WORK NEEDED / PENDING ADVICE / REQUESTS		
SUMMARY OF FINDINGS / DIAGNOSIS / COMMENTS		
Pathologist:	Date:	

References

Berkovitz B, Shellis P (2018a) The teeth of mammalian vertebrates. Elsevier, Academic Press, Amsterdam

Berkovitz B, Shellis P (2018b) The teeth of non-mammalian vertebrates. Elsevier, Academic Press, Amsterdam

Bierens J (ed) (2014) Drowning. The World Congress on Drowning 2002. Springer, Berlin, Heidelberg

Blood DC, Studdert VP (1988) Saunders comprehensive veterinary dictionary. W.B. Saunders, London

Bortolotti F, Tagliaro F, Manetto G (2004) Objective diagnosis of drowning by the "diatom" test – a critical review. Forensic Sci Rev 16:135–148

Brownlie HB, Munro R (2016) The veterinary forensic necropsy: a review of procedures and protocols. Vet Pathol 53(5):919–928

Byrd JH, Castner JL (eds) (2011) Forensic entomology: the utility of arthropods in legal investigations. CRC Press, Boca Raton, FL

Capula M, Rugiero L, Capizzi D, Milana G, Vignoli L, Franco D, Petrozzi F, Luiselli L (2014) Long-term, climate change-related shifts in monthly patterns of roadkilled Mediterranean snakes (*Hierophis viridiflavus*). Herpetol J 24:97–102

Cheesbrough M (2005) District laboratory practice in tropical countries. Cambridge University Press, Cambridge

Cooper ME (1996) Community responsibility and legal issues. Semin Avian Exotic Pet Med 5(1):37–35

Cooper JE (2008) Methods in herpetological forensic work—post-mortem techniques. Appl Herpetol 5:351–370

Cooper JE (2012) The estimation of post-mortem interval (PMI) in reptiles and amphibians: current knowledge and needs. Herpetol J 22:91–96

Cooper JE (2013a) What is fieldwork? J Exotic Pet Med 22(1):7–16

Cooper JE (2013b) Field techniques in exotic animal medicine. J Exotic Pet Med 22(1)

Cooper ME (2013c) Legal, ethical and practical considerations of working in the field. J Exotic Pet Med 22(1):17–33

Cooper JE (2019) Nutrition. In: Girling SJ, Raiti P (eds) BSAVA manual of reptiles, 3rd edn. BSAVA British Small Animal Veterinary Association, Quedgeley, Gloucester

Cooper JE, Cooper ME (2007) Introduction to veterinary and comparative forensic medicine. Blackwell, Oxford

Cooper JE, Cooper ME (2013) Wildlife forensic investigation: principles and practice. Taylor & Francis/CRC Press, Boca Raton, FL

Cooper JE, Cooper ME (2016) Forensic investigations in avian medicine. In: Samour J (ed) Avian medicine. Elsevier, Amsterdam, pp 582–585

Cooper JE, Cunningham AA (1991) Pathological investigation of captive invertebrates. Int Zoo Yearb 30:137–143

Cooper JE, Hull G (2017) Gorilla pathology and health, with a catalogue of preserved materials. Academic Press, New York

Cooper JE, Samour JH (1997) Portable and field equipment for avian veterinary work. Proceedings European Committee of AAV (Association of Avian Veterinarians), London

Cooper JE, Dutton CJ, Allchurch AF (1998) Reference collections: their importance and relevance to modern zoo management and conservation biology. Dodo J Jersey Wildl Preserv Trust 34:159–166

Cooper JE, Cooper ME, Budgen P (2009) Wildlife crime scene investigation: techniques, tools and technology. ESR 9:229–238

Cooper JE, Deacon AE, Nyariki T (2014) Post-mortem examination and sampling of African flamingos (Phoenicopteridae) under field conditions. Ostrich J Afr Ornithol 85(1):75–83

Cork SC, Halliwell RW (2002) The veterinary laboratory and field manual: a guide for veterinary laboratory technicians and animal health advisors. Nottingham University Press, Nottingham

Cummings P, Trelka DP, Springer KM (2011) Atlas of forensic histopathology. Cambridge University Press, Cambridge

Cyrus C (2005) The Dr Cecil Cyrus Museum: public attitudes to tissue donation for display in St. Vincent. Bull R Coll Pathol 131:44–47

DiMaio VJM, DiMaio DJ (2001) Forensic pathology, 2nd edn. CRC Press, Boca Raton, FL

Dolinak D, Matshes EW, Lew EO (2005) Forensic pathology principles and practice. Academic Press/Elsevier, Amsterdam

Dorion RBJ (2004) Bitemark analysis. CRC Press, Boca Raton, FL

Elkan E (1981) Pathology and histopathological techniques. In: Cooper JE, Jackson OF (eds) Diseases of the reptilia. Academic Press, London

Fowler ME (2014) Various dates, numerous volumes. In: Zoo and wild animal medicine and Fowler's zoo and wild animal medicine. Academic Press, London

Frye FL (1999) Establishing the time of death in reptiles and amphibians. Proc Assoc Reptilian Amphibian Vet 23–25

Frye FL, Cooper JE, Keymer IF (2001) Outfitting and employing a compact field laboratory. Zoo Med Bull Br Vet Zool Soc (BVZS) 1:28–36

Haggblade MK, Smith WA, Noheri JB, Usanase C, Mudakikwa A, Cranfield MR, Gilardi KV (2019) Outcomes of snare-related injuries to endangered mountain gorillas (*Gorilla beringei beringei*) in Rwanda. J Wildl Dis 55(2):298–303

Haigh J (2013) Fieldwork in a cold climate. In field techniques in exotic animal medicine. J Exotic Pet Med 22(1):51–57

Hanley CS, Hernandez-Divers SJ (2003) Practical gross pathology of reptiles. Semin Avian Exotic Pet Med 12:71–80

Huffman JE, Wallace JR (2012) Wildlife forensics: methods and applications. Wiley Blackwell, Oxford
Jones DM (1982) Biological council lecture. The veterinary surgeon and wildlife conservation. Vet Rec 111 (19):427–431
Kalema-Zikusoka G, Rubanga SV (2013) The establishment and use of field laboratories: lesion from the CTPH Gorilla research clinic, Uganda. J Exotic Pet Med 22:34–38
Miles AEW, Grigson C (1990) Colyer's variations and diseases of the teeth of animals, Revised edn. Cambridge University Press, Cambridge
Munro R (1998) Forensic necropsy. Seminars Avian Exotic Pet Medicine 7(4)
Munro R, Munro R (2013) Some challenges in forensic veterinary pathology: a review. J Comp Pathol 149 (1):57–73
Munson LL (1990) Future directions for zoological pathology. J Zoo Wildl Med 21:385–390
Nielsen SW (1988) Importance of wildlife pathology. Comp Pathol Bull XIX(5):1–4
Parry NMA, Stoll A (2020) The rise of veterinary forensics. Forensic Sci Int 306:110069
Pessier AP (2003) Practical gross necropsy of amphibians. Semin Avian Exotic Pet Med 12:81–88
Pollanen MS (1997) Forensic diatomology and drowning. Elsevier, Amsterdam
Rotstein DS (2008) How to perform a necropsy if a toxin is suspected. J Exot Pet Med 17:39–43
Scott KR, Morgan RM, Jones VJ, Cameron NG (2017) The transferability of diatoms to clothing and the methods appropriate for their collection and analysis in forensic geoscience. Forensic Sci Int 241:127–137
Shepherd R, Simpson K (2003) Simpson's forensic medicine. Arnold, London
Simpson VR, Fisher DN (2017) A description of the gross pathology of drowning and other causes of mortality in seabirds. BMC Vet Res 13(302):1–13
Smith DA, Barker IK, Allen BO (1988) The effect of ambient temperature on healing cutaneous wounds in the common garter snake (*Thamnophis sirtalis*). Can J Vet Res 52:120–128
Stephenson L, Van den Heuvel C, Byard RW (2019) The persistent problem of drowning - a difficult diagnosis with inconclusive tests. J Forensic Legal Med 66:79–85. https://doi.org/10.1016/j.jflm.2019.06.003
Taylor JH (2011) Mummy: the inside story. NHM, London
Thomson RG (1978) General pathology. W.B. Saunders, Philadelphia
Williams DJ, Ansford AJ, Priday DS, Forrest AS (1998) Forensic pathology. Churchill Livingstone, Edinburgh and London, UK
Wiseman J (1993) SAS Survival Guide. Harper Collins, Glasgow
Wobeser G (1996) Forensic (medico-legal) necropsy of wildlife. J Wildl Dis 32:240–249
Woodford M (1965) The role of the veterinarian in wildlife conservation. Vet Rec 77(45):1311
Wyllie I, Newton I (1999) Use of carcasses to estimate the proportions of female sparrowhawks and kestrels which bred in their first year of life. Ibis 141:504–506

The Role of Conservation Dog Detection and Ecological Monitoring in Supporting Environmental Forensics and Enforcement Initiatives

Ngaio L. Richards, Jennifer Hartman, Megan Parker, Lauren Wendt, and Chris Salisbury

Abstract

The unique capabilities of dogs to further conservation efforts, harnessed in partnership with proficient handlers, continues to gain recognition but also remains under-utilized in many relevant sectors. This chapter shares firsthand knowledge acquired by conservation detection dog handlers in pursuit of related initiatives worldwide. Ecological monitoring applications at the interface of environmental forensics and enforcement are discussed around ways that dog-handler teams could be further incorporated therein. The notion of proactive monitoring, to facilitate sustainable preservation of organisms and ecosystems, and to lessen the strain on enforcement resources, is a prevailing theme. Recognizing the restricted resources generally available to address our most pressing conservation issues, limitations of dog-handler teams are outlined alongside strengths. Common misconceptions around who (dogs and handlers) can do this work, and the possibility of undermining dog team effectiveness through erroneous perceptions of their use as a monitoring tool, are also detailed. Considerations regarding the feasibility of implementing new international detection dog programs relative to those already in place and to additional resources and capacity, are also offered.

Keywords

Canine · Endangered species · Invasive species · Environmental contaminants · Pesticide poisoning · Poaching · Fisheries · Wildlife

Introduction

Much of what has been written about conservation detection invariably begins with a remark that humans have partnered with dogs for approximately 100 years on various forms of ecological monitoring and in environmental protection programs. Recent publications highlight the increasing recognition of the unique capabilities of dogs, focusing on olfactory abilities, and the ways these are harnessed in partnership with proficient handlers to acquire indispensable information that may otherwise be challenging or

N. L. Richards (✉) · M. Parker
Working Dogs for Conservation, Missoula, MT, USA
e-mail: ngaio@wd4c.org

J. Hartman
Rogue Detection Teams, Rice, WA, USA

L. Wendt
(Formerly with) Washington Department of Fish and Wildlife, Olympia, WA, USA

C. Salisbury
(Former) SAR & HRD K9 Handler, Conservation Detection Dog Handler, Patagonia Employee Environmental Internship Program Intern, Ventura, CA, USA

S. C. Underkoffler, H. R. Adams (eds.), *Wildlife Biodiversity Conservation*,
https://doi.org/10.1007/978-3-030-64682-0_11

impossible to gather (e.g., Bennett et al. 2020). And yet, despite a growing body of compelling evidence that Conservation Detection Dog and Handler Teams (CDDHTs) offer a powerful method and versatile tool in the field of conservation (see e.g., Hurt and Smith 2009; Dahlgren et al. 2012; Woollett et al. 2014), they remain under-utilized.

As conservationists with field experience acquired around the world, we have firsthand knowledge of the ways through which anthropogenic activities and human avarice is undermining ecosystems that people and wildlife rely upon for survival. In our respective capacities as professionally trained conservation detection dog handlers, we are keenly aware too of specific ways that CDDHTs could be further and more effectively utilized in a largely ecological monitoring context to lend concrete support at the interface of environmental forensics and enforcement. Intensive habitat encroachment, unsustainable harvest, and consumption of imperiled organisms and natural resources demand restoration and human behavioral change. These bestow upon us an even greater sense of urgency to impart the information in this chapter.

We are also concerned that the severity of illegal and/or unsustainable activities is overshadowing an equally critical but thus far less globally highlighted reality that is unfolding: the widespread decline of "common" or "abundant" species. Some losses are ostensibly subsistence-based (e.g., bushmeat harvest), with corresponding constriction and decimation of habitats from human encroachment. Ongoing declines in both biodiversity and ecosystem integrity, often in parallel with incursion by invasive species, is occurring to such an extent that species and habitats previously assumed to be ubiquitous are slipping away before sufficient baseline data can be obtained or necessary protections can be provided. At current rates, the harvest and consumption of these rapidly dwindling species casts the responding actions somewhere between ecological monitoring and enforcement. Without adequate data, it is impossible to determine which category species fall into, so we remain at an impasse of how to accord them appropriate attention.

In this chapter, we outline ways that CDDHTs can be incorporated in ecological monitoring amidst present rates of extirpation. Essentially proactive tools, their use can facilitate the long-term safeguarding of various organisms and ecosystems in order to lessen the strain on conservation enforcement efforts tasked with responding in the aftermath of their loss. Ecological monitoring and use of CDDHTs in these contexts may facilitate what Massé (2019) has referred to as "holistic understandings of conservation and poaching" as a means of protecting so-called "mundane ecological and biological management of protected areas" from oversimplification of the root issues and solutions. Many prior publications have already described *how* dogs are able to assist us in the ways that will be discussed in this chapter. We refer the reader to these sources, which also recount in intricate detail the canine olfactory system, the superior abilities of dogs to detect an array of target scents to trace levels and recount from historical to contemporary ways dogs have partnered with people in a range of detection disciplines, including that of conservation detection (Syrotuck 1972; Johnston 1999; Matre 2003; Helton 2009; Dahlgren et al. 2012; Jezierski et al. 2016).

Our narrative serves as a snapshot of CDDHT applications that hover around the edges of forensic, enforcement, conservation and environmental monitoring principles, which have been effective in addressing the issues in question. As such, it is meant to be illustrative rather than exhaustive. The applications we consider do not tidily fit into one category, and, therefore, many might not know of their existence. In addition to generating awareness of further possibilities for incorporation of this method and for collaboration, our aim is to share frank perspectives regarding the strengths and limitations of the use of dog-handler teams. This is because we fully recognize that limited resources are available to address the most pressing conservation issues and feel it important to outline instances where use of dogs may not be the most cost-effective or optimal of the available approaches (see also

Coppolillo/WD4C unpublished data, Long et al. 2007). We also discuss common misconceptions around the types of dogs and handlers engaged in this unique intersection of disciplines, and how erroneous perceptions can undermine the overall effectiveness of the teams as a monitoring tool. We conclude with recommendations and considerations regarding the feasibility of implementing new international detection dog programs relative to those already in place, as well as additional resources and capacity.

Detection Dog Applications at the Interface of Environmental Forensics, Enforcement, and Conservation Monitoring

There is already a general awareness of dogs used in the disciplines of narcotics and explosives detection work (e.g., GICHD 2004). As such, their enlistment in forensic and conservation enforcement-based initiatives may be considered an extension of the former. In contrast, the role of dog-handler teams in helping conservation managers and researchers acquire information, integral to early-stage and long-term species monitoring and habitat conservation efforts, is less known.

Contemporary environmental forensic and enforcement detection dog applications include: the physical tracking of poachers for apprehension and interrogation, tracking and accompaniment of rhinoceros for *within-habitat* protection from poachers (and broader conservation efforts); and the inspection of vehicles, buildings, shipping containers and individuals, as well as items in their possession (e.g., contraband). Likewise, targets to which dogs have been trained in this context range from bushmeat (i.e., wildlife that is illegally hunted), wire snares and traps, gun powder and associated illicit weaponry, elephant ivory and rhinoceros horn, bear gall bladders, animal pelts (e.g., cheetah), lion bones, pangolin scales, shark fin, and clandestinely harvested lumber, to name but a few (Fig. 1a, b).

Dog-handler teams around the globe are stationed at ports, airports and along transit ways, in an effort to intercept the smuggling and trade of flora and fauna (Fig. 2a, b). Any information-sharing and coordination networks that are in place help those teams to better identify and focus on region-specific targets. Because the presence of dog-handler teams may deter the use of certain exit/entry points for trade and shift smuggling routes, knowledge of alternate routes and increased vigilance is also necessary. The teams may work to a single target that has conservation priority or have strategically cross-trained to a

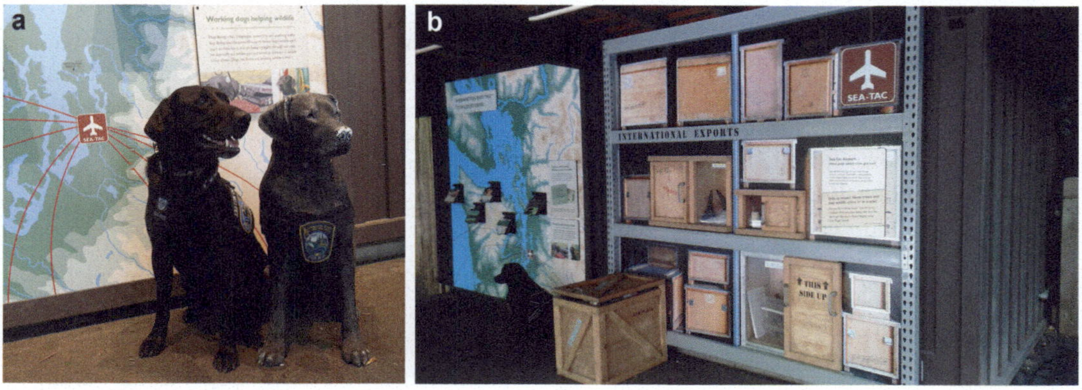

Fig. 1 K9 Benny inspects a statue made in his likeness, part of an exhibit at Woodland Park Zoo (WA, USA) on wildlife trafficking and the role of detection dogs in combatting it. (**a**) Credit: Isabelle Groc, copyright owner. (**b**) Credit: Katura Reynolds/Woodland Park Zoo, copyright owner

Fig. 2 Dog-handler teams around the globe are stationed at ports, airports, and along transitways, in an effort to intercept the smuggling and trade of flora and fauna. (**a**) Credit: WD4C, copyright owner. (**b**) Credit: Joel Caldwell, copyright owner

range of targets, carefully selecting them to complement the scents to which the dogs are already trained. In principle, judiciously layering on additional targets to a team's existing scent portfolio can increase the scope of their work and maximize their conservation impact. But doing so is not always possible. Challenges around procuring requisite samples for training (e.g., ivory) can limit a dog-handler team's scent portfolio (as detailed in Braun et al. 2018). Finally, it is important to note that the search/inspection authority of, and mandated response by, "conventional" enforcement teams (e.g., narcotics) is usually incompatible with those relative to environmental forensics and enforcement targets, which prevents these teams from cross-training (further discussed in Information Box 1).

Other important applications of dog-handler teams at the conservation-enforcement interface that are lesser-known include: detection of retaliatory or unintentional/secondary poisonings due to human–wildlife conflict, interception of prohibited or excessive marine organism harvest, inspection of watercraft for the presence of invasive flora and fauna, terrestrial surveys for invasive plants, and monitoring of environmental contaminants through noninvasive sampling of indicator (aka sentinel) species. Here, we briefly describe these applications in turn, emphasizing where they offer truly unique opportunities—beyond what people can do without dogs—for better safeguarding individuals and species, habitats and ecosystems, and hence, how this can be an invaluable preventive approach.

INFORMATION BOX 1. *The role of jurisdiction, inspection authority, and mandated response in cross-training to environmental forensics and conservation targets by K9 enforcement teams*

At the international level, dog teams may be trained to and deployed on both environmental forensics targets and narcotics. Alternatively, they may previously have been trained in narcotics or ballistics and now only work on environmental forensics targets. For example, in central Asia, the State Customs Service of the Republic of Kyrgyzstan has trained operational narcotics dogs to targets from five wildlife species: Saker Falcon (*Falco cherrug*), snow leopard (*Panthera uncia*), Argali sheep (*Ovis ammon*), Siberian ibex (*Capra sibirica*), and Saiga antelope (*Saiga tatarica*). Narcotics dog teams from Kazakh law enforcement agencies tasked

(continued)

with customs and border inspections are currently also used to intercept illegally transported wildlife targets. Going forward, during and post-2020, dogs to be deployed on wildlife will solely be trained to those targets. Not surprisingly, the line for explosives detection dog teams is firmly drawn to that single target.

In the United States, the Federally operated Department of Homeland Security (DHS) is akin to the interior ministries of other countries. DHS is mandated to maintain public security against threats. Customs and Border Protection (CBP) is the largest federal law enforcement agency of the DHS and the main United States border control entity. Several different CBP K9 teams exist, for the purpose of detecting: narcotics, people, currency/weapons, and explosives, and for agricultural inspections (US Department of Homeland Security, Canine Program 2020). Border search authority is granted to CBP K9 teams, wherein under Section 19 C.F.R. 162.6a, a team has "carte blanche" to search any person or property entering or leaving the country, without needing probable cause or a warrant to do so (US Department of Homeland Security, CBP Search Authority 2020). Subsequent detection of items under the team's remit leads to a series of responses including: seizure, arrest, and penalization.

Despite the breadth of their search authority, a CBP dog would *never* be trained in BOTH narcotics or agriculture, or have conservation targets layered on; the teams are trained only to a single CBP grouping of targets. In this context, cross-training could actually hamper the efficacy of the teams, given that the nature and/or severity of the penalties conferred in each case can differ significantly. Interestingly, CBP does have within its mandate the enhancement of ways to intercept contraband used to fund terrorist and criminal activities. While the illegal trade in flora and fauna has been linked to the financing of these sectors (United Nations Office on Drugs and Crime, UNODC 2016), this premise has not been incorporated into CBP dog team training to date.

Absent broad search authority, it would be highly problematic to have a dog trained to a range of targets, some but not all requiring a warrant for inspection following an alert. In true "chicken and egg" form, the officer or agent would not know which of these targets the dog was alerting to without conducting an inspection, for which they might require—but not have—permission.

At the State level, the Department of Natural Resources (DNR) is primarily responsible for enforcing environmental and conservation laws and policies. As such, DNR dog-handler teams could be trained to a range of conservation targets like dreissenid mussels for watercraft inspections and related contraband (e.g., illegal firearms, venison outside the hunting season). In this case, caution would be needed around the other targets of the mussel inspection dog, since an alert would initiate a time-consuming boat decontamination process. Alternately, an inspection or search initiated because of a dog's alert might uncover additional items—to which the dog is not trained—but that do fall under the officer/agent's jurisdiction, justifying seizure and apprehension or other appropriate penalties.

Detecting Deliberate Wildlife Poisoning

The use of dogs to combat deliberate wildlife poisoning was pioneered in Eastern Europe, in response to the extensive level of poisoning historically encountered there (Fajardo et al. 2012). Trained to find commonly used substances, often restricted-use pesticides, detection dog-handler teams are now routinely called out to suspected scenes of wildlife crime across numerous European countries and territories (e.g., Kret et al. 2015; Vavylis et al. 2018). They are also increasingly lending support and capacity to their counterparts in Africa.

In many parts of Europe, Africa, and increasingly Latin America, it is common to place baits and lures laced with toxic substances such as organophosphorus or carbamate insecticides, strychnine or cyanide, in the environment (e.g., Fajardo et al. 2012; Ogada 2014; Richards et al. 2017; Alarcón and Lambertucci 2018). The aim of this practice is to poison wildlife perceived as competition for food and hunting quarry, or in retaliation against damage to agricultural crops and/or predation on livestock. Bodies of water may also be tainted in order to precipitate a mass poisoning event. This is not to be confused with the practices of "pesticide hunting" or "pesticide fishing," wherein poisons are used to capture wild animals for human consumption (Odino 2012). Occasionally, retaliatory poisoning also arises from disputes between people (e.g., neighbors poisoning one another's herds, working animals or pets). In Africa, rangers have noted that poachers of large animals like elephants and rhinoceros deliberately poison carcasses in an effort to reduce vultures circling overhead, which could have acted as a beacon to alert park rangers to the scene of the crime (Ogada et al. 2015). For instance, in June of 2019, over 530 vultures were poisoned in Botswana in association with the poaching of three elephants (BirdLife Africa 2019). The impact of this loss was further magnified because it coincided with the post-fledgling period for the vulture species, leaving behind vulnerable young. Similarly, in Argentina, more than 90 condors were poisoned with the carbamate pesticide carbofuran in 2018, with 34 found at a single poisoned carcass—of which 30 were adults of breeding age (Alarcón and Lambertucci 2018).

Regardless of the underlying motivations, the acute toxicity and non-discriminate nature of poisons and the extent to which poisoning occurs is decimating global populations of avian and mammalian scavengers (e.g., vultures, hawks, lions, hyenas, jackals). In addition, micro-scavengers that provide critical ecosystem services within these landscapes are frequently encountered as casualties at or near poisoned carcasses (Martins 2012; Fernández et al. in review).

Working alongside conservation law enforcement agents (in Europe), as per strict protocols and typically in accordance with the issuance of a warrant by a judge, the teams are directed to search property, buildings, and vehicles. Dispatched to conduct outdoor/area searches, always accompanied by an agent, a single dog-handler team may locate poisons and a range of peripheral evidence such as animal carcasses, baits, and items left by potential perpetrators (e.g., cigarette butts, gum wrappers, gloves) from which DNA and fingerprints may be obtained (Fig. 3a, b). Often, these searches uncover additional clandestine practices, such as placement of snares and traps, which are also confiscated. Once the dog indicates a find, agents document the evidence in situ and collect it for analysis at an environmental forensics laboratory. Intermittent but reliable patrols by dog-handler teams, with impromptu follow-up visits in known poisoning hotspots, have also been found to deter poisoning (Fajardo et al. 2012). Wherever incidents arise, poison detection dog teams can also be used to sweep and clear an area *after* a wildlife crime scene has ostensibly been decontaminated. This enables the removal of any remnants of poisons or poisoned carcasses that could cause further mortalities (i.e., through secondary poisoning) or contaminate the water table and ambient environment. Although this follow-up application may be overlooked, its importance cannot be overstated. When interdicted, poachers can be inspected for poisons they may be carrying on their person or traces remaining on their clothing. Confiscation and penalization may in some instances be an effective deterrent for those who are considering the use of poisons in their illegal activities.

Although poisoning is usually directed at a specific animal group, e.g., predators or perceived "nuisance" animals, the practice indiscriminately

Fig. 3 Alongside conservation law enforcement agents, poison detection dog teams are directed to search property, buildings, vehicles, and outdoor areas. A single dog-handler team may locate poisons and a range of peripheral evidence (e.g., carcasses, baits, items left by suspects) from which DNA and fingerprints may be obtained. (**a**) Credit: Iñigo Fajardo, copyright owner. (**b**) Credit: Iñigo Fajardo, copyright owner

affects a much broader swath of species. However, the penalties or procurement of penalties for poisoning may be more weighted according to the protected status of a given species. For example, regardless of whether they were the intended target, evidence that a single Eurasian lynx (*Lynx lynx*) or Egyptian vulture (*Neophron percnopterus*) perished during a poisoning incident may ultimately prove more compellingt than tens or even hundreds of dead foxes or griffon vultures (*Gyps fulvus*) recovered from the same scene.

Thus, identifying an imperiled species amidst mass casualties can strenghten the rationale or even provide the necessary impetus for bringing a case before a judge. However, there must also be evidence that poisoning caused the death in that protected species. In one particular case (summarized in Richards et al. 2015), the carcasses of several species of vultures of varying protected statuses were recovered during a search made in conjunction with a dog-handler team from the canine unit. Initially, evidence of poisoning was only found in the more abundant (i.e., unprotected) species. The forensics laboratory personnel then considered the chemical properties of the suspected poisons and the feeding behavior of the vultures and decided to take different additional samples from all of the carcasses, focusing on the palate and tongue. This novel approach, focusing on two sample types not typically analyzed, led to poison detection in samples from the carcass of one of the endangered vulture species, thereby further compelling the case to proceed in court.

The discovery of the vulture carcasses as a whole is credited to the entire team rather than being specifically attributed to the dog-handler duo. However, there is no doubt that the involvement of the canine units significantly increases the probability of retrieving viable evidence from a putative scene of wildlife poisoning. This case reflects that fighting wildlife poisoning is very much a collaborative undertaking comprised of many dedicated participants, with poison detection dog-handler teams at the core of such efforts.

Intercepting Overharvest and Illegal Take of Marine Organisms

During lobster-fishing season, a detection team formerly operating with the Department of Fisheries and Oceans inspected hundreds of wharves across the Maritime Provinces of Canada (Fig. 4a, b). Lobster crates were inspected as they were unloaded from fishing boats. As fisheries officers checked crates for infractions

Fig. 4 A fisheries officer/handler and dog team can inspect approximately 100 crates full of lobsters in under 10 min, whereas the same could take a couple of hours for a pair of fisheries inspectors. Reproductively viable female lobsters can promptly be released for a chance to spawn and contribute to the long-term stability of the population, and the fisheries resource. (**a**) Credit: DFO Canada, copyright owner. (**b**) Credit: DFO Canada, copyright owner

such as the presence of undersized lobsters, the dog-handler/fisheries agent team searched for spawning females, including those whose eggs had been removed or "scrubbed." The team also participated in outreach at schools in fishing communities and to stakeholder groups, at times with suspected poachers in their midst (and/or potentially their children). Evidence of their impact was made clear when fishermen in attendance recalled their child mentioning a visit from the team, thus bringing the outreach effort full circle.

The repercussions of overharvesting and illegal take to the industry cannot be overstated. For example, the lobster fishery off the northeastern shores of North America is highly lucrative, bringing in billions of dollars in revenue annually. The lobstering season spans the spring and autumn months, which also coincide with the crucial egg-bearing period for the Atlantic lobster (*Homarus americanus*). Whereas a one-pound female might bear some 8000 eggs, she could conceivably carry more than 100,000 eggs if allowed to reach a weight of nine pounds. And yet, out of every 50,000 eggs, only one or two lobsters are expected to survive to legal capture size (NOAA, https://www.nefsc.noaa.gov/faq/faq-archive/fishfaq7.html).

Given these impressive figures and their implications for the stability of the population and resource, fishermen are forbidden from returning to shore with egg-bearing lobsters; all must immediately be returned to the water if caught. The penalty for illegal capture is approximately $1K (CAD) for a single egg-bearing female, with a maximum of $100K. Coupled with how few enforcement officers there are, the penalties do not necessarily offset how lucrative the illegal trade in egg-bearing lobster is on the Asian market. Indeed, the estimated value of one egg-bearing lobster shipped clandestinely overnight is approximately five times more per pound (than a non-egg bearing female) for the harvesting of lobster caviar.

This application of detection dogs has the potential to save considerable money and time for fishermen and fisheries agents. For example, a former fisheries officer/canine handler estimates that a dog-handler team can inspect 100 crates full of lobsters in under 10 minutes, which would take a pair of fisheries inspectors a couple of hours. There is the added benefit of this being a preventive measure, enabling rapid identification of reproductively viable female lobsters that might swiftly be returned to the water and, ideally, allowed a chance to spawn and contribute to the long-term stability of the population. Further discussion of dog-handler teams for the broader purpose of intercepting practices harmful to species and populations, e.g., illegal abalone fishing, illicit clam digs, and shark fin harvest and smuggling can be found in Braun et al. (2018) (and see also, e.g., https://www.sowetanlive.co.za/news/2016-11-15-police-dog-sniffs-out-r3m-in-illegal-abalone/).

Watercraft Inspection to Prevent the Entry or Spread of Invasive Species into Aquatic Ecosystems

The role of dogs to combat the spread and infiltration of invasive species can generate awareness in the public, and compliance, in a way few other measures can. Invasive *Dreissenid* mussels are a case in point. Zebra and quagga mussels can quickly establish themselves in new habitats, as a single female can produce up to one million offspring per reproductive cycle (Sawchuk 2018). Mussels are prodigiously adaptable and highly tolerant to temperature change; thus, they are able to outcompete established native species. Invasive mussel infestations alter entire aquatic ecosystems and foul water quality, requiring aggressive and expensive mitigation treatments including pesticides, which can be harmful to native flora and fauna. The sudden eruption of razor-sharp shells that adhere to many surfaces, not to mention the pervasive odor, can transform beaches from tourist havens into very unpleasant or outright hazardous places to visit (ibid). Infrastructure such as turbines or irrigation pipes can clog rapidly, which is a major problem and economic burden for the stakeholders reliant on water outflow. Reactive rather than preventive remediation is costly, time-consuming, and onerous (Vyn 2019; Sawchuk 2018).

Boaters may inadvertently carry mussels or microscopic larvae on watercraft, infecting clean watersheds. One might expect that the tax burden from the mitigation costs would be sufficient to induce practices to clean boats and inspire boater cooperation. Unfortunately, boaters can instead perceive having multiple inspections on a single journey as a nuisance to be avoided if possible, especially when infractions for failing to stop at a check station may be negligible. Part of their frustration could stem from the belief that the odds of a human inspector actually detecting a contaminated boat are miniscule and therefore that inspections are of limited value. However, the involvement of a detection dog can significantly increase confidence in the effectiveness of the process and in willingness to submit to it.

Working with Inspectors, mussel detection dog-handler teams can efficiently inspect a boat, sometimes in less than 5 minutes - their speed appreciated, especially during times of high boat traffic. Despite having submitted to prior inspections along their route, members of the public are usually quite amenable to further scrutiny from a dog-handler team. Curious and engaged, they will exit their vehicle to watch the process and take photographs. Meanwhile, as an Inspector narrates the process and answers their questions, an opportunity has been created to impart the importance of inspections, of adhering to thorough cleaning protocols, and the damage wrought by an infestation. In the event that mussels are detected and a boat must be decontaminated, discovery by a dog-handler team can make owners more agreeable to an enforced interruption in their journey (Sawchuk 2018). Here, the teams play a dual part in helping to streamline inspections for maximum efficiency and in generating interest in and awareness of an issue that is—candidly—not terribly attractive or compelling for the public to learn more about (Fig. 5a, b). In so doing, they increase the chances of keeping un-infested water bodies, beaches, and infrastructure free of zebra and quagga mussels.

Terrestrial Monitoring and Eradication of Non-native Fauna, Noxious Weeds, and Invasive Insect Pests

The trade of flora and fauna may give way to invasive species infestations (Hulme 2009). This can be further facilitated by the pet industry. For example, Burmese pythons have been released into protected areas by people who initially got them as pets, then released them into landscapes where they are not native and into which they subsequently wrought havoc on those ecosystems. Dogs have been trained as one of

Fig. 5 Boaters may inadvertently infest clean watersheds by spreading invasive mussels or microscopic larvae. In-season, dog-handler teams work with Inspectors to maximize the efficiency of watercraft inspections. Year-round, they attend outreach events to generate public interest and compel boaters to learn more and take precautions to combat invasive mussel incursions. (**a**) Credit: WD4C, copyright owner. (**b**) Credit: Cindy Sawchuk, copyright owner

the tools employed to locate and remove pythons (e.g., Avery et al. 2014). Another example is feral swine, also unwelcome in non-native landscapes, wherein dog-handler teams have been enlisted to assess the effectiveness of eradication efforts, via the absence of deposited feces (Jackling et al. 2016). There is increasing concern about keeping national parks and other important habitats free of invasive species or dealing with their encroachment into such areas (e.g., Dayer et al. 2019 and Novoa et al. 2016).

Dog-handler teams have been used to monitor and combat the spread of terrestrially invasive species, sometimes in parallel with efforts to protect native species and grassland ecosystems (Fig. 6). For example, since 2010, dog-handler teams have been central to efforts to contain and eradicate the noxious weed dyer's woad (*Isatis tinctoria*) in Montana (Hurt et al. 2016a). Other noxious weeds to which dogs have been trained and deployed include spotted knapweed (*Centaurea maculosa*; Goodwin et al. 2010). At a wildlife refuge in Iowa, teams have been employed to survey for Chinese bush clover (*Lespedeza cuneata*) several years running, and most recently the dogs were also there to simultaneously seek native whorled milkweed (*Asclepias verticilliata*). This enabled tandem mapping of viable monarch butterfly habitat, denoted by the presence of the milkweed, with assessment of bush clover removal efforts (WD4C unpublished data). Similarly, in a proof of concept study, dogs were able to distinguish native Kincaid's lupine (*Lupinus sulphureus*) from other types of co-occurring lupine. Their ability to detect this particular lupine offers an opportunity for assessing habitat viability for the endangered Fender's blue butterfly (*Icaricia icarioides fenderi*), which uses it as a larval host, without necessarily having to train directly to the butterfly itself (Vesely 2008).

Dogs have also been trained to detect a variety of invasive insect pests, including the Asian longhorned beetle (*Anoplophora glabripennis*; Hoyer-Tomiczek et al. 2016) and the emerald ash borer beetle (EAB; *Agrilus planipennis*). In the case of EAB, a subset of the dogs were trained

Fig. 6 Dog-handler teams can help monitor terrestrially invasive species, and CDDHT surveys are sometimes used in parallel with efforts to protect native species and grassland ecosystems. Credit: WD4C, copyright owner

to discriminate ash from non-ash wood and infested versus uninfested wood matter (e.g., chips). Although these species are known to have devastating and costly impacts on forests and urban treescapes (e.g., RPWCO 2019), the subject may nonetheless be considered rather underwhelming by the general public. However, here again, the involvement of dogs can generate much-needed public engagement. When several news pieces outlining the use of detection dogs to monitor and halt the spread of EAB-infested ash wood in Minnesota ran within a single week in 2012, extensive household viewing estimated at a value of one million dollars (USD) of free media outreach was achieved in support of an ongoing awareness campaign (WD4C unpublished data).

Noninvasive Monitoring of Environmental Contaminants Through Indicator Species

The ability of dogs to locate the fecal matter (often referred to as "scat") of wildlife species can be incorporated with great effectiveness into monitoring environmental contaminants in terrestrial and aquatic ecosystems. As one of the greatest threats to modern conservation, the prevalence of environmental toxin inputs is now acknowledged, but the study of the long-term impacts on the environment and/or wildlife remains in its infancy (Richards et al. 2018). The degree to which contaminants like human and veterinary pharmaceuticals and industrial compounds (e.g., polybrominated flame retardants and heavy metals) are present in the environment, primarily in aquatic ecosystems, has been well-documented in some parts of the world but not in others. Exposure has been investigated in a range of avian and mammalian indicator species. However, it has primarily entailed blood or tissue sampling, which may necessitate direct contact (and the associated required permits), locating nesting sites (and potentially climbing to nests), culling animals, or opportunistic carcass collection. Further challenges arise when the species of interest is imperiled, reclusive, trap-savvy, or occurs at low density.

CDDHTs have been used to seek fecal matter of River otter (*Lontra canadensis*) and American mink (*Mustela vison*), i.e., spraints and scats, respectively, both having a preponderance of fish and other aquatic organisms in their diet (Richards et al. 2018; Sutherland et al. 2018). Exclusive to riverine habitats, these two apex predators are notoriously elusive and therefore difficult for researchers to capture and sample directly. Instead, fecal samples can be analyzed for many different types of environmental contaminants (e.g., heavy metals, flame retardants, and pharmaceuticals; and, separately, provide information on the animal's diet and

Fig. 7 Dogs can find fresh and aged samples across landscapes. Finding and mapping bleached shell remains from otter fecal matter at a former communal "latrine" relative to more recently deposited samples can provide insight into habitat occupancy and species co-occurrence. Samples of a certain freshness can be collected for genetics analysis and contaminants monitoring, among other informative parameters. (**a**) Credit: WD4C, copyright owner. (**b**) Credit: WD4C, copyright owner

overall health, see, e.g., La Guardia et al. (2020)). It should be emphasized that in these cases detection dogs were *not* specifically trained to detect environmental contaminants, but rather to the scats that may or may not contain them. The reader is directed to Reynolds and Reynolds (2018) for a description of using detection dogs to locate illicit sewage input and contamination, e.g., from faulty or aging infrastructure.

Although the notion of training dogs to detect (and confining detection to) samples only containing contaminant residues might seem more efficient, directing dog-handler teams to find a subset of a target is in reality *more* challenging. In principal it also removes objectivity by imposing an assumption of contamination, assigns unequal value to presence versus absence of residues, and assumes there is no gradient of contamination. Instead, searching unrestrictedly for fecal matter of the focal species capitalizes on a dog's ability to detect a representative range of samples occurring across the landscape, whether freshly deposited or months, even years, old (Fig. 7a, b). These finds proffer a treasure trove of complimentary information regarding presence, (co-)occupancy, and habitat usage that can be examined relative to observed residues and levels. When specifically targeting scats for analysis, different levels of sample freshness may be acceptable, depending upon the type of contaminant sought and its associated molecular properties, and rates of persistence. Similarly, hormonal or genetic information (e.g., sex, individual level) may also be acquired from fecal samples for consideration with observed residues. Here, the dog is free to find the fecal matter of the target species, while the handler can be trained to choose and collect a subset of detected samples for a range of desired analyses based on specific viability criteria. Alternately, the descriptor of "old" or "fresh" and other observations provided by handlers can inform laboratories seeking to prioritize samples for analysis in the face of a limited budget.

In summary, training dogs to the fecal matter of indicator species may enable concurrent examination of exposure to contaminants, genetic analysis, and a big picture assessment of habitat usage and occupancy, as well as patterns of species co-occurrence. Much of the information obtained could in turn be used to pinpoint contaminant input sources for mitigation purposes.

Incentives for Increased Use of Ecological Monitoring Applications of Detection Dogs Within a Preventive and Supportive Framework

The examples provided in the previous section describe environmentally-related variations of forensics and enforcement initiatives to illustrate the unique role that trained CDDHTs play within these realms. This section will focus on conservation detection. Here, we outline how detection dogs can be used preventively, at the ecological monitoring and community engagement level of the spectrum, to survey habitats, and identify species occurrence and status. We aim to focus on ways to keep organisms alive and protect ecosystems while enforcement and forensics are dealing with individual people and entities who have committed offenses against wildlife and the environment, to prevent adding to their workload.

Ecologists and landscape managers have a veritable arsenal of data-collection tools at their disposal. Traditional ecological monitoring methods used to gather information on animal species include setting out camera traps, hair snares, baited track plate stations, and live animal traps to collar, band, or tag individuals. Auditory surveys may also be conducted. Similarly, botanical inventories and surveys are carefully conducted to determine the presence and health of a complement of plant species and communities, just as environmental scientists can sample plant life, microorgamisms, sediment/soil, water, and/or sentinel species to gain broader insight on ecosystem health.

Within this framework, CDDHTs may be a noninvasive, comprehensive, and cost-effective means of surveying and sampling—providing that their unique benefits are fully acknowledged from the outset, and that prospective partners have a clear understanding of when and how best to use them in relation to other available information-gathering tools (See Common Misconceptions below for further discussion). At various stages of discovery and investigation, the support of CDDHTs may be particularly indispensable when the species of interest is wide-ranging, nocturnal, cryptic, hidden, stowed away, or the information sought is otherwise incredibly difficult, time-consuming or laborious for people to acquire (Figs. 6, 7a, b, 8a, b, 9a–b). A determination that key baseline information might not be obtained or augmented without the assistance of a CDDHT may justify a higher cost expenditure in order to facilitate later, comparatively less costly measures (see also the discussion below on biological significance).

The previous section elaborated on some of the reasons why dog-handler teams engaged in forensics and enforcement work may only be trained to a minimal or singular number of targets. Conversely, while still bearing in mind any potential incompatibilities, it is *ideal* that conservation detection dogs be trained to many different types of conservation/environmental targets (e.g., scat, living animals, refugia, plants). Typically, an individual dog in our profession will acquire a vast catalog of scents (e.g., 31 and counting) over their working life (Table 10.1; Woollett et al. 2014). Accompanied by a trained handler, these dogs have the ability to seek multiple, compatible conservation targets at the same time. This further magnifies the cost- and time-effectiveness of this method and an individual team's impact. Working as one, the duo can comprehensively survey the designated search area, which can be an essential time-saver and catalyst, for example at the initial stages of discovery where next steps hinge on ascertaining presence/absence within a given area. Having found recent evidence of the animal(s) of interest (e.g., via detection of fresh scat) additional dog-handler surveys can be conducted, or tools like camera traps can then be selectively sited within an appropriate defined area for follow-up monitoring. Similarly, pinpointing the location of even a single plant, or a small patch of plants, of particular ecological significance within a given habitat can compel land acquisition and inform the delineation of buffer zones. As described in the previous section, fecal samples (aka scat or dung) collected from indicator species can provide complimentary, synchronized information on natural history, ecology, and the general health of an ecosystem.

Another possibility is to use a target as a "proxy" for the focal species. For example, the

Fig. 8 Getting to work alongside a highly capable dog is a privilege. (**a**) Credit: WD4C, copyright owner. (**b**) Credit: Jennifer Hartman, copyright owner

presence of certain key habitat components (e.g., plants) can provide more readily obtained interim information about an elusive or endangered species (e.g., the previously discussed Kincaid's lupine and Fender's blue butterfly case; Vesely 2008). Likewise, when it is particularly difficult to gather direct samples related to the focal species, it might be possible instead to train a dog to other targets closely associated with it to increase the chances of finding that species or samples upon which they can then be trained.

When a dog has been trained to a target that is present within an environment yet not of apparent value to an investigator, initial interest can quickly be extinguished when the dog realizes no reward will be forthcoming for finding it *during these specific surveys.* This notion is further confirmed when they alert to the desired targets and are handsomely rewarded for it (i.e., positive reinforcement). With the aid of the handler, the dog understands to switch the focus to the target that *does* confer a reward in this instance.

There are many reasons for deploying dog-handler teams that transcend cost considerations or time constraints. Fundamentally, dog team surveys often challenge and identify underlying fallacies in our prior perceptions and preconceived ideas relative to the biota we study. Data acquired and observations made during surveys, led by olfaction rather than vision or auditory cues, often reveal the ways in which we may unwittingly have constrained ourselves, by providing answers to questions we had not thought to ask. This is one of the defining ways that only dogs can instruct us. Searching primarily by scent, the teams operate in a manner that inherently minimizes the influence of surveyor bias—i.e., steering a search according to where finds have previously been made, or where there is an expectation of making a find. When, in addition to the expected locations, the dog also finds the target (s) in a new and unforeseen area, it refreshes our own search image and reinvigorates both our way of seeing the landscape and our imagining of how species operate within it. Although this can be surprising and even a little disconcerting to partners initially, it is vital to gaining a true understanding of the focal species and to ensuring that follow-up monitoring and protective measures are appropriate and successful.

Common Misconceptions About Dogs, CDDHTs and Handlers in Conservation Detection

Although there is overlap with many detection disciplines, conservation detection is a distinct

Fig. 9 The olfactory abilities of dogs are harnessed in partnership with proficient handlers to acquire indispensable information that may otherwise be challenging to gather. (**a**) Mustelid scat. Credit: Jennifer Hartman, copyright owner. (**b**) Bat guano. Credit: Rogue Detection Teams, copyright owner. (**c**) Fisher and marten scat. Credit: Jennifer Hartman, copyright owner. (**d**) Moose scat. Credit: WD4C, copyright owner

field with standalone operating procedures, associated methodologies, and underlying principles. In this section, we identify misconceptions and erroneous expectations that are commonly encountered. These are noteworthy in that they can and often do prevent CDDHTs from being utilized in ways that capitalize on the benefits and versatility of dogs as an information-gathering tool. As such, these are a barrier to more completely, often cost-effectively, applying CDDHTs to ecological monitoring and related enforcement and forensic endeavors. Discussions held early on between conservation dog professionals that are experienced/established and prospective project partners and stakeholders can alleviate to an extent certain misconceptions and/or erroneous expectations, though some may still persist and reemerge once on the ground and in the field.

CDDHTs typically operate via "fee for service," which may perpetuate an assumption that they are short-term contractors first and foremost. While this more transactional partnership is applicable in certain cases and might be appropriate—even preferable—in others, many conservation dog handlers are authorities in this field, professionals and researchers on an equal footing, and even potential stakeholders with a long-term vested interest in (and commitment to) supporting the accomplishment of project objectives. The outcome of this fundamental misunderstanding can be dog-handler teams not being used to their fullest potential. This is ultimately counterproductive, might be wasteful of resources, and may create damaging, erroneous perceptions about the viability of CDDHTs.

In actuality, conservation detection dog handlers are dedicated professionals who possess a rare and extensive combination of skills and expertise, including the ability to:

- Assess potential candidate dogs
- Train and handle dogs that make it through a rigorous trial period
- Judiciously rehome dogs that do not pass the trial period
- Draw from backgrounds in biology, ecology, and environmental science—among others—or possess the natural desire to gain the requisite knowledge to become a detection dog handler
- Partner with different dogs to find a varied range of targets in multiple types of terrains and landscapes
- Assess the viability and feasibility of using dogs to find certain targets
- Make informed decisions about which new targets to train a dog to given their scent repertoire and prospective future project deployment
- Work in precarious/challenging work environments
- Collect and process samples, gather/analyze/interpret data—including mapping and weighing in with recommendations
- Prepare successful grant applications, to assist prospective partners in finding funding support
- Contribute to or lead in authoring reports and preparing manuscripts for scientific publication and dissemination of information through a variety of venues and formats

Different Types of Dogs Can Be Suited to Conservation Detection

While by no means a new field, conservation detection is still overcoming misperceptions or misconceptions regarding the types of dogs that can do this work. In certain forensic or enforcement contexts, a dog's physical appearance and demeanor may be used—either to reinforce authority or, conversely, to put people at ease. Predominantly seeing certain breeds of dogs in enforcement applications (e.g., Shepherds, Malinois, Labrador Retrievers) may generate the notion that only these breeds, or by extension only purebred dogs, can effectively serve in scent detection. Yet numerous dogs employed in the field of conservation have been adopted from shelters, or "changed careers" (e.g., started in Schutzhund as sport dogs but were redirected for conservation work, to which the individual dog is ultimately better suited). Whether of uncertain lineage or a known breed, these individuals possess undeniable potential and a deeply ingrained need for a job that gives them a distinct purpose. Their drive, intense exuberance, and toy/play obsession prevents them from partaking in the kind of life that satisfies a pet dog. Such traits may also prevent them from excelling at other jobs that require a different temperament and/or focus (e.g., guide dog for the blind, therapy dog). Additional information about the important and distinctive traits that may be required of conservation dogs, regardless of breed, can be found in Mackay et al. (2008) and Hurt et al. (2016b).

Multiple Dogs and Handlers May Work Together as a Team

Dogs operating in conservation detection groups may be paired with a primary handler or may work with multiple handlers. Within each pairing, the detection dog and the handler must coalesce as a solid unit. Their joint efficacy and capabilities are contingent upon training, both apart and together, as well as upon sufficient time spent getting comfortable and familiar with one another. With the proper training, it is entirely possible for a complete novice who has never before laid hands on a dog, but nonetheless possesses the requisite skills and attributes, to become a skilled handler. However, pairing a trained dog (regardless of their versatility or skill) with someone unfamiliar with conservation handling and expecting the duo to adjust as they go can entirely invalidate the reliability and effectiveness of this method.

Human and Dog Team Surveys Gather Information in Vastly Different Ways

The misperceptions around the breeds of dogs that are able to do conservation detection work can also extend into how the teams should look when on-duty. First, a CDDHT at work does not resemble a single human surveyor or survey crew in action, each team instead using a tailored search strategy that requires different senses and skills. Thus, a first glimpse of a dog and handler searching for conservation purposes may not correspond with expectations. People who accompany dog-handler teams to observe or assist (e.g., as orienteers, or PIs—primary investigators) sometimes express concern when, for example, they observe what is interpreted as a dog ignoring, or failing to indicate on a known or visible target. What is misunderstood, however, is the dog's method of searching, which utilizes an invisible odor signature with a characteristic pattern of diffusion and dispersion to locate the source of the odor, i.e., where it is strongest. Alternately, the dog may have detected another as yet unseen target in the vicinity and could be working to that. Humans rely so much on sight and sound to negotiate their environment, and as a result, can initially be very resistant to scent-focused navigation.

It is somewhat ironic that the sight of an experienced team at work may appear quite unremarkable to onlookers—like a person out for a hike with their dog. In reality, numerous elements are unfolding and being managed in relation to one another inside the handler's brain: they are conceptually mapping the terrain in relation to the survey requirements (e.g., grid lines), often with GPS in hand; maintaining visual contact of their dog's movements within the terrain; being mindful of habitat features, natural history traits of the focal species and target, behavioral indicators that the dog is on the scent, and the role of environmental factors in the dispersal of scent—all to support the dog in the search as needed. They are also tasked with ensuring adequate hydration, nourishment, safety, and maintaining good spirits for both themselves and the dog. Additionally, they must: confirm the dog's find and reward them appropriately, which requires a certain degree of experience and dexterity in itself (see, e.g., Fig. 10c); absent an assistant or orienteer, collect samples per the requirements unique to each project; make field notes; and move in an efficient and timely manner so that adequate coverage for the study area is obtained throughout the duration. Finally, at the end of each workday, they must: attend to their dog through rigorous body checks; feed and rest the dog; process samples and enter data - all typically before attending to their own needs.

In summary, conservation detection rests upon a foundation of conservation science, and, when correctly applied, is an acknowledged tool through which to reliably and noninvasively gather a range of crucial ecological monitoring data. At the heart of this discipline lies not an automated machine, but rather a living being, which introduces an element of art and intuition in complement to the science of odor detection. Since this important element does not lend itself easily to precise explanation, it carries the inherent risk of initial skepticism and possibly mistrust from casual observers who may at first be uncomfortable operating outside the scope of conventional sciences and methodologies (e.g., visual surveys, collaring, camera traps).

It may run counter to the expectation that the people who excel at this type of work do not necessarily know a great deal about dog breeds or have multiple training certificates to their names. Rather they are inquisitive, driven, self-directed, conscientious, able to troubleshoot, open to constructive criticism, and willing to trudge through inhospitable, albeit beautiful, terrains (Fig. 10a–k). At all times, they maintain mental and physical focus, keeping themselves safe while simultaneously attending to their canine partner's well-being (Fig. 11a–g). The types of people who commit to this field and the sacrifices entailed largely do so out of an abiding love of the natural world and its inhabitants, and a genuine desire to tangibly contribute to conservation, quality research outcomes, and sound science. They understand that achieving any of these outcomes with a highly capable dog by their side is a privilege (Fig. 8a, b).

Interpreting Seemingly Low or Variable Detection Performances, Factoring in Biological Versus Statistical Significance and Context

There is an increasing movement for parameters observed around conservation detection to be reported (see, e.g., Bennett et al. 2020). The aim is to enhance overall quality control and to help potential partners more readily assess efficacy, timing, and cost-effectiveness around deploying CDDHTs. Here, we identify several caveats, in order for such transparency to accurately reflect the context of the parameters reported upon.

Despite the obvious role and recognized importance of statistical significance, a certain erroneous emphasis may still be placed on it in relation to the quality or import of the data acquired. Yet for many species under consideration, what matters most, and sets apart the value of CDDHTs, lies in the realm of biological/ecological/conservation significance. Sometimes, the question best addressed with CDDHTs is whether *any* evidence of the species of interest can be gleaned. Thus, it is critical to hold early and detailed discussions of the study objectives prior to entering the field to survey, in order to achieve clarity on what can and cannot be realistically achieved through detection dog-handler surveys and ensure that is reflected in study design. This also helps to better frame the meaning of the data and results achieved once the field component of a study is completed.

Not all targets are equally detectible or can be located using the same search strategy or intensity, as exemplified in scat work (e.g., Statham et al. 2020). Nor are all reported or observed conservation detections comparable - even when concerning the same target - due to survey execution, accuracy in detection, and environmental conditions, among others. True survey effort, an important consideration for assessing cost-effectiveness and efficacy, is often masked in the way that find rates are reported. For example, an average scat find rate reported as, e.g., 5 scats/km, can actually encompass a broad range of raw find rates, from 200/40 km to 10/2 km and, by extension, an array of survey efforts. Additional context would be provided by explicitly stating the range of raw find rates. Further discussion of this can be found in Richards et al. (2018) where, in seeking to measure the performance of dog-teams in finding otter feces against prior human efforts, this very phenomenon of masked survey effort in reporting was encountered.

Sometimes less is more. For example, within a given season, a total find rate of 4/200 acres relative to 25/200 the prior year is good news when the target is a noxious weed being monitored for containment and eradication. Alternately, 25/200 acres may be a preferable scenario to 4/200 if the former were emergent seedlings found at one seed bank in a single location, as opposed to a single plant being

Fig. 10 (continued)

Fig. 10 (continued)

found at four disparate locations, especially if one or all had reached the flowering or seeding stage. In the context of noxious weed management and invasive species eradication, dogs may be especially useful when a target is thought to occur at low density (Hurt et al. 2016a; Goodwin et al. 2010). At higher density, the expectation in using CDDHTs should not be to find every single target sample within the surveyed area, but rather to (a) increase the *probability* of finding samples while (b) detecting a representative number of those present (Fig. 12).

Fig. 10 Conservation detection handlers get to work in many beautiful, remote and often inhospitable landscapes. (**a**) Credit: WD4C, copyright owner. (**b**) Credit: Rogue Detection Teams, copyright owner. (**c**) Credit: WD4C, copyright owner. (**d**) Credit: Jennifer Hartman, copyright owner. (**e**) Credit: Marci Johnson, copyright owner. (**f**) Credit: Jennifer Hartman, copyright owner. (**g**) Credit: WD4C, copyright owner. (**h**) Credit: Michael Westphal, US Bureau of Land Management, copyright owner. (**i**) Credit: Rogue Detection Teams, copyright owner. (**j**) Credit: Mark Jackling, USDA-APHIS, Wildlife Services, copyright owner. (**k**) Credit: Casey MCcormack, copyright owner

Suppose that over 2 weeks, a single CDDHT found three invasive snakes after meticulously searching a 10-acre area. On the face of it, this rate of detection may not seem a viable use of dogs. However, if one clarifies that the species was thought to have been eradicated within that area after a team of human surveyors swept and cleared it, suddenly the finds take on more value. Similarly, a dog-handler team may have a find rate comparable to humans, but in considerably less time. Or, they may have a single find after expending a huge amount of energy, where human surveyors had been unable to make any finds. Sometimes a single or low number of finds is sufficient to catalyze a range of efforts, justifying the investment. In other cases (e.g., Arandjelovic et al. 2015) the expense and effort may be deemed prohibitive relative to the judicious use of other available tools and methods over a greater period of time.

A fine-scale examination of reported rates may parse out:

(a) Total number of finds relative to
(b) Number of finds that were viable for the desired analyses and then
(c) Those subsequently determined to be of the targeted species

Fig. 11 Conservation detection work is intensely hands-on, requiring a significant time commitment as well as mental and physical stamina in a range of challenging environments and conditions. At all times, handlers are

Fig. 12 Dogs have proven useful in noxious weed management and invasive plant monitoring, often serving to confirm the effectiveness of eradication efforts, when the focal species is ostensibly present at lower density. Credit: WD4C, copyright owner

When samples are not readily visually distinguishable as that of the target species, which is often the case, they cannot be identified to species without genetic analysis, which may require the sample to be of a certain freshness. In this respect, the handler plays a central role in the determined level of accuracy, by ensuring to the best of their ability that viable target samples are collected (or, alternately, that the dog is not erroneously rewarded). These contingencies are considered and addressed in Statham et al. (2020) and in Richards et al. (2018).

In summary, it is important for prospective partners to be explicitly aware of find rate context and biological/conservation significance in reported detection rates when considering the use of CDDHTs to seek a particular target.

Dogs Are Not Always the Answer

Just as some may feel skeptical about the abilities of dogs and CDDHTs, so too can people hold unreasonable expectations of what can be realistically achieved using dogs. Certain conditions must prevail in order for dogs to be successfully employed. Some targets or circumstances simply do not lend themselves to detection by dogs and, when this is the case, it is important to acknowledge this and seek alternate methods. In the simplest of terms, time is running out for many species and there are only limited resources to draw from. Species once categorized as ubiquitous are now trending downward. As we have previously discussed, there are many ways and cases where CDDHTs could be more strategically incorporated to aid in ecological monitoring. However, there are other pressing cases where we remain unsure of how, or even whether, dogs have a significant role to play. The pangolin, one of the world's most poached animals, serves as a poignant example. Information Box 2 discusses the plight of this small, unobtrusive animal about which researchers still know alarmingly little, how CDDHTs are being enlisted, and whether it is feasible or practical to try and incorporate them any further as part of ongoing efforts underway.

Fig. 11 (continued) responsible for ensuring their own well-being and that of their canine partner. While not nearly as glamorous as the public may imagine it to be, the work and accompanying life is immensely rewarding. (**a**) Credit: Casey MCcormack, copyright owner. (**b**) Credit: WD4C, copyright owner. (**c**) Credit: Jennifer Hartman, copyright owner. (**d**) Credit: Rogue Detection Teams, copyright owner. (**e**) Credit: Jesika Reimer, copyright owner. (**f**) Credit: Bryan Peterson, copyright owner. (**g**) Credit: WD4C, copyright owner

INFORMATION BOX 2. *The plight of the pangolin (Manidae): Can CDDHTs be enlisted to help?*

Contributed in part by Katie Wells, former intern with the IUCN Pangolin Specialist Group

Image adapted from a Wildlife Alliance photo, https://creativecommons.org/licenses/by-sa/2.0/

Conservation enforcement initiatives are meant to apprehend perpetrators and halt the associated repercussions of illegal species take. However, penalties rarely deter many would-be perpetrators on the ground, or hamper purveyors higher up the chain. Nor do they sate the wider cravings of consumers who continue to relentlessly drive the demand for wildlife contraband. The pangolin provides a case in point.

Covered in hundreds of hard keratinous scales, pangolins (Manis spp.) are reportedly the most trafficked wild mammals in the world. These unassuming animals are sought predominantly for their meat, considered a delicacy, and for their scales—believed to have curative properties and used in "traditional medicines." Pangolins are worth far more dead than alive, for their perceived medicinal benefits, and this has led all eight Manis spp. being listed as Appendix 1 (i.e., greatest level of threat/protection). Four pangolin species are found in Africa, and four occur in Asia (see Table 1, below). As CITES Appendix I species, they are protected from commercial international trade of wild taken specimens.

Here we broadly consider ways that dog-handler teams have been enlisted to protect pangolins, by helping to acquire information, and in enforcement initiatives. But first, to present these applications in context, we consider what is currently known about these animals and their plight.

The name "pangolin" originates from the Malay word "*penggulung*," which translates to "roller," in reference to their ability to roll into a tight ball when threatened. Coiled up in this way, the pangolin's scales protect it from the bites of their natural predators, including lions and leopards. Despite their similarity in appearance and ecology to armadillos and anteaters, pangolins are more closely related to animals in the order "Carnivora."

Most of the genera is nocturnal. During the day they shelter in burrows approximately 1.5 m deep or in tree hollows. Pangolins have well-adapted front claws for digging, though they are known to opportunistically occupy the empty burrows of other animals such as aardvarks and aardwolves. Their prehensile tail allows them to easily climb trees, though only some pangolin species (e.g., *Phataginus tricuspis*) live arboreally.

Pangolins do not have teeth. All eight species are myrmecophagous, i.e., exclusively feeding on ants and termites. With an ability to forage ant and termite nests using their sharp claws and long, saliva-covered tongue, pangolins help to regulate social insect populations across their habitat range and within their ecological niche as a result of their highly specialized diet. In order to digest ants and termites, part of their stomach is thickened and muscular with keratinous spines, and they ingest small amounts of dirt and grit to further assist with the digestion process. Estimates

(continued)

Table 1 Various parameters which may be drawn from to inform coordinated CDDHT and community efforts to research and conserve the pangolin

Information that could be used in fundraising campaigns

Data that could steer use of CDDH teams

Appropriate for public outreach

Species	Geographic range	IUCN Red List status	Length (cm)	Weight (kg)	Habitat	Identification	Did you know?
White-bellied (*Phataginus tricuspis*)	Central and West Africa, including; Guinea, Angola, Uganda, and Central African Republic	V	60 - 105	1.6 – 3.0	Arboreal; tropical forests, dense woodland, agricultural land. Particularly around watercourses	Distinctive white underbelly. Scales are keeled with three points (hence 'tri-cuspis')	The most frequently encountered pangolin species in Africa
Black-bellied (*Phataginus tetradactyla*)	Central and West Africa, including; Sierra Leone, Democratic Republic of the Congo, and Cameroon	V	85 - 120	2.0 – 3.0	Arboreal; riverine and swamp forests dominated by palms	Black underbelly. Scales are dark brown with yellow-orange colouration at tips	Have the longest tail to body ratio. With tails that can be more than twice the length of their body, they are often called 'long-tailed pangolin'
Giant (*Smutsia gigantea*)	Central and West Africa, including; Democratic Republic of the Congo, Guinea Bissau, and Uganda	V	125 - 170	20 - 35	Fossorial (aka ground-dwelling) savanna, lowland tropical moist forest, and swamp forest	Distinguished from the white and black bellied pangolins by its fossorial lifestyle, large size and regular scale arrangement	The largest extant species of pangolin
Temminck's ground (*Smutsia temmincki*)	Central and southern Africa, including; Chad, Tanzania, Namibia and South Africa	V	65 - 111	7.0 – 18	Fossorial; arid savanna and desert, rocky terrain, and floodplain grassland		The only species found in southern Africa
Philippine (*Manis culionensis*)	Endemic to four islands of the Philippines, including Palawan and Culion	E	85 - 115	5.0	Semi-arboreal; forests, grassland, agricultural and disturbed land	This species has the greatest number of scale rows of the Asian pangolins	Has only recently been described as a separate species distinct from the Sunda pangolin
Chinese (*Manis pentadactyla*)	Southeast Asia, including; Nepal, China, Taiwan, Vietnam and Myanmar	CE	70 - 96	4.0 – 9.0	Predominantly fossorial. Occupies a broad range of habitats, including; agricultural and grassland, as well as tropical, bamboo, deciduous and coniferous forests	Small scales, large external ear and a 'helmeted' scale formation around the head	In Cantonese, the name pangolin translates to 'animal that bores through the mountain'. In Chinese legend, pangolins are said to travel all around the world underground
Indian (*Manis crassicaudata*)	Bangladesh, India, Pakistan and Sri Lanka	E	78 - 120	8.0 - 17	Fossorial; urban areas, tropical forests, shrub and grassland	Largest scales of the Asian species	The most western ranging of the Asian species
Sunda (*Manis javanica*)	Southeast Asia, including; Cambodia, Myanmar, Singapore, Thailand, Indonesia and Vietnam	CE	75 - 122	4.0 - 14	Semi-arboreal; forests and agricultural areas - including palm oil and rubber plantations	Can have distinctive white scales on the lower half of their tail. This species differs from the Philippine pangolin as it has a longer head and body to tail length ratio	The most widely distributed Asian pangolin species

V vulnerable, *E* endangered, *CE* critically endangered

suggest that one adult pangolin can consume more than 70 million insects in a year.

Little is known about pangolin reproduction, and gestation periods vary by species. In general, one offspring is produced per year. Pangolins are solitary, thus males and females meet only to mate. Males mark their location using both urine and feces. A young pangolin stays with its mother for 3–4 months, clinging to or riding on her tail for protection when away from the burrow.

Could detection dogs help enhance knowledge and datasets about pangolins?

All eight species of pangolin are affected by habitat destruction and fragmentation. Ground-dwelling African species are additionally threatened by land-management practices such as electric fencing and pesticide use. The most severe threat to pangolin populations is poaching, particularly in Asia. Emerging trends over the past decade indicate that African pangolins are now targeted for intercontinental trade to meet the demands of Asian markets, as Asian populations are depleted. The International Union for the Conservation of Nature (IUCN) estimates that over one million pangolins were trafficked between 2000 and 2013, though given the difficulty of obtaining true trade data, this estimate is considered conservative. Over 120 tons of whole pangolins and pangolin parts were seized from international trade between 2010 and 2015.

Detection dogs are already used at ports of entry and during inspections to intercept concealed pangolins and scales. Several CDDHT surveys have been undertaken in hopes of contributing to the meager ecological data pool for pangolins. However, in the wild, it has proven challenging for dog teams to detect evidence of live pangolins or their sign to date. As alluded to elsewhere in the chapter, in order for dog-handler surveys to be effective, there must be a certain level of sample availability and general natural history knowledge to steer initial detection efforts, which is currently lacking for pangolins - at least from conservationists and researchers. Scat is a conventional ecological monitoring target, which may explain its focus in numerous prior surveys. Yet, procuring pangolin scat for training is difficult. And in the wild, scat may rapidly degrade, which reduces a person's ability to confirm finds (and hence, reward opportunities for dogs) and sample viability for genetic testing to species. However, in theory, there is an abundance of better types of training samples available (e.g., confiscated scales and live animals). Does this perhaps mean that project leaders currently do not have at their disposal the means or contacts to procure access to such alternative samples?

The pangolin is currently facing unsustainable pressure from poaching, habitat loss, and fragmentation. In concert with enforcement and forensic efforts, more research is needed on the ground and within communities to more fully understand the ecology and behavior of wild pangolins and to begin to protect them. A variety of survey techniques implemented in an attempt to learn more about their habits such as diet, preferred habitat, and social ecology have yielded variable or limited data. Much like fingernails, the pangolin's keratinous scales grow continuously and are smooth in texture, which can cause radio transmitters to slip off in the field and provides only temporary windows of data collection. Unlike leopards or whales, which can be identified individually by their unique spot patterns or scars, respectively, pangolins do not have physical features that lend themselves to individual visual identification, making camera trap identification virtually impossible.

With targets like this about which so little is known, CDDHTs could in principle be used to counter survey bias, detecting

(continued)

free-ranging pangolins in unexpected places, that might not otherwise be searched. On the one hand, the current dearth in available knowledge also confers *biological significance*, wherein detecting even a few individuals would benefit conservationists who have little information on wild pangolins. Unfortunately, the very same scarcity of information can also limit the feasibility and the efficacy of CDDHTs or prohibit their use altogether, since a certain amount of capacity and support is required, including access to trusted samples (i.e., genetically confirmed fecal matter, known burrows, or live animals) for training, fielding, and general troubleshooting. But is information truly that scarce? What if a central part of our inability to reliably find evidence of pangolin stems from lack of communication and insufficient information gathering from those operating on the ground, with the most direct and best knowledge of pangolin habitat and the species itself?

Three fundamental questions to consider in relation to general ecological monitoring and CDDHT implementation within ongoing efforts are:

1. *How are poachers able to find so many pangolins where others seem to falter?*
2. *Can the knowledge of poachers and traffickers be harnessed in some way, drawing from other forms of intelligence gathering, to alleviate the strain on the pangolin populations before it is too late?*
3. *How would any information gained through CDDHT surveys be used advantageously to confer tangible benefits upon ongoing pangolin conservation efforts?*

If decisions around the training and use of CDDHT surveys are indeed being driven by assumptions that do not reflect the conditions and realities on the ground, and without greater communication and inclusion of local knowledge in study design and execution, it will be difficult to move forward productively.

Conservation Detection Dogs as Ambassadors

The complimentary role that conservation detection dogs play as ambassadors—drawing positive attention, sparking excitement, and creating unique openings for dialogue between and amongst stakeholders and communities—cannot be overstated. However, we note that in some places there can be a large disconnect between the work and presence of CDDHTs and the wider community. Dogs are feared in some areas, but a first interaction with them in a safe, controlled context can be an uplifting experience and change an individual's views of dogs, and by extension, of conservation and its importance (Fig. 13a–c). As with the boating public, passengers aboard buses in Africa appreciate that submitting to dog team rather than "by hand" inspections at checkpoints means they can be on their way in a matter of minutes rather than hours. The mere presence of the teams can also be a powerful deterrent.

In areas where species are reviled and persecuted, engaging with a dog and handler can plant a seed in children that there are many different ways to be a scientist or biologist aside from wearing a white coat and working in a laboratory with test tubes and beakers. It also gives children who like hunting, fishing, and other outdoor pursuits an opportunity to see ways to make use of their skills and passions while simultaneously protecting the resources that bring them enjoyment (Fig. 14).

In terms of generating public awareness and support, the charismatic nature of dogs is second to none. Using the previously described example of invasive mussels, the involvement of dogs in

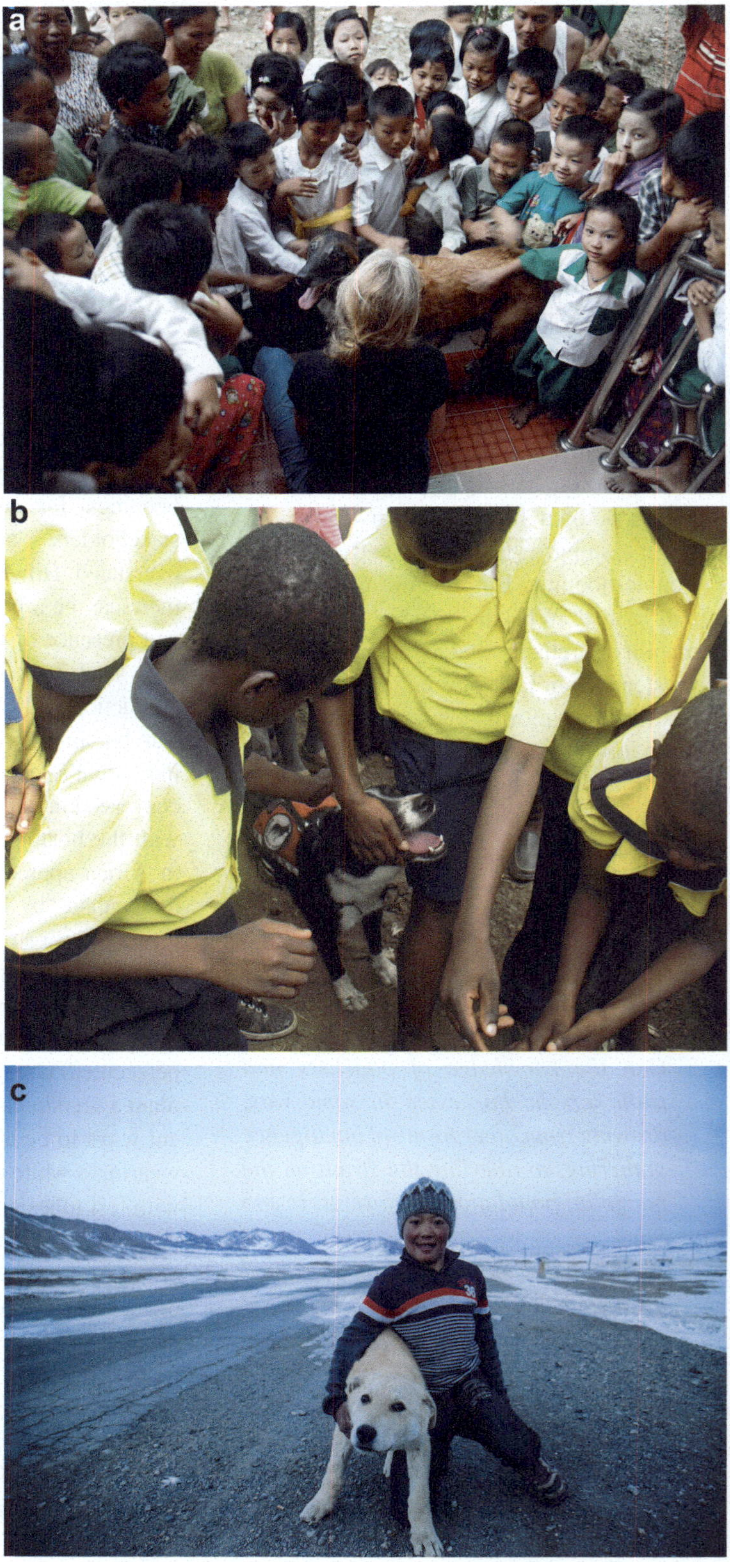

Fig. 13 Interacting with dogs in a safe, controlled space can be an uplifting experience. It may change an individual's views of dogs, and by extension, of conservation and its importance. (**a**) Credit: Simon Hedges/WCS, copyright owner. (**b**) Credit: WD4C, copyright owner. (**c**) Credit: Joel Caldwell, copyright owner

Fig. 14 Educational outreach material created by Woodland Park Zoo (USA) and distributed in their Spring 2018 members bulletin

inspections may tip the scales toward compliance. This has also been observed with a suite of harmful species nonetheless considered underwhelming by the general public in North America, a case in point being the emerald ash borer beetle, also discussed earlier.

These animals need Benny's help!

The animals below are often victims of wildlife trafficking.

Can you help Benny identify each species?

Unscramble the letters and write out the name of each animal. Then copy the letters from the color coded boxes into the same colored box below to reveal the mystery message.

1. DCGKOEU  _ _ _ _ _ _ _ K

2. ESA URBCMCEU  _ E _ _ _ C _ _ _ _ _

3. GANNOLIP  _ _ _ G _ _ _ _

4. ALKCB EBRA  _ _ _ _ _ B _ _ _

5. SCHIOERNRO  _ _ _ _ O _ _ _ _

6. KRHSAS DAN YARS  S _ _ _ _ _ _ _ _ _ _ _

The mystery message is…

_ _ _ _ _ _ _, _ _ _ _ _ !

Part of Benny's training involves a reward: time with his favorite ball!

Dogs can smell separately with each nostril. WHOA.

Answers, page 28 :

Geoducks—some people sell these sea creatures illegally

Black bear parts—poaching black bears is against the law

Tiger pelt—tigers are endangered

Pangolin scales—pangolins are the most trafficked *animal*

Sea cucumbers—these creatures from the Pacific North*west are vulnerable to illegal trade*

Answers, page 29 : *1. Geoduck, 2. Sea cucumber, 3. Pangolin, 4. Black bear, 5. Rhinoceros, 6. Sharks and rays.*

Mystery message: Good Boy, Benny!

Fig. 14 (continued)

An Opportunity to Pool Knowledge and Share Resources

Ultimately, the degree to which any monitoring program can be conducted, and any lasting protection measures can be successfully implemented, depends to a large extent on the resources which stakeholders, governments, and other assorted partners have at their disposal. The ways such resources are disbursed are also key. Conservation dog professionals can and do advise on how best to allocate or use scant resources to optimize both detection efficacy outcomes and conservation efforts.

Training Samples and Multi-target Training

The nature of the conservation issues we seek to address and the protected status accorded a given species often imposes an inherent difficulty in acquiring viable training samples. In this chapter, we have highlighted a selection of targets to which detection dogs have been trained for ecological and environmental monitoring, and for parallel forensics and enforcement initiatives. Pooling available samples among groups or entities to facilitate training can help immeasurably. Alternatively, the reality may simply be that much of the training must be conducted in-situ, for which a suitable period of acclimation and troubleshooting will be required. Consideration could also be given to the appropriateness of more easily obtained "proxy" targets, which enables complimentary information to be obtained without having to train directly to a given species, for which samples may be too challenging to obtain.

Detection Dog Programs

The prospect of starting one's own canine program may seem tantalizing. However, existing regional detection dog capacity ought to be considered first, in conjunction with potential resource pooling and networking to meet common goals. Programs should be undertaken when the requisite long-term support, infrastructure, and capacity has been secured. In this regard, the acquisition of dogs and suitable facilities to house them is just the beginning, with success contingent upon continual, sustained support (veterinary care; housing; dietary concerns; daily fitness and well-being). We reiterate here that the use of CDDHTs is optimized when there is acknowledgment and understanding of when and how best to use them in relation to other existing and available information-gathering tools. This critical consideration may be overlooked at the exploratory stages of a program since many times dogs are seen as a "silver bullet," without the understanding that the rest of the pieces and resources need to be in place to support the canine team(s) and utilize them in their fullest capacity. Further details related to logistics and requirements, factoring in that standards and guidelines may differ by region and country, have been considered in a series of reports and audits on African programs (unpublished data, available from WD4C by request).

Development of In-house Dog-Handler Teams

It may be tempting and even considered fiscally responsible or good for community relations, to contemplate the use and viability of biologists paired with their own pet dogs, or of hunters paired with their personal hunting dogs. On the surface, these may seem economical as opposed to developing the capacity for bringing on a trained conservation detection dog-handler team, flying in a dog from an established program and arranging in-person training and support for the dog with prospective handlers, or bringing in a conservation dog professional for in-person assistance and capacity development. However, as we have outlined throughout this chapter, opting to forego professional support and consultation, which includes pre-vetting of dogs and prospective handlers, is often a false economy. Attempts to avoid professional training may put both wildlife and the program itself at risk. CDDHTs come with built-in quality control protocols and continual risk assessment and mitigation standards, in order to maintain the safety of the team at all times and ensure their presence is both as noninvasive and respectful to the environment in which surveys are taking place as possible. A valid concern is that, separate from the scent target sought, focal and non-focal animals may be physically harmed if dogs without the requisite temperament and traits are deployed on surveys. Additionally, sample integrity may be compromised, and/or the quality of samples and observations may be called into question, which is in opposition to the objectives of conservation and sets the detection field back. In a more strictly forensic or enforcement realm, this could nullify or jeopardize evidence recovered by the team.

Coordinated strategies and networks, forensic techniques and tools are continually being developed and adapted for use against the unrelenting menace of poaching, to intercept smuggling and trade of living animals, animal parts, and contraband items, and to suppress wildlife poisonings, among many other actions. It is worth reiterating here that, providing there is no interference or incompatibility with the primary purpose or mandate of a given dog, additional targets can potentially be layered on to nearby existing teams (including those operating in a large enforcement capacity) to maximize their impact and distribute available resources.

Conclusions

As conservation detection dog handlers who have conducted conventional ecological monitoring and participated in some of the previously described overlapping applications, we have seen firsthand how judiciously implemented and strategically executed detection dog-handler team surveys generate information and insights that could not have been obtained through other means. By helping to demarcate the boundaries of our knowledge, detection dogs, aided by their handlers, further understanding regarding the needs of, and the pressures being exerted upon, species and ecosystems.

We engage with residents of affected communities and with government representatives and operate within the sphere of non-governmental organizations and conservation affiliates worldwide. We are also eminently familiar with the limited resources typically available for conservation needs. Bearing all these factors in mind, it is our assessment that within this realm we could be of much greater assistance to those laboring at the wildlife forensics and enforcement end of the spectrum—by according a larger supporting role to, and extensively coordinating and corralling, existing ecological and environmental monitoring efforts and entities. The overriding aim is to strengthen our ability to *prevent* the demise of species and habitats through the acquisition of scant baseline and existing dataset information, thereby lessening the weight of the response at the opposite end of the spectrum. Information acquired can also be used to nurture and strengthen intermittent measures, e.g., where still-living organisms have been confiscated for immediate or eventual release.

The strategic incorporation of CDDHTs may extend a lifeline to researchers and conservation groups eager to learn more about specific species, especially those that are rare or elusive. However, this in no way means dog-handler teams are always the best and only means of gathering data. Sometimes at certain stages of discovery, this is true, however, overwhelmingly, judicious timing in the use of dog-handler teams can provide pivotal, cost-cutting (or, alternately, cost-justifying) insights and inform further study design. In this regard, conservation dog professionals have a comprehensive sense of how dog surveys can be used to best advantage, in relation to other available techniques, or even whether the use of dogs is warranted. If consulted early in the planning process, e.g., at the brainstorming and grant-writing stage, their knowledge and perspectives may be invaluable to potential partners or would-be funders in optimizing study design and data-gathering. Similarly, when results are to be disseminated as a publication, allowing conservation dog professionals to contribute to the study design and supporting methodology, and later to articulate its implementation in their own words, may further limit the perpetuation of misconceptions. They may also have valuable networks through which information and messaging can be further and more effectively disseminated.

Extending lasting protections to the biota and ecosystems of any region spans a range of tools, strategies, and partners, including local communities. The continuum begins with the acquisition of a profound understanding of populations and occupancy through concerted ecological/environmental monitoring and culminates in protective forensic and enforcement measures. The illegal trade and traffic of flora and fauna is a multi-billion-dollar industry with many resources and creative outlets to avoid detection. It takes years to enact laws to provide protection for at-risk species and involves the collaboration and coordination of multinational agencies in order to be successful. While resources allocated for conservation are few—especially in contrast to the coffers of those operating clandestinely—and can be difficult to secure, collaborations and resource-sharing among agencies, organizations and communities would benefit the preservation of species. Through it all, conservation dog professionals and the detection dogs they partner with stand ready, willing, and able to assist.

Summary and Recommendations:

Conservation detection dog-handler teams are a unique, standalone monitoring method and should be recognized as such. Study design and objectives must reflect optimal usage of the teams for data-gathering to ensure successful, efficient, and cost-effective outcomes.

Where sampling is concerned, clear communication between investigators and laboratory partners (e.g., for genetic and toxicology analyses, among others) is required in order to ensure the best and most up-to-date protocols for correct sample collection, storage, and transfer.

Deployment on multiple complimentary targets is one way to get more value from surveys and teams and may lead to unexpected and useful findings, even potentially to additional partnerships if additional data interpretation is sought.

Drawing upon the ambassador role of the dogs whenever possible can significantly heighten stakeholder engagement and create unique opportunities for dialogue.

Common misconceptions and erroneous perceptions reduce the effectiveness of conservation detection dog-handler teams. These can be cleared through consultation with established conservation dog professionals, starting from the study design phase and following through to the field and beyond, to data analysis, interpretation of results and their dissemination. Handlers and related detection dog professionals are not only service providers but potential stakeholders with insight into how dogs can be most effectively used. These individuals also harbor a genuine desire to make a meaningful contribution to conservation efforts and to securing the success of a project, for its own sake, and to benefit the conservation detection field as a whole.

The views put forward in this chapter reflect the authors' common and distinct experiences. Other conservation dog professionals may have contrasting perspectives, and we welcome their additional input to this evolving discussion.

Acknowledgments This chapter is dedicated to all those working on the very front lines to protect our wild places and species, whether they be two-legged or four-pawed. We extend our appreciation and gratitude to the people who make the sacrifices and commitment to embark on partnerships with dogs and all that this work entails. Many of our working dogs have been given a second chance, rescued from an uncertain or short future, and are now engaged in a job they relish and are uniquely qualified to do. The exuberance with which all of them approach the work is a veritable joy to behold.

Earlier versions of the chapter manuscript were greatly improved by input from André Botha (Endangered Wildlife Trust); Sergio Lambertucci, Pablo Plaza and Lucía Zamora (Instituto de Investigaciones en Biodiversidad y Medio Ambiente, National Scientific and Technical Research Council, Argentina); Nick Ahlers (TRAFFIC) and Nancy Harrison (formerly Anglia Ruskin University). Many thanks to these individuals for taking the time to

discuss the topics and issues outlined in the chapter with us and to provide their frank, invaluable feedback.
All photos have been used with permission, and we salute the photographers who strengthened our words and sentiments through their images. Special thanks to Isabelle Groc (www.isabellegroc.com), Joel Caldwell (www.joelcaldwell.com), Iñigo Fajardo, and the Department of Fisheries and Oceans (DFO) for permission to use their beautiful photographs.
We remember Pepin, Wicket, Steven, Denali, Maxwell, and Scooby. They taught us everything we thought we should know, and so much more. This chapter honours the memory of our colleague and dear friend Christine Salisbury.

Bibliography

Aisher A (2016) Scarcity, alterity and value: decline of the pangolin, the world's most trafficked mammal. Conserv Soc 14(4):317–329

Akrin F, Mahmood T, Hussain R, Qasim S, Zangi I (2017) Distribution pattern, population estimation and threats to the Indian pangolin Manis crassicaudata (Mammalia: Pholidota: Manidae) in and around Pir Lasura National Park, Azad Jammu and Kashmir, Pakistan. J Threat Taxa 9(3):9920–9927

Alarcón PAE, Lambertucci SA (2018) Pesticides thwart condor conservation. Sci Mag 360(6389):612. https://science.sciencemag.org/content/360/6389/612.1. Accessed 04 December 2019

Arandjelovic M, Bergl RA, Ikfuingei R, Jameson C, Parker M, Vigilant L (2015) Detection dog efficacy for collecting faecal samples from the critically endangered Cross River gorilla (*Gorilla gorilla diehli*) for genetic censusing. R Soc Open Sci 2:140423. https://doi.org/10.1098/rsos.140423

Avery ML, Humphrey JS, Keacher KL, Bruce WE (2014) Detection and removal of invasive Burmese pythons: Methods development update. In: Timm RM, O'Brien JM (eds) Proceedings of the 26th Vertebrate Pest Conference. University of California, Davis, Davis, CA, pp 145–148

Bennett EM, Hauser CE, Moore JL (2020) Evaluating conservation dogs in the search for rare species. Conserv Biol. 34(2):314–325

BirdLife International (2019) BirdLife calls for urgent high-level support for African vultures; BirdLife International strongly condemns the recent poisoning of vultures in the country, and calls on the government to take action to prevent further similar incidents https://www.birdlife.org/worldwide/news/birdlife-calls-urgent-high-level-support-african-vultures. Accessed 04 Dec 2019

Boakye MK, Pietersen DW, Kotzé A, Dalton D-L, Jansen R (2014) Knowledge and uses of African pangolins as a source of traditional medicine in Ghana. PLoS One 10(1):e0117199

Braun B, Stuart J, Chhabra DP, Choudhary AN, Wendt L (2018) A glimpse into the use of dogs to address global poaching, overharvesting, and trafficking of aquatic species. In: Richards NL (ed) Using detection dogs to monitor aquatic ecosystem health and protect aquatic resources. Palgrave Macmillan, Cham, Switzerland, pp 263–286

Challender D (2009) Asian Pangolins: how behavioural research can contribute to their conservation. In: Proceedings of the workshop on Trade and Conservation of Pangolins Native to South and Southeast Asia, pp 95–102

Challender DWS, Waterman C (2017) Implementation of CITES Decisions 17.239b) and 17.240 on Pangolins (Manis spp.). Prepared by IUCN for the CITES Secretariat. https://cites.org/sites/default/files/eng/com/sc/69/E-SC69-57-A.pdf

Challender DWS, Baillie JEM, Waterman C, Pietersen D, Nash H, Wicker L, Parker K, Thomson P, Nguyen TV, Hywood L, Shepherd CR (2016) On scaling up pangolin conservation. TRAFFIC Bull 28(1):19–21

Dahlgren DK, Elmore RD, Smith DA, Hurt A, Arnett EB, Connelly JW (2012) Use of dogs in wildlife research and management. In: Silvy NJ (ed) Wildlife techniques manual: research, vol I. John Hopkins University Press, Baltimore, MA, pp 140–153

Dayer AA, Redford H, Campbell KJ, Dickman CR, Epanchin-Niell RS, Grosholz ED, Hallac DE, Leslie EF, Richardson LA, Schwartz MW (2019) The unaddressed threat of invasive animals in U.S. National Parks. Biological Invasions, Published Online

Fajardo I, Ruiz A, Zorrilla I, Valero A, Fernández I, Sáez E, Molino FM, Olivares J (2012) Use of specialized canine units to detect poisoned baits and recover forensic evidence in Andalucía (Southern Spain). In: Richards NL (ed) Carbofuran and wildlife poisoning: global perspectives and forensic approaches. Wiley-Blackwell, London, pp 147–155

Fernández I, Zorrilla I, Richards NL Detection of illegally used pesticides in entomofaunal specimens recovered from suspected environmental scenes of crime in southern Spain – dual applications for wildlife forensics and complimentary ecological monitoring. Submitted to Forensic Science International. (In review)

Geneva International Centre for Humanitarian Demining (GICHD) (2004) Training of mine detection dogs in Bosnia and Herzegovina (NPA Global Training Centre). Switzerland, Geneva. https://www.gichd.org/fileadmin/GICHD-resources/rec-documents/Training_of_MDD.pdf. Accessed 19 Feb 2019

Goodwin KM, Engel RE, Weaver DK (2010) Trained dogs outperform human surveyors in the detection of rare spotted knapweed (Centaurea stoebe). Invas Plant Sci Manag 3:113–121

Heinrich S, Wittmann TA, Ross JV, Shepherd CR, Challender DWS, Cassey P (2017) The global trafficking of pangolins: a comprehensive summary of seizures and trafficking routes from. TRAFFIC,

Southeast Asia Regional Office, Petaling Jaya, Selangor, Malaysia, pp 2010–2015
Helton WS (ed) (2009) Canine ergonomics: the science of working dogs. CRC Press, Boca Raton, FL
Hoyer-Tomiczek U, Sauseng G, Hoch G (2016) Scent detection dogs for the Asian longhorn beetle, Anoplophora glabripennis. EPPO Bull 46:148–155
Hulme PE (2009) Trade, transport and trouble: managing invasive species pathways in an era of globalization. J Appl Ecol 46:10–18
Hurt A, Smith DA (2009) Conservation dogs. In: Helton WS (ed) Canine ergonomics: the science of working dogs. CRC Press, Boca Raton, FL, pp 175–194
Hurt A, Guscio D, Tirmenstein DA, Richards N, Burch A, Marler M (2016a) Using search dogs for biological eradication programs – a tale about dyer's woad (*Isatis tinctoria* L.). In: Schwarzländer M, Gaskin JH (eds) 2014 proceedings of the 3rd Northern Rockies Invasive Plants Council Conference February 10–13, 2014. Airway Heights, WA. USDA Forest Service, Forest Health Technology Enterprise Team, Morgantown, West Virginia. FHTET-2016-03, pp 73–86
Hurt A, Woollett (Smith) DA, Parker M (2016b) Training considerations in wildlife detection. In: Jezierski T, Ensminger J, Papet LE (eds) Canine olfaction science and law: advances in forensic science, medicine, conservation, and environmental remediation. CRC Press, Taylor and Francis Group, Boca Raton, FL, pp 139–153
Ingram DJ et al (2017) Assessing Africa-wide pangolin exploitation by scaling local data. Conserv Lett. https://doi.org/10.1111/conl.12389
Jackling ME, Gansowski JT, Hojnacki DM, Gosser AL (2016) Overview of Feral Swine Management and Elimination Efforts in New York (2008-2016). In: Timm RM, Baldwin RA (eds) Proceedings of the 27th Vertebrate Pest Conference. University of California, Davis, Davis, CA, pp 140–146
Jezierski T, Ensminger J, Papet LE (eds) (2016) Canine olfaction science and law: advances in forensic science, medicine, conservation, and environmental remediation. CRC Press, Boca Raton, FL
Johnston JM (1999) Canine detection capabilities: operational implications of recent R & D findings. Institute for Biological Detection Systems. Auburn University, Auburn
Kret E, Vavylis D, Saravia V, Ntemiri K (2015) Poison bait detection with specially trained dogs in Thrace and Central Greece, Annual report 2014. In: Technical report under action C1 of the LIFE+ project "The Return of the Neophron" (LIFE10 NAT/BG/000152). Hellenic Ornithological Society & WWF-Greece, Athens, 41 p
La Guardia M, Richards NL, Hale RC (2020) A noninvasive environmental monitoring tool for brominated flame-retardants (BFRs) assisted by conservation detection dogs. Chemosphere 260:127401
Long RA, Donovan TM, Mackay P, Zielinski WJ, Buzas JS (2007) Comparing scat detection dogs, cameras, and hair snares for surveying carnivores. J Wildl Manag 71:2018–2025
MacKay P, Smith DA, Long RA, Parker M (2008) Scat detection dogs. In: Long RA, MacKay P, Zielinski WJ, Ray JC (eds) Noninvasive survey methods for carnivores. Island Press, Washington, DC, pp 183–222
Martins DJ (2012) Repercussions of pesticides (including carbofuran) on nontarget, beneficial insects and use of insects in forensic analyses in Kenya. In: Richards NL (ed) Carbofuran and wildlife poisoning: global perspectives and forensic approaches. Wiley-Blackwell, London, UK, pp 81–84
Massé F (2019) Anti-poaching's politics of (in)visibility: representing nature and conservation amidst a poaching crisis. Geoforum 98:1–14
Matre PJ (2003) Training, organisation and skills: case studies of practice using mine detection dogs. In: McLean IG (ed) Mine detection dogs: training, operations, and odour detection. Geneva International Centre for Humanitarian Demining, Geneva, pp 139–163
National Oceanic and Atmospheric Administration (NOAA) (2020) New England Fishery Science Center Fish FAQ. https://www.nefsc.noaa.gov/faq/faq-archive/fishfaq7.html. Accessed 19 Feb 2020
Novoa A, Kaplan H, Wilson JRU, Richardson DM (2016) Resolving a prickly situation: involving stakeholders in invasive cactus management in South Africa. Environ Manag 57:998–1008
Odino M (2012) Measuring the conservation threat that deliberate poisoning poses to birds in Kenya: the case of pesticide hunting and fishing with Furadan in the Bunyala Rice Irrigation Scheme. In: Richards NL (ed) Carbofuran and Wildlife poisoning: global perspectives and forensic approaches. Wiley-Blackwell, London, UK, pp 53–70
Ogada D (2014) The power of poison: pesticide poisoning of Africa's wildlife. Ann NY Acad Sci 1322:1–20. https://doi.org/10.1111/nyas.12405
Ogada D, Botha A, Shaw P (2015) Ivory poachers and poison: drivers of Africa's declining vulture populations. Oryx. https://doi.org/10.1017/S0030605315001209
Regional Public Works Commission of Ontario (RPWCO) (2019) Estimated expenditures on invasive species by Ontario municipalities and conservation authorities. Infographic report prepared by RPWCO. https://yorkpublishing.escribemeetings.com/filestream.ashx?DocumentId=8474. Accessed 23 Feb 2021
Reynolds K, Reynolds S (2018) In situ detection of sewage pollution and its sources in aquatic ecosystems. In: Richards NL (ed) Using detection dogs to monitor aquatic ecosystem health and protect aquatic resources. Palgrave Macmillan, Cham, Switzerland, pp 167–192
Richards NL, Zorrilla I, Fernandez I, Calvino M, Garcia J, Ruiz A (2015) A preliminary assessment of the palate and tongue for detecting organophosphorus and carbamate pesticide exposure in the degraded carcasses of vultures and other animals. Vulture News 68:32–51

Richards NL, Ogada D, Buij R, Botha A (2017) The killing fields: The use of pesticides and other contaminants to poison wildlife in Africa. Encycl Anthropocene 5:161–167

Richards NL, Tomy G, Kinney CA, Nwanguma FC, Godwin B, Woollett DA (2018) Using scat detection dogs to monitor environmental contaminants in sentinel species and freshwater ecosystems. In: Richards NL (ed) Using detection dogs to monitor aquatic ecosystem health and protect aquatic resources. Palgrave Macmillan, Cham, Switzerland, pp 193–262

Sawchuk C (2018) Intercepting invasive invertebrate species before they infest waterbodies: the inception and implementation of Alberta's dedicated Canine Mussel Program. In: Richards NL (ed) Using detection dogs to monitor aquatic ecosystem health and protect aquatic resources. Palgrave Macmillan, Cham, Switzerland, pp 119–163

Statham MJ, Woollett (Smith) DA, Fresquez S, Pfeiffer J, Richmond J, Whitelaw A, Richards NL, Westphal MF, Sacks BN (2020) Noninvasive identification of herpetofauna: pairing conservation dogs and genetic analysis. J Wildl Manag 84:66–74. https://doi.org/10.1002/jwmg.21772

Sutherland C, Fuller AK, Royle JA, Hare MP, Madden S (2018) Large-scale variation in density of an aquatic ecosystem indicator species. Sci Rep 8(8958). https://doi.org/10.1038/s41598-018-26847-x

Syrotuck WG (1972) Scent and the scenting dog. Arner Publishing, Rome

United Nations Office on Drugs and Crime, UNODC (2016) World Wildlife Crime Report: Trafficking in protected species

US Department of Homeland Security, Canine Program (2020). https://www.cbp.gov/border-security/canine-program. Accessed 24 Feb 2020

US Department of Homeland Security, CBP Search Authority (2020) https://www.cbp.gov/travel/cbp-search-authority. Accessed 19 Mar 2020

Vavylis D, Kret E, Ntemiri K, Saravia V (2018) Dogs for conservation: results of the antipoison dog units in central Greece and Thrace during 2014-2018. In: 9th Congress of the Hellenic Ecological Society. Heraklion, Crete, pp 113–114

Vesely D (2008) Training of conservation detection dogs to locate Kincaid's Lupine (*Lupinus sulphureus ssp. kincaidii*). USFWS report

Vyn RJ (2019) Estimated expenditures on invasive species in Ontario: 2019 Survey Results. https://www.invasivespeciescentre.ca/Portals/0/Documents/Economic%20Report/Final%20Report%20-%202019%20Survey%20Results%20V2.pdf?ver=2019-10-09-174853-000%3E%3Cspan%20style= . Accessed 12 Feb 2020

Woollett D, Hurt A, Richards NL (2014) The current and future roles of free-ranging detection dogs in conservation efforts. In: Gompper ME (ed) Free-ranging dogs and wildlife conservation. Oxford University Press, Oxford, pp 239–264

Processing the Wildlife Crime Scene and Evidence of Forensic Importance

Roderick B. Potter and Susan C. Underkoffler

Abstract

Wildlife crime is increasingly recognized as a growing threat to wildlife species. It can take many forms, from illegal wild capture, trading, and selling online for the exotic pet trade, to harvesting protected species for traditional medicine and ornamental products, to poaching for various reasons. Investigations into these illicit activities are often nonexistent or inferior at best, as many wildlife officials, game wardens, conservation professionals, and veterinarians do not have an adequate background in crime investigation or any familiarity with forensic science and its applicability to wildlife, environmental, or conservation violations. Additionally, wildlife crime scenes present unique challenges and thus may not resemble the preconceptions of crime scenes as portrayed in movies and on television, resulting in confusion for investigators. A lack of awareness and training may: compromise an investigation; destroy a crime scene; render evidence or items of importance unrecognizable; and may ultimately allow the perpetrator(s) to escape justice. It is vitally important that all participants involved be aware of the proper procedures and actions to employ in order to avoid such mistakes, as all actions taken by participants at a suspected crime scene will have either a positive or negative impact on the outcome of the investigation. The goal of this chapter is to present an overview of wildlife crime scene investigation, issues, and problems with this discipline, general considerations and principles of crime scene management that are common to all crime scenes, and some specific actions and evidence that are distinct for crimes involving animals.

Keywords

Crime scene · Wildlife forensic science · Investigation · Wildlife crime · Evidence · Forensic investigation

Introduction

What is a crime scene? All people who deal with wildlife matters must be able to recognize a possible crime scene. Wildlife crime scenes can take many forms. Some examples include crime scenes which are: spread over several kilometers in densely wooded areas; several fathoms deep in the ocean; encompassing the bed of a lorry, or pickup truck (Fig. 1), or a stand-alone garage freezer; in an outbuilding where makeshift

R. B. Potter (✉)
Wildlife Crime Investigations, Pietermaritzburg, South Africa

S. C. Underkoffler
William R. Maples Center for Forensic Medicine, University of Florida, Gainesville, FL, USA
e-mail: sunderkoffler@ufl.edu

S. C. Underkoffler, H. R. Adams (eds.), *Wildlife Biodiversity Conservation*,
https://doi.org/10.1007/978-3-030-64682-0_12

Fig. 1 Marine poachers intercepted with a load of lobsters. This truck may constitute an initial crime scene

butchering or taxidermy takes place; a market stand displaying bushmeat or in shipment crates confiscated at ports of entry. The myriad of possible scenarios can be confusing, and for professionals without a background in forensic science can prove daunting—how should this be documented? What evidence needs to be collected? A crime scene is any area where a crime may have been committed or where there may be evidence of a crime, irrespective of location or size. Having a standard methodology for determining what constitutes a scene, for preserving and documenting the scene, and for collecting evidence regardless of scenario will alleviate confusion and enable an investigator to effectively walk step by step through the procedure, eliminating as many errors as possible (Fig. 2).

Crime scene investigation has often been compared to solving a jigsaw puzzle—"clues," or physical, tangible evidentiary items and testimonial statements from witnesses or others involved constitute the puzzle pieces that may fit together or maybe discarded as irrelevant. As an investigator systematically searches for the truth about an incident, these puzzle pieces, whether from people or objects, will guide the investigation. Clues that are collected by the investigator in accordance with strict rules can become evidence. This evidence can be used to present a version of what happened at the scene to the court. But if these clues are unrecognized or ignored, there will be no case at all. The purpose of this chapter is not to serve as a step-by-step processing guide, as there are many such resources available (see Gardner and Krouskup 2019; Saferstein 2019; Ogle and Plotkin 2018; Cooper and Cooper 2013; Cooper et al. 2009; Fisher and Fisher 2012). However, some of the primary steps will be discussed. This chapter is mainly designed as an overview and summary of the current state of the discipline, a description of the applicability and utility of scene investigation techniques to wildlife crime, and a brief introduction to scene concerns and items of forensic importance associated with wildlife crime scenes. The goals are to unite the fields of wildlife and conservation forensics with crime scene investigation and to provide a familiarity with techniques that may assist a professional in the investigation of a potential crime.

Problems with the Wildlife Crime Scene Investigation Discipline

There are many issues slowing the progress of the overall acceptability and utilization of crime scene investigation techniques in wildlife cases. Many are obvious, but some are worth exploring in more detail. It is important to note that all of the following challenges/problems listed are considered to be interrelated.

Fig. 2 Illegal hunting trophy; sometimes a crime scene may involve a single shipping crate and its contents

Perceptions

In the authors' experience, traditionally, and in regular practice, crime scene investigation pertaining to animal or environmental crime is not taken seriously or considered worth much effort. It is often overlooked or practiced in a haphazard matter, with disregard for many of the steps and processes utilized in human crime scene processing. Often, the practice of crime scene investigation is not utilized at all. This can be due to many factors including: a lack of awareness, knowledge, or training; lack of manpower or personnel; a belief that it will not make much difference to the case outcome; pressure from outside agencies to wrap the matter up quickly; or general malaise or feeling that animal or conservation crimes are not as important as human crimes.

This is to the detriment of all of those involved, as well as the entire field of wildlife forensic science, because, as will be pointed out in later sections, proper investigation begins at the crime scene. Although in some circumstances cases may make it to court and even be adjudicated without following proper crime scene procedures, this is due in part to a lack of awareness on the part of attorneys and judges as to the correct protocols, forensic science techniques, and even the law as it pertains to animal crimes. It may be only a matter of time before the prosecution of wildlife crime becomes more commonplace and those in the legal profession become more experienced and astute. A clever defense attorney exposing mistakes made or shortcuts taken on the crime scene will also expose investigators previously emboldened by successes in court to professional discrediting and complete failure of their cases.

Lack of Awareness, Training, and Knowledge

In many if not most countries, basic forensic science principles are not routinely taught, especially in conservation or veterinary programs, and the only knowledge of forensics comes from dramatized television shows. The use of forensic science in anything but human crime is relatively new, albeit increasing. However, while true that human crimes take precedence, more pressure is being put on conservation and animal management and advocacy organizations to take crimes against animals and the environment seriously, pass more restrictive laws, increase the amount of prosecutions, and enforce harsher penalties. "Many law enforcement agencies in range, transit and destination states require additional technical and financial assistance to address capacity gaps, including help to better prevent, detect and investigate wildlife trafficking, and better training and protocols on how to secure and deal with specimens once seized or confiscated." (UNODC, World Wildlife Crime Report). In order to do these things, more emphasis will need to be placed on quality investigations, and professionals working in these fields will need to have a familiarity with forensic science procedures.

In many STEM (science, technology, engineering, and math) curricula across North America, Europe, Asia, and Australia, forensic science and its disciplines are introduced as early as middle school, and even elementary school, with special activities designed to highlight forensic techniques applied in various scenarios, and there are innumerable undergraduate- and graduate-level forensic science programs at universities (Horan and Rothchild 1980; Houck and Siegel 2010). But these are typically stand-alone programs. Forensic theories and principles, even the most basic, need to be incorporated routinely into other disciplines, as they apply not only to criminal justice and law enforcement, but also to many other subjects including law, computer science, veterinary science, ecology, conservation biology, and wildlife management.

Many who encounter wildlife or conservation crimes will not be familiar with even basic forensic science protocols such as: wearing gloves and other protective equipment; not smoking or eating near a crime scene; not entering a crime scene or touching/moving any items; how to recognize a crime scene; and knowledge that a suspected crime scene must be approached with a strategic protocol. While the likelihood that livestock owners, game managers, wildlife reserve and park personnel, and zoo staff may be faced with a crime scene is small, having a basic understanding of forensic practices and skills could prevent the inadvertent destruction of evidence, and improve the chances of a positive legal outcome should an incident occur. It has been the authors' experience that small, in-person forensic trainings in the field are well received, not only by law enforcement, but also ranchers, game reserve staff, and individuals responsible for public and private land and national park management. More of these combined practical field- and classroom-based trainings should be emphasized as worthwhile or even required continuing education by organizations. And it should be made known that individuals working for themselves and not a company or formal association can benefit. The hallmarks of effective investigations using the forensic science doctrines of meticulousness, thoroughness, attention to detail, and an open-minded approach, can apply to any profession and any situation.

Although many who are the first to encounter potential wildlife and environmental crime scenes may not have a background or education in forensic science, it is vital to understand the importance of scene integrity and proper preservation.

Everything that occurs at a crime scene is destructive in nature; some of the damage that occurs may not be avoided, e.g., inclement weather events, or when life-saving measures necessitate a veterinarian to enter the scene and possibly disturb items of evidence, and most definitely impact the original condition of the scene. It is critical in such emergency circumstances, that the work carried out on the scene is properly documented by the person involved in order to clearly explain their impact on the scene to the investigator, as the ultimate objective is to minimize the damage and preserve as much as possible in as pristine a condition as possible. In the authors' experience, it is not uncommon to find those who are either first on the scene of a wildlife crime, or those arriving later whose task it is to investigate that crime, running into the middle of the scene to capture an excited "selfie" with the deceased animal. Not only does this potentially destroy any evidence along with the very context of the scene itself, but it undermines the integrity of investigations and the seriousness with which they need to be considered. Perhaps the most important concept in crime scene processing methodology is the knowledge that, at the very instant one realizes a crime may have occurred, he or she must take all possible precautions not to alter or move any item in the scene or change the context of the scene in any way.

Misunderstanding of What Is Important

It can be said that "forensic science begins at the crime scene," (Saferstein, Criminalistics, 30) not at the veterinary hospital or even at the laboratory. The use of DNA evidence is often touted as a golden standard; many people, including forensic science professionals, believe that without DNA evidence there is no case, what is often referred to as the "CSI effect"—an inaccurate perception of forensic investigations based on a misguided belief (encouraged by television shows) that all cases end in court with the presentation of irrefutable, conclusive evidence (DNA) that was obtained in a process lasting no longer than 50 min without commercials (Dowler et al. 2006). While no one can dispute the importance of DNA in crime solving (see chapter "Wildlife Forensic Genetics and Biodiversity Conservation: The Intersection of Science, Species Management, and The Law"), it is only one piece of the puzzle. In fact, without proper crime scene investigation, there would be no usable DNA evidence at all. If a crime scene is not properly secured and the correct processing procedures are not followed, DNA evidence can be collected improperly, destroyed, or missed entirely. If it is collected and then mishandled or not documented accurately, an attorney or judge can question the integrity of the evidence presented in court. And this is true of all potential evidence items, not just DNA. Essentially, a crime scene presents the one and only chance to get it right.

In addition, it is not necessary to have expensive supplies to properly process a crime scene; many of these items can be inexpensively improvised (see section "Evidence Supplies: Substitution and Improvisation") But, while having the proper materials for scene processing at hand (whether commercially purchased or improvised) is vital, a demonstrated competency with their use is equally essential. Knowing how to use a camera and all of its functions, for example, is paramount to taking quality crime scene photographs. If all one has available is the camera embedded within a cell phone, then by all means utilize it, but ensure that there is a proficiency in its use and with its features. If someone is uncomfortable with his or her expertise, they should pass the task on to someone more comfortable. Personal pride has no place in matters of crime investigation and working with others is essential (see section "Working with Others"). It is exceedingly difficult, if not impossible, to be an expert in all forensic disciplines, nor is it necessary. In addition to recognizing one's strengths and abilities, it is crucial that participants recognize their own limitations, in order to strengthen the team as a whole.

Responsibility

Also common is the confusion that can arise when deciding what agency or individuals are responsible for crime scene processing and evidence handling. In many instances, it is not a law enforcement officer who first discovers a crime scene or encounters an animal deceased under mysterious circumstances. It may be a conservation officer or game manager out on routine patrol. Or it may even be a cattle rancher checking his or her herd. This person is more than likely unfamiliar with crime scene processing techniques and has no knowledge of what to do. He or she may seek help from someone equally unfamiliar with what constitutes proper procedures in these instances.

In the authors' experience, many game professionals, including those tasked with ensuring the wellbeing of wildlife such as conservation officers, are uneasy about performing any duties associated with crime scenes, as they fear retribution from the police, whose job is to handle these responsibilities. However, many countries, states, or local municipalities do not have police specifically designated for animal crimes; handling cases involving animals may be a minor off-shoot of their normal professional obligations, and one they have little to no experience with or interest in. In addition, it is not uncommon for police to arrive at a potential wildlife or livestock crime scene, especially one in inhospitable or unnavigable terrain, days or weeks after the crime occurred. More pressing human-related crimes or understaffing may have left them without available personnel, or their perceptions of the lack of importance make it less of a priority. By that point, any evidence of value will have been lost, and those who initially arrived were crippled by a lack of knowledge or a fear of overstepping boundaries.

Even if the only thing that an individual can do, while waiting for police or another party to arrive and process the scene, is secure and protect the scene and capture a few photos with a cell phone, it is better than doing nothing.

Working with Others

Knowing who to call when someone encounters a possible wildlife crime scene is not intuitive as it is with human crimes. It may not be as simple as grabbing a cell phone and dialing a three-digit number for emergency personnel. Who is responsible for handling animal cases in a particular location—animal services? A fish and wildlife office? Game commission? Police criminal investigative division? If a person encounters a possible crime scene through the general course of his or her job duties, even if it is not likely but merely a possibility, having this information in advance saves wasted time.

Just as in human forensic cases, animal cases require the assistance of people from many different professional backgrounds. An investigator should not be afraid to utilize the assistance and experience of others. It is not uncommon for seasoned forensic investigators to be unfamiliar with a specific technique or process, whether collecting impression evidence or botanical samples. Investigation of a crime scene is rarely a solitary endeavor. Even if there is only one investigator, specialist, or veterinarian on the scene, testing of evidence samples will most likely be performed at a later time and off scene by laboratory personnel or a forensic examiner (Gardner and Krouskup 2019). Compiling a list of appropriate laboratories for testing, and having knowledge of sampling protocol is necessary for proper evidence collection. For example, some labs may require the use of particular swabs for DNA collection, or mandate the use of certain packaging. Additionally, it is wise to consult with professional forensic references based on specialty (i.e., forensic entomologists, geologists, botanists, toxicologists), thus having a list at the ready allows a more efficient investigation.

Animal crimes necessitate working closely with veterinarians. The presence of a veterinarian in a wildlife case is not only likely but essential. Since most veterinarians are unfamiliar with law enforcement and/or forensic protocol, he or she will need to be guided by an investigator. Veterinarians can also help recognize shortcomings of the work of the law enforcement officers or investigators who may not be very familiar with wildlife characteristics, traits, behaviors, and subtle species differences. An investigator should have a list of veterinarians that he or she can call in the event of a wildlife case; these individuals must be willing to perform forensic necropsies and live animal examinations, whether in the field or at the clinic, and must be willing to testify in court.

Working with other professionals is not always easy. In forensic work and especially on a crime scene, people are tense and edgy, and often we see the worst in someone's personality. Egos and proprietary stances are common, as some feel it is "their" scene, only they can process it correctly. Police may be resentful of the crime scene team or vice versa, veterinary professionals may refuse badly needed assistance or may need guidance that no one is willing to provide. This is counterproductive and can result in critical mistakes. The situation must be approached with professionalism. Complete focus must be on the task at hand and all entities must work together and remain objective. It must be remembered that everyone shares the common goal of seeking truth and the administration of justice. This can only occur if everyone works together.

Safety

Wildlife crime scenes can be inherently dangerous and there can be a resulting reluctance on the part of investigators to get involved. In 2018, it was reported that over 100 wildlife rangers were killed in Asia and central Africa in 1 year, nearly half killed by poachers (WWF 2018). The International Ranger Federation stated that from 2009 to 2018, 871 rangers were killed in the line of duty (WWF 2018). Hazards include not only the threat of poachers/hunters still in the vicinity, but also the presence of loaded weapons left behind by fleeing suspects, poisons in the soil, water, or air, and the risks associated with the animals themselves. Such problems are exacerbated exponentially if the incident occurs at night, as often the case.

Measures must be taken to ensure the safety of all those involved in the investigation and law enforcement work, as well as the general public. Anticipation of possible dangers based on information provided by dispatch or the person calling to report the incident, prior to arrival at the scene, and advanced preparation of materials such as personal protective equipment (PPE) is crucial (Cooper and Cooper 2013; Gardner and Krouskup 2019; Saferstein 2019). Always arrive with a group, never alone, and have backup support personnel at the ready. Always assume the possibility of suspects still lurking in the area. Arrival in coordination with a veterinarian prepared to sedate an injured and aggressive or fearful animal is recommended. The safety of everyone must be the first priority at all stages of the investigation.

Evidence

There are many definitions of evidence, but it can be summarized as anything used to prove or disprove that which is being questioned (Gardner and Krouskup 2019). It can be in the form of a physical object, a "context" provided by a physical item, or an oral or written statement (i.e., a witness or a victim). There are various admissibility rules that govern court presentation of evidence, and they differ depending on jurisdiction. More credence is given to physical evidence than testimonials, as physical evidence is viewed as definite and absolute, whereas testimonial evidence is subjective. Physical evidence by its very nature is indisputable and can either strengthen or weaken testimonial evidence. A *lack* of physical evidence can also be significant and telling. The context of physical evidence items is as important as the items themselves. An investigator must consider how they all relate to one another, and what their presence in the scene may indicate (Gardner and Krouskup 2019). Awareness of context also assists in determining what may be pertinent to the particular situation at hand. Investigators are often confused about what constitutes pertinent evidence, what to mark and collect, and what items to exclude. It is not effective to simply vacuum up every item present in a scene; one must be able to evaluate how each item relates to the overall picture and its context and function (Gardner 2012). Wildlife

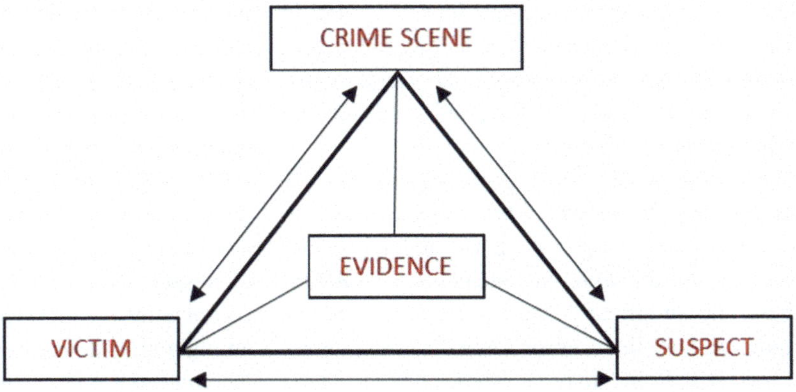

Fig. 3 Linkage triangle

crime scenes often occur outdoors, where there may be a multitude of items that on initial inspection seem to be important, and possible evidence, but on closer examination can be excluded. For example, there may be a great deal of trash and other random debris at the scene of a wildlife poaching that occurs near a busy highway. A shell casing discovered partially hidden in vegetation not far from the body could be considered critical evidence given the situation, but a scrap of yellowed, weather-beaten newspaper partly stuck under a log is probably not related to the case; it has most likely been there for some time. A certain amount of flexibility and judgment need to be utilized in considering how items are related to the scene and situation as a whole. Arbitrarily including every single thing encountered in a crime scene is ineffectual and a waste of time. An investigator must use discretion, intuition, and good decision-making skills when searching for possible evidence—the totality of the circumstances must be considered. It can be helpful to evaluate each item's value and merit as evidence in terms of the forensic linkage triangle (Gardner and Krouskup 2019) (Fig. 3); this triangle demonstrates that the purpose of evidence is to link it to the victim, suspect, and crime scene. Items that do not satisfy this requirement can be disregarded.

Another potential problem for investigators is evidence that changes in form or appearance, or disappears altogether. The odor of cigarette or cigar smoke, a smoldering campfire, the sound of pebbles falling down a ridge, or footprints visible only when the sun hits the morning dew on grass at a certain angle are all examples of transitory, changeable forms of evidence, but evidence nonetheless. Their presence must be recorded as quickly as possible, either via photographs or descriptive notes (Gardner and Krouskup 2019). Evidence recognition requires the use of all senses at times, not just the sense of sight.

Physical evidence can have differing characteristics, termed either class or individual characteristics, that help determine what information a particular item of evidence may present (Gardner and Krouskup 2019). Class characteristics allow an item of evidence to be identified back to and compared with a group of items, i.e., weapon type or paint color, while individual characteristics allow evidence to be compared with a specific source, such as blood type or fingerprints. Individual characteristics carry more weight than class characteristics. An item found on a crime scene and simply identified as a certain type of object is not going to assist an investigator as much as positively identifying the origin of that object through the process of individualization. For example, an investigator would first need to know that a particular piece of evidence is actually a *Cerotatherium simum simum* (white rhino) horn, and then he/she would attempt to identify the specific carcass from which that horn originates. Using DNA the investigator could "individualize" that rhino horn as having originated from that animal only and no other animal. But class characteristics can help avoid needless further processing of evidence; if the evidence item is compared with a known item and found not to match at the class characteristic level, then more extensive processing is not necessary—the item can be excluded. The goal in wildlife or environmental crime scene investigation is certainly to locate as much individualizing evidence as possible, but sometimes enough evidence with only class characteristics, when the entire scene context is considered, is sufficient to link a suspect with a crime scene.

Evidence Supplies: Substitution and Improvisation

For many involved in processing wildlife crime scenes, the cost of commercially made forensic supplies is prohibitive. However, inexpensive substitutes do exist. For instance, in place of paper or plastic evidence bags with preprinted sections for case number and item information, such as those from forensic supply companies, a clean, unused paper or plastic lunch bag and a permanent marker for hand-writing will suffice. Instead of a roll of specially designed crime scene barrier tape, any type of barrier material will work, including rope, for all that is needed is boundary demarcation. Plastic food storage containers can be used for temporary storage and transport of evidence, but if an item is wet or if it may contain DNA, it must be removed immediately after arrival at a lab or storage facility (biological evidence cannot be stored in plastic because of the likelihood of mold or bacterial growth which will degrade samples). Commercial tamper-proof forensic seals/tape are available from forensic supply companies, but plain plastic tape will suffice. Evidence tags to affix to large items can be made from strips of thin cardboard and fastened via wire or plastic zip ties. Instead of the heavy plastic yellow evidence placards with preprinted numbers, evidence can be marked using cardboard folded into a tent shape or simple wire construction flags with the corresponding evidence number handwritten in permanent marker. For examples of other inexpensive equipment substitutes, see Cooper and Cooper (2013) and chapter "Wildlife Forensic Pathology". It is advisable for anyone who may, however unlikely, encounter a wildlife crime scene, to keep essential equipment in a kit that can be stored in his or her vehicle; see Appendix 1 for a suggested list. There are, of course, some items that cannot be substituted, for example, proper DNA collection swabs, blood sample vials, gloves, and the like. Before collecting evidence, always consult with any outside laboratories as to which samples will be sent, in order to follow any mandated laboratory procedures.

Evidence Particular to Wildlife Cases

While some of the evidence encountered on wildlife crime scenes will be similar to that found on crime scenes involving humans, there are many items discovered in the course of an investigation that are wildlife- or conservation-specific (see Table 1). While there are many more wildlife and plant parts involved in trade than can be listed here, Table 1 lists some of the more common items. Space does not allow for detailed descriptions; however, more information can be found in Cooper and Cooper 2013; Huffman and Wallace 2012; Walker and Adrian 2012.

In many instances an entire animal will not be discovered; there may be only parts of an animal due to field butchering, trophy-taking, treatment such as processing, tanning or carving, decomposition, or scavenging by other animals. If the animal product is encountered in the course of transport, it may have already been turned into the product for which it was destined, e.g., handbags or shoes, leaving an investigator to struggle to identify the species (Figs. 4 and 5). It is crucial to determine whether the products encountered are legal to possess and/or trade and all permitting requirements were satisfied, as in some cases it may be perfectly acceptable to find these items in transit. Identification of species is imperative as this will determine legality; this can be accomplished through various means including genetic and morphological analysis. In some cases, it is possible to individualize an animal, either by the recovery of a microchip, DNA comparison with a database, i.e., the Rhino DNA Index System (RhODIS) at the University of Pretoria, South Africa, or comparison of identifiable

Table 1 Evidence particular to wildlife cases

Evidence item	Examples	Possible purpose
Animal victim/organs of an animal	Bear gall bladder, tiger penis, pangolin scales; bear or gorilla paws/hands	Traditional medicine, hunting trophies
Smoked or cured meats	Bats, porcupine, reptiles, duiker, primates	Bushmeat
Fur	Individual hairs or clumps of hairs found as trace evidence	N/A
Feathers	Migratory bird feathers; feathers from birds of prey or other large birds; feathers found as trace evidence	Ceremonies; rituals; jewelry; costumes; dance fans; headdresses; hats
Bird feet	Feet of birds of prey	Ceremonies; rituals
Bird pellets/castings	Undigested regurgitates of birds of prey	N/A
Eggs	Eggs of rare or endangered species	Collectors; breeders
Claws	Tiger claw, bear claw	Collectibles; jewelry
Reptiles/skins/products	Snake, crocodile, cayman skin	Accessories; shoes; clothing
Fish/fillets/organs/eggs	Totoaba swim bladders; sturgeon eggs	Food; caviar; traditional medicine
Plants or plant material	Endangered orchids; ginseng; mushrooms	Collectors; traditional medicine
Wood products	Rosewoods, teaks, ayugue	Furniture; decorative items; floorings; building materials; instruments
Elephant ivory products		Jewelry, carvings
Rhino horn		Traditional medicine
Animal bones/teeth	Tiger and leopard bones	Traditional medicine
Animal hides/pelts	Leopard, cheetah, giraffe	Rugs, clothing, pelts themselves
Antlers/horns	Mule deer, white-tailed deer, elk	Trophies; traditional medicine

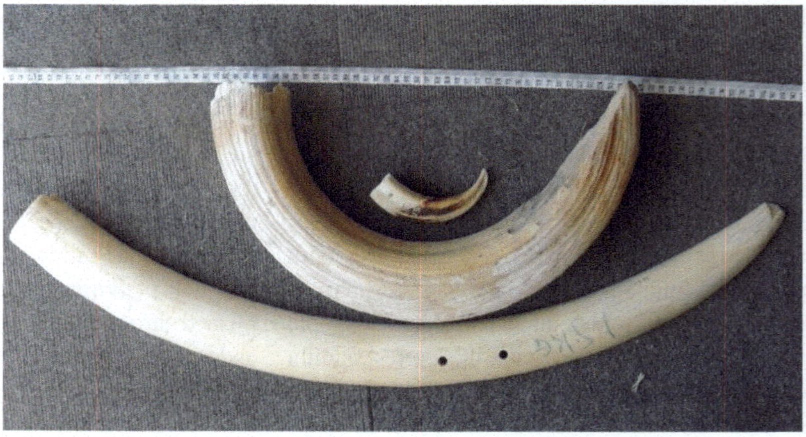

Fig. 4 Ivory: top to bottom: warthog, hippopotamus, African elephant

Fig. 5 Carved rhino horn

characteristics such as coat pattern, unusual horn shapes, scars, ear tears, or ear notchings, with a managed database of photographs and records depicting morphologic characteristics for all animals of a particular species within a national park or reserve.

Even seemingly insignificant items can prove very useful: trace evidence items such as hairs can be microscopically examined and compared with known animals, feathers can give an indication as to the health of the birds, and pellets can demonstrate diet, which can suggest the birds' habitat and location. Fish fillets and even leather goods may be able to be identified via morphological traits such as scale counts or patterns. Recognizing that what seems inconsequential is actually extremely important as evidence is crucial. And it must always be remembered that whatever part is found, it is evidence and must be handled accordingly; the animal itself, whether alive or deceased, is evidence. Under all but the most unusual circumstances, no changes may be made to evidence or its value as such is destroyed. Any alterations to an animal that occur during the course of a live examination or necropsy must be documented in detail with descriptive notes and photographs. Additional wildlife crime evidence will be discussed in later sections.

Chain of Custody

The integrity of evidence items can be questioned at any point in an investigation. Defense attorneys may ask questions such as, "how do I know this evidence has not been tampered with?" Establishing a "chain of custody" or documentation of the location and possession of every item of evidence is essential. The principle of the chain of custody can easily be explained as "who-had-what-when-and-where," and this needs to be recorded from the time an item was discovered on scene through its documentation, collection, and either storage in an evidence room or submission to another facility or lab for further processing. If there is any question as to the location of an evidence item or who had possession of it at any point during the course of the investigation, the integrity of that item can be called into question and the evidence may be thrown out of court. Chain of custody forms the focal point in many legal arguments by defense attorneys when taking on the merits of what appears to be a strong case against their client; this is often done in an attempt to discredit the investigator and his/her procedures on scene and even later when the evidence is in storage. The defense lawyers often get this right, and poor documentation enables them to do so.

Following a standard crime scene processing procedure with regard to evidence collection and documentation ensures that the disposition of the evidence will survive any courtroom questioning (Saferstein, 45). To ensure the correct chain of custody and prevent contamination between samples, gloves must be worn at all times and changed between the collection of each item of evidence. Evidence must be collected (it is preferable that only one person collect all evidence) and placed into a container or bag labeled with the collector's name or initials, date and time of collection, case number (if known), and the location where found. The container or bag must then be sealed with tamper-proof evidence tape (evidence tape is specially designed to indicate tampering; if evidence tape is unavailable, an investigator can use regular plastic tape, but ideally only temporarily until proper evidence tape can be obtained). The person collecting the evidence must sign or place his or her initials and the date across the container tape or seal and onto the container. A separate chain of custody form with the collector's name, date, time collected, case number, and location must accompany all evidence items and can be attached to the evidence container. The evidence can then be transported to a storage facility where it will be entered in a logbook or sent on to a laboratory for testing. If arriving at a lab, the person receiving the evidence must sign the chain of custody form, enter the date and time, avoid destroying the original seal (another opening must be made), and all laboratory personnel handling that item must do the same. When the item is returned from the lab it must be resealed in a new location on the original bag or container, signed and dated across the seal, and the chain of custody form filled out again. The form is to be signed and dated by the person releasing it, be signed for and dated at the place of storage by the receiver, entered again in the logbook with date and time, and placed in a recorded location. If an evidence item is removed from its storage location for any reason, i.e., to be shipped to a different laboratory for testing purposes, the chain of custody form needs to be sent with it and updated with the name of the person signing the item out of storage, date and time, the purpose of removal and location to which the item is being sent, and so on. This may appear to be an extremely arduous process, but one mistake caught by a perceptive attorney can have disastrous results for a case. For a sample chain of custody form, see Appendix 2.

Scene Processing Procedures and Special Considerations

There is no one correct way to process a crime scene. The steps and techniques utilized will differ depending on the circumstances. However, there are concepts and principles that are absolutely necessary regardless of the situation. The processing methodology and proper steps to follow in wildlife and conservation crime scene investigation are essentially the same as those prescribed for human crime scene analysis, albeit with some special concerns. We will attempt to elaborate on these and summarize and briefly describe some of the steps involved in scene processing, focusing on some of the factors pertinent to wildlife and conservation crime scenes.

Initial Notification

In all cases, the investigation begins with the receipt of initial information from the person suspecting the presence of a crime scene. This may happen through any number of scenarios: a person encountering an injured or deceased animal with a suspicious wound, who then notifies a state game and fish department; an inspector discovering a shipping container concealing live animals or plants at a transit port, who then contacts the federal investigative division; an observant tourist who spots ivory or coral jewelry in a marketplace and calls the ministry or department of natural resources. The person initially notified, be it an investigator, veterinarian, or law enforcement official, must record any and all details of initial contact: date and time of notification; name of person providing the notification; how that individual obtained his or her information; and any pertinent details. This begins the process of crime scene documentation, which encompasses a few main aspects: (1) notes, logs, and reports; (2) sketches; and (3) photographs/video; having written records of every action taken ensures the case will stand up to rigorous cross-examination by legal professionals, and it also maintains the chain of custody.

Written Documentation

Detailed notes at every stage of the investigation are mandatory. They are used to explain everything, such as: the time an investigator was first contacted and all relevant information; how the scene appeared on first arrival; all investigatory steps taken and evidence found; all personnel present; witness statements; analysis of evidence; all the way through to the final report. The date and time of each action or observation must be recorded. It is important that notes are written in a contemporaneous manner. They should be descriptive in nature and must include details of not only where evidence was found but also where it was not found. For instance, if an area was searched for fingerprints but none were obtained, this must be noted. One cannot simply leave out negative results of search attempts (Gardner and Krouskup 2019). Descriptive notes are used along with photographs, sketches, and logs as supplementary material—together they help paint a clear picture to a judge and/or jury of exactly what was encountered at a crime scene—all steps taken, all techniques utilized, and all results found.

Logs are necessary throughout the crime scene investigation as they strengthen and verify the notes taken. At a minimum: crime scene entry and exit logs are vital to record data such as the name, time entering/leaving the scene, and purpose for the presence on the scene; photography logs record time, location, and subject of each photo taken; evidence logs provide locations and descriptions of all evidence items, as well as time collected; and measurement logs contain the measurements of all pieces of evidence; sample log forms can be found in the appendices.

The final report is not only a compilation of all contemporaneous notes, sketches, logs, and photographs obtained during scene processing, it is a chance for the investigator to depict, in detail, every action taken, procedure followed, technique utilized, and test carried out. Descriptive in nature, it encompasses scene and environmental conditions, information about the victim(s), explanations of all the areas of the scene(s) and what techniques were utilized, the results of any tests performed (presumptive tests for blood, fingerprinting, etc.), whether those results were positive or negative, and any findings from further laboratory evidence testing (Gardner and Krouskup 2019; Cooper and Cooper 2013). It should also contain the veterinarian's notes, photographs, and other documentation. It is a direct reflection of the investigator's professionalism; it must be clean, free from grammatical and/or typographical errors and formatting mistakes. Any mistakes made on a crime scene will have to be declared in the report; attempts should not be made to try to conceal errors.

Crime Scene Assessment, Personnel, and Roles

If there is someone assigned to serve as an initial responding officer (IRO) from the investigative department involved (a particular game warden, fish and wildlife officer, criminal investigative division member, etc.) when a wildlife or environmental crime occurs, that person is usually first on the scene (the role of the IRO is not to be confused with the person who makes the initial discovery and calls in a possible crime scene). Regardless of whether there is an assigned person or not, the first person on the scene after initial notification by the person who made the discovery should assume this role. This person functions as scene control, evaluates the overall situation to determine what is needed, identifies any possible safety hazards, and assigns tasks to additional personnel. An IRO should try to ascertain as many facts about the situation as possible prior to arriving on the scene, as this will help with determining what supplies are required, how many additional investigators may need to be called, whether a veterinarian is needed immediately (in the case of live animal victims) or whether the veterinarian's arrival can be delayed (in the case of deceased animal victims), as well as whether the veterinarian may require additional staff to assist with his or her work at the scene (especially true if the animal is large and/or the scene is not accessible by vehicle).

In larger departments, the IRO rarely processes the crime scene but instead hands the scene over to crime scene investigators or a crime scene controller (CSC) after securing the area. In many wildlife crime scene scenarios, however, the IRO will fill multiple, if not all, roles him- or herself due to a lack of staff assigned to animal crime investigations. Once on the scene, the IRO must determine whether there are any safety hazards present, any life-saving measures necessary, and document any initial observations including temperature and weather conditions and particularly any transitory evidence (evidence that may disappear over time) such as odors of cigars/cigarettes, footprints in the dew on the grass, moisture rings from a drinking glass, or puddles resulting from drips from the undercarriage of a car. The IRO should also determine if there is sufficient daylight remaining to initiate the investigation (there is little investigation work that can be done on the scene, at night, in the bush). If life-saving measures are needed, either for humans or animals, those take precedence, and a veterinarian or emergency medical technicians must be called. As their priority on scene is to tend to the injured, they will most likely not be concerned with the preservation of any evidence in the area. It is advisable to take photos of the scene prior to their arrival to at least have some recording of the original scene appearance. The IRO must then determine the crime scene boundary and cordon off any witnesses or onlookers and maintain control of the scene at all times until additional personnel arrive. The IRO should record details for each person found on the scene upon his or her arrival and brief notes to accompany the names and contact details should include what each of these persons did or observed at the scene prior to the arrival of the IRO. A specific "staging" area, set up within the scene boundary but outside the scene itself (and one that has been confirmed free of potential evidence), should be established so that crime scene personnel and veterinarians have an area in which to store their processing equipment and materials that allows for easy access while still maintaining the integrity of the scene (Fig. 6).

Once additional personnel arrive to process the scene, the IRO can release the scene to them, and the two entities should compare notes in a debriefing. All information regarding everything the IRO did and observed should be conveyed to the arriving staff and the IRO can take on a support role and help with scene processing tasks (Gardner and Krouskup 2019).

Fig. 6 Staging area established directly outside of crime scene boundaries to hold equipment and minimize contamination from multiple scene entries and exits

One concern with wildlife crime scenes is that there are rarely sufficient personnel to which to designate duties and process the scene. It is rare for a department to have the ability to dispatch multiple people to a single scene; often it is only one or two people serving as IRO, crime scene "team," and security all at once, which is fine if the scene is small and confined, i.e., just one house or boat or outbuilding. But having limited personnel can be problematic for actual large-scale scenes, which can take hours if not days to process thoroughly. If all that is available in this situation are one or two individuals, they must be familiar with crime scene investigation and forensic science techniques, and they must still follow proper procedures to the best of their ability. It will take longer to execute, but it is certainly still possible, and one must ensure that if pressured to finish quickly, he or she emphasizes the seriousness of the situation and the fact that there is only one chance to do things correctly.

Ideally, if there are multiple people present, each person should be assigned a particular task. This reduces the chances that something is missed. For example, having one person to take photographs and another to follow along and record the time taken and descriptive information about each photo ensures that essential information will be recorded for future reference. And at all times there should be someone stationed at one crime scene entry/exit point (usually at or close to the "staging area") from which all access to the scene itself is controlled; this person must document on a log sheet (see Appendix 3 for example) every individual entering or leaving the scene and for what reasons. Crime scene boundaries may change as additional evidence is found, but regardless of where the boundary tape is placed, one point of entry and exit should be utilized to minimize the chances of scene contamination.

Veterinarians and Medical Personnel

While veterinary/medical staff is tending to any injured parties (remember, there may be human as well as animal victims), all scene investigation must stop (Fig. 7). This may be a good time for an investigator to speak with witnesses and record their statements.

It may be necessary for an investigator to guide the veterinarian regarding forensic procedures, as the veterinarian may not be aware of proper forensic techniques. An investigator can also serve as a scribe and record any additional notes necessary, as well as take photographs during every stage of the necropsy or animal

Fig. 7 Veterinarian treating injured rhino

examination. All dead animals should be considered illegally killed until confirmed otherwise by a qualified individual (veterinarian). If an animal has been injured and is still mobile, it may need to be darted and anesthetized for examination and recovery of evidence; meticulous notes must be taken regarding darting and anesthesia, including dart and anesthetic information and dosing amounts. The animal may be ear notched or microchipped at this time for future identification, and blood, tissue, or other samples for DNA data banking can be taken as well, but these procedures must be recorded.

Information on procedures for the veterinarian at the scene regarding evidence collection and animal examination can be found in Cooper and Cooper 2007; Cooper and Cooper 2013; Brooks 2018; Rogers and Stern 2018). The veterinarian must document (written notes and photographs) everything done during the examination, the general condition of the animal, all wounds, and any normal and abnormal findings (Fig. 8).

When feasible, an animal carcass or remains should be removed from the scene in order to examine them in detail in a laboratory or clinic setting. Radiographs of the entire body or parts are recommended for the determination of embedded bullet fragments, bone fractures, or other trauma not initially seen. Bones can be examined for artifacts resulting from instruments used either during the killing or during the butchering process (be sure to distinguish between trauma to bone and the gnaw marks of rodents and other scavenging animals; this is often difficult and requires significant examination).

Plastic coated numbered cards can be used to number wounds found on the carcass, similar to marking various evidence items with flags or placards on a crime scene (Fig. 9). Each animal should have its own reference number. It is useful to refer to the live animal or carcass as "1," with each wound on the animal numbered "1.1," "1.2," and so on. Overall photographs of the animal should be taken with and without wound indicator cards, and both midrange and close-ups should be taken of every wound, both with and without indicator cards and with and without a scale or ruler to demonstrate the size of the wound.

All photographs taken by the veterinarian should be retained and safely filed, preferably cut onto a CD immediately upon returning to the

Fig. 8 Veterinarian recording contemporaneous notes and GPS coordinates

Fig. 9 Veterinarian recording bullet wounds with numbered rods

office. No photograph should be deleted, be it a poor photograph or a duplicate, as all photographs may be requested by the court. The metadata behind each photo will indicate a continuous series, and missing photographs could be called into question by a defense lawyer.

Scene Processing: Photographs, Sketching, and Evidence Documentation

Processing a wildlife crime scene should proceed from the least intrusive step to the most intrusive, in order to preserve as much of the scene as possible in its original condition.

General Scene Considerations

If sufficient staff are present, appointing someone to serve as a lead investigator or CSC can be extremely helpful, as this person is ultimately responsible for overall scene oversight, determination of roles, ensuring everyone has what they need to do their jobs effectively, and finally, that everything is conducted properly and accounted for. If only one or two people are available to process the scene, it will be incumbent on those individuals to ensure these things are accomplished.

As mentioned previously, there may not be a scene per se, as in the discovery of beached marine mammals injured by watercraft (see chapter "Forensic Science in Marine Mammalogy: Applications and Limitations"), or an animal carcass could be discovered in the bed of a vehicle but was most likely killed elsewhere. An animal carcass discovered in a field could have been moved or dragged from another location. These situations indicate a secondary or even tertiary scene that will also need thorough investigation. The techniques used for investigation and processing will be scene- and situation-dependent, but whatever the situation, in ***every*** case, the investigation must be meticulous and thorough.

As mentioned in prior sections, whichever crime scene processing methodology is chosen must be carried out in a sequential order, as some steps are destructive and thus can only follow certain others, e.g., photography must precede evidence collection. Box 1 provides the foundational questions that must be asked for every crime scene processed. Additionally, each crime scene and its context must be taken into consideration in order to choose the appropriate stepwise strategy. Reconsider what may be needed for scene processing—are all necessary supplies present? Is there fragile evidence that could be disturbed by wind or rain? What noninvasive, nondestructive steps should be carried out before the more intrusive steps that will alter the scene?

Box 1: Basic Questions to Ask Before Initiating Crime Scene Processing

1. **Was a crime committed?**—By this point in most cases the answer will be yes, but it is entirely possible that what first appeared to be a criminal act is, in fact, not; consider pertinent facts, e.g., if the crime is a possible poaching, is the animal a protected species? If not, what might the hunting season for that species be?
2. **What crime was committed here?**—Like the first question, determination of this answer will be based on a knowledge of the laws applicable in that jurisdiction; as in the above scenario, consider whether it could be an actual poaching, or could it be a retaliatory killing? Contemplation of this question could provide for a structured search for clues to prove an illegal activity.
3. **If a crime was committed, how might it have happened?** Think of possibilities, paths of entry/exit, whether a vehicle was used, etc. This "thinking like a criminal" can help direct the evidence search but one must be careful to avoid bias or tunnel vision and remain open to all possibilities.

Again, there is no one-size-fits-all definitive method for wildlife or conservation crime scene investigation. Because these crime scenes are situationally variable, it is best to approach them methodically, utilizing relevant steps for individual cases (detailed processing steps can be found in previously mentioned sources). There are, however, certain concepts that must be adhered to in order to conduct a proper investigation, as there is only one chance to get it right. An investigator can never go back later and do it over. There is no way to put evidence items back in the same, original location, and in the same condition. Once someone enters a scene, it has been altered in an irreparable way, as he or she will inadvertently move evidence, destroy evidence, or somehow add materials to a scene. Thus, following a methodical procedure enables an investigator to limit the damage and capture as much of the original context as possible.

Utilizing checklists can assist with ensuring that no steps are missed; however, allowance must be made for flexibility in their use. Merely checking off boxes does not always prevent mistakes and can, in some circumstances, cause an investigator to overlook important factors pertinent to a particular situation (Gardner 2012). Depending on the situation, the order of scene processing steps followed may need to be changed, i.e., if a pending storm threatens to destroy fragile items, rather than photograph an entire scene prior to collecting evidence, an investigator may need to photograph and collect those items first (Gardner 2012).

Whether the processing methodology utilized has been created and enforced by an agency or by an individual, it must be carried out in a methodical fashion, with the primary goals to capture the scene in its original condition, minimize any/all disturbance or alteration, and to record every detail. The purpose is to demonstrate for those who cannot visit the crime scene, namely judge and jury, precisely what was encountered and what was done.

Photographic Documentation

It is good practice to format the memory card in the camera before embarking on the task of photographing a scene. It is recommended to use a new memory card for photography, and package and store separately in case of use in court.

Complete documentation of the overall crime scene must occur before any collection of evidence or even marking of evidence. Photography is carried out first, working from outside to inside the scene, in three iterations: (1) overall photographs, which capture the scene in its entirety or as much as possible, taken from all four cardinal directions; (2) midrange or evidence-establishing photographs, which capture the location of evidence items by including in the photo a landmark of some kind, and (3) close-up photographs of evidence items,

Fig. 10 Photographic improvisation: using the "flashlight" feature on a cell phone to provide adequate oblique lighting in order to capture the details of a footprint

which capture detail and are typically taken with and without the use of a scale or ruler. But these are not the only photographs that should be taken; photos should ideally be taken during every step of processing in order to provide a photographic record of what was done. Typically, overall scene photographs are shot first; next, a search for evidence is conducted and items are marked; and finally evidence-establishing photographs are shot, followed by close-up photographs. However, this procedural order may change depending on circumstances. An investigator can never take too many photos. The use of a flash is also recommended to reduce dark shadows in the photographs and is essential in overcast conditions. Improvisation may be necessary, as shown in Fig. 10. A description of all photographs taken along with the date, time, and name of the photographer should be entered into a photography log (see Appendix 4 for example).

Thorough descriptions of crime scene photography techniques can be found in other sources (Robinson 2016; Duncan 2015; Gardner and Krouskup 2019).

If all that is available for capturing images is a cell phone camera, by all means, use it. Obviously, better quality photographs will be obtained when using a DSLR camera, but in many small countries with understaffed, underfunded police or investigative departments, expensive cameras are not feasible. An investigator must use what he or she has, while still striving for quality. It is often stated as true that poor quality photos are better than no photos. As more emphasis is placed on forensic investigations as applied to wildlife and environmental crimes and given the importance of photographic documentation, it is the authors' hope that departmental budgets will allow for the purchase of high-caliber equipment; at the very least, a top-notch camera.

Important to note is that photographs can be altered to a certain extent, e.g., clarity, lighting, or contrast can be improved using digital techniques, but all originals must be saved in order to demonstrate to the court that the photographs were not intentionally misrepresented or doctored in any way.

Video

In some cases, it is advisable to supplement crime scene photographs with video. Video may be recorded along with the initial overall scene photographs, and again once all evidence items have been marked. Videography is also extremely helpful to assist in documenting evidence that cannot be captured with a camera, such as animal behavior. Recording a lame animal or an animal in distress can convey the gravity of an injury or suffering to a court much more effectively than can photos. It can also be used as part of an animal's medical record, to demonstrate its state at the time of seizure or arrival at a clinic, which can then be compared with follow up video taken at intervals throughout treatment to document improvement.

Ensure that consideration is given to focus and movement throughout the video process; all movement must be slow and deliberate. And while at times capturing sound may be helpful, e.g., the vocalizations of birds seized as evidence or distress calls of injured animals, be cautious of what is said by investigators at the time of recording; avoid subjective statements, and/or suggestive comments. Unless attempting to capture these vocalizations specifically, it is advisable to disable audio recording.

Be mindful of what appears in the background of your photographs or video. Try to avoid capturing faces if possible and avoid the common error of including the toes of shoes in downward-focused shots. Any person appearing in the picture may be called as a witness or items not disclosed in the written report may be questioned by an observant Defense Council. Plan your photographs and video clips carefully, and always remember that you may not delete any photographs or footage which you have recorded.

Evidence Search

After overall scene photographs/video are taken, an investigator can then move carefully through the scene to search for evidence. The technique used will be dictated by the location and condition of the scene. For example, if it is an outdoor scene spread over many acres, it may be best to use a line search method, in which multiple searchers move through the scene in lanes at the same pace and formation, carefully searching his or her lane (the size of the lane can be designated based on what is reasonable for a single individual to inspect thoroughly) (Fig. 11). A smaller scene may dictate that a strip search be used, where fewer investigators are needed to move down lanes, reversing direction on reaching the end of the perimeter to evaluate the next adjacent lane. The investigators may then choose to search in lanes approaching the scene from an opposing direction, effectively dividing the scene into a grid. Another option is the spiral search; this is best utilized in smaller interior spaces. Using this search technique, investigators move through the

Fig. 11 Using the line search method to look for evidence

Fig. 12 Broken branches should be marked as evidence and collected; these can indicate paths of entry/exit, especially if there is fresh sap present

scene either inward or outward, in a spiral pattern, searching from high surfaces to low (or vice versa) (Gardner 2012; Girard 2015; Gardner and Krouskup 2019).

Evidence may be anywhere within a wildlife crime scene. Look up, not just on the ground directly ahead. Look for disturbed or broken vegetation that could indicate a path of entry or exit (Fig. 12). Look for items that may have been cast into bushes or shrubs as suspects fled. Pay special attention to footprints and animal tracks. Local wildlife trackers are often extremely skilled at identifying animal markings—do not be afraid to consult with them. Poachers have become increasingly sophisticated, as they are aware that footprint/footwear impressions can be used for identification, and in some cases, they have begun wrapping their feet or shoes in cloth to avoid leaving discernible prints (authors' personal experience).

Any evidence located during the search should be marked or flagged (Figs. 13 and 14), and once

Fig. 13 Improvised cardboard evidence marker

Fig. 14 Flag with evidence number written in marker

all items are located and marked, a second round of overall scene photographs is taken. Then midrange photographs can be taken of each item.

Evidence Measurements

After the initial search is completed and evidence has been marked, a second set of overall photographs have been taken, and evidence-establishing and close-up photographs of each evidence item have been obtained, investigators need to measure every piece of evidence to fixed points in the scene (Gardner 2012; Gardner and Krouskup 2019). Measurements are another form of supporting documentation that allows the judge and/or jury to better visualize a scene and are essential if scene recreation is required. Proper measurements may help to explain unknowns, such as: the actions and movements of the perpetrator(s); the actions and movements of the victim; and such actions and movements in relation to one another.

There are several measurement techniques that can be utilized based on scene conditions. Whatever method is chosen, the evidence must be measured to immobile, permanent structures, such as corners of buildings or utility poles. Additionally, measurements should never be only from one item of evidence to another; all items must be fixed in the scene on their own. As many wildlife crime scenes occur outdoors, the most feasible methods include triangulation, in which four straight measurements are taken from fixed reference points to specific points on the evidence (for items with fixed shapes that do not change with movement, i.e., guns, knives) and two straight measurements from the center of mass of an irregularly shaped evidence item (i.e., clothing) to two reference points (Figs. 15 and 16) (Gardner 2012; Gardner and Krouskup 2019).

If there are no structures to use, as might be encountered in outdoor scenes, reference points must be manually fixed in the scene so that they can be readily identified later. Using trees, rocks or shrubs is not ideal, as it would be very difficult for another investigator to try to recreate the scene and know exactly what point on the tree/shrub was used as the actual reference point. A better solution is to use metal stakes placed on the ground near these types of landmarks (trees, rocks); once the geographic positioning system (GPS) coordinates of the stakes are recorded, they can serve as extremely effective points for measurements. The stakes can then be hammered into the ground when measuring is complete and can be relocated later by the use of a metal detector.

Measurements can also be taken using the baseline technique (Fig. 17), where a datum point is set from which the baseline will be extended in a straight line along a cardinal direction. The datum point itself is fixed by triangulation to a permanent structure (metal stakes or rebar placed at landmarks can also be utilized for this). Evidence is then measured (from its center of mass or fixed points on the item) at right angles to the baseline, and the point at which the evidence measurement line intersects the baseline is marked on the baseline, and

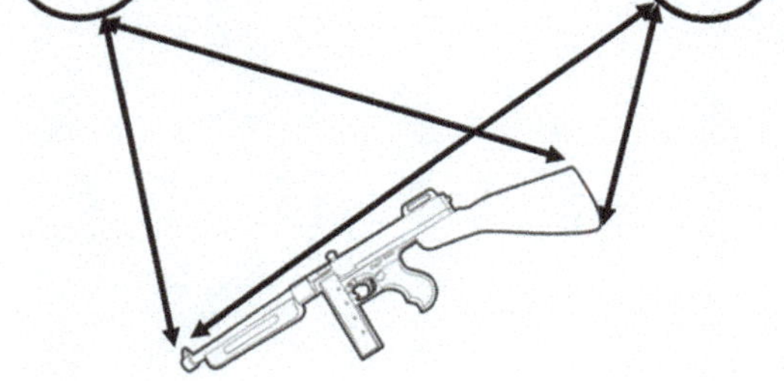

Fig. 15 (Left) Triangulation (regularly shaped item)

Fig. 16 Triangulation (irregularly shaped item). Measurements taken from fixed reference points (RPs)

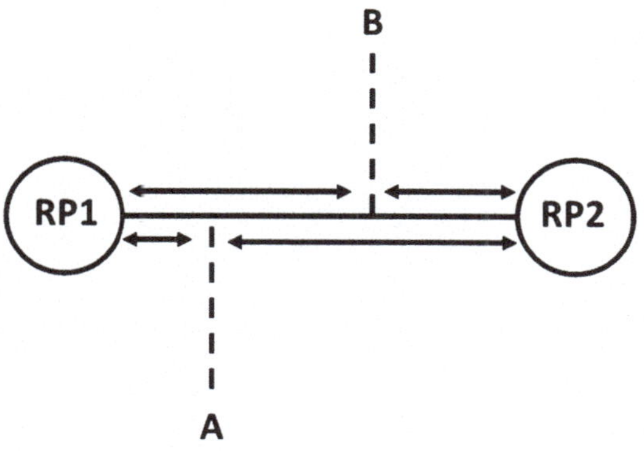

Fig. 17 Baseline technique: baseline established between fixed datum point, RP 1, to RP 2; evidence items A and B are measured in straight lines to baseline, and measurements are taken from the RPs to the points at which the evidence measurement intersects with baseline

measurements are taken from the datum to this point (Gardner 2012; Girard 2015).

All evidence measurements must be recorded on a measurement log or form (see Appendix 5 for example form).

There are other measuring techniques that can be utilized depending on what the scene dictates; although space does not allow for detailed explanations, these can be found in other sources (Cooper and Cooper 2013; Gardner 2012; Gardner and Krouskup 2019; Girard 2015; Saferstein 2019).

Sketching

Sketches can be useful for complicated scenes, and when combined with other forms of documentation such as notes, photographs, and logs, may add to the overall processing effort, and may enhance the understanding of the judge and jury. They are especially effective for outdoor wildlife crime scenes where photographs may not do a complete job of capturing the scene's overall structure and appearance (Fig. 18).

Crime scene sketches need not be works of art, and unless the scene mandates extraordinary investigatory efforts (high-profile cases, etc.) they needn't require exorbitant amounts of time. They are merely ways to combine overall scene elements into one document. Multiple sketches are usually made, as attempting to combine complete landmarks and all items of evidence with their measurements can make the sketch too crowded. A rough sketch of major landmarks, landscape features (or for an interior scene, interior features: doors, windows, furniture, etc.), and items of evidence could be made at the scene, and later combined with a sketch detailing evidence items and their measurements. These sketches can in turn be followed by more refined, final sketches. Typically sketches depict a birds-eye view, where the viewer is looking down onto the scene, although there are other variations (Gardner and Krouskup 2019).

Sketches should include a title block that lists: the purpose of the sketch (preliminary/rough sketch, final sketch, depiction of certain areas or items, etc.); name of the person making the sketch; case number; address or location of the scene; date and time of sketch; a legend if necessary; directional indication; and if the sketch is not made to scale, the phrase "not to scale" should be written on the document (Gardner and Krouskup 2019). As with notes and photographs, all sketches should be saved for the final report, or for legal purposes.

It is useful to combine the measuring and sketching of a scene into a single report (see Appendix 7 "Scene Summary Report" for an example of how to include a sketch in the case report). A rough sketch is compiled at the scene and the necessary measurements are taken. These are combined in a neater sketch at the office which, while not usually drawn to exact scale, is a reasonable representation of the crime scene. This can be coupled with an aerial photograph and a topographical map in order for judicial officers to confirm remote locations (Fig. 19).

Scene Rechecks

The crime scene should be rechecked multiple times throughout processing to ensure no evidence items were missed and to determine whether secondary and/or tertiary scenes exist. If additional evidence is found, the crime scene boundary may need to be adjusted, and these items will need to be flagged or marked and photographed.

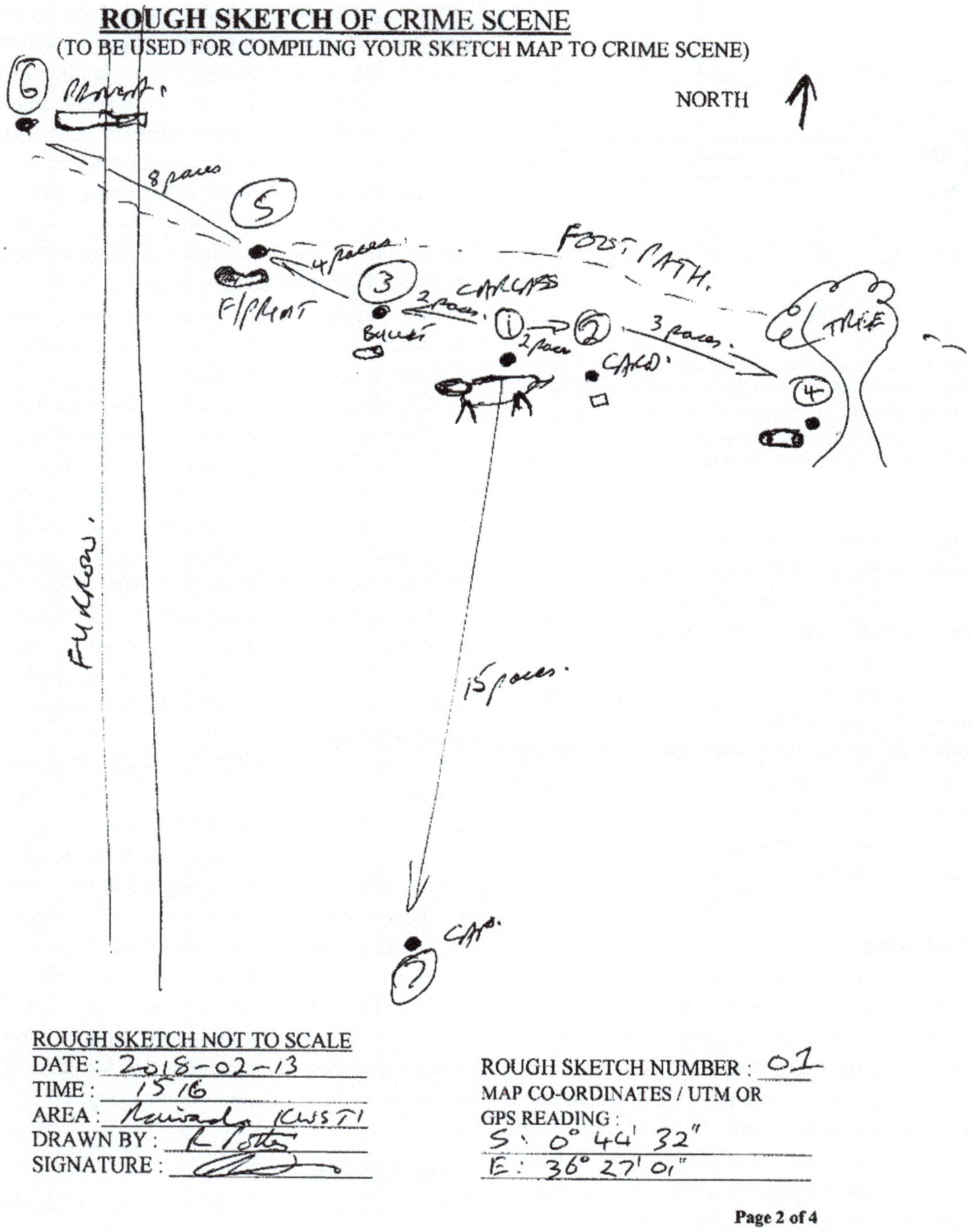

Fig. 18 Rough scene sketch

Evidence Collection

Once all documentation steps have been completed, and another group of overall and possibly midrange/evidence-establishing photographs have been taken, individual evidence items may be collected. The collection,

Fig. 19 Aerial photograph of the crime scene that can be combined with sketches and topographical maps as part of scene documentation

packaging, marking, transport, and storage of samples taken from a crime scene is as important as the samples themselves and critical in professional investigations. It is preferable that only one person be assigned to collect and package evidence in an effort to eliminate contamination and to adhere to the proper chain of custody. Collection priority is given to fragile or degradable evidence items—these are collected first. Once an evidence item has been collected, additional photographs should be taken of the area under and around this item (photographs should be taken of the area under the body as well, once it has been removed). As each item is collected, the date, time, location, and item description are entered on an evidence log or form, along with the name of the person collecting the item (see Appendix 5 for sample form).

Individual evidence items should be collected with caution so as not to destroy additional evidence, whether on the item itself or the surrounding area. For example, a handgun should be picked up by the textured grip, where fingerprints are least likely to be found (never insert anything into the barrel as this can damage the lands and grooves inside and introduce artifacts).

Most evidence items can be placed into paper bags or envelopes. Be sure to select the correct size of bag or envelope to fit the evidence; small items must be placed in small packaging. Write any relevant information (name, date, time collected, and brief item description) on the outside of the packaging before placing the item inside. Seal them with plastic or evidence tape and sign and date over the edges of both sides of the sealing tape. Large evidence items can be stored separately with cardboard or plastic labels attached by wire or plastic cable ties (Fig. 20). There is no need to package these objects, although care should be taken to secure them during transport to protect from any damage, tampering, or contamination.

Only nonbiological items should be placed into plastic for long-term storage. Although blood-stained or soiled items may need to be placed into plastic for transport, due to the potential for leakage through paper, they must be removed as soon as possible, allowed to dry, preferably by hanging in a drying cabinet with paper placed beneath them to catch any trace evidence that may fall, and then folded carefully and stored in a paper bag or box. Items with biological material on them cannot be

Fig. 20 Large skull marked with individualizing metal evidence tag

haphazardly stuffed into a plastic bag, they must be carefully folded in paper sheets, so as to avoid one stained area coming into contact with a clean area, or another wet area, as this can create artifacts or contaminate other evidence.

Evidence from an Animal Victim

Unlike emergency treatment of a live animal, which takes precedence over scene processing, necropsies that need to be performed in the field should be conducted following the crime scene investigation to avoid scene alteration and disturbance or loss of evidence.

If projectiles can be removed during the necropsy, they should be rinsed free of debris, allowed to air dry, wrapped in tissue, and placed into individual small boxes (Figs. 21 and 22).

The veterinarian should also look for and document evidence indicating that a weapon was fired in close proximity to or against the hair or skin of an animal. This may include gunshot residue or an "abrasion ring"—a dark circle around the entrance wound (gunshot residue may also be found on the hands or face of suspects, although it is easily washed off, and the clothing of any suspects should be tested for residue by using commercially available testing kits or by using a length of sticky tape to press against skin or clothing. The piece of tape can then be placed across the mouth of an open bottle

Fig. 21 Projectile recovered from carcass

Fig. 22 Projectile rinsed, air-dried, and ready for packaging

with a screw-on lid with the test area of the sticky tape facing into the clean bottle. The lid or cap is then replaced, and any necessary details are noted on a label placed on the outside of the container (Potter and Steyl 2020).

Reflecting back on the skin of an animal during necropsy can reveal not only the extensive trauma caused by bullet wounds (Fig. 23) but may also indicate lesions hidden under fur. This should all be documented.

Animals should be scanned with a microchip scanner (the entire animal should be scanned, as microchips may vary as to placement and can move underneath the skin), which may help in identification (Fig. 24). A microchip is a 10- or 15-digit alphanumeric passive transponder manufactured by different companies, with an indefinite lifespan, and can be implanted under the skin of an animal or into various animal parts, e.g., a rhino's horn.

Microchips are frequently used for domestic animals but can also be inserted into the wildlife that has been sedated for a routine veterinary or management purpose. A microchip scanner/reader is a handheld device that when waved above an animal will automatically display the alphanumeric code of the embedded chip. Not all microchip readers will read both 10- and 15-digit microchips and different microchip readers have different strengths, thus be aware of the capabilities and specifications of the reader being used. By ensuring that a reader who can interpret both the 10- and the 15-digit microchips is employed, the chances of recovering a microchip implanted into an animal is vastly improved. In order to be of benefit, accurate records must be kept of the serial numbers and locations used for

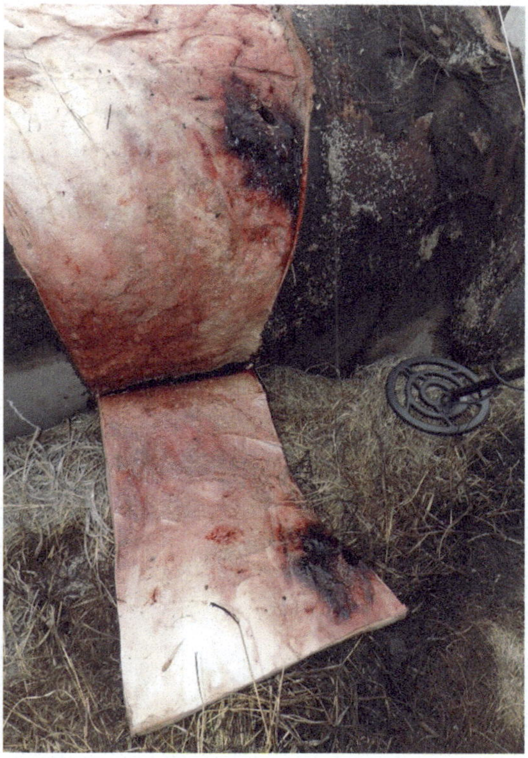

Fig. 23 Extensive hemorrhage from bullet wound visible on skin reflection

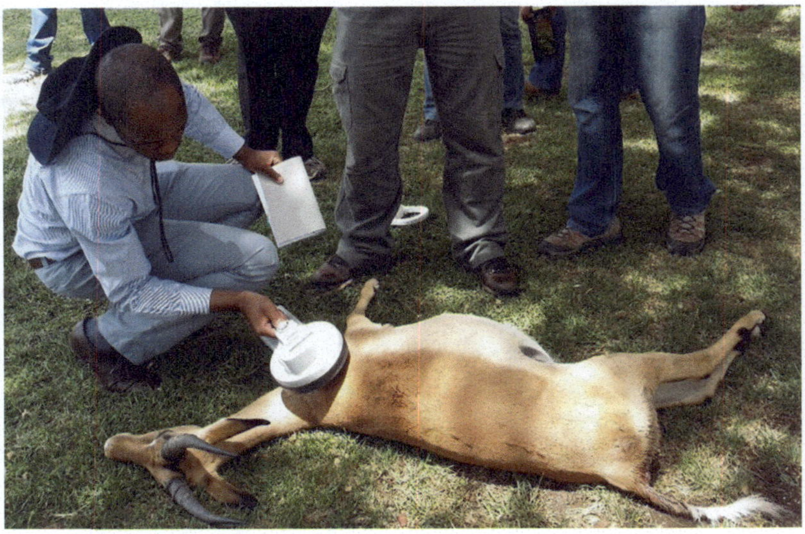

Fig. 24 Using a microchip scanner

the chosen system. For example, rhino horns taken from a rhinoceros with microchips fitted to the horns and later recovered will be conclusively proved to have originated from an individual rhinoceros—this will only happen if proper records are kept, preferably in a centralized database (Fig. 25).

Metal detectors can be used to scan an animal for the presence of firearm projectiles and are especially useful with large animals (Fig. 26). Depending on the skin and tissue thickness of the animal, the metal detector may not penetrate deep enough to detect projectiles.

DNA from Animal Victims or Other Evidence

Items suspected to have DNA should be packaged carefully to avoid contamination, i.e., water bottles or weapons. These items will often contain other forms of evidence such as fingerprints, and if testing is to be done by a laboratory, the items should be labeled as potentially containing multiple sources of evidence. Always consult with the lab to which samples will be sent to in order to

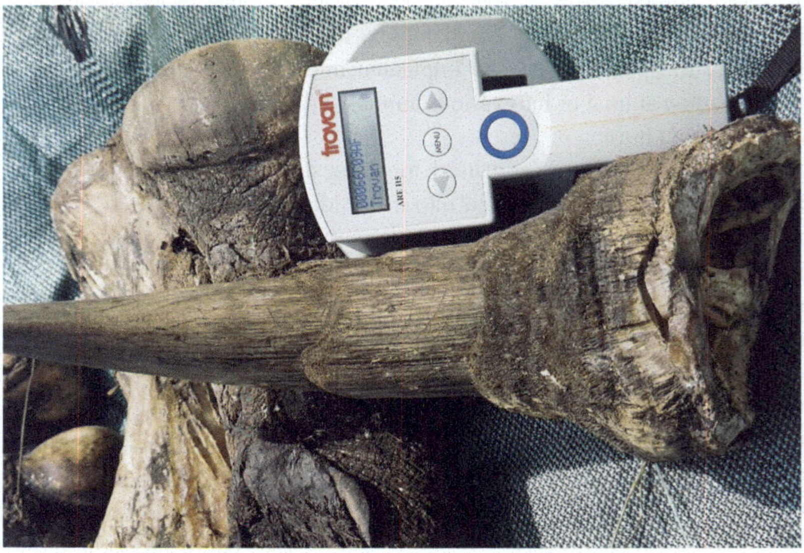

Fig. 25 Microchip identified in a rhino horn

Fig. 26 Searching for evidence using a metal detector

ascertain their sampling requirements. There are presumptive blood test kits commercially available, such as Hexagon OBTI, that can indicate whether a suspected bloodstain is of human or animal origin.

If DNA sampling is conducted on the scene, bloody items or other evidence should be wiped with sterile swabs in the area suspected of containing DNA. The swabs should then be allowed to air dry, and each swab placed in its own cardboard swab box and appropriately labeled. For dried blood stains, moisten the end of the swab with sterile or bottled water and rub the swab against the stain. Concentrate the sample at one point on the swab rather than rolling the swab over the sample area.

If animal tissue is to be used for DNA testing, most any tissue can be used (even teeth, horn, hooves, or tusks). Preferred samples are from heart muscle or other muscle tissue, brain, or hide that is not completely desiccated (Clark, personal communication). Tissue samples need not be larger than marble size. Ideally, they should be placed into a sealable container and frozen, or put into containers with SED (Saturated NaCl; 250 mM EDTA pH 7.5; 20% DMSO) buffer (see Appendix 8 for instructions). However, if buffer is not available and there is no access to freezers, very thinly sliced samples can also be air-dried and placed in salt (table salt is acceptable). If dried and then placed in salt, the dried tissue can be wrapped in paper and the packet covered in salt or the tissue and salt can be wrapped together in paper. The salt must be replaced regularly to keep it dry. Tissue samples can also be smoked thoroughly to dry, and then wrapped in paper (with or without) salt (Clark, personal communication).

Genetic as well as isotopic analysis of ivory may be used to ascertain an elephant's geographic location. This can be of great assistance in investigations resulting from ivory seizures, and it can also be used to determine the approximate age of ivory (Wasser et al. 2004, 2007, 2008; Ishida et al. 2013; Ziegler et al. 2016).

Releasing and Leaving the Scene

When all evidence has been collected and packaged and a thorough final recheck of the scene and surrounding areas has been conducted, all team members should compare notes to ensure that all protocols were followed, and nothing was missed. The scene should be released to appropriate personnel (property owner, etc.) At the lab or evidence storage facility, evidence items that were hastily packaged for transport can be repackaged properly and labeled. Additional close-up photographs can be taken at this point as well, in conditions that allow for more appropriate lighting. Consult regularly with all team members and the examining veterinarian as the final report is prepared.

Additional Evidence

Although space does not allow for descriptions of every type of evidence that may be encountered, there are some additional items worth elaboration.

Poisoning and Water/Soil Samples

In many areas, departments of environmental health will be able to presumptively test for toxins using a variety of sensors and other tools, but regardless of the findings, actual substrate samples must be obtained and sent for confirmatory testing. Always consult with the laboratory to which suspected poison samples will be sent as to their required or preferred collection protocols, including the amount of substrate to collect and submit. Evidentiary samples must be taken from nearby water bodies, food sources, local vegetation, or soil within the vicinity of where the animal(s) was found. Water and/or soil samples from suspected poisoning scenes should typically be placed into glass vials or bottles. Take caution when using plastic vials, as the chemicals in plastic may leach into an evidence sample. Confirm with the lab whether plastic or metal utensils should be used to obtain samples, as the chemical components of each can impact the sample in different ways. Relevant sample information can be handwritten on the outside of the vial, using a permanent marker on self-adhesive labels or tape affixed to the outside of the container. Place tape across the lid, extending onto the sides of the bottle, and sign across the seal. Ensure that containers of liquid are tightly closed before placing them in a vehicle for transport, and always ensure that packages are secure—use crumpled up paper or a clean cloth in between glass containers to keep them upright and protected from breakage. Keep detailed notes of the locations sampled and amounts of substrates taken, along with how they were collected.

One of the most important factors when dealing with potential hazards such as poison is the health and safety of the investigators and the general public. Poisons are often decanted into unmarked containers and sold in smaller quantities through informal and sometimes illegal methods and these containers can occasionally be found left at the scene. Such ubiquitous containers may be easily disregarded or mishandled. When working with the recovery of poisons, experts trained in hazardous materials should be called in to supervise the collection and handling of the evidence. The following safety guidelines should also be observed:

- Wear prescribed PPE as recommended by law enforcement or HAZMAT department.
- Do not allow skin contact with suspected poisons.
- Do not enter enclosed rooms where poisons are stored without first ventilating the room.
- Mark all evidence with "Hazardous" or "Poison" evidence labels.
- Cordon the area and do not allow the public into the vicinity of unknown chemicals.
- Do not open containers—if poison is suspected, let the laboratory do the analysis.

Poisoning and Darts

When poisoning is suspected, it is essential to obtain at least a basic knowledge of what type (s) of poisons are likely to have been used. This can be done by talking with local people to gather background information, ascertaining the availability and usage of poisons (for any reason), as well as any recent incidents of human–wildlife conflict.

There are a variety of dart guns that may resemble rifles, have a fitted scope, operate from compressed gas or a ballistic round, or may even consist of a simple hollow tube containing a projectile. All present safety hazards and require certain precautions regarding sampling and

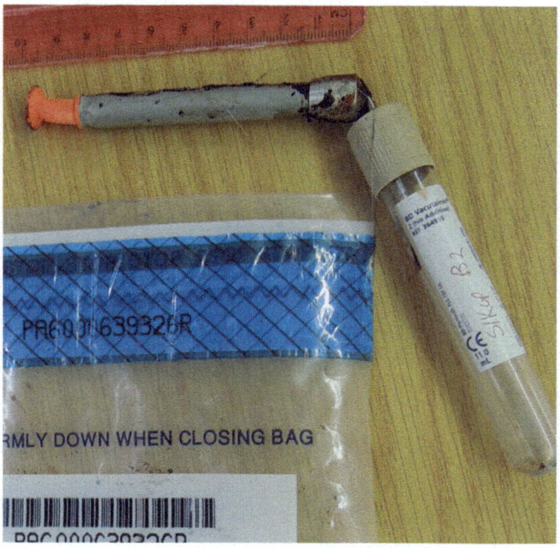

Fig. 27 Dart with a bent needle (as found) inserted into cap of vacutainer (note: needle was not straightened)

packaging. Any darts or arrows must be packaged appropriately, ideally within cardboard boxes or tubes.

When collecting darts with needles, use a vacutainer that contains no additives (usually those with a tan-opaque-colored cap [this color indicates that it is lead-free with a normal vacuum]) and pierce the rubber cap or stopper of the vacutainer with the dart needle (Fig. 27). Leave the needle in the cap at all times, including while packaging and transporting. Insert the needle into the rubber cap of the vacutainer immediately when removing a dart from an animal to protect anyone handling the dart after collection, or when packaging or swabbing the dart for human DNA. This will also prevent the liquid from possibly escaping after the dart is withdrawn from the skin. When withdrawing a dart from the skin, wear goggles or a face shield and mask, and always wear clean gloves.

The needle of the dart may be manufactured with or without a barb, it may or may not have a side port, and it may have one of various shapes of collars (where the needle joins the body of the dart). The needle itself, as well as the body of the dart, can vary in length and material. All of these factors can indicate a particular type of dart and the possible origin of manufacturer and distributor. When the dart is not found in the animal, the dart site must be carefully examined for traces of the barb or small rubber O-ring used to cover the side port on some darts. The O-ring originates from around the needle and covers the ejection port on the side of the needle. Not all darts have an O-ring; some only eject from the end of the needle while others eject from both the end and the side port. Recovery of the rubber ring at the dart site may indicate a particular brand of dart.

The outside of the dart should be handled as little as practically possible, as there may be fingerprints on the dart surface or human DNA from handling. Gasp the dart where it is not normally held in order to preserve fingerprints, e.g., on the flight (part of the dart that attaches to the shaft; these can vary in structure, shape, and material) or just behind the needle. As there may be a need to attempt to recover DNA from the dart to link it to a specific animal, swab the area of the dart most likely in contact with the skin of the animal at the time of impact (Fig. 28). Darts can also have a tissue plug from the animal skin, or there may be blood in the dart which has pushed back into the dart via the needle after the chemical has been ejected, and this material can be used for DNA analysis. Upon collection, darts should be frozen if possible, to slow any chemical reaction which would break down the poison. Testing must be done as soon as possible, preferably within 24 h.

Detailed procedures for taking forensic samples of animals suspected to be the victims of poisoning can be found in Cooper and Cooper (2013), and in veterinary medical texts.

Botany and Entomology

Botanical evidence can be exceptionally helpful in wildlife forensic cases, not only for indicating possible paths of the perpetrators, as mentioned previously, but also for linking suspects with crime scenes. For example, certain plants found growing at a crime scene may only grow in that particular area; their presence on vehicles or suspects found far from the scene could connect

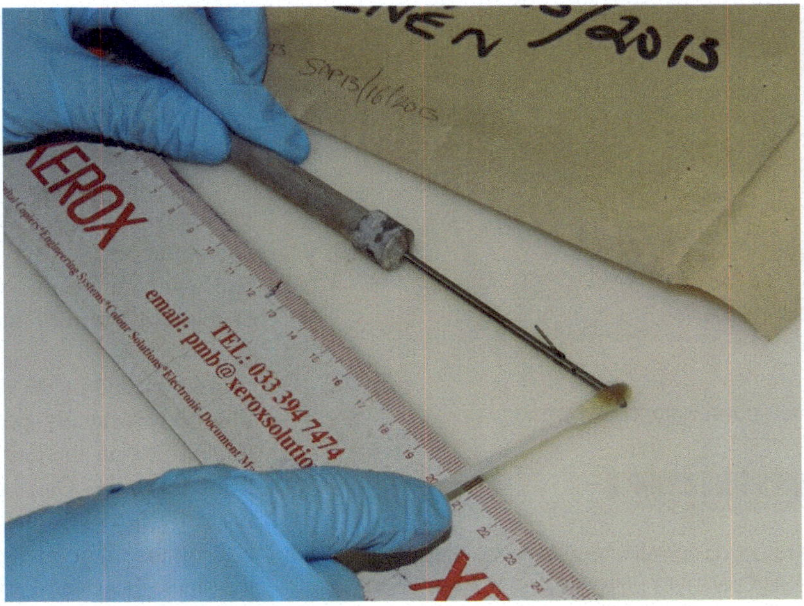

Fig. 28 Swabbing a dart found in the field for animal DNA

them with the crime. It is imperative that vehicle wheel wells, chassis, grills, radiators, and interior mats and upholstery are searched thoroughly, along with the clothing and shoes of suspects, and any plant material collected appropriately. Small plant fragments can be placed into paper envelopes; larger items need to be placed into a commercially available or hand-made plant press, allowed to dry, and sent to a forensic botanist as soon as possible. Broken branches should be cut a few inches away from the break and placed into a plant press; botanists may be able to determine the approximate age of the break by examining the break and any sap present. Similarly, pollen, spores, fungi, and algae can be obtained from people and objects and used to ascertain possible presence in a certain location, and certain types of algae, i.e., diatoms, can assist in the possible determination of drowning. Further information on constructing a plant press, proper botanical sampling, and collection of plant material can be found in Coyle (2004) and Hall and Byrd (2012).

Entomological evidence can be of great value in a wildlife case to assist in the determination of postmortem interval or the time between death and discovery of the body. When combined with environmental conditions and weather data obtained for the area where the body is found, insect succession has a fairly accurate rate of time assessment (Anderson 2010). Wildlife trackers and livestock owners local to the area may have a fairly good idea of the rate of decomposition of a large carcass in that location. However, an investigator must obtain independent professional assessments that can be combined with the opinion of the attending veterinarian. Insects tend to colonize a carcass in well-documented stages (for some regions insect succession has not been well-studied, and all species of insect have not been identified; the use of forensic entomology in these areas therefore will require significantly more time). Having a forensic entomologist present on the scene to collect samples is ideal; however, if this is not possible, the investigator should obtain samples (both live and killed specimens) of the Diptera (fly) species on or around the carcass (both aerial and in maggot [larvae] form) as well as Coleoptera (beetle) species. These samples should be sent via express courier to a qualified forensic entomologist (notify the forensic entomologist prior to shipment and enquire as to any specific sampling methods he or she requires). Detailed information on carrion ecology and arthropod communities can be found in chapter "Carrion Ecology", and Benbow et al. (2016). Proper entomological sampling techniques can be found in Gennard (2012); Byrd and Tomberlin (2020), and many publications within the scientific literature.

Conclusion

Successful prosecutions and hefty sentences acting as, effective deterrents against wildlife and conservation criminal activity can only occur if crime scenes are investigated and processed correctly and if everyone works together toward a common goal. It is the authors' hope that the field of wildlife forensic science continues to expand and strengthen and that more attention will be paid to the importance of crime scene investigation techniques as applied to wildlife crime. We need to teach, encourage and empower wildlife investigators because the successful application of forensic science at wildlife crime scenes will close investigation loopholes, replace speculation with fact and go a long way toward countering the worldwide illegal trafficking of wildlife products in order to preserve biodiversity and our wildlife heritage.

Acknowledgments The authors would like to thank and dedicate this chapter to all of the wildlife rangers and conservation officers who daily put their lives on the line to protect and conserve the wildlife that remains. You are the true heroes.

Appendix 1

Recommended Basic Crime Scene Kit Supply List

Ballpoint pens/permanent markers/pencils (no gel ink pens); clipboards
Gloves (latex/nitrile)
Shoe covers; goggles; face shields; masks; protective suits
Paper bags—assorted sizes
Paper envelopes—assorted sizes
Plastic storage containers with lids—assorted sizes
Heavy-duty tubes and boxes
Plastic and Glass bottles/vials with lids—assorted sizes
Plastic reclosable storage bags—assorted sizes
Measuring tapes (of various lengths)
Rulers—assorted lengths
Evidence tent-shaped placards or construction flags (improvisation: cardboard)
Tags and wire attachments/Zip ties
Evidence tape
Plastic tape for trace evidence collection; white card stock or cardboard to use as backing
Barrier tape or rope to cordon off crime scene
Forceps/tweezers (plastic and stainless steel)
Scissors (stainless steel)
Magnifying glass
Binoculars
All-purpose tool
Camera with extra SD cards
Flashlight/torch; spare batteries
Metal rods or steel Rebar and hammer
GPS device
Insect collection net
Plant press
Printed logs and forms
Dental stone (impression collection casting material); water and plastic sealable bag for mixing
Hexagon OBTI blood testing kit
Swabs and swab boxes for DNA collection
SED buffer for tissue samples
Sterile water
Scalpels and blades/knives
Syringes and needles
Vacutainers (various)
Metal detector
Microchip scanner
Peel-off labels

Appendix 2

<table>
<tr><th colspan="5">CHAIN OF CUSTODY RECORD</th></tr>
<tr><td colspan="3">DATE, TIME AND PLACE OF SEIZURE:</td><td colspan="2">EVIDENCE/PROPERTY SEIZED BY:</td></tr>
<tr><td colspan="3">SOURCE OF EVIDENCE/PROPERTY (PERSON AND/OR LOCATION):
☐ TAKEN FROM:
☐ RECEIVED FROM:
☐ FOUND AT:</td><td colspan="2">CASE TITLE AND REMARKS:</td></tr>
<tr><td>ITEM NO.</td><td colspan="4">DESCRIPTION OF EVIDENCE/PROPERTY (INCLUDE SEIZURE TAG NUMBERS AND ANY SERIAL NUMBERS):</td></tr>
<tr><td rowspan="2">ITEM NO.</td><td>FROM: (PRINT NAME AND RANK)</td><td>RELEASE SIGNATURE:</td><td>RELEASE DATE</td><td rowspan="2">DELIVERED VIA:
☐ POST
☐ IN PERSON
☐ OTHER:</td></tr>
<tr><td>TO: (PRINT NAME AND RANK)</td><td>RECEIPT SIGNATURE:</td><td>RECEIPT DATE</td></tr>
<tr><td rowspan="2">ITEM NO.</td><td>FROM: (PRINT NAME AND RANK)</td><td>RELEASE SIGNATURE:</td><td>RELEASE DATE</td><td rowspan="2">DELIVERED VIA:
☐ POST
☐ IN PERSON
☐ OTHER:</td></tr>
<tr><td>TO: (PRINT NAME AND RANK)</td><td>RECEIPT SIGNATURE:</td><td>RECEIPT DATE</td></tr>
</table>

CHAIN OF CUSTODY RECORD (CONTINUED)

ITEM NO.	FROM: (PRINT NAME AND RANK)	RELEASE SIGNATURE:	RELEASE DATE	DELIVERED VIA: ☐ POST ☐ IN PERSON ☐ OTHER:
	TO: (PRINT NAME AND RANK)	RECEIPT SIGNATURE:	RECEIPT DATE	
ITEM NO.	FROM: (PRINT NAME AND RANK)	RELEASE SIGNATURE:	RELEASE DATE	DELIVERED VIA: ☐ POST ☐ IN PERSON ☐ OTHER:
	TO: (PRINT NAME AND RANK)	RECEIPT SIGNATURE:	RECEIPT DATE	
ITEM NO.	FROM: (PRINT NAME AND RANK)	RELEASE SIGNATURE:	RELEASE DATE	DELIVERED VIA: ☐ POST ☐ IN PERSON ☐ OTHER:
	TO: (PRINT NAME AND RANK)	RECEIPT SIGNATURE:	RECEIPT DATE	
ITEM NO.	FROM: (PRINT NAME AND RANK)	RELEASE SIGNATURE:	RELEASE DATE	DELIVERED VIA: ☐ POST ☐ IN PERSON ☐ OTHER:
	TO: (PRINT NAME AND RANK)	RECEIPT SIGNATURE:	RECEIPT DATE	
ITEM NO.	FROM: (PRINT NAME AND RANK)	RELEASE SIGNATURE:	RELEASE DATE	DELIVERED VIA: ☐ POST ☐ IN PERSON ☐ OTHER:
	TO: (PRINT NAME AND RANK)	RECEIPT SIGNATURE:	RECEIPT DATE	
ITEM NO.	FROM: (PRINT NAME AND RANK)	RELEASE SIGNATURE:	RELEASE DATE	DELIVERED VIA: ☐ POST ☐ IN PERSON ☐ OTHER:
	TO: (PRINT NAME AND RANK)	RECEIPT SIGNATURE:	RECEIPT DATE	

Appendix 3

CRIME SCENE LOG (Version 3)

STATION & CASE #		DATE	LOCALITY	
LOG KEEPER		Name:		
Details of persons entering the scene				
NAME	DESIGNATION	TIME IN	TIME OUT	ANYTHING REMOVED? SPECIFY

Log Keeper signature: ______________________

Appendix 4

CRIME SCENE
Photo Log

Case Number:______________________ Page:______
Primary Investigator:_______________________________________
Photographer:____
Date:__________________ Time: ___________
Location:____

Photo #	Description: Items-Location

Appendix 5

Evidence Measurements/General Information

Area/Activity

#	Item	Size	Ht	Measure-ments	Remarks	L*	E*

Officers	Date

Comments	*Use different pages of this form for different areas of investigation. *L – LAB WORK NEEDED *E – EVIDENCE RETAINED

Appendix 6

Evidence Log

Case No.:

Item Number	Date	Location Found	Collected/Packaged By	Description/Notes

Appendix 7

WILDLIFE CRIME SCENE SUMMARY REPORT

DATE :________________ **NAME** (Of person making this report) : ______________________

SIGNATURE : ________________________ **CONTACT No** : _____________________

LOCATION OF CRIME SCENE (Address/Grid Ref/UTM/GPS Ref/Description/Directions)
____
____
____

POLICE STATION : _____________________ CASE No : _____________________

OFFENCE DESCRIPTION:____
____
____

COMPLAINANT : Name : ____
Address : ____
_________________ Tele numbers Work :______________ Mobile : ______________

POLICE MEMBERS PRESENT : YES ☐ NO ☐
Detective :____ Phone : _______________
Scene of Crime Officer : ___________________________ Phone : _______________
Other : ____Phone : _______________
Other : ____Phone : _______________

INVESTIGATIONS MEMBERS PRESENT : YES ☐ NO ☐
Name : ____ Phone : _______________
Name : ____ Phone : _______________

PHOTOGRAPHS TAKEN : YES ☐ NO ☐
Cameraman : ______________________________________ Phone : _______________
Cameraman : ______________________________________ Phone : _______________

DATE : _________________ TIME : _______________SIGNATURE : ___________

ATTACH GOVERNMENT TOPOGRAPHICAL MAP TO THIS REPORT, TO INDICATE LOCATION OF CRIME SCENE ON 1:50 000 SCALE MAP, IF AVAILABLE.

REFER TO ROUGH SKETCH ON REVERSE

ROUGH SKETCH OF CRIME SCENE

(TO BE USED FOR COMPILING YOUR SKETCH MAP TO CRIME SCENE)

NORTH

ROUGH SKETCH NOT TO SCALE

DATE : ____________________________

TIME : ____________________________

AREA : ____________________________

DRAWN BY : ____________________

SIGNATURE : ____________________

ROUGH SKETCH NUMBER : _____

MAP CO-ORDINATES / UTM OR

GPS READING (TAKEN AT POINT _):

____

____

KEY TO **ROUGH SKETCH** AND **SKETCH MAP**

POINT **DESCRIPTION (of item at point)**

1 - ____________________
2 - ____________________
3 - ____________________
4 - ____________________
5 - ____________________
6 - ____________________
7 - ____________________
8 - ____________________
9 - ____________________
10 - ____________________

DISTANCE IN PACES

POINT - DISTANCE		POINT - DISTANCE	
1 to 2	________	4 to 5	________
1 to 3	________	4 to 6	________
1 to 4	________	4 to 7	________
1 to 5	________	4 to 8	________
1 to 6	________	4 to 9	________
1 to 7	________	4 to10	________
1 to 8	________		
1 to 9	________	5 to 6	________
1 to10	________	5 to 7	________
		5 to 8	________
2 to 3	________	5 to 9	________
2 to 4	________	5 to10	________
2 to 5	________		
2 to 6	________	6 to 7	________
2 to 7	________	6 to 8	________
2 to 8	________	6 to 9	________
2 to 9	________	6 to10	________
2 to10	________		
		7 to 8	________
3 to 4	________	7 to 9	________
3 to 5	________	7 to10	________
3 to 6	________		
3 to 7	________	8 to 9	________
3 to 8	________	8 to10	________
3 to 9	________		
3 to 10	________	9 to10	________

COMPILED BY : ____________________ SIGNATURE : ____________________

SKETCH MAP OF CRIME SCENE
(TO BE COMPLETED BACK AT THE OFFICE)

NORTH

SKETCH MAP NOT TO SCALE
APPROX SCALE : _______________
DATE : _________________________
TIME : _________________________
AREA : _________________________
DRAWN BY : ___________________
SIGNATURE : ___________________

SKETCH MAP NUMBER : _____
MAP CO-ORDINATES / UTM OR
GPS READING :
____
____

Appendix 8

UF | Maples Center for Forensic Medicine
UNIVERSITY *of* FLORIDA

4800 S.W. 35th Drive
Gainesville, FL 32608
352-294-4487

PROTOCOL: COLLECTING TISSUE WITHOUT REFRIGERATION FOR FORENSIC ANALYSIS

Mailing and Freight Address: Maples Center for Forensic Medicine, University of Florida, 4800 SW 35th Drive, Gainesville, FL 32608 *Advanced notification of shipment is required!*

Phone: 352-294-4487

MATERIALS FOR COLLECTING TISSUE

- SED buffer: Saturated NaCl; 250 mM EDTA pH 7.5; 20% DMSO
- Screw-cap tubes or other sealed storage containers
- Razor blade or scalpel (sterile is ideal but clean will suffice)
- Disposable gloves (highly recommended)

TISSUE SOURCE

Almost any tissue will suffice, but the preferred sources are (1) muscle tissue, (2) brain or heart muscle, (3) hide that has not desiccated. We have a separate protocol for blood, which can also be preserved without refrigeration. Tissues that have been previously frozen are acceptable.

METHOD FOR COLLECTING TISSUE

- Wearing gloves and using a clean razor blade, scalpel, or sharp knife collect a piece of tissue the size of a large marble from area not exposed to the environment or insects.
- Chop the tissue a few times with the blade to increase penetration of buffer.
- Add the tissue to a properly, uniquely labelled tube containing SED buffer.
- Change gloves between samples or wash and dry gloves and knife before the next sample.
- Samples can be stored at room temperature for several months or in a refrigerator for at least a year. Avoid extended exposure to heat or sunlight.
- Prior to shipping, reseal the tubes carefully and double wrap in an airtight plastic bag (ziploc or equivalent) to prevent leakage.

Ship samples by air freight or express mail. *Advanced notification of shipment is required!*

Note 1: SED buffer is nontoxic, nonflammable, and can be stored indefinitely at room temperature. Since this buffer contains saturated salt (NaCl), you may find a white precipitate in some tubes. This does not affect the ability of the buffer to preserve tissue.

Note 2: Handling SED buffer without gloves may result in exposure to DMSO. This substance soaks into skin very rapidly and is commonly used as a carrier for medications to alleviate muscle aches. DMSO will produce a vague garlic/oyster taste in the mouth along with a comparable breath odor. While not damaging, it is recommended that gloves be worn to prevent exposure.

Note 3: To make SED buffer:

- Dissolve 95 g tetrasodium EDTA in 700 ml distilled water (1 liter)
- pH to 7.5 with glacial acetic acid
- Saturate with NaCl, about 200 g. Allow salt to dissolve completely
- Add 200 ml DMSO, bring up to 1 liter with distilled water

*protocol modified from 1) Amos, B. and A.R. Hoelzel. 1991. Long-term preservation of whale skin for DNA analysis. Rep. Int. Whal. Comm. Special Issue 13:99-103; and 2) Proebstel D S R.P Evans, D.K. Shiozawa, and R.N. Williams, 1993. Preservation of nonfrozen tissue samples from a salmonine fish Brachymystax lenok (Pallas) for DNA analysis. Journal of ichthyology 9:9-17.

References

Anderson G (2010) Factors that influence insect succession on carrion. In: Forensic entomology: the utility of arthropods in legal investigations, pp 201–250

Benbow ME, Tomberlin JK, Malone AM (2016) Carrion ecology, evolution, and their applications. CRC, Boca Raton, FL

Brooks JW (2018) Veterinary forensic pathology, vol 1 & 2. Springer, Cham

Byrd JH, Tomberlin J (2020) Forensic entomology: the utility of arthropods in legal investigations, 3rd edn. CRC, Boca Raton, FL

Clark AM. Maples Center for Forensic Medicine, University of Florida, Gainesville, FL. Personal communication

Cooper JE, Cooper ME (2007) Introduction to veterinary and comparative forensic medicine. Wiley-Blackwell, West Sussex, UK

Cooper JE, Cooper ME (2013) Wildlife forensic investigation: principles and practice. CRC, Boca Raton, FL

Cooper J, Cooper ME, Budgen P (2009) Wildlife crime scene investigation: Techniques, tools and technology. Endangered Species Res 9:229–238

Coyle HM (2004) Forensic botany: principles and applications to criminal casework. CRC, Boca Raton, FL

Dowler K, Flemming T, Muzzatti SL (2006) Constructing crime: Media, crime, and popular culture. Can J Criminol Crim Justice 4(6):838–850

Duncan CD (2015) Advanced crime scene photography, 2nd edn. CRC, Boca Raton, FL

Fisher BAJ, Fisher DR (2012) Techniques of crime scene investigation, 8th edn. CRC, Boca Raton, FL

Gardner RM (2012) Practical crime scene processing and investigation, 2nd edn. CRC, Boca Raton, FL

Gardner RM, Krouskup D (2019) Practical crime scene processing and investigation, 3rd edn. CRC, Boca Raton, FL

Gennard D (2012) Forensic entomology: an introduction, 2nd edn. Wiley-Blackwell, West Sussex, UK

Girard JE (2015) Criminalistics: forensic science, crime and terrorism, 2nd edn. Jones & Bartlett Learning, Burlington, MA

Hall DW, Byrd JH (2012) Forensic botany: a practical guide. Wiley-Blackwell, West Sussex, UK

Horan J, Rothchild R (1980) Introduction to forensic science: an advanced elective for high schools. J Chem Educ 57(8):599–600

Houck MM, Siegel JA 2010 Fundamentals of forensic science, 2nd edn. Elsevier, Oxford

Huffman JE, Wallace JR (2012) Wildlife forensics: methods and applications. Wiley-Blackwell, West Sussex, UK

Ishida Y, Georgiadis NJ, Hondo T, Roca AL (2013) Triangulating the provenance of African elephants using mitochondrial DNA. Evol Appl 6(2):253–265

Ogle RR, Plotkin SL (2018) Crime scene investigation and reconstruction, 4th edn. Pearson Education, New York

Potter RB, Steyl J (2020) Wildlife crime scene investigations. Unpublished training manual

Robinson EM (2016) Crime scene photography, 3rd edn. Elsevier, London

Rogers ER, Stern AW (2018) Veterinary forensics: investigation, evidence collection, and expert testimony. CRC, Boca Raton, FL

Saferstein R (2019) Forensic science: from the crime scene to the crime lab, 4th edn. Pearson, New York

UNODC, World Wildlife Crime Report 2020, United Nations Office on Drugs and Crime, 2020. https://www.unodc.org/unodc/en/data-and-analysis/wildlife.html

Walker DN, Adrian WJ (2012) Wildlife forensic field manual, 4th edn. Association of Midwest Fish and Game Law Enforcement Officers, Fort Collins, CO

WWF (2018) Over 100 wildlife rangers died on duty in past year. Retrieved August 2020 from https://phys.org/news/2018-07-wildlife-rangers-died-duty-year.html

Ziegler S, Merker S, Streit B, Boner M, Jacob DE (2016) Towards understanding isotope variability in elephant ivory to establish isotopic profiling and source-area determination. Biol Conserv 197:154–163

Part III

Species-Specific Methodologies and Special Topics in Conservation Forensics and Biodiversity Protection

Gorilla Conservation and One Health

Gladys Kalema-Zikusoka, Alex Ngabirano, and Stephen Rubanga

Abstract

Disease is one of the key threats to gorillas along with high human population growth rates, habitat loss, and poaching of wild animals in their habitat. Though mountain gorillas were once critically endangered, a One Health approach has contributed to reversing the trend and increasing the number from 700 to just over 1000 in the past 17 years, leading to a downgrading of the IUCN status of this subspecies to endangered. Conservation Through Public Health (CTPH) founded in 2003, establish a One Health or integrated Population Health and Environment (PHE) approach that addresses human, animal, and ecosystem health together in order to reduce threats to wildlife and fragile ecosystems. CTPH established integrated programs that improve wildlife health, conservation attitudes and practice, community health, and alternative livelihoods. This includes regular gorilla health monitoring and comparative pathogen analysis and support to community volunteers including Village Health and Conservation Teams (VHCTs) and Human and Gorilla Conflict resolution (HUGO) teams; as well as, providing premium prices for good coffee from farmers bordering Bwindi Impenetrable National Park. This One Health approach to gorilla conservation has contributed to reduced disease incidences in the gorillas, increased adoption of family planning methods, improved hygiene and attitudes toward conservation, and reduced conflict between people and gorillas. CTPH has built upon this award winning model that addresses 10 out of the 17 Sustainable Development Goals (SDGs) to mitigate the impact of COVID-19 on gorilla conservation.

Keywords

Mountain gorillas · Conservation · One Health · Zoonoses · Disease · Sustainable development goals

Introduction

According to the IUCN, mountain gorillas (*Gorilla beringei beringei*) are classified as an endangered species, with an estimated population of 1063 free-ranging individuals. An estimated 459 individuals inhabit Bwindi Impenetrable

G. Kalema-Zikusoka (✉)
Conservation Through Public Health, Entebbe, Uganda

Gorilla Conservation Coffee, Entebbe, Uganda
e-mail: gladys@ctph.org

A. Ngabirano
Conservation Through Public Health, Entebbe, Uganda

Bwindi Development Network, Kanungu, Uganda
e-mail: alex@bwindidevelopmentnetwork.org

S. Rubanga
Conservation Through Public Health, Entebbe, Uganda
e-mail: rubanga@ctph.org

S. C. Underkoffler, H. R. Adams (eds.), *Wildlife Biodiversity Conservation*,
https://doi.org/10.1007/978-3-030-64682-0_13

Fig. 1 Photograph of a mountain gorilla at Bwindi Impenetrable National Park, Uganda

National Park in southwestern Uganda, with the remainder inhabiting the Virunga Volcanoes Mountain Range that spans Rwanda, the Democratic Republic of Congo (DRC), and Mgahinga National Park in Uganda. The IUCN status of the mountain gorilla was downlisted from critically endangered to endangered in 2018 due to positive growth trends over the past two decades. The mountain gorilla is also the only gorilla subspecies in recorded history to demonstrate a rise in population this century, from 650 in 1998 to 1063 in 2018 (Hickey et al. 2018) (Figs. 1 and 2).

Threats to Gorillas

Habitat loss, poaching, competition for food, poverty, high human population growth, human–wildlife conflict, and the spread of zoonotic diseases between people, wildlife, and livestock all threaten the continued survival of endangered mountain gorillas and other gorilla subspecies.

Gorillas and other nonhuman great apes are particularly vulnerable to zoonotic diseases because they share 98.4% genetic material with humans, and are thus highly susceptible to

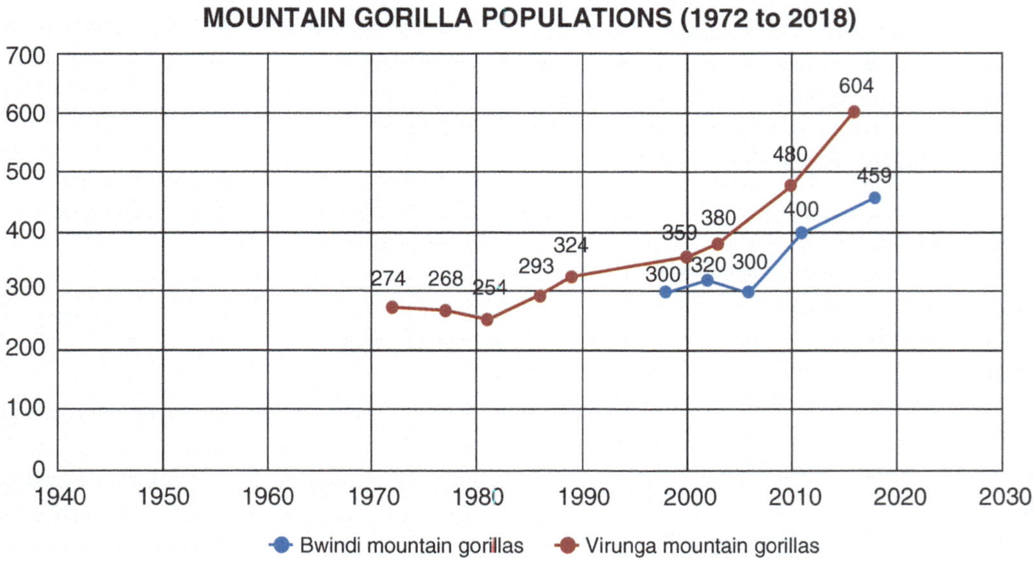

Fig. 2 Mountain gorilla population census results from 1970 to 2020

virulent pathogens spreading between humans and nonhuman great apes (Köndgen et al. 2011; Scully et al 2018; Coscolla et al. 2013, Patrono et al. 2018; Grutzmacher et al. 2018; Negrey et al. 2019). Pathogens of concern include those of global pandemic scale such as SARS (Severe Acute Respiratory Syndrome), novel coronaviruses (COVID-19), and Ebola, as well as endemic human pathogens such as measles, and scabies, a contagious skin disease spread via direct contact with *Sarcoptes scabiei* mites (Graczyk et al. 2001; Kalema-Zikusoka et al. 2002; Walsh et al. 2003; Melin et al. 2020; Damas et al. 2020). Outbreaks have occurred in mountain gorillas, from contact with the local communities, resulting in morbidity in many gorilla groups, with a few mortalities. Tourists have also been implicated as potential reservoirs of disease transmission. In 2009, an outbreak of human metapneumovirus, a respiratory illness was traced to a tourist outside Rwanda, and resulted in the death of a mother and infant gorilla (Palacios et al. 2011).

One Health

One Health is described as an approach that promotes human, animal, and ecosystem health as a whole, and thus encourages veterinarians, medical practitioners, and conservationists to work together to address issues that require a multidisciplinary solution, such as zoonotic disease prevention and control, and promotion of environmental health.

This chapter will focus on how the One Health approach has been implemented in order to achieve conservation outcomes. We will illustrate this with a case study of Conservation Through Public Health (CTPH), a grassroots nongovernmental and nonprofit organization established in 2003 to prevent and control disease transmission at the human/wildlife/livestock interface, while cultivating a winning attitude to conservation and public health in local communities. The mission of CTPH is to promote biodiversity conservation by enabling people, gorillas, and other wildlife to coexist through improving health and community livelihoods in and around protected areas in Africa. CTPH's goal is for people, wildlife, and livestock to live together in balance, health, and harmony, with local communities acting as stewards of their environment. CTPH envisions a world where gorillas live in a secure environment, coexisting with healthy communities. To achieve these goals, CTPH has three main integrated programs: wildlife health, community health, and alternative livelihoods. CTPH aims to:

- Promote biodiversity conservation by reducing threats to gorillas and their habitat through proactive health monitoring of gorillas; preventive care for gorillas, people, and livestock; and emergency care for animals.
- Provide an early warning system for disease outbreaks through regular analysis of gorilla fecal samples, and comparative pathogen analyses of people, gorillas, and livestock, in order to minimize cross-species disease transmission.
- Promote community hygiene and sanitation and reduce infectious disease transmission by encouraging health-promoting behaviors.
- Improve community reproductive health outcomes using an integrated approach to population, health, and the environment (PHE).
- Improve community attitudes toward conservation practices.

This chapter describes how a One Health approach contributes to gorilla conservation. Direct impacts include the reduction of anthropozoonotic disease transmission. Indirect impacts include improved public health and family planning outcomes for local communities, and improved attitudes toward gorilla conservation, which in turn may reduce dependence on the utilization of vital gorilla habitat for food and firewood.

Assessment of Disease Risks to Gorillas

Gorillas come into close contact (less than 7 m) with humans through the following ways:

foraging outside their protected habitat, research, and tourism (Kalema-Zikusoka and Rwego 2016; Weber et al. 2020). This close proximity increases the risk of transmission. Bwindi Impenetrable National Park currently has 18 habituated gorilla groups utilized for tourism and research. Approximately half of these gorilla groups also periodically leave the park to forage on community land an estimated 30% of the time.

Outbreaks of scabies have occurred among mountain gorillas in 1996 (Kalema-Zikusoka et al. 2002), and again between 2000 and 2001 (Graczyk et al. 2001). The outbreaks were traced to local communities with limited access to basic health care and adequate hygiene practices. This is an example of anthropozoonotic disease transmission, or transmission of a pathogen from humans to animals. The 1996 outbreak resulted in a high morbidity rate of all gorilla group members who displayed symptoms of alopecia, pruritus, and scaling and crusting of the skin, and one fatality of an 8-month old in the Katendegyere group. The severity of the disease was age and size related, with the infant being the worst affected, and the lead silverback gorilla exhibiting only scratching with no hair loss (Kalema-Zikusoka et al. 2002). Three of the four individuals in the Katendegyere gorilla group only recovered after being treated with ivermectin using a remote injection (dart).

In 2000, another scabies outbreak affected 17 gorillas who also developed the same symptoms, and recovered after treatment with ivermectin (Graczyk et al. 2001). The source of the outbreak was again traced to the local human community surrounding Bwindi Impenetrable National Park, Uganda (Graczyk et al. 2001). Gorillas have opportunities for close contact with communities surrounding the Park when they range outside of park boundaries to forage on banana plants and other appealing crops. Gorillas are curious animals, and it is theorized that they may have touched infected clothing placed on scarecrows to deter wildlife from eating crops (Kalema-Zikusoka et al. 2002).

Bwindi Impenetrable National Park is surrounded by some of the most economically depressed regions in Africa, where access to proper hygiene, sanitation, and basic health services is limited. There is significant spread of communicable disease among the human population, and the rate of chronic immunosuppressive diseases such as tuberculosis and HIV/AIDS is high (https://www.cdc.gov/globalhivtb/cdcs-impact/cdconthefrontlines.html).

Bwindi Impenetrable National Park is 321 square kilometers and considered by many conservationists as an island of biodiversity in Uganda amidst a sea of human habitation. The human population surrounding Bwindi has one of the highest global densities, at 200–600 people per square kilometer (Kalema-Zikusoka 2013). When CTPH programs began, the average family size was 10, and higher than the Ugandan national average of 7 (Uganda Demographic and Health Survey 2016). Thirteen years after CTPH implemented a community-based family planning program, the fertility rate has reduced to 4.2 children per family in Kanungu District with the two parishes where the program was implemented. Additionally, Uganda has one of the highest global population growth rates, at 3.1% in the latest Demographic and Health Survey (UDHS 2016). These figures contribute to the increasing human pressure for land utilization, resulting in a hard edge with no buffer zone around much of the park, exacerbating human, gorilla, and other wildlife conflicts.

Promoting Health and Conservation Using a Community-Based Model

CTPH's community health program is centered around strengthening community-based health care by working with Village Health Teams (VHTs) established by the Ministry of Health (MOH). The field program began in 2004 in two high human–gorilla conflict parishes, Mukono and Bujengwe, in Kanungu District, where CTPH implemented ongoing educational programs for local communities. Trusted members of the community were selected in 2007, based on their ability to read and write, and were trained by CTPH to conduct both health and conservation outreach, creating a model of Village Health and Conservation Teams

(VHCTs). Thus, CTPH links the community to local health centers through the VHCTs who are trained in the promotion of hygiene and sanitation; infectious disease prevention, referral, and control with a focus on zoonoses; family planning promotion, including the administration of family planning measures such as the Depo-Provera injection for women; advancement of good nutrition; and the referral of children with malnutrition to the nearest health centers. The VHCTs additionally promote gorilla conservation, forest conservation, and sustainable agriculture. The program expanded to the neighboring Kisoro District in 2010, and the most active VHTs were trained to become VHCTs.

In addition to the risk of anthropozoonotic disease transmission, gorillas may also transmit emerging infectious diseases such as Ebola to humans. The Ebola case fatality rate in humans is as high as 90%, and often associated with consumption of meat from monkeys and close contact with infected people. However, at Bwindi Impenetrable National Park, people do not have a culture of consuming monkeys though they hunt duiker and bush pigs in the forest bringing them into closer contact with gorillas. VHCTs place an emphasis on decreasing such risks through the avoidance of bushmeat consumption, which may include great apes in other countries.

Evaluating Monthly Data on Positive Behavior Change Within the Local Communities

From the outset, CTPH established a system to monitor and evaluate human behavioral change each month. After external and internal reviews of the methodology, a decision was made to develop a simpler data collection tool, which mirrored MOH methodology, namely the ticking of indicators, and individual pages for each household. This system minimizes errors in data entry by simplifying collection and reducing the use of subjective assessment, facilitating the ease of both monitoring and evaluation. CTPH made a strategic decision in 2018 to engage all VHTs as VHCTs, based on a recommendation from Travers et al. (2019) to improve conservation outcomes. This reduced the number of homes each VHCT was responsible for, enabling them to visit homes more frequently and facilitated data sharing with MOH, who is ultimately responsible for the VHTs and other relevant health NGOs.

Results from a comparison between 60 households frequently visited by the VHCTs with a control group of 60 non-visited or much less frequently households over a duration of 10 years revealed that VHCT-visited homes had better markers for health and conservation, including hand-washing facilities outside of pit latrines, greater knowledge of human and gorilla disease transmission, knowledge of family planning, less human–gorilla conflict, and reduced dependence on the forest to meet their basic needs (Ainerukundo et al. 2019). This information was subsequently used to demonstrate the success of VHCTs and community outreach efforts by CTPH.

Village Health and Conservation Teams

In order to support the efforts of community health volunteers, CTPH has invested in a number of measures to increase their health, overall livelihood, and boost morale. CTPH encouraged the volunteers to establish group livestock income-generating projects. Income from these projects is reinvested into VHCT Village Saving and Loan Associations (VSLAs), and members meet monthly to lend and borrow money, as well as to exchange knowledge related to their work within the communities served. Additionally, CTPH conducts routine quarterly engagements with the groups, as well as on an individual basis, for progress review and continuing education in methodology and emerging issues. As a result of these efforts, there has been no volunteer loss or turnover among the 29 volunteer workers during the first 10 years of the program.

Training Park Staff and Community Volunteers to Monitor Gorilla Health

CTPH trains park staff, who are the first line of contact with the mountain gorillas, to report gorillas showing overt signs of disease and to conduct daily observations for clinical signs of abnormal appearance or behavior, as well as to collect routine fecal samples from each habituated gorilla group on a monthly basis. Human and Gorilla Conflict resolution (HUGO) teams are community volunteers trained and supervised by the Uganda Wildlife Authority (UWA), with support from International Gorilla Conservation Programme (IGCP) and CTPH to safely herd gorillas and other wildlife back to the park when found foraging in community land (Kalema-Zikusoka and Rwego 2016). In addition to reducing conflict between people and wildlife, CTPH also trains HUGOs to monitor the health of gorillas when the risk of disease transmission is highest, at the interface between gorilla habitat and community lands. Both park staff and HUGO members receive training in One Health. Park staff are also trained to manage tourists during gorilla viewing in order to enforce rules such as: maintaining a minimum distance of 7 m during visits to the gorillas; encouraging people to turn away to cough or sneeze, proper disposal of human defecation while in gorilla habitat; and the refraining of eating, drinking, smoking, or flash photography while in the presence of the gorillas (Hanes et al. 2018; Weber et al. 2020). The HUGOs, like the VHCTs, are a volunteer organization, and are given the same support of group livestock income-generating projects by CTPH as described above for the VHCTs.

Building a Gorilla Health and Community Conservation Center for Analysis of Pathogens from Wildlife, People, and Domestic Animals

In 2005, the first initiative for CTPH's wildlife health program was to build a field laboratory that would facilitate proactive gorilla health monitoring through regular analysis of normal gorilla fecal samples, as well as abnormal gorilla fecal samples from gorillas showing clinical signs of disease. The Gorilla Research Clinic enabled monthly laboratory analysis of fecal samples from each habituated gorilla group, gorillas showing clinical signs, and unhabituated gorillas during the gorilla census, conducted once every 4–5 years (Rubanga and Kalema-Zikusoka 2013). In 2015, the clinic was upgraded and expanded to a permanent Gorilla Health and Community Conservation Centre. Disease investigations to date at the Gorilla Health Centre field laboratory include testing for: intestinal helminths and other parasites; protozoa, including entamoeba, giardia, and cryptosporidium; and bacteria, including *Salmonella* and *Shigella* (Kalema-Zikusoka 2015; Nolan et al. 2017).

During the first year, data collected from CTPH identified the Nkuringo gorilla group to have the highest rate of parasitic infections, which likely correlated with their preference for regular foraging in community lands. This prompted UWA to recruit and train additional HUGO community volunteers to herd the Nkuringo gorillas from community land, in an effort to reduce their risk of disease transmission from humans. A few years later the parasite infection rate was reduced and was comparable to other gorilla groups.

Linking with Local Human Health Centers Through Comparative Disease Investigations

A study implemented in 2010 investigated the prevalence of cryptosporidium and giardia species in people, gorillas, and livestock that interface with one another (Kalema-Zikusoka et al. 2018). Samples collected from habituated and non-habituated gorillas, community-owned livestock herds, and humans around Bwindi were screened for cryptosporidium and giardia using the ImmunoSTAT Commercial Field Kit (Johnstone et al. 2003; Scheffler and Van Etta 1994), and dubious samples confirmed with Direct Fluorescence Antibody Test (DFA) (Chalmers et al. 2011).

Giardia was detected in 5.5% of livestock, 40% of symptomatic humans from the local hospital, and 9.5% of asymptomatic park staff, but not in gorillas. *Cryptosporidium* was detected in 3.1% of habituated gorillas, 4.7% of livestock, and 62.4% of park staff. Unlike previous studies in Bwindi and Virunga National Parks, no giardia species were detected in gorillas.

The Village Health and Conservation Teams (VHCTs) may have contributed to the decreased prevalence of giardia in the mountain gorilla population through the implementation of improved hygiene and sanitation of local communities living at the interface with gorillas.

Cryptosporidium species found in the habituated gorillas may have been associated with human interaction, similar to previous studies (Graczyk et al. 2001) where a VHCT was selected for each village with positive human samples, and where gorillas often range. Local health centers were mobilized to educate patients on the health risks of collecting water from unprotected sources, and cattle water troughs were built to prevent cattle from defecating and contaminating water that is shared by humans and gorillas

Linking with Local Human Health Centers Through Community Based Health Promotion

Over the past 10 years of work at the Gorilla Research Clinic, we have contributed to the identification of some of the primary causes of diarrhea in the community, such as giardia, *E. coli*, and other coliforms. CTPH encourages communities to better manage their water resources in order to reduce contamination with diarrhea-causing pathogens. Prevention education includes: teaching farmers to dig water troughs for their cattle in order to reduce defecation in streams where water is collected for human use; collecting water with clean containers from protected water sources; and boiling drinking water prior to consumption.

In 2015, CTPH built a permanent home by expanding the Gorilla Research Clinic to a larger Gorilla Health and Community Conservation Centre, with more rooms for laboratory work and a community resource meeting room. Through partnerships with UWA, local health centers, and local veterinary centers, the new One Health Center conducts a regular analysis of both normal and abnormal gorilla, human, and livestock fecal samples. Results from gorilla fecal sample analysis help to guide teams in how often and where to deworm people, livestock, and in extreme circumstances, gorillas that have a high parasite burden, or abnormal parasites from people and livestock causing clinical signs.

Conclusion: Impact on Gorilla Conservation

Since 2003, CTPH has contributed to an increase in the overall mountain gorilla population, from 700 to an estimated 1063, and at Bwindi specifically, from 320 to an estimated 459 gorillas.

There has been a reduction in human-related disease outbreaks in the gorillas, with no scabies outbreaks since 2002, and reduced exposure of gorillas to human and livestock diseases such as giardia, which lowered to non-detectable levels in the mountain gorillas (Kalema-Zikusoka et al. 2018). This outcome is attributed to a combination of increased referrals of people with infectious diseases such as tuberculosis, and improved health and hygiene in homes bordering park boundaries. Data collected from CTPH has revealed a three to seven fold increase in handwashing facilities outside the pit latrine, from 10% to between 30 and 75%, as well as an increase in homes with pit latrines, clean water storage containers, and drying racks (Ainerukundo et al. 2019).

Additionally, CTPH's family planning program has contributed to a significant increase in women who are using modern contraceptives from 22 to 67% of women in the communities of Mukono and Bujengwe parishes. This increase falls above the national Ugandan average in rural areas from 30 to 47% in the same time period (National Population Council 2019). Behaviour Change Communication has resulted in an increase in the inter-birth interval, contraceptive prevalence rate, and a reduction in the rate of conception in the number of children for each woman in Kanungu District from 6.7 to 4.2, which is now among the lowest in Uganda (National Population Council 2019). When a family reduces their number of children, they are better able to provide good health care and education, thus reducing the threat of disease transmission between people and gorillas, and helping to break the poverty cycle. An unintended positive outcome has been the reduction of gender disparity, as women are more involved in conservation, and men are more involved in family planning and health care.

CTPH has seen similar gains within the marginalized Batwa pygmy community, where data has shown that when the program began 1% of households had hand-washing facilities compared to 10% among the majority Bakiga community, and overall hygiene and sanitation measures are poor at best. Batwa communities rely heavily on natural resources from within Bwindi Impenetrable National Park, and efforts by CTPH to support this community may contribute to a decrease in such dependence, and subsequent drain on gorilla habitat.

CTPH's One Health program has resulted in a better quality of life for Bwindi local communities, which in turn has led to more positive attitudes toward conservation. Examples include: the local community protecting an aged silverback gorilla from Mubare gorilla group who could not keep up with the rest of his family group, and chose to settle in community lands rather than the forest (Kalema-Zikusoka 2019) and the VHCT community volunteers promoting sustainable agriculture, which utilizes less land, and adhering to improved soil and water conservation methods, resulting in less pressure on fragile gorilla habitat.

Future Recommendations for Model Sustainability and Scalability

CTPH has established one of the first One Health field programs in the world, in communities surrounding Bwindi Impenetrable National Park. This program is also being scaled up to Virunga National Park in the Democratic Republic of Congo (DRC) in order to serve communities surrounding the DRC populations of mountain and eastern lowland or Grauer's gorillas (*Gorilla beringei graueri*) in DRC. This One Health model is being sustained through alternative livelihoods at the community level, which have been expanded to include reformed poachers, who are being given similar group income-generating projects as the VHCTs and HUGOs. The One Health model is also being supported externally through a social enterprise, Gorilla Conservation Coffee, that provides above-market prices for coffee farmers in the Bwindi community, reducing their dependence on gorilla habitat for food and fuelwood. The farmers are trained in sustainable agricultural practices that conserve soil and water, enabling ecological sustainability. Additionally, a portion of the proceeds from coffee sales is funneled into CTPH's community health, gorilla health, and conservation education programs.

CTPH's programs embody a One Health or integrated Population Health and Environment (PHE) approach that addresses human, animal, and ecosystem health together in order to reduce threats to wildlife and fragile ecosystems. This is accomplished through the prevention and control of cross-species disease transmission, as well as the promotion of: family planning, hygiene and sanitation, nutrition, sustainable agriculture, wildlife and forest conservation, agribusiness, ecotourism, and the reduction of conflict between people and wildlife. An analysis by CTPH has revealed that this work began at Bwindi Impenetrable National Park, home to 43% of the world's remaining mountain gorillas, addresses 10 of the 17 UN Sustainable Development Goals (United Nations 2015). The SDGs are important in bringing countries together to holistically achieve shared goals where the work of CTPH is mainly contributing to the goals of conservation, health, development, gender, and sustainable consumption. The model of CTPH can be extrapolated to other species, ecosystems, or areas of concern, by addressing the human side of the equation in conservation, namely by addressing the needs of the people in order to indirectly support conservation and biodiversity. The 10 UNDP sustainable development goals that CTPH contributed to are:

SDG 1. No Poverty
SDG 2. Zero Hunger
SDG 3. Good health and well-being
SDG 5. Gender equality

SDG 6. Clean water and sanitation
SDG 8. Decent work and economic growth
SDG 10. Reduced inequalities
SDG 12. Responsible consumption and production
SDG 15. Life on land
SDG 17. Partnerships for the goals

A One Health model has the potential to help reverse the negative growth trend of other gorilla subspecies, including eastern lowland gorillas (*Gorilla beringei graueri*) western lowland gorillas (*Gorilla gorilla gorilla*) and cross river gorillas (*Gorilla gorilla diehli*).

CTPH has also tested this One Health model in savannah ecosystems such as Queen Elizabeth National Park and Pian Upe Wildlife Reserve, where the primary threat to conservation is disease transmission at the interface between wildlife and livestock. Many of these diseases are zoonotic, including brucellosis, tuberculosis, and Rift Valley Fever. This work has helped to improve community attitudes toward park management by focussing on improving livestock health and husbandry through facilitating the formation of Community Conservation Animal Health Workers. CTPH has also tested this model in a non-great ape mountainous ecosystem, Mount Elgon National Park, where the primary threat to conservation is habitat encroachment through planting of crops and cattle grazing leading to cracks in the mountain and landslides (UWA 2019). The model has reduced conflict between local communities and park management by focusing on improving community health and family planning. Finally, this model has been implemented in a freshwater ecosystem at Lake Victoria Basin through Ecological Christian Organization, Osienala and Pathfinder International to reduce overfishing while improving community health, and to slow the region's high human population growth rates.

Donors

Since our inception in 2003, CTPH is grateful to many donors for ongoing support to our programs, including Population Connection, Tusk Trust, Darwin Initiative, Whitley Fund for Nature, Development Cooperation Ireland, Critical Ecosystem Partnership Fund, Global Development Network, Mulago Foundation, Disney Conservation Fund, National Geographic, USAID, United States Fish and Wildlife Service, John D and Catherine T MacArthur Foundation, Wildlife Conservation Society, African Wildlife Foundation, FHI360, Population Reference Bureau, Ashoka, North Carolina Zoological Park, San Diego Zoo Global, Cotswold Wildlife Park, Forix Foundation, The Gorilla Organization, Zoos Victoria, Primate Conservation Inc., Webinar Vet, Wildlife Conservation Network, Wildlife Direct, Rain Forest Trust, Rare Species Fund, Empowers Africa, Intervet Netherlands/Merck Animal Health, World Wildlife Fund for Nature, Growth Africa, Solidaridad, Arcus Foundation, Ernest Kleinworth Charitable Trust, Babel Foundation, Saint Andrews Prize for the Environment and several individual donors, who have enabled our success thus far. Government partners in Uganda include the Uganda Wildlife Authority, Kanungu and Kisoro District Local Governments, Ministry of Health, Ministry of Agriculture Animal Industry and Fisheries, and in Democratic Republic of the Congo include the Institute Congolais de Conservation de la Nature (ICCN) and health zones at Mikeno and Mount Tshiabirimu sectors of Virunga National Park. Finally, this work would not be possible without the support of many additional NGOs.

References

Ainerukundo E, Gaffikin L, Kalema-Zikusoka G (2019) Evaluation of a community-based health and conservation model at Bwindi Impenetrable National Park. In: Conference proceedings of the 2nd African Primatological Society congress

Chalmers RM, Campbell BM, Crouch N, Charlett A, Davies AP (2011) Comparison of diagnostic sensitivity and specificity of seven Cryptosporidium assays used in the UK. J Med Microbiol. 60:1598–1604. https://doi.org/10.1099/jmm.0.034181-0

Coscolla M, Lewin A, Metzger S, Maetz-Rennsing K, Calvignac-Spencer S, Nitsche A, Dabrowski PW, Radonic A, Niemann S, Parkhill J, Couacy-Hymann E,

Feldman J, Comas I, Boesch C, Gagneux S, Leendertz FH (2013) Novel *Mycobacterium tuberculosis* complex isolate from a wild chimpanzee. Emerg Infect Dis 19(6):969–976

Damas J, Hughes GM, Keough KC, Painter CA, Persky NS, Corbo M, Hiller M, Koepfli K-P, Pfenning AR, Zhao H, Genereux DP, Swofford R, Pollard KS, Ryder OA, Nweeia MT, Lindblad-Toh K, Teeling EC, Karlsson EK, Lewin HA (2020) Broad host range of SARS-CoV-2 predicted by comparative and structural analysis of ACE2 in vertebrates. BioRxiv preprint. https://doi.org/10.1101/2020.04.16.045302v1

Graczyk TK, Mudakikwa AB, Cranfield MR et al (2001) Hyperkeratotic mange caused by *Sarcoptes scabiei* (Acariformes: Sarcoptidae) in juvenile human-habituated mountain gorillas (*Gorilla gorilla beringei*). Parasitol Res 87:1024–1028. https://doi.org/10.1007/s004360100489

Grutzmacher KS, Keil V, Metzger S, Wittiger L, Herbinger I, Calvignac-Spencer S, Matz-Rensing K, Haggis O, Savary L, Kondgen S, Leendertz FH (2018) Human respiratory syncytial virus and *Streptococcus pneumoniae* infection in wild bonobos. Ecohealth 15(2):462–466

Hanes AC, Kalema-Zikusoka G, Svensson MS, Hill CM (2018) Assessment of health risks posed by tourists visiting mountain gorillas in Bwindi Impenetrable National Park, Uganda. Primate Conservat 2018 (32):10 pp

Hickey JR, Uzabaho E, Akantorana M, Arinaitwe J, Bakebwa I, Bitariho R, Eckardt W, Gilardi K, Katutu J, Kayijamahe C, Kierepka EM, Mugabukomeye B, Musema A, Mutabaazi H, Robbins MM, Sacks BN, Kalema-Zikusoka G (2018) Bwindi-Sarambwe 2018 surveys monitoring mountain gorillas, other select mammals, and human activities. Unpublished Report to Uganda Wildlife Authority

Johnstone SP, Ballard MM, Beach MJ, Causer L, Wilkins PP (2003) Evaluation of three commercial assays for detection of Giardia and Cryptosporidium organisms in fecal specimens. J Clin Microbiol. 41:623–626. https://doi.org/10.1128/JCM.41.2.623-626.2003

Kalema-Zikusoka G (2013) New security beat. 13 Mar 2013. In Uganda, integrating population, health, and environment to meet development goals. https://www.newsecuritybeat.org/2013/03/uganda-integrating-population-health-environment-meet-development-goals/

Kalema-Zikusoka G (2015) Chapter 13. Special considerations and scenarios. Mountain gorilla disease: implications for conservation. In: Cooper JE, Cooper ME (eds) Wildlife forensic investigations. CRC Press, pp 455–457

Kalema-Zikusoka (2019) From human-wildlife conflicts to a human-gorilla friendship, 25 November. http://www.birdlife.org/africa/human-wildlife-conflicts-human-gorilla-friendship

Kalema-Zikusoka G, Rwego IB (2016) Mountain gorillas, tourism, and conflicts with people living adjacent to Bwindi Impenetrable National Park. In: Aguirre A, Sukumar R (eds) Tropical conservation, perspectives on local and global priorities. Oxford University Press, pp 136–139

Kalema-Zikusoka G, Kock RA, Macfie EJ (2002) Scabies in free-ranging mountain gorillas (*Gorilla beringei beringei*) in Bwindi Impenetrable National Park, Uganda. Vet Rec 150(1):12–15

Kalema-Zikusoka G, Rubanga S, Mutahunga B, Sadler R (2018) Prevention of Cryptosporidium and GIARDIA at the human/gorilla/livestock interface. Front Public Health. Brief Research Report. published: 14 December 2018. https://doi.org/10.3389/fpubh.2018.00364

Köndgen S, Leider M, Lankester F, Bethe A, Lubke-Becker A, Leendertz FH, Ewers C (2011) *Pasteurella multocida* involved in respiratory disease of wild chimpanzees. PLoS One 6(9):e24236

Melin AD, Janiak MC, Marrone F, Arora PS, Higham JP (2020) Comparative ACE2 variation and primate COVID-19 risk. BiorxiV. The Preprint server for Biology. Cold Spring Harbor Laboratory. https://www.biorxiv.org/content/10.1101/2020.04.09.034967v2

National Population Council (2019) https://www.npc.gov.gh

Negrey J, Reddy R, Scully E, Phillips-Garcia S, Owens L, Langergraber K, Mitani J, Thomson ME, Wrangham R, Muller M, Otali E, Machanda Z, Hyeroba D, Grindle K, Pappas T, Palmenberg A, Gern J, Goldberg T (2019) Simultaneous outbreaks of respiratory disease in wild chimpanzees caused by distinct viruses of human origin. Emerg Microbes Infect 8:139–149

Nolan MJ, Unger M, Yeap Y-T, Rogers E, Millet I, Harman K, Fox M, Kalema-Zikusoka G, Blake DP (2017) Molecular characterisation of protist parasites in human-habituated mountain gorillas (Gorilla beringei beringei), humans and livestock, from Bwindi Impenetrable National Park, Uganda. Parasit Vectors 10:340. https://doi.org/10.1186/s13071-017-2283-5

Palacios G, Lowenstine LJ, Cranfield MR, Gilardi KVK, Spelman L, Lukasik-Braum M, Kinani J-F, Mudakikwa A, Nyirakaragire E, Bussetti AV, Savji N, Hutchison S, Egholm M, Lipkin WI (2011) Human metapneumovirus infection in wild mountain gorillas, Rwanda. Emerg Infect Dis 17(4):711–713. https://doi.org/10.3201/eid1704.100883

Patrono LV, Samuni L, Corman VM, Nourifar L, Röthemeier C, Wittig RM, Drosten C, Calvignac-Spencer S, Leendertz FH (2018) Human coronavirus OC43 outbreak in wild chimpanzees, Côte d'Ivoire. Emerg Microbes Infect 7:118. https://doi.org/10.1038/s41426-018-0121-2. www.nature.com/emi

Rubanga SV, Kalema-Zikusoka G (2013) The establishment and use of field laboratories: lessons from the conservation through Public Health Gorilla Research Clinic, Uganda. J Exotic Pet Med 22(1):34–38

Scheffler EH, Van Etta LL (1994) Evaluation of rapid commercial enzyme immunoassay for detection of Giardia lamblia in formalin-preserved stool specimens. J Clin Microbiol. 32:1807–1808

Scully EJ, Basnet S, Wrangham RW, Muller MN, Otali E, Hyeroba D, Grindle KA, Pappas TE, Thompson ME, Machanda Z, Watters KE, Palmenberg AC, Gern JE,

Goldberg TL (2018) Lethal respiratory disease associated with human rhinovirus C in wild chimpanzees, Uganda, 2013. Emerg Infect Dis. https://doi.org/10.3201/eid2402.170778. https://wwwnc.cdc.gov/eid/article/24/2/17-0778_article

Travers H, Milner-Gulland EJ, Roe D, Kalema-Zikusoka-G, Ngabirano A (2019) Monitoring and evaluation for non professionals—how to ensure quality in data collection processes. Published by IIED, March, 2019. ISBN: 978-1-78431-679-2. http://pubs.iied.org/17647IIED

Uganda Demographic and Health Survey (2016) https://dhsprogram.com/pubs/pdf/FR333/FR333.pdf

Uganda Wildlife Authority (UWA) (2019) https://www.ugandawildlife.org/news-events/news/landslides-strike-southern-part-of-mt-elgon

United Nations (2015) https://www.un.org/sustainabledevelopment/blog/2015/12/sustainable-development-goals-kick-off-with-start-of-new-year/

Walsh PD, Abernethy KA, Bermejo M, Beyers R, De Wachter P, Akou ME, Huijbregts B, Mambounga DI, Toham AK, Kilbourn AM, Lahm SA, Latour S, Maisels F, Mbina C, Mihindou Y, Obiang SN, Effa EN, Starkey MP, Telfer P, Thibault M, Tutin CEG, White LJT, Wilkie DS (2003) Catastrophic ape decline in western equatorial Africa. Nature 422:611–614

Weber A, Kalema-Zikusoka G, Stevens NJ (2020) Lack of rule-adherence during Mountain Gorilla Tourism encounters in Bwindi Impenetrable National Park, Uganda, Places Gorillas at risk from human disease. Front Public Health. Original Research published: 13 February 2020. https://doi.org/10.3389/fpubh.2020.00001

Forensic Science in Marine Mammalogy: Applications and Limitations

Megan Stolen

Abstract

Marine mammals are unique in many ways including their aquatic adaptations, vulnerability to threats, legislation and laws protecting them, and their popularity with humans. Despite their reputation as charismatic megafauna, marine mammals still suffer injuries and death due to anthropogenic causes. Some are on the brink of extinction due to negligence and direct harm. Strong laws have made improvements in the management of marine mammal populations but every year, cases of injury and death warrant forensic investigation. These include vessel strikes, entanglement, drowning, shootings, mutilation, and illegal trade. Around the world, teams of stranding responders, fishery observers, veterinarians, and researchers investigate cases of human interaction with marine mammals. Often, the work is difficult as many obstacles make the full examination and forensic investigation challenging. This chapter will briefly review the laws in place to protect marine mammals, common causes of injury and death, limitations with some examinations, and recommendations in the application of forensic techniques.

M. Stolen (✉)
Hubbs-SeaWorld Research Institute, Melbourne Beach, FL, USA
e-mail: mstolen@hswri.org

Keywords

Marine mammals · Human interaction · Wildlife forensics

Laws Protecting Marine Mammals

Marine mammals are not a strict taxonomic group in the traditional sense. On the one hand, marine mammals do share some common traits; most notable, by definition, is the fact that they are all mammals, and all rely on aquatic environments for their survival. However, the term "marine mammal" includes several taxonomic groups that have converged and specialized in habitats that include not just saltwater oceanic environments but also riverine systems and estuaries that are, in fact, not strictly marine at all. Marine mammals include all seals, sea lions, walruses (all species in order Pinnipedia), marine and sea otters (select genera in the family Mustelidae), porpoises, dolphins and whales (order Cetacea), manatees and dugongs (order Sirenia), and polar bears (single species in family Ursidae). This somewhat artificial construct of assigning these taxonomically disjunct species to a group, at least in the United States, was a decision to grant them protection under the Marine Mammal Protection Act of 1972 (MMPA, most recently amended in 2015) as an ecosystem approach to conservation. It passed as a reaction to public concern over risks to marine

S. C. Underkoffler, H. R. Adams (eds.), *Wildlife Biodiversity Conservation*,
https://doi.org/10.1007/978-3-030-64682-0_14

mammals from human causes and a growing concern of threats to the marine environment (Lavigne et al. 1999). The result is a broad approach to conservation for all mammals that inhabit the vast marine environment as well as those species that have made their way into rivers and estuaries where they have specialized to feed and avoid predation in unique ways.

Some species of marine mammals are afforded additional protection in the United States under the Endangered Species Act (ESA) depending on their population status. Internationally, laws protecting marine mammals vary by country but may also include conservation measures under international treaties and agreements such as the Convention on International Trade in Endangered Species of Wild Fauna and Flora (CITES) and regional agreements between neighboring countries (Baur et al. 1999).

In the United States, the authority to enforce the MMPA and other international treaties falls under the jurisdiction of the Department of Commerce, National Oceanic and Atmospheric Administration, National Marine Fisheries Service (NMFS), and the Department of the Interior, United States Fish and Wildlife Service (USFWS). NMFS has the authority to protect cetaceans and pinnipeds (except walruses) and USFWS enforces laws protecting walruses, otters, polar bears, and sirenians and they have the authority to regulate trade (both legal and illegal) of all marine mammals under CITES (Baur et al. 1999).

The MMPA is primarily built on the pillar which states that it is illegal to "take" a marine mammal, which is defined in the Act as: "*to harass, hunt, capture, collect, or kill or attempt to harass, hunt, capture, or kill any marine mammals, including, without limitation, any of the following: the collection of dead animals or parts thereof; the restraint of detention of a marine mammal, no matter how temporary; tagging a marine mammal, or the negligent or intentional operation of an aircraft or vessel, or the doing of any other negligent or intentional act which results in the disturbing or molesting of a marine mammal.*" (Baur et al. 1999). This sweeping definition of take underscores the broad effects that the law has had on marine mammal protection and the vast scope of the actions needed by law enforcement to enforce it. By protecting the animals and their parts as well as any human effect that may negatively influence their behavior, the MMPA touches all aspects of human influence on these species from driving boats and dredging canals to collecting beachcast bones and shooting a marine mammal. It defines legal and illegal take in fisheries and describes how and when a marine mammal can be caught for research or public display in a zoological park. The law describes limits for approaching whales during an ecotour and hunting by indigenous communities. Thus, the MMPA protects marine mammals in all aspects from birth to death and even after death with respect to trade in their parts.

To implement the MMPA with all its pieces, like most wildlife protection plans, there must be involvement of countless people in multiple agencies including law enforcement at the local, state, national, and international levels, wildlife protection and trade agencies, tribal governments, animal health responders, and fishing industries, just to name a few. There are personnel solely dedicated to help understand and mitigate risks to marine mammals, such as observers on fishing vessels, but there are also volunteers working for stranding programs and experts looking for instances of wildlife trafficking as part of numerous global cybercrime initiatives. At each level, there are challenges of funding the work, keeping up with gear modifications and changes in fishing pressures, and understanding new threats as novel technologies such as remote operated vehicles and sonar arrays are developed. Effects on marine mammals are constantly under scrutiny as changes to the environment, economic development, and military operations have advanced significantly since the MMPA was first enacted. Amendments to the MMPA and other laws protecting marine mammals are needed to keep up with the times so that the original goals and mandates can be achieved.

Marine mammals by their very nature often travel between states, countries, and even continents as they navigate rivers, nearshore environments, and vast expanses of open ocean. Because of this, strong international agreements and monitoring must be in place to protect marine mammals in all parts of their natural range. Likewise, these species are often the targets of wildlife traffickers, so strong import and export laws must be enforced. When a crime or violation is suspected, the investigator must notify the appropriate law enforcement agency early in the investigation so that there will be appropriate sample collection and preservation as well as interviews with appropriate people involved in the event. Often teams of scientists, forensic analysts, and on-the-ground responders are needed to fully investigate a marine mammal case. Fines may be levied, or, in very rare and egregious cases, perpetrators may be sentenced to time in probation or jail.

Classifying Anthropogenic Effects on Marine Mammals

Humans affect marine mammals in a myriad of negative ways and this may include criminal offenses in direct defiance of conservation laws such as the MMPA. These include direct lethal and nonlethal take of a marine mammal by means of shooting or other projectiles, drowning (usually associated with fishing gear), mutilation (such as a knife or gaff), vessel strike, entrapment, entanglement, and toxin exposure (e.g., oil spill). It is important to recognize that there may be chronic effects on marine mammals by all of these sources resulting in a reduction in lifespan or fecundity or an increase in pain and suffering as well as susceptibility to infectious and noninfectious disease. Other effects leading to changes in behavior or movement are more subtle and therefore fall outside the discussion of wildlife forensics that we will consider in this chapter.

Wildlife conflict and cases of anthropogenic effects on marine mammals fall into two general categories depending on the extent and type of harm. The first category involves cases with several marine mammals over time and often with some consistent input over time (chronic effects). Examples of these include post-oil spill assessments of marine mammal stocks in an affected area (part of the legal responsibility of a National Resource Damage Assessment) or chronic species level concerns for an endangered animal such as prolonged and widespread vessel strikes. These large population or species level concerns are often built upon data collected from individual cases, but the goal of managers is often to understand long-term effects on the population or species. The second category falls under what might be considered the acute result on an individual marine mammal after the interaction or crime occurs. Examples of these cases could include gunshot mortality, stabbing, or an unusual entanglement in fishing gear. In many instances, reliance on evidence and data is based on stranded animal recovery and examination.

As part of the MMPA, the United States established a National Marine Mammal Stranding Network with a responsibility to investigate causes for strandings (animals who have come ashore alive or dead or who may be injured or killed at sea) and to collect data on anthropogenic effects that may have contributed to the stranding event (Wilkinson and Worthy 1999). Often, chronic effects of long-term entanglement, toxin exposure, and nonlethal vessel strikes are also apparent during the investigation and these data likewise contribute to the understanding of harmful human–marine mammal interactions. Over time, the stranding network has developed procedures and protocols to ensure a consistent and objective examination of these cases. Currently, active and permitted stranding responders are required to complete a standardized form used by all responders in the United States which includes basic data (total length, sex, etc.) as well as more detailed information such as a description of their assessment of human interaction (HI form). Data fields on this form require a yes, no, or cannot be determined (CBD) answer to several gross examination questions and then a final determination to the main question "was there evidence of human interaction?". In order to answer yes, the examiner must have conducted an examination and must have found some evidence that is consistent with a known human interaction effect. Examples of evidence include the presence of fishing gear or plastic ingestion but may also include evidence of interactions even in the absence of the gear. Other examples include lacerations or tissue damage consistent with a known type of trauma that can be reasonably assigned to a human interaction. Often these assessments must rely on the expertise and experience of the examiner or another expert relying on photographs and other types of evidence collected at the scene or during the examination. The question of human interaction can be placed into the "no" category when, after a full exam and based on reasonable expertise, there appears to be no evidence that a human interaction occurred with the animal. This is considered the most difficult determination to make because so many factors can affect the external or internal appearance of the evidence. These include decomposition, scavenger damage, incomplete examination as a result of the large size of the carcass or the inability to perform a complete necropsy due to weather or time constraints, lack of equipment, or proper tissue storage, etc. The third possible determination for the data field is "cannot be determined" (CBD) which is the most common answer because it accounts for all the difficulties outlined above as well as the possible inexperience of the examiner. Therefore, the evidence is deemed inconclusive (see review by Moore and Barco 2013).

Evidence of human interaction is categorized based on the type of injury (laceration, abrasion, amputation, ingestion, entanglement) and the object causing the injury (projectile, net, propeller, etc.). It is also important to note that evidence of human interaction with a stranded animal (dead or alive) does not always conclude with a determination that the interaction caused the death or stranding of the animal. For example, plastic or rope may be found inside the stomach of a whale but it may not have been the ultimate cause of the stranding. Similarly, evidence of human interaction does not necessarily imply criminal intent. However, all evidence and descriptions from human interaction cases are reviewed by teams at the regulating agency who then make determinations about the extent of the injury and make conclusions that may result in changes in maritime rules or charges being brought against a known offender. Evidence considered includes: gross examination, photographs, measurements of wounds, pathology reports, close examination of gear or other physical evidence by an expert familiar with the type, and eyewitness statements.

While a considerable amount of information regarding wildlife conflicts and marine mammals comes from stranded animal investigations, there are numerous studies analyzing data from field studies of marine mammals using techniques such as photo-identification to determine nonlethal interactions. However, these cases rarely result in a court case or investigation because the strong evidence needed is often inconclusive or missing. Exceptions to this are rare cases of harassment (considered a take and therefore a crime) that are posted on social media or the Internet and can be investigated by law enforcement.

Challenges of Investigating Marine Mammal Cases

Body Size

All wildlife forensic cases pose some unique challenges often not found in human crime investigations. These have been summarized well by Cooper and Cooper (2013) and Huffman and Wallace (2012). In general, wildlife crime is difficult to monitor and respond to because of several factors including proper identification of the species, lack of resources, the difficulty of securing or processing the crime scene, and lack of adequate comparative data, just to name a few. Many additional challenges affect marine mammal investigations. The first considerable challenge is the large body size of many marine mammals, with baleen whales being one extreme, but even moderately sized marine mammals such as dolphins and large pinnipeds often weigh several hundred pounds. Large body size is problematic because it makes a full forensic necropsy difficult as the carcass often requires heavy equipment to lift and turn so that all sides of the animal can be evaluated. For example, many of the largest marine mammals require large knives, large hooks, and several hours of work just for the removal of the blubber or fat layer. This blubber layer also contributes to the increase of internal body temperature which increases the rate of decomposition. Relatedly, large amounts of pressurized gases inside the body cavities resulting from decomposition will displace organs and can lead to difficulty when attempting to identify sources of trauma.

Due to their large size and equipment challenges, marine mammal examinations are often performed in-situ on the beach, rocky cliff, or in water. This leads to a series of additional challenges, not only with decomposition but also because these issues make it difficult to collect and preserve samples without cross-contamination. Moreover, radiographs (for ballistic or skeletal trauma analysis) and other tests used in small and moderate-size animal veterinary forensic exams are often not available to prosecutors working up marine mammal cases, even in many laboratories. In a few cases, parts of the animal such as the head or flippers may be removed and can be examined or scanned back at a specialized laboratory or veterinary hospital but again, these may require large coolers, large vehicles, and appropriate types of testing equipment not often available to local stranding program managers.

Aquatic Environments

The second consideration is the difficulty presented by the nature of the marine (or aquatic) environment itself. Carcasses often initially sink below the surface of the water making them difficult to detect until gases build up in the body cavities, whereby the carcass will begin to float (see review by Liebig et al. 2003). Once floating, the animal can then drift for several miles over several days resulting in an expansive search area. Floating animals are essentially removed from the crime scene in the traditional sense, so investigators often rely on techniques such as hindcasting (using oceanic currents and weather as a model) to place the animal back at the site where it was injured or killed (Peltier et al. 2012). This level of modeling requires expertise outside the scope of most teams and is normally only used in endangered species cases. Investigations of carcasses at sea are difficult and dangerous for the investigator so carcasses are often towed to shore requiring adequate equipment and experienced mariners. The very nature of the saltwater and shorelines that comprise the site also make examinations difficult due to saltwater intrusion into the body, effects of weather and heat/cold on the external surface, and significant opportunities for scavengers from small crabs to large terrestrial carnivores and most notably sharks (Moore and Barco 2013). All of these effects lead to difficulties in trace evidence collection. For example, traditional evidence such as fingerprint and blood splatter analysis are often not appropriate or possible to collect. It is important to note that most marine mammal deaths are never recovered as the carcasses sink, become scavenged, or float away from human detection (Wells et al. 2015; Williams et al. 2011).

Taphonomy

The third consideration for applying forensics to marine mammal cases is the lack of comparative taphonomy information such as postmortem change details and marine-specific forensic entomology and scavenger assessment. While some data do exist for proxy animals such as pigs and humans in marine environments (Anderson and Bell 2014), there are few published studies on marine mammals using experimental design to test effects such as saltwater immersion, aquatic temperature changes, or scavenger marks (Peltier et al. 2012). Carcasses in aquatic environments require an advanced understanding of aquatic (marine and freshwater) organisms that can colonize, scavenge, and live as epibionts on marine mammals. Although deepwater whale falls are well researched and continue to be a focus of researchers with advanced technology (remote and human-piloted submersibles), less research has been conducted in shallow water environments (Anderson and Bell 2014).

The best use of epibionts in marine mammal forensics has been made in large whale entanglement and ship strike cases. Whale lice (family Cyamidae) have been used as indicators of long-term, chronic injuries and serve as evidence of chronic injury and lack of vigor (Pettis et al. 2004; Sharp et al. 2019). As with terrestrial forensic cases, parasites and epibionts can also indicate the origin of the carcass which is especially important in aquatic environments as floating marine mammals can move great distances due to oceanic currents and offshore and nearshore winds.

Confounding Effects

Care should be taken in all wildlife forensic cases to understand confounding effects. One important type in marine mammal cases is the effect of scavengers on carcasses. Scavenger damage can range from a small opening such as a bird peck to complete loss of major organs, skeletal parts, and even complete consumption of the body. Understanding and identifying scavenger marks is critically important in marine mammal cases as some are often misidentified as anthropogenic mutilations and projectile cases. For example, small diameter, penetrating wounds to the body often result from bird pecks (antemortem and postmortem). In Florida, stingray barb stab wounds often present similar to bullet wounds and can include an associated path of migration. After all, a dolphin killed by a stingray is essentially a non-anthropogenic interspecies stabbing case (Fig. 1).

Other confusing external lesions that can be misinterpreted as human-made marks such as those resulting from net entanglement can occur before and after death (Moore et al. 2013) so the examiner should document all marks well with sketches and photographs in situ if possible and consult other experts when marks are present (Table 1).

Postmortem Investigation of Marine Mammal Carcasses

Marine mammal case investigations do not generally allow for a true crime scene investigation for obvious reasons, such as the aquatic environment and drift as noted earlier, although underwater investigations should be approached with traditional crime scene processing techniques if possible and warranted (Byrd and Sutton 2012). Once landed or recovered, the carcass can be treated as a typical wildlife forensic case and should be secured and treated as evidence when a crime or violation is suspected. This includes a full chain of custody for biological samples, suspected fishing gear, projectiles, photographs of the body, and standard forms with notes, measurements, and gross necropsy results.

Standardized forms and protocols are easily accessible for use in marine mammal necropsies. These include standard data such as total length and sex (Level A data in the United States) and human interaction investigation forms. Standard procedures and protocols are readily available (Geraci and Lounsbury 2005). Carcasses are categorized by decomposition code from 1 (alive) to 5 (mummified, skeletal). This allows investigators to gauge the need and applicability of the sample modality and to inform the pathologist and other consulting scientists about the limitations of the investigation at the time of recovery. For example, it is important to note if the carcass had been frozen prior to necropsy because freezing may produce artifacts that change an interpretation during examination both grossly and in histopathology. Condition codes and appropriate sample collection methods are provided by Geraci and Lounsbury (2005) and Moore et al. (2013). Level B data are collected upon exam. These include photographs, sketches, and morphometric data. Level C data include pathology interpretations, advanced sample results such as DNA analyses, viral screenings, toxicology screening, etc.

Marine mammal forensic necropsies rely on universal techniques common to veterinary forensic science (Cooper and Cooper 2013, Huffman

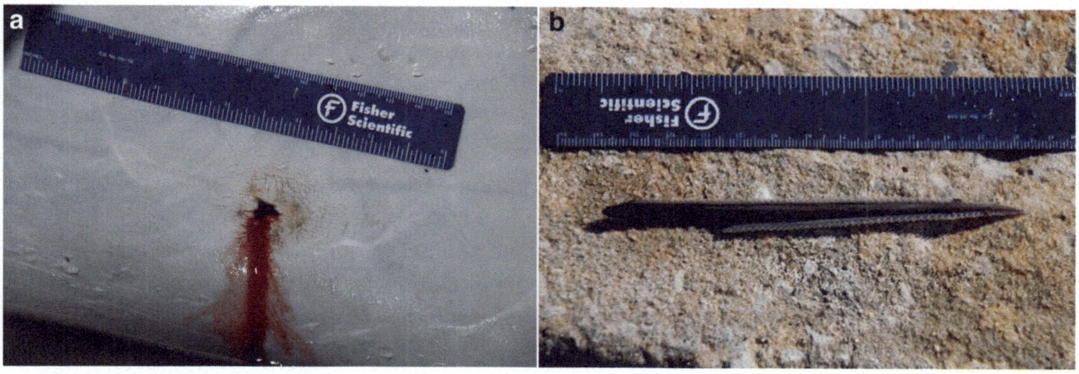

Fig. 1 (**a**) Ventral body of a bottlenose dolphin (Hubbs-0903-Tt) with stab wound resulting from a stingray. (**b**) Barb from the stingray found inside the dolphin. Photo credit: Hubbs-SeaWorld Research Institute

and Wallace 2012). The recent review of marine mammal cases by Moore et al. (2013) best summarized the appropriate approach to many forensic marine mammal case investigations including sharp and blunt force trauma, entrapment, gear entanglement, and gunshot cases.

Table 1 Types of marks commonly found on marine mammals with potential anthropogenic sources and confounding factors

Gross description	Common anthropogenic source	Confounding causes
Open circular wound of small diameter	Bullet hole	Antemortem stingray barb sharp injury
	Sharp instrument (e.g., Screwdriver)	Bird peck
		Other scavengers (invertebrates etc.)
		Postmortem carcass collection (e.g., Gaff use)
		Cookie-cutter shark (*Isistius*) bites
Parallel impressions, abrasions, lacerations	Net or line entanglement	Conspecific tooth "rake" marks
	Debris entanglement	Scavenger marks
		Predation tooth marks (sharks, other marine mammals)
		Freezer artifacts
Cross-hatching or cross-shaped marks	Net entanglement	Postmortem artifacts (e.g., Truck bed liners, body bags)
		Scavenger marks
Encircling marks around head and appendages	Rope or debris entanglement	Marks resulting from securing and towing carcass
Head or body part missing including teeth	Mutilation (e.g., trophy collection)	Antemortem predation
	Entanglement	Postmortem scavenging (shark, alligator)
		Decomposition
Diffuse subdermal blood staining	Blunt force trauma from watercraft	Conspecific aggression
	Entanglement or entrapment	Predation
Broken and luxated bones	Watercraft strike	Conspecific aggression
		Predation and large scavengers
		Mechanical disruption during stranding (alive or dead)

Likewise, individual case reports such as Byard et al. (2001, 2013) provide details for conducting thorough necropsies when an anthropogenic injury or criminal case is suspected. Generally, marine mammal forensic cases fall into several common types: watercraft strike, underwater entrapment, entanglement in gear or debris, ingestion of gear or debris, projectiles/bullets, mutilation (often postmortem) including to obtain trophies and parts, collection/sale crimes including cybercrime, toxic spills, and underwater sound emissions. Brief reviews of each are provided here as a summary.

Watercraft Strike

Virtually any type of vessel moving at speed on the water has the potential to injure or kill a marine mammal. Naturally, the larger the vessel and the faster the speed, the more potential for harm to marine mammals. Large ships are known to kill and seriously injure large whales at sea and even small vessels such as personal watercraft can injure and kill smaller marine mammals. Ships and smaller vessels cause harm in two ways: sharp injuries from propellers and blunt force trauma from a strike from the hull or skeg. In some cases, the animal can suffer from both types of injury. Sharp injuries have the potential to cause acute death by exsanguination, lethal mutilation and amputation, and fatal laceration and puncture to vital organs, as well as chronic mortality through infection, loss of mobility leading to reduction in feeding, swimming ability, or predator avoidance, issues with buoyancy and effects on the natural physiology of the animal. Examples of watercraft injury and death are well documented but perhaps two stand out more than others as they continue to significantly contribute to mortality in two ESA-listed species: North Atlantic right whales (*Eubalaena glacialis*) and West Indian manatees (*Trichechus manatus*).

North American right whales (NARWs) are the most endangered large whales on the planet and are protected by both the MMPA and the ESA in the United States. These whales were nearly extinct by the early 1900s as a result of several hundred years of commercial whaling (Kenney 2009). A recent study by Sharp et al. (2019) of right whale mortalities found that 42% of deaths during the study period could be attributed to vessel strikes. Right whales migrate from northern latitudes to Florida, close to shore, and straight through many commercial shipping areas. Despite large-scale efforts to decrease their susceptibility (e.g., mariner warning systems, protected status, education campaigns) NARWs continue to be killed by vessels. Gross descriptions and pathology and a review of cases by Sharp et al. (2019) are illustrative of the type of injuries, both acute and chronic, that result from vessel strikes.

Similarly, it is well known that boat strikes are a common source of mortality for manatees, another species at risk in the southeastern United States. These slow-moving sirenians have difficulty maneuvering away from approaching vessels especially when boats are moving quickly. Speed limits for boats have been in place for a number of years and failure of boaters to comply with these "manatee zone" enforcement regulations can result in fines, probation, and imprisonment for repeat offenders. Because of the high mortality rate from this trauma and the legal requirements of correctly identifying the cause of death in these cases, manatee experts with the Florida Fish and Wildlife Conservation Commission in Florida have developed detailed protocols for each manatee examined by the agency. Fresh carcasses are transported to a laboratory where they are systematically examined, and detailed notes and photographs are collected. Highly decomposed animals may be examined by regional field biologists but again, using a stepwise protocol that is built upon a strong foundation of forensic evidence collection. This allows the agency to reasonably assign carcasses, even when decomposed, to the watercraft death category (Fig. 2). Results from Rommel et al. (2007) show the benefits of taking a stepwise forensics approach to these cases resulting in accurate and defendable cause of death determinations.

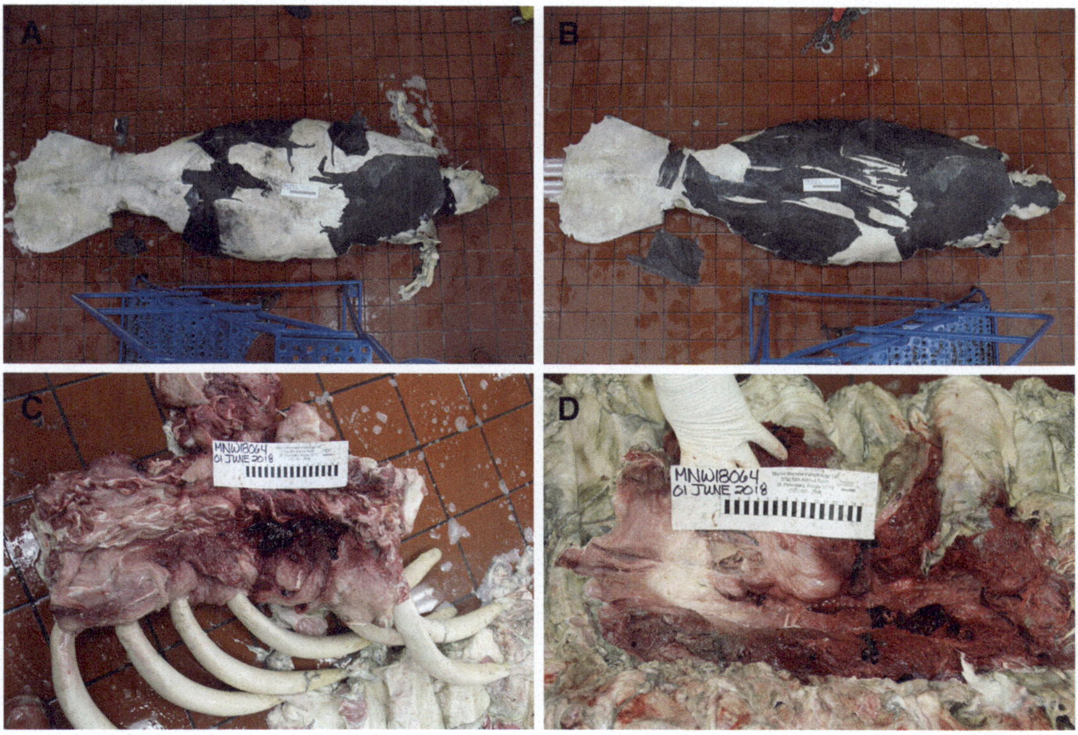

Fig. 2 (**a**) Decomposed body of a Florida manatee (MNW18064) with sloughing epidermis. (**b**) Reconstruction of the epidermis from the manatee showing marks consistent with propeller strike. (**c**) Dissected thoracic area of the manatee showing hemorrhage. (**d**) Dissected body of the manatee showing a broken rib and large blood clot. Photo credit: Florida Fish and Wildlife Conservation Commission

Important features of these sharp and blunt force trauma watercraft cases include broken or luxated ribs, broken vertebrae and skulls, propeller lacerations, large blood clots in body cavities, evidence of hemothorax and pneumothorax, contusions, organ rupture and emaciation (Lightsey et al. 2006; Rommel et al. 2007; Sharp et al. 2019; Moore and Barco 2013). Examiners involved in vessel strike cases should familiarize themselves with associated gross lesions and take adequate histological samples to confirm injury and cause of death. As always, photographs showing the location of the wound and close-up images of wound/lesion sampling should be included as well as sketches, measurements, and trace evidence collection (e.g., hull/skeg paint on carcass).

Underwater Entrapment/Drowning

While all marine mammals are dependent on and spend much of their time in or around the water and are, in fact, the most aquatically adapted mammals in the world, nevertheless they are still air breathers and have limited survival time when trapped underwater. Entrapment is typically a result of being caught in fishing gear but can also result from human-made structures such as construction equipment, underwater machinery, or loch gates (Osinga and Morick 2008; O'Shea et al. 1985; Reynolds et al. 2018). While entrapment in fishing gear and entanglement are linked, we single out this type of entanglement here to highlight the cause of death in these cases as asphyxiation or capture myopathy rather than death as a result of gear wounds, which we discuss in the following section on entanglement. Drowning in marine mammals is often difficult to interpret without expert veterinary pathology opinion, especially in cases where there is no direct evidence of gear on the animal. Some evidence of drowning in marine mammals includes edematous lungs, froth or large amounts of fluid in the airways, dark red lungs, and gas bubbles in tissues (Jepson et al. 2013). Examiners in these cases should make detailed descriptions of opened airways and photograph evidence as early in the necropsy as possible. Appropriate histopathology samples of the entire body with emphasis on the respiratory system are important.

Entanglement

Entanglement in gear and debris is a leading cause of anthropogenic mortality in marine mammals and occurs in virtually all areas where marine mammals are found, including shallow bays, estuaries, and rivers as well as pelagic environments (Read 2008). International laws and treaties include regulations for fishing gear to mitigate the effects of bycatch on marine mammals. Entanglement in human-made materials results in both acute and chronic health effects and is caused by the obvious sources such as fishing gear (Cassoff et al. 2011; Moore and Van der Hoop 2012; Moore et al. 2013) as well as debris not always easily associated with marine waters such as plastic waste, clothing, and other household items (Laist 1987; Moore and Barco 2013). Injuries from entanglement vary by the associated species but generally result in lacerations, abrasions, mutilation, and amputations of appendages and impressions from gear or debris (Moore and Barco 2013; Moore et al. 2013). Examiners should be familiar with fishing practices in their study area and spend some time understanding the signs of entanglement in such gear for local species (Moore and Barco 2013). Marine mammals are susceptible to entanglement as they swim through a three-dimensional space where floating debris is often essentially invisible to them and accumulates throughout foraging areas. Gear marks are commonly found around appendages especially the insertion or leading edge of the dorsal fins, flippers, and flukes of pinnipeds and cetaceans and the mouth, cervical regions, and peduncles of cetaceans and pinnipeds (Moore et al. 2013). Large whales may carry gear from several interactions and gear may remain embedded for years often leading to death years after the initial entanglement event (Fig. 3).

The cause of death in these cases may include hypoxia if the drag of the gear prevents the animal from surfacing adequately, exsanguination when gear severs major blood vessels, systemic infection, emaciation, and shock. Important sampling considerations for suspected entanglements include a full suite of histological samples with lesions and associated photographs, general tissue collection to identify any underlying health effects, frozen samples to identify potential toxicology that may be involved, sketches and photographs of gear and photographs of identifiable areas of the body such as dorsal fin and flukes to individually identify the animal for prior history in areas with ongoing population studies.

Ingestion of Gear and Marine Debris

Just as marine mammals become entangled in gear and debris so will they ingest pieces of these items. Plastic debris ingestion cases are becoming more common (Butterworth 2016; Jacobsen et al. 2010; Laist 1997) but other items such as recreational and commercial fishing gear ingestion are also well described in the literature. While ingestion of gear and debris does not always lead to death in many cases (Beck and Barros 1991; Stolen et al. 2013), the effects of indigestible trash are not well understood or studied (Simmonds 2012). Some cetaceans, such as dolphins, are unfortunately uniquely maladapted to the ingestion of fishing gear by their anatomy. Modified laryngeal cartilages (goosebeak) allow for the passage of air between the external nares (blowhole) and the trachea leading to the lungs. Ingested gear (often with a hook or lure attached) may become lodged around the cartilages and may encircle the structure. Multiple wraps of line around the goosebeak may lead to

Fig. 3 (**a**) At sea photo of decomposed North Atlantic right whale (VAQS20081005) showing rope entanglement. (**b**) Pectoral flipper of right whale showing multiple wraps of rope. (**c**) Open oral cavity of right whale showing intertwined rope through the baleen. Photo credit: Virginia Aquarium and Marine Science Center

asphyxiation and aspiration of prey items into the respiratory system (Stolen et al. 2013). While some cases of debris ingestion are a result of non-targeted prey selection, some types of plastic debris mimic normal prey for some marine mammals such as beaked whales and other pelagic squid eaters. Large baleen whales that skim the surface of the water are also susceptible to the ingestion of floating debris. Examiners of all marine mammal carcasses should take care to open the stomachs and intestines of all carcasses, even of decomposed animals, to look for debris and should attempt to understand the proximate effects of the ingested items on the area of the body. Ingested gear and debris should be photographed in situ and collected using appropriate forensic techniques including chain of custody if warranted.

Projectiles

Domestic, agricultural, and wildlife veterinary forensic science commonly includes the study of bullets and other projectiles in cases of animal cruelty and death. Marine mammals are no exception, but such cases are less likely to be: first, identified, and second, reasonably proven to have contributed to the death of the animal. For reasons that we have already discussed, finding a projectile in a large marine mammal, especially one that is decomposed, is difficult. The victim is often too large to be easily examined using scanning equipment and small wounds are often difficult to identify or may be misinterpreted as from another source such as a bird peck. Following the path of a projectile through a large carcass is also challenging. However, in some populations where death from gunshot is more common, researchers have been able to develop systems for identifying and describing such cases. For example, in the western United States, sea lions are often the victims of aggression from humans as they are competitors for the same prey species (Goldstein et al. 1999; Stroud and Roffe 1979; Würsig and Gailey 2002). However, small cetaceans have also been the targets of shootings using firearms as well as other projectiles and sharp weapons (Byard et al. 2001; Vail 2016). Examiners of such cases should, if possible, use scanning equipment such as CT and radiography, even if only for a portion of the body (the suspected area of concern); this portion must be removed and scanned separately. Careful examination of external wounds and the track of the projectile or weapon and a detailed collection of photographs and histology samples should be made (Fig. 4). Prosecutors unfamiliar with these types of cases should refer to published protocols in standard forensic references (Cooper and Cooper 2013; Moore et al. 2013; Moore and Barco 2013). Of course, law enforcement should be notified immediately and should help coordinate chain of custody and evidence handling.

Illegal Mutilation and Trophy Collection

Governments and regulating agencies around the world including the United States recognize the long history of indigenous peoples' interaction with several marine mammal species especially in the high arctic. As such, the MMPA and other international treaties allow for exemptions for such groups and provide a framework for the legal collection and use of marine mammal parts and have set up safeguards to ensure regulations are followed (Burn 1998; Hovelsrud et al. 2008). While hunting or harvest of a marine mammal is legal in such circumstances, it remains a controversial subject as many marine mammals that are targeted are still declining. In this forensic context, we are referring to the illegal collection of marine mammal parts. Several marine mammal species are commonly involved in mutilation and trophy collection either to retain the artifact in its original form (e.g., teeth) or for it to be modified for art or souvenirs such as figurines or scrimshaw.

Mutilation may be a secondary human interaction when the primary injury/death was caused by a fishery interaction (entrapment or entanglement). In some cases, once the discovery of the animal with attached gear is made, there may be an attempt by the fishery personnel to sink the body or otherwise disguise the evidence. In these cases, the ventral body may be incised with a knife or appendages may be removed (Moore and Barco 2013). Mutilation may also occur when someone removes a portion of the body for fishing bait or to collect the body part as a "souvenir." In some of these cases, the person may not know that doing so violates the MMPA so public education is paramount. Examiners of carcasses that have been mutilated should recognize the possibilities for such in the context of local fishery interactions and both legal and illegal collection of parts. Careful collection of samples including the remainder of skeletal parts is important in the forensic investigation to determine how and when (antemortem or postmortem) the part was removed.

Illegal Sale of Parts

While some legal collection of marine mammal parts is permitted both by the United States and international law, illegal collection, and sale of parts continues to be a significant issue. There are generally two categories of marine mammal part sales/trades. The first is collection and trade for art or handicrafts and household goods. Prior to 1972 when the MMPA was enacted, marine mammal parts could be collected and used legally in such industries. It is therefore important to identify the proper documentation such as sale receipts and identifying marks on the piece to provide the best evidence of a legal take and sale. Likewise, some parts (or derivations of parts) may be legally sold by indigenous tradespeople provided they follow the requirements of the laws in the countries where the animal was killed and laws regarding the trade of the pieces

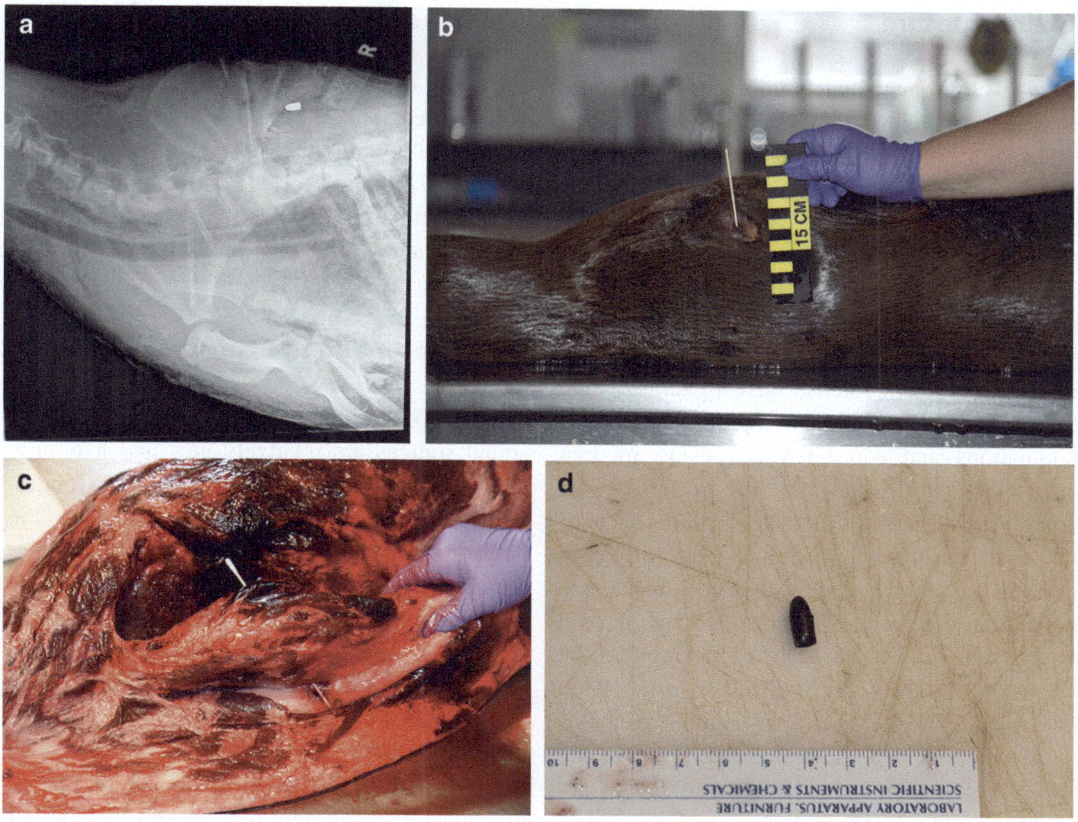

Fig. 4 (**a**) Radiograph of California sea lion (LMLZC2018OCT01-2) showing bullet lodged near the scapula. (**b**) Penetrating wound on the lateral body of a sea lion with a swab inserted to show wound track. (**c**) Dissection of sea lion showing internal wound track from bullet. (**d**) Bullet removed from the body of sea lion during necropsy. Photo credit: UCSC Long Marine Laboratory

through international laws and agreements such as CITES (Baur et al. 1999).

The second category refers to soft tissues (usually blubber or meat) from harvested marine mammals. In some countries, it is legal to kill and sell marine mammals for consumption. This trade takes place under the oversight of the International Whaling Commission (IWC) with international agreements (Gambell 1993). Several investigations and published research have shown how difficult it is to regulate such trade due to the difficulty in correctly identifying the tissues/meat at the market or in a restaurant. Breakthroughs in DNA collection and analyses including portable test kits have provided researchers and investigators invaluable devices in the forensic toolbox (Baker et al. 2007, 2010; Palumbi and Cipriano 1998). This genetic expertise remains a vital piece of forensic investigations and the continual effort to conserve cetaceans and other marine mammals through surveillance of products sold in markets and through Internet suppliers (Baker et al. 2003). Note that although technology has vastly improved the efficiency and accuracy of parts identification, the ability to do so remains with genetic and morphology experts such as those in dedicated forensic laboratories at NOAA/NMFS, USFWS, and academic institutions.

Toxin Exposure

It is well-known that marine mammals are exposed to toxins throughout their lifetime. Many of these toxins bioaccumulate in their tissues over time and cause harm in various ways. While these findings are anthropogenic in nature, the long-term nature of the exposure does not often lead to forensic investigations. However, acute toxic exposure, most notably in the form of oil spills are indeed some of the most important types of forensic investigations as the effects often take hold on multiple species or populations and can be traced using modern chemistry even when the effects persist for many years. The full extent of harmful effects of oil spills on marine mammals is still being investigated. Unfortunately, we do not have to go back in time very far to see this play out on a large scale; it can be seen as recently as the Deepwater Horizon Oil Spill (DWH) of 2010 in the Gulf of Mexico. Scientists continue to monitor the populations of marine mammals in the Gulf to determine what long-term effects are being experienced by populations of bottlenose dolphins and other pelagic species (Takeshita et al. 2017; Schwacke et al. 2013). Effects of oil spills on pinnipeds and sea otters (as well as cetaceans) are also well documented from the 1989 Exxon Valdez Spill in Alaska (Peterson et al. 2003).

The DWH spill investigation continues as part of the Natural Resource Damage Assessment, which is the legal framework for identifying impacts to natural areas and their biological components. Thus, it could reasonably be argued that the resulting investigation including collection and sampling of dead and sick marine mammals during and since the spill constitutes one of the largest and most comprehensive forensic investigations in history (Helm et al. 2015; Wallace et al. 2017). Examiners and investigators involved in oil spills (and other toxic spills) must be familiar with standard protocols for collecting, archiving, and analyzing samples and data. Often this type of investigation requires large teams working together (Takeshita et al. 2017; Wilkin et al. 2017). Workers on site must be properly trained as hazardous waste operators/responders and chain of custody for collected samples must be followed. Therefore, prior training and drills are critically important. When at all possible, caches of supplies for the initial collection of samples and response should be ready for deployment in staging areas near possible spill sites.

Underwater Sound Exposure

Threats to marine mammals from underwater sound sources are relatively new and it is only recently that researchers have begun to understand the impact that sound can have on the behavior and bodies of marine mammals. There are numerous types of anthropogenic sounds now ubiquitous in the marine environment including ship noise, military sonar, pile driving associated with industry and construction, seismic surveys, and even noise from underwater wind turbines and other energy exploration. All of these have the potential to disrupt the behavior of marine mammals for short and long periods of time but many also have the potential to cause hearing damage (Ketten 1995; Madsen et al. 2006; Moore et al. 2012) and there is mounting evidence that sonar events can change dive behavior and lead to serious, even fatal physiological changes such as fat embolism and gas bubble lesions (Fernández et al. 2005; Jepson et al. 2003). Generally, gross descriptions and examinations of suspect sound trauma in marine mammals will not be conclusive. Advanced sampling, specialized clinical equipment, and expert opinion are necessary in such cases (De Quirós et al. 2011; Ketten 2014; Van Bonn et al. 2011). It is not reasonable to expect that most stranding responders or even pathologists will have all the tools and supplies needed to follow exact protocols for sound-related trauma events, but examiners are encouraged to have protocols in hand and to be familiar with in situ sampling procedures. As with any forensic investigation, photographs are essential as well as histology, blood samples, and frozen tissues. Notification should be made to the proper regulating body so that potential sources of sound or other factors can be investigated with the potential to suspend operations if warranted.

Final Thoughts

Perhaps no marine mammal on the planet epitomizes the need for more attention on conservation and the focus of law enforcement, regulators, forensic investigators, and the public than the vaquita (*Phocoena sinus*). The vaquita is the most endangered marine mammal in the world. A small porpoise with a very limited range in waters of the Gulf of California in Mexico, the species has been in decline since the 1950s. The population has been decimated as bycatch in gillnets and trawls set for fish and shrimp (D'Agrosa et al. 2000). The most recent and deadliest threat to this endemic cetacean is the illegal gillnet fishery for one particular fish (*Totoaba macdonaldi*) and the trade of their swim bladders which are sent to China and Hong Kong as a food delicacy and for their perceived medicinal properties (Martínez and Martínez 2018; Rojas-Bracho et al. 2006). Recent efforts to capture and breed these porpoises in an ex situ facility failed with one animal dying in the attempt to save the species. International laws and enforcement have proven insufficient to curb the high price that fishermen can demand for the illegal product (Rojas-Bracho et al. 2019). This case illustrates the role of forensic science, conservation biology, law enforcement, local people, and multilateral efforts in a global conservation crisis.

Conservation measures, forensic investigations, technology, and field efforts must be improved if we are going to help the remaining vaquitas, North Atlantic right whales, Hawaiian monk seals, and all other marine mammals at risk. Unfortunately, we are running out of time as the effects of climate change and political and cultural unrest only fuel the conflicts, illegal trade, and poverty that is at the heart of many conservation crises. One part of a global strategy to halt the decimation of marine mammal populations should include an increase in the number of qualified forensic investigators, advances in portable field equipment, and wider dissemination of forensic approaches to wildlife conservation.

References

Anderson GS, Bell LS (2014) Deep coastal marine taphonomy: investigation into carcass decomposition in the Saanich Inlet, British Columbia using a baited camera. PLoS One 9(10):e110710

Baker CS, Lento GM, Dalebout ML, Pichler F (2003) DNA surveys and surveillance of marine mammals: species identification, discovery and management. Marine Mammals and Humans CSIRO, Melbourne, pp 366–382

Baker CS, Cooke JG, Lavery S, Dalebout ML, MA YU, Funahashi N, Carraher C, Brownell RL Jr (2007) Estimating the number of whales entering trade using DNA profiling and capture-recapture analysis of market products. Mol Ecol 16(13):2617–2626

Baker CS, Steel D, Choi Y, Lee H, Kim KS, Choi SK, Ma Y-U, Hambleton C, Psihoyos L, Brownell R (2010) Genetic evidence of illegal trade in protected whales links Japan with the US and South Korea. Biol Lett 6 (5):647–650

Baur DC, Bean MJ, Gosliner ML (1999) The laws governing marine mammal conservation in the United States. Conservation and management of marine mammals. Smithsonian Institution Press, Washington, DC, pp 48–86

Beck CA, Barros NB (1991) The impact of debris on the Florida manatee. Mar Pollut Bull 22(10):508–510

Burn DM (1998) Estimation of hunter compliance with the marine mammal marking, tagging, and reporting program for walrus. Wildl Soc Bull 68–74

Butterworth A (2016) A review of the welfare impact on pinnipeds of plastic marine debris. Front Mar Sci 3:149

Byard RW, Gilbert JD, Kemper CM (2001) Dolphin deaths: forensic investigations. Med J Aust 175(11–12):623–624

Byard RW, Machado A, Woolford L, Boardman W (2013) Symmetry: the key to diagnosing propeller strike injuries in sea mammals. Forensic Sci Med Pathol 9 (1):103–105

Byrd, JH, Sutton, LK (2012) Defining a crime scene and physical evidence collection. In: Wildlife forensics: methods and applications, 6

Cassoff RM, Moore KM, McLellan WA, Barco SG, Rotstein DS, Moore MJ (2011) Lethal entanglement in baleen whales. Dis Aquat Organ 96(3):175–185

Cooper JE, Cooper ME (2013) Wildlife forensic investigation: principles and practice. CRC

D'Agrosa C, Lennert-Cody CE, Vidal O (2000) Vaquita bycatch in Mexico's artisanal gillnet fisheries: driving a small population to extinction. Conserv Biol 14 (4):1110–1119

De Quirós YB, González-Díaz Ó, Saavedra P, Arbelo M, Sierra E, Sacchini S, Jepson PD, Mazzariol S, Di

Guardo G, Fernández A (2011) Methodology for in situ gas sampling, transport and laboratory analysis of gases from stranded cetaceans. Sci Rep 1:193
Fernández A, Edwards J, Rodriguez F, De Los Monteros AE, Herraez P, Castro P, Jaber J, Martin V, Arbelo M (2005) "Gas and fat embolic syndrome" involving a mass stranding of beaked whales (Family Ziphiidae) exposed to anthropogenic sonar signals. Vet Pathol 42 (4):446–457
Gambell R (1993) International management of whales and whaling: an historical review of the regulation of commercial and aboriginal subsistence whaling. Arctic 97–107
Geraci JR, Lounsbury VJ (2005) Marine mammals ashore: a field guide for strandings. In: National Aquarium in Baltimore
Goldstein T, Johnson S, Phillips A, Hanni K, Fauquier D (1999) Human-related injuries observed in live stranded pinnipeds along the. Aquat Mamm 25:43–51
Helm RC, Costa DP, DeBruyn TD, O'Shea TJ, Wells RS, Williams TM (2015) Overview of effects of oil spills on marine mammals. In: Handbook of oil spill science and technology. Wiley Online Library, pp 455–475
Hovelsrud GK, McKenna M, Huntington HP (2008) Marine mammal harvests and other interactions with humans. Ecol Appl 18(sp2):S135–S147
Huffman JE, Wallace JR (2012) Wildlife forensics: methods and applications, vol 6. Wiley
Jacobsen JK, Massey L, Gulland F (2010) Fatal ingestion of floating net debris by two sperm whales (Physeter macrocephalus). Mar Pollut Bull 60(5):765–767
Jepson P, Arbelo M, Deaville R, Patterson I, Castro P, Baker J, Degollada E, Ross H, Herráez P, Pocknell A (2003) Gas-bubble lesions in stranded cetaceans. Nature 425(6958):575
Jepson PD, Barbieri M, Barco SG, Bernaldo de Quirós Y, Bogomolni AL, Danil K, Rowles T (2013) Peracute underwater entrapment of pinnipeds and cetaceans. In: Moore MJ, Van Der Hoop J, Barco SG, Costidis AM, others (eds) Criteria and case definitions for serious injury and death of pinnipeds and cetaceans caused by anthropogenic trauma. Dis Aquat Org, vol 103, pp 235–239
Kenney RD (2009) Right whales: Eubalaena glacialis, E. japonica, and E. australis. In: Encyclopedia of marine mammals. Academic, pp 962–972
Ketten DR (1995) Estimates of blast injury and acoustic trauma zones for marine mammals from underwater explosions. Sens Syst Aquat Mamm 391–407
Ketten DR (2014) Sonars and strandings: are beaked whales the aquatic acoustic canary. Acoust Today 10 (3):46–56
Laist DW (1987) Overview of the biological effects of lost and discarded plastic debris in the marine environment. Mar Pollut Bull 18(6):319–326
Laist DW (1997) Impacts of marine debris: entanglement of marine life in marine debris including a comprehensive list of species with entanglement and ingestion records. In: Marine debris. Springer, pp 99–139
Lavigne DM, Scheffer VB, Kellert SR (1999) The evolution of North American attitudes toward marine mammals. In: Conservation and management of marine mammals, pp 10–47
Liebig PM, Taylor T-SA, Flessa KW (2003) Bones on the beach: marine mammal taphonomy of the Colorado Delta, Mexico. Palaios 18(2):168–175
Lightsey JD, Rommel SA, Costidis AM, Pitchford TD (2006) Methods used during gross necropsy to determine watercraft-related mortality in the Florida manatee (Trichechus manatus latirostris). J Zoo Wildl Med 37(3):262–276
Madsen PT, Wahlberg M, Tougaard J, Lucke K, Tyack P (2006) Wind turbine underwater noise and marine mammals: implications of current knowledge and data needs. Mar Ecol Prog Ser 309:279–295
Martínez IA, Martínez ER (2018) Trafficking of totoaba maw. In Green crime in Mexico. Palgrave Macmillan, Cham, pp 149–170
Moore KT, Barco SG (2013) Handbook for recognizing, evaluating, and documenting human interaction in stranded cetaceans and pinnipeds
Moore MJ, Van der Hoop JM (2012) The painful side of trap and fixed net fisheries: chronic entanglement of large whales. J Mar Biol 1–4
Moore SE, Reeves RR, Southall BL, Ragen TJ, Suydam RS, Clark CW (2012) A new framework for assessing the effects of anthropogenic sound on marine mammals in a rapidly changing Arctic. BioScience 62 (3):289–295
Moore MJ, van der Hoop J, Barco SG, Costidis AM, Gulland FM, Jepson PD, Moore KT, Raverty S, McLellan WA (2013) Criteria and case definitions for serious injury and death of pinnipeds and cetaceans caused by anthropogenic trauma. Dis Aquat Organ 103(3):229–264
O'Shea TJ, Beck CA, Bonde RK, Kochman HI, Odell DK (1985) An analysis of manatee mortality patterns in Florida, 1976-81. J Wildl Manag 1–11
Osinga N, Morick D (2008) By-catch and drowning in harbour porpoises (Phocoena phocoena) stranded on the northern Dutch coast. Eur J Wildl Res 54 (4):667–674
Palumbi A, Cipriano F (1998) Species identification using genetic tools: the value of nuclear and mitochondrial gene sequences in whale conservation. J Hered 89 (5):459–464
Peltier H, Dabin W, Daniel P, Van Canneyt O, Dorémus G, Huon M, Ridoux V (2012) The significance of stranding data as indicators of cetacean populations at sea: modelling the drift of cetacean carcasses. Ecol Indic 18:278–290
Peterson CH, Rice SD, Short JW, Esler D, Bodkin JL, Ballachey BE, Irons DB (2003) Long-term ecosystem response to the Exxon Valdez oil spill. Science 302 (5653):2082–2086
Pettis HM, Rolland RM, Hamilton PK, Brault S, Knowlton AR, Kraus SD (2004) Visual health

assessment of North Atlantic right whales (Eubalaena glacialis) using photographs. Can J Zool 82(1):8–19

Read AJ (2008) The looming crisis: interactions between marine mammals and fisheries. J Mammal 89 (3):541–548

Read AJ, Murray KT (2000) Gross evidence of human-induced mortality in small cetaceans

Reynolds III JE, Powell JA, Diagne LWK, Barton SL, Scolardi KM (2018) Manatees: Trichechus manatus, T. senegalensis, and T. inunguis. In: Encyclopedia of marine mammals. Elsevier, pp 558–566

Rojas-Bracho L, Reeves RR, Jaramillo-Legorreta A (2006) Conservation of the vaquita Phocoena sinus. Mamm Rev 36(3):179–216

Rojas-Bracho L, Gulland F, Smith C, Taylor B, Wells R, Thomas P, Bauer B, Heide-Jørgensen M, Teilmann J, Dietz R (2019) A field effort to capture critically endangered vaquitas Phocoena sinus for protection from entanglement in illegal gillnets. Endangered Species Res 38:11–27

Rommel SA, Costidis AM, Pitchford TD, Lightsey JD, Snyder RH, Haubold EM (2007) Forensic methods for characterizing watercraft from watercraft-induced wounds on the Florida manatee (Trichechus manatus latirostris). Mar Mamm Sci 23(1):110–132

Schwacke LH, Smith CR, Townsend FI, Wells RS, Hart LB, Balmer BC, Collier TK, De Guise S, Fry MM, Guillette LJ Jr (2013) Health of common bottlenose dolphins (Tursiops truncatus) in Barataria Bay, Louisiana, following the Deepwater Horizon oil spill. Environ Sci Technol 48(1):93–103

Sharp SM, McLellan WA, Rotstein DS, Costidis AM, Barco SG, Durham K, Pitchford TD, Jackson KA, Daoust P-Y, Wimmer T (2019) Gross and histopathologic diagnoses from North Atlantic right whale Eubalaena glacialis mortalities between 2003 and 2018. Dis Aquat Organ 135(1):1–31

Simmonds MP (2012) Cetaceans and marine debris: the great unknown. J Mar Biol. https://doi.org/10.1155/2012/684279

Stolen M, Durden WN, Mazza T, Barros N, St Leger J (2013) Effects of fishing gear on bottlenose dolphins (Tursiops truncatus) in the Indian River Lagoon system, Florida. Mar Mamm Sci 29(2):356–364

Stroud RK, Roffe TJ (1979) Causes of death in marine mammals stranded along the Oregon coast. J Wildl Dis 15(1):91–97

Takeshita R, Sullivan L, Smith C, Collier T, Hall A, Brosnan T, Rowles T, Schwacke L (2017) The Deepwater Horizon oil spill marine mammal injury assessment. Endangered Species Res 33:95–106

Vail CS (2016) An overview of increasing incidents of bottlenose dolphin harassment in the Gulf of Mexico and possible solutions. Front Mar Sci 3:110

Van Bonn W, Montie E, Dennison S, Pussini N, Cook P, Greig D, Barakos J, Colegrove K, Gulland F (2011) Evidence of injury caused by gas bubbles in a live marine mammal: barotrauma in a California sea lion Zalophus californianus. Dis Aquat Organ 96(2):89–96

Wallace BP, Brosnan T, McLamb D, Rowles T, Ruder E, Schroeder B, Schwacke L, Stacy B, Sullivan L, Takeshita R (2017) Effects of the Deepwater Horizon oil spill on protected marine species. Endangered Species Res 33:1–7

Wells RS, Allen JB, Lovewell G, Gorzelany J, Delynn RE, Fauquier DA, Barros NB (2015) Carcass-recovery rates for resident bottlenose dolphins in Sarasota Bay, Florida. Mar Mamm Sci 31(1):355–368

Wilkin SM, Rowles TK, Stratton E, Adimey N, Field CL, Wissmann S, Shigenaka G, Fougères E, Mase B, Network SRS (2017) Marine mammal response operations during the Deepwater Horizon oil spill. Endangered Species Res 33:107–118

Wilkinson D, Worthy, GA (1999) Marine mammal stranding networks. Conserv Manag Mar Mamm (2):396–411

Williams R, Gero S, Bejder L, Calambokidis J, Kraus SD, Lusseau D, Read AJ, Robbins J (2011) Underestimating the damage: interpreting cetacean carcass recoveries in the context of the Deepwater Horizon/BP incident. Conserv Lett 4(3):228–233

Würsig B, Gailey G (2002) Marine mammals and aquaculture: conflicts and potential resolutions. In: Responsible marine aquaculture. CAP, New York, pp 45–59

Contaminants as a Conservation Threat to Marine Mammals

Marie Noël and Tanya M. Brown

Abstract

Persistent Organic Pollutants (POPs) and inorganic elements like mercury (Hg) are ubiquitous environmental pollutants. Their physico-chemical properties allow them to be transported over great distances, deposited, and incorporated into aquatic food webs. As long-lived species, many high trophic level marine mammals bioaccumulate high levels of certain contaminants, and a variety of contaminant-related health effects have been reported in populations around the world. Contaminants have been identified as one of the top threats to over 80% of the marine mammal populations globally, including two of Canada's most endangered cetacean populations, Southern Resident killer whales (*orcinus orca*) from the northeastern Pacific, and St Lawrence Estuary beluga whales (*Delphinapterus leucas*). It is therefore critical that we continue to improve upon our understanding of contaminant sources, their fate, and effects in order to better inform on marine mammal conservation measures. The collection of samples from marine mammals poses a number of challenges, however, over the past few decades novel approaches have been developed that allow for minimally invasive collection of samples (e.g., fecal). As well, the development and use of new laboratory techniques (e.g., omics) has allowed for the assessment of contaminant-related impacts at different levels of biological organization. Despite these advances, there is much to be learned about contaminant-mediated effects in marine mammals at both the individual and population levels. There is a need for long-term monitoring programmes that use multi-disciplinary approaches to assess regional contaminant trends and effects in marine mammals in the context of other environmental stressors, including climate change.

Keywords

Persistent organic pollutants · Mercury · Bioaccumulation · Southern Resident killer whales · St Lawrence Estuary belugas · Marine mammals · Toxicology · Marine conservation

Contaminants of Concern in Marine Mammals

Marine mammals face a number of threats globally. Contaminants have been identified as the second most reported threat for over 80% (99 of the 121) of marine mammal species globally

M. Noël (✉)
Ocean Wise Conservation Association, Vancouver, BC, Canada
e-mail: Marie.noel@ocean.org

T. M. Brown
Fisheries and Oceans Canada, West Vancouver, BC, Canada

S. C. Underkoffler, H. R. Adams (eds.), *Wildlife Biodiversity Conservation*,
https://doi.org/10.1007/978-3-030-64682-0_15

(Avila et al. 2018). This threat is particularly concerning for toothed whales because they feed higher up in the food web, are long-lived, and show a reduced metabolic ability to eliminate some persistent environmental pollutants (Sonne et al. 2018). Persistent organic pollutants (POPs) and inorganic elements like mercury (Hg) are ubiquitous environmental pollutants that readily enter aquatic food webs and bioaccumulate and biomagnify to concentrations that exceed thresholds of known or potential adverse effects. POPs include a number of man-made organohalogen compounds (OHCs) such as polychlorinated biphenyls (PCBs), polybrominated diphenyl ethers (PBDEs), and poly- and perfluoroalkyl substances (PFASs) that have been released globally and exhibit a wide range of physico-chemical properties (Wania 2003). In contrast, Hg occurs both naturally and anthropogenically and is present in several forms in the environment, including its most toxic form, monomethylmercury (MMHg). This chapter provides an overview of legacy and emerging POPs and mercury (Hg) which pose a conservation threat to marine mammal populations globally, a summary of novel tools and approaches that have been used to evaluate contaminant levels and related health effects in marine mammals, and highlights two of Canada's most endangered marine mammal populations (Southern Resident killer whales and St Lawrence Estuary beluga whales), of which contaminants is one of the top threats impeding both their survival and recovery.

Sources

Legacy and Emerging Persistent Organic Pollutants (POPs)

Polychlorinated biphenyls (PCBs) and polybrominated diphenyl ethers (PBDEs) are two contaminant classes that are persistent, bioaccumulative, and toxic. Both classes of chemicals tend to degrade slowly and persist in the environment for extensive periods of time. PCBs are legacy contaminants that were commercially manufactured from about 1930 to the late 1970s (Tanabe 1988) when production was banned in North America. The production of PBDEs started in the 1960s (USEPA 2010), with two (pentaBDE and OctaBDE) of the three main commercial components being phased out in North America by 2005 (Government of Canada 2006), and the third, DecaBDE, following thereafter. International action on POPs culminated in the signing of the Stockholm Convention in 2001. The Convention was enacted in 2004 and identified 12 chlorinated chemical groups for global bans, including PCBs (Muir and Howard 2006). PentaBDE and OctBDE were added to the Stockholm Convention in 2009, followed by DecaBDE in 2017.

PCBs were used extensively in a variety of industrial applications, such as coolants and lubricants in capacitors and transformers, as well as in hydraulic fluids, heat transfer fluids, and plasticizers (Wu et al. 2008). PBDEs were used as flame retardants in a wide array of products, including furnishings, electronics, foams, textiles, and construction materials, in order to comply with fire safety standards (Mack 2004). Despite the banning of production and use of these two contaminant classes, PCBs contained in products manufactured prior to the ban will continue to represent a potential source of environmental contamination until the complete elimination of equipment containing PCBs is required by 2025 (Fauchald and Hunter 2008). Similarly, an estimated thousands of tons of PBDEs are still present in products currently being used (Abbasi et al. 2015).

Bioaccumulation and Biomagnification of Contaminants

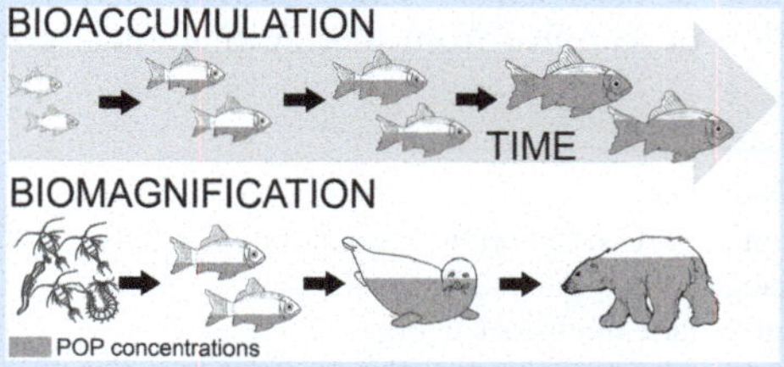

Poly- and perfluoroalkyl substances (PFASs) are a group of emerging organic contaminants of concern and include two well-studied groups: perfluoroalkyl sulfonates (PFSAs) and perfluoroalkyl carboxylates (PFCAs). Their chemical structure allows them to have both lipid and water-repellent properties, making them ideal for several commercial uses. Production of PFASs began in the 1950s and currently there are more than 200 industrial and consumer applications, including liquid repellents for paper, food packaging, carpets, textiles, floor polishes, insecticides, and aqueous fire-fighting foam (Renner 2006). They can be released directly as manufacturing by-products, or indirectly through the degradation of precursors. Given their versatility, they often end up in landfills and wastewater treatment plants, and ultimately into the marine environment. It has been estimated that over 95% of PFAS emissions are directly released into the aquatic environment, with limited atmospheric emissions (Prevedouros et al. 2006; Ahrens and Bundschuh 2014). Major manufacturers from the United States (US) phased out the production of perfluorooctanoic acid (PFOA) and perfluorooctanesulfonic acid (PFOS) in the early 2000s amidst growing concern over the potential for toxic effects. In 2009, PFOS, its salts, and perfluorooctane sulfonyl fluoride were added to the Stockholm Convention Annex B substance list, restricting the production and use of these chemicals. In 2017, PFOA, its salts, and PFOA-related compounds were added to the Annex A substance list, whereby parties are advised to eliminate the production and use of these chemicals. Currently, Perfluorohexane sulfonic acid (PFHxS), its salts, and PFHxS-related compounds have been proposed for listing and are under the Stockholm Convention review.

Mercury

Mercury (Hg) can be found in the environment from both natural and anthropogenic sources. Natural sources account for approximately 60% of total atmospheric Hg emissions and include releases from volcanoes, forest fires, contaminated soils in ancient mining industrial areas or geological units rich in Hg, and surface waters (Pirrone et al. 2010). Since the Industrial Revolution, human activities have substantially increased the amount of Hg released into the environment, resulting in two to three times more atmospheric and oceanic mercury. While electric power generation facilities using coal contribute to more than 50% of the total anthropogenic Hg sources, there are a number of other potential sources, including ferrous and non-ferrous metal smelters, Chlor-alkali plants, waste incinerators, and small-scale gold mining (Pirrone et al. 2010).

In the USA, Canada, most European countries, and Japan, there are regulations to limit mercury emissions from coal-fired power plants. However, regulations are lacking in Asia, the major emitter of Hg. This raises a global concern, as emissions from China are predicted to increase, thus adding to the global environmental burden of Hg (Pacyna et al. 2010). The Minamata Convention, signed in 2013, provided the means for international legislation to mitigate Hg emissions through the use of a legally-binding treaty. The Convention aims to control emissions from products such as thermometers and energy-saving light bulbs, as well as from industries such as mining, cement production, and coal-fired power (www.unep.org).

Incorporation into Marine Food Webs

Legacy and Emerging POPs

POPs physico-chemical properties drive their ability to be transported into the environment, broken down, or deposited and incorporated into aquatic food webs. Once emitted into the atmosphere, gas/particle partitioning is the first key process that influences the environmental fate of POPs. Because of their moderate vapour pressure, most POPs exist in a gaseous state (Manchester-Neesvig and Andren 1989), allowing them to be transported over great distances, where they have been documented in remote regions such as the Arctic. Most POPs enter the marine environment through the gas exchange, dry, and wet deposition (Cotham and Bidleman 1991). For example, atmospheric PCBs have been estimated to represent 93% of the total PCB inputs to the North Sea (Duce 1998), and 60–90% of the PCB burden in the Great Lakes may originate from the atmosphere (Eisenreich et al. 1981). Whereas, for some of the more dominant PFAS compounds (PFOA and PFOS), oceanic transport appears to be the dominant source of these chemicals to remote Arctic environments (Stemmler and Lammel 2010; Wania 2007).

Once in the marine environment, most POPs including PCBs and PBDEs will partition out of the water onto organic matter, due to their lipophilic nature. At the bottom of the food chain, there is rapid adsorption of POPs to phytoplankton and/or zooplankton surfaces, followed by diffusion through the membrane into the plankton matrix (Del Vento and Dachs 2002; Swackhamer and Skoglund 1993). From there, accumulation up the food chain occurs through a series of solvent switching and solvent depletion, where POPs move through different organismal compartments and deposit into tissues with a high lipid content (Macdonald et al. 2002). Because they are long-lived, feed at the top of the food chain, and have significant blubber storage, marine mammals are reported to have significant POP burden (Ross et al. 2000; Ross 2006; Lebeuf et al. 2014; Marsili et al. 2018; Nyman et al. 2002). Unlike PCBs and PBDEs, PFASs are relatively water soluble, non-lipophilic, and tend to accumulate more in protein rich tissues such as the liver and blood. Short-chain PFASs are more hydrophilic, and therefore more mobile, while long-chain PFASs are more hydrophobic and tend to bioaccumulate, and thus are of particular concern for marine mammals (Ahrens and Bundschuh 2014).

Mercury

Atmospheric transport and deposition are responsible for approximately 90% of the oceanic accumulation of Hg (Outridge et al. 2008). Once deposited in the ocean, Hg can be deposited into the sediment layer, transformed into methylmercury (MeHg), or reduced via microbial reduction and/or photoreduction into elemental Hg that is recycled back into the atmosphere. It has been estimated that 24–36% of deposited Hg is photoreduced and returned to the atmosphere (Schroeder and Munthe 1998).

With regard to contamination of the food chain, methylation is key, as MeHg bioaccumulates and is a toxic form of Hg. Microorganisms such as methane- and sulphate-producing bacteria are key components in the formation of MeHg (Barkay et al. 2003). In addition, certain environmental conditions such as low dissolved oxygen content, low pH, and high concentrations of organic matter can increase the formation of MeHg. Once in the marine environment, MeHg can be demethylated through photodegradation or microbial pathways, or incorporated into the food web.

At the bottom of the food chain, Hg binds to the membrane of diatoms, while MeHg is associated with the soluble fraction of the organism. Copepods feed on diatoms and thus assimilate MeHg, while inorganic Hg will be excreted along with the membranes. There is still much uncertainty surrounding the transfer of mercury through the food chain, however, it appears to be the relative liposolubility of MeHg that allows it to be partly retained in the fatty tissues (Morel et al. 1998). MeHg in fish and top predators is mostly associated with proteins, suggesting that factors other than its relative liposolubility may explain MeHg bioaccumulation.

Tools to Measure Contaminant-Related Health Effects in Marine Mammals

The effects of anthropogenic contaminants on marine mammals may manifest at different levels of biological organization, including altered population dynamics, behavioural, and physiological changes of individual organisms, as well as altered molecular biological pathways. While investigating abundance informs on the cumulative anthropogenic impacts to populations, assessing physiological or molecular level responses lends insight into the biological effects individual and or combined stressors (e.g., contaminants, climate change, noise) may have on marine mammals. Over recent decades several non-lethal, minimally invasive approaches have been developed to collect samples from cetaceans, pinnipeds, and polar bears, which have allowed for a variety of physiological and molecular techniques to be used in the laboratory.

Marine Mammal Sample Collection in Support of Physiological and Molecular Investigations

Early studies have used marine mammals in captivity to provide important insight into the impacts of contaminants on marine mammal health (Ross et al. 1996; De Swart et al. 1996; Thomas et al. 2005), however current ethical standards prevent such studies to be undertaken. Opportunistic sampling from carcasses or stranded marine mammals is possible and allows for a variety of tissues to be sampled, however collection of these samples can introduce confounding factors such as tissue necrosis, degradation of contaminants, and degradation of RNA for transcriptomic (gene expression) analyses. Several research projects have also benefited from the collection of samples through traditionally harvested marine mammals (Loseto et al. 2008; Ostertag et al. 2013; Noël et al. 2014; Brown et al. 2017). In such studies, a variety of samples were collected, including muscle, liver, brain, and claws, and could be used for contaminant analyses and related health effects. Samples obtained from live free-ranging marine mammals are desirable, however, sample collection, especially from large cetaceans, is difficult due to the inherent logistical challenges in obtaining adequate samples (Hunt et al. 2013). Blood is acknowledged as being the preferred tissue for physiological or molecular analyses in marine mammals (Hunt et al. 2013), however, collection of blood from marine mammals, and in particular from cetaceans, is often not feasible. Alternative, non-lethal, minimally invasive sampling techniques to investigate contaminant levels and health include (Fig. 1):

- Blubber/skin biopsies are the most widely used sample in marine mammal toxicology. The subcutaneous fatty layer is the main site for the storage of lipids and has great potential for providing information about stored lipophilic substances, thermoregulation, energetics, fasting metabolism, seasonal patterns of dietary intake, and hormones. Blubber/skin biopsies have therefore been collected from a number of live free-ranging cetacean species using projectile darts, and have provided a wide array of toxicological information, such as the concentration of lipophilic contaminants of concern (Ross et al. 2000; Ross 2006; Hobbs et al. 2003; Fossi et al. 2000), hormone levels (Loseto et al. 2017; Galligan et al. 2019; Hayden et al. 2017; Boggs et al. 2017), as well as transcriptomic information (Buckman et al. 2011; Noël et al. 2014; Brown et al. 2017; Simond et al. 2019).
- Faecal samples have been widely used in dietary studies, and more recently have been used to investigate contaminants concentrations in marine mammals. Lundin et al. (2016) were the first to measure a series of organic contaminants (PCBs, PBDEs, DDTs, and HCB) in killer whale scats collected from endangered Southern Resident killer whales (*Orcinus orca*). In addition, reproductive and stress hormones have been measured in scat samples from the same population of killer whales (Wasser et al. 2017) and from North Atlantic right whales (*Eubalaena glacialis*) (Rolland et al. 2017).
- Blow (exhaled droplets of condensed respiratory vapour) samples have become valuable physiological samples (Hunt et al. 2013). While blow condensate is not likely an appropriate matrix for the determination of contaminant burden, it can provide valuable insight into the health of individuals. Steroid hormones have been measured successfully in blow samples from North Atlantic right whales (*Eubalaena glacialis*) and beluga whales (*Delphinapteurs leucas*) (Hunt et al. 2016; Richard et al. 2017; Burgess et al. 2018). In addition, recent studies demonstrated

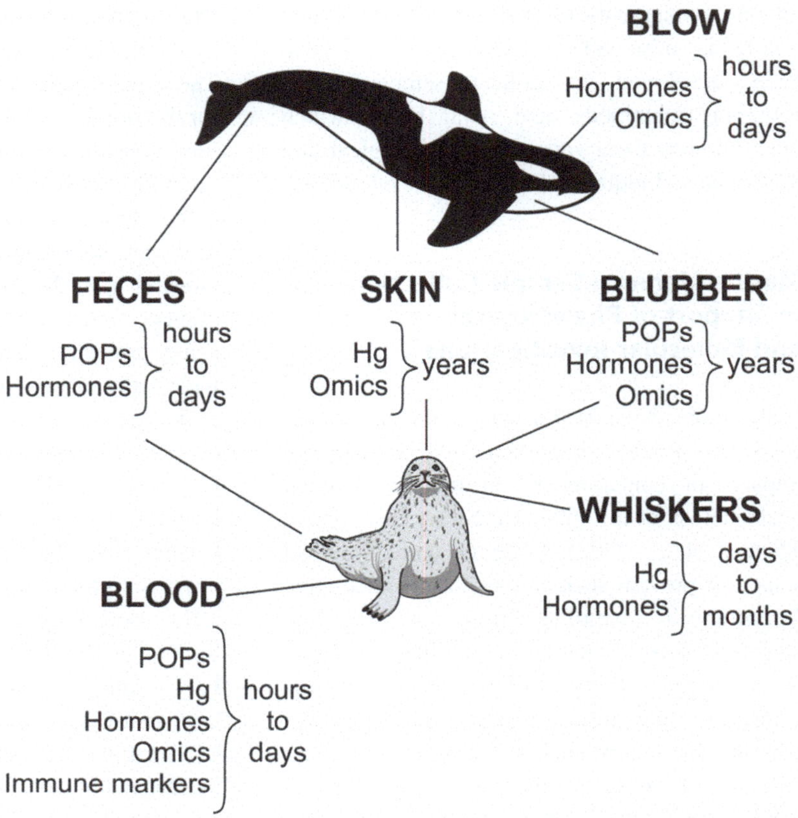

Fig. 1 A variety of samples can be collected from pinnipeds and cetaceans in order to investigate contaminant levels and their related health effects

that in common bottlenose dolphins (*Tursiops truncatus*), there was a correlation between breath metabolites and conventional health measures conducted in the blood (Pasamontes et al. 2017; Borras et al. 2017). Given that mitochondrial and microsatellite DNA have been isolated from cetacean blow (Frère et al. 2010; Richard et al. 2017), there is also the potential for transcriptomics methods to be applied to blow samples (Hunt et al. 2013).

Contaminants and related health effects may be affected by biological variables such as age, sex, molting, fasting, and lactating. Accounting or controlling for a number of these confounding variables is important when designing marine mammal toxicology studies. When working with cetaceans, logistical challenges usually lead to opportunistic sampling rather than randomized, stratified sampling in a more controlled setting. Sampling pinnipeds tends to be easier, as they spend some of their time on land. Various techniques have been developed over the years to catch and handle them safely (Gentry and Holt 1982; Jeffries et al. 1993), allowing for the collection of a variety of samples and biological data to support contaminant and health effect analyses. Further, it allows for the sampling of a specific age class or sex which reduces the number of variables to account for in a study. Similar to cetaceans, blubber/skin biopsies are also collected from pinnipeds and other types of samples are additionally accessible (Fig. 1).

– Blood samples have allowed for the measurement of a wide variety of contaminants and their metabolites (Sandala et al. 2004; Griesel et al. 2008; Weijs et al. 2009; Kunisue and Tanabe 2009), as well as hormones (Simms and Ross 2000; Braathen et al. 2004; Tabuchi et al. 2006; Mos et al. 2007), and parameters of the immune system (Lie et al. 2004, 2005; Lahvis et al. 1995; Mos et al. 2007; Dupont et al. 2013) in a number of pinniped species and polar bears (*Ursus maritimus*). Blood has direct contact with target organs and is likely to reflect bioavailable contaminant concentrations. On the other hand, blood will only reflect a relatively short exposure time, from a few hours to a few days, as opposed to other tissues like blubber, hair, or whiskers, that will reflect a much longer time period, from months to years.

- Hair samples have been collected from a variety of pinniped species, as well as from polar bears. Hair is usually the matrix of choice when monitoring heavy metals, especially toxic methylmercury in marine mammals (Brookens et al. 2008; Rea et al. 2013; Bechshoft et al. 2015). More than 90% of Hg present in keratinized tissues such as hair or whiskers exists as the toxic methylated form (Kehrig et al. 1997; Voegborlo et al. 2010). Unlike blood or other biological tissues that are used in toxicology studies which have transient pharmacokinetic properties, hair is resistant to hydrolysis and/or enzyme activity, and is considered to have a stable composition (Hopps 1977). Incorporation of Hg in hair occurs via blood circulation during the formation of the hair shaft and therefore acts as a record of mercury intake over the time of hair growth. Steroid hormones have also been measured successfully in polar bear hair, with the advantage of expressing chronic stress rather than short-term hormonal fluctuations (Bechshøft et al. 2011, 2012, 2015).
- Whisker samples have been used in several pinniped studies to provide insight into the temporal accumulation of Hg (Ferreira et al. 2011; Noël et al. 2016). Similar to hair, whiskers are considered metabolically inert and can provide temporal record of metal accumulation over the past year for phocids (Hall-Aspland et al. 2005; Beltran et al. 2015), and several years for otariids (Stricker et al. 2015; McHuron et al. 2016). Recently, Karpovich et al. (2019) demonstrated that phocid whiskers could also be a valuable tool for physiological investigations, as they successfully measured cortisol along the whiskers of ringed (*Phoca hispida*), spotted (*Phoca largha*), and harbour seals (*Phoca vitulina*).

While in vitro toxicology studies can be conducted using cells derived from the tissues collected from free-ranging animals and exposing them to specific contaminants, it does not capture the contaminant kinetics that takes place within the body (Weijs and Zaccaroni 2016). In vitro studies and modelling efforts are discussed in detail in Weijs and Zaccaroni (2016). In the present chapter, we focus on 'natural exposure' studies (in vivo) used to evaluate the potential impacts of contaminants on marine mammal health, using samples collected from free-ranging marine mammals.

Measurements of Key Hormones and Vitamins

Hormones are biochemicals produced by specific glands or organs that can influence cellular functions in order to achieve the optimal internal environment (St. Aubin et al. 2001) whereas vitamins are essential nutrients that are not produced endogenously and are generally acquired through diet.

Methods have been developed to measure vitamin A and E in a variety of marine mammal tissues such as blubber, liver, and plasma using high performance liquid chromatography (HPLC) (Simms and Ross 2000; Debier et al. 2005; Mos et al. 2007; Routti et al. 2010; Desforges et al. 2013; Pedro et al. 2019). Vitamin A and E are essential for a number of biological functions including growth, development, reproduction, and protection against tissue damage, and have been shown to be disrupted by PCBs in Arctic belugas, northeastern Pacific harbour seals, California sea lions (*Zalophus californianus*), Baltic ringed and grey seals (*Halichoerus grypus*), and Greenland killer whales (Desforges et al. 2013; Nyman et al. 2003; Mos et al. 2007; Debier et al. 2005; Routti et al. 2010; Pedro et al. 2019), and PBDEs in grey seals from Scotland (Berghe et al. 2013).

Hormone levels vary constantly in order to adjust to internal and external stimuli. The measurement of hormone levels in marine mammals has been used over the years to provide information on the activity of specific endocrine organs and to shed light on the potential endocrine disruptive effects of contaminants. Numerous enzyme- or radio-immunoassays kits are available commercially for hormone analysis. However, a significant amount of sample is required for the determination of an individual hormone. Additionally, there is variability in test output between kit manufacturers, and limited ability for absolute quantification when using biopsies (Hayden et al. 2017). Techniques have therefore been developed to measure hormones in a variety of other matrices by either modifying steroid isolation techniques and using the commercially-available immunoassays (Kellar et al. 2015; Champagne et al. 2017), or using advanced equipment such as HPLC or high performance liquid chromatography tandem mass spectrometry (LC/MS/MS) (Desforges et al. 2013; Loseto et al. 2017; Hayden et al. 2017; Boggs et al. 2017). In addition, LC/MS/MS methods were adapted recently to simultaneously evaluate the levels of progesterone, 17-hydroxyprogesterone, testosterone, androstenedione, cortisol, 11-deoxycortisol, cortisone, and 11-deoxycorticosterone in a blubber sample as small as 50 mg (Hayden et al. 2017; Boggs et al. 2017). Thyroid hormones, involved in the regulation of metabolic processes, growth, and differentiation of the developing animal, have also been reported to be affected by PCB and PBDE in grey seals, harbour seals, beluga whales (Debier et al. 2005; Sørmo et al. 2005; Tabuchi et al. 2006; Hall and Thomas 2007; Villanger et al. 2011a), as well as polar bears (Skaare et al. 2001; Villanger et al. 2011b). Reproductive hormones have also been shown to be associated with POP levels. For example, testosterone was negatively associated with PCBs in polar bears (Oskam et al. 2004), and it was suggested that PCBs may affect progesterone levels in female polar bears (Haave et al. 2003). Plasma thyroid hormone levels were correlated with POP concentrations in Baltic ringed seals (Routti et al. 2010). Finally, cortisol, an important stress-related hormone involved in the regulation of energy metabolism, maintenance of growth, and development were altered by PCB exposure in polar bears (Oskam et al. 2004).

Immunotoxicity Measurements

The immune system is comprised of a variety of cells, tissues, and chemical messengers that allow an organism to prevent and eradicate infections. Several disease outbreaks in marine mammal populations have been reported since the 1970s, including epizootic outbreaks of morbillivirus that killed thousands of European harbour and grey seals (Dietz et al. 1989), and an epidemic of cancer among the St Lawrence beluga whale population (De Guise et al. 1995a, b; Martineau et al. 1994). A combination of in vitro laboratory studies, captive feeding experiments, and field studies suggest that exposure to POPs and Hg has the potential to affect the immune system, and therefore reduce the ability of individuals to fight infectious microorganisms and malignant diseases (Ross et al. 1996; Ross 2002). Desforges et al. (2016) provided an extensive review of immunotoxic effects in marine mammals, in which they combined all marine mammal data available in the literature to create dose-response relationships for a variety of immunotoxic endpoints. They concluded that both the in vivo and in vitro experiment results on lymphocyte proliferation were consistent with one another, supporting the use of extrapolation of in vitro dose-response data in wildlife risk assessment.

When investigating potential contaminant-related immunotoxicity in live marine mammals, blood (plasma or serum) is often the preferred matrix, as it contains white blood cells, which are key players of the immune response. White blood cells are comprised of T- and B-lymphocytes, natural killer cells, macrophages, and neutrophils, which are all involved in an organism's first line of defence. The simplest approach to investigating immune imbalances is the evaluation of clinical blood parameters such as the number and relative proportion of white blood cells in circulation. While studies have reported an association between the number of various types of white blood cells and contaminants such as PCBs, Hg, and PFCs (Beckmen et al. 2003; Mos et al. 2006; Fair et al. 2013), results are usually difficult to interpret due to natural fluctuations in such parameters. A variety of assays have been developed over the years to evaluate various components of the immune system, with B- or T-lymphocyte proliferation being by far the most commonly used in marine mammal toxicology. B- and T-lymphocytes are the major cellular components of the adaptive immune response and are responsible for recognizing foreign entities entering the body. After the collection of fresh blood samples, they can be isolated and exposed to proteins (mitogens) that induce their proliferation. The majority of studies suggest that PCBs and Hg can reduce the proliferative ability of lymphocytes (Lahvis et al. 1995; Levin et al. 2005; Kakuschke et al. 2005; Mos et al. 2006), therefore preventing the development of a proper immune response. On the other hand, Fair et al. (2013) demonstrated that PFCs were positively associated with lymphocyte proliferation in bottlenose dolphins.

Neutrophils are also impacted by contaminants. By incubating neutrophils with fluorescent beads, and using flow cytometry to evaluate the ability of these cells to engulf the beads (phagocytosis), in vivo studies demonstrated that PCBs and PFCs were affecting the ability of marine mammal white blood cells to engulf pathogens (Mos et al. 2006; Schwacke et al. 2012; Fair et al. 2013). After phagocytosis, the engulfed material is destroyed via biochemical processes of the respiratory burst, leading to the creation of reactive oxygen species. In harbour seals and beluga whales, PCBs have been associated with high respiratory burst (De Guise et al. 1995a, b, Mos et al. 2006), therefore increasing the risk of oxidative stress and tissue damage. Transcriptomics is an emerging tool that has been used in a few immunotoxicity studies looking at the expression of cytokines, important signalling proteins involved in the modulation and mediation of immune response. For example, the expression of IL-1 was associated with PCBs in ringed seals from northern Europe and Labrador (Routti et al. 2010; Brown et al. 2014), and IL-10 mRNA levels were positively correlated with PCB concentrations in northeastern Pacific killer whales (Buckman et al. 2011).

Omics

Omics refers to the quantification of biological molecules that translate into the structure, dynamics, and function of an organism. With technological advances allowing for cost-efficient and high-throughput analysis of biological molecules, omics studies have been increasingly used in the medical field (Hasin et al. 2017). While there are a variety of omics technologies available, genomics (identification of genetic variants), transcriptomics (quantification of RNA levels), proteomics (quantification of protein abundance, modification, and interaction), and metabolomics (quantification of small molecules that are substrates or end products of cellular regulatory processes) are the most commonly used (Cavill et al. 2015; Hasin et al. 2017). The majority of molecular profiling studies published in marine mammal toxicology involve transcriptomics, with only a few recent ones using metabolomics in polar bears (Tartu et al. 2017; Morris et al. 2018, 2019) and bottlenose dolphins (Aksenov et al. 2014).

Transcriptomics

One of the first components of biological response to environmental change includes altered expression of mRNA, followed by an adjustment in transcriptome profile in a given tissue (Veldhoen et al. 2012). Transcriptomics studies may be targeted whereby RNA is extracted from the samples, complementary DNA (cDNA) synthesized, and the mRNA levels of specific genes quantified using polymerase chain reaction (PCR). This technique requires the design of primers specific to the genes of interest. mRNA transcript abundance levels of genes involved in growth, development, reproduction, and metabolism in killer whales, ringed seals, harbour seals, beluga whales, bottlenose dolphins, and fin whales have been associated with PCB and POP exposure (Fossi et al. 2010; Buckman et al. 2011; Noël et al. 2014, 2016; Brown et al. 2014; Routti et al. 2010, Lunardi et al. 2016; Simond et al. 2019).

Other transcriptomic techniques such as microarrays and RNA-sequencing (RNA-Seq) provide information on thousands of genes at once, allowing a deeper understanding of the molecular responses to environmental stressors, as well as the identification of new genes of interest. Mancia et al. (2014, 2015) developed a common bottlenose dolphin microarray that identified altered expression of genes involved in contaminant metabolism, development, differentiation, and oncogenic pathways in response to PCB exposure. While microarrays can only detect transcripts that correspond to existing genomic sequencing information, RNA-Seq techniques are able to explore both known and new transcripts. Additionally, with RNA-Seq, the transcriptome is evaluated using high-throughput sequencing methods that are sensitive and allow for the quantification of RNA levels expressed at very low levels. The application of RNA-Seq to marine mammals is a recent and emerging approach that can generate vast amounts of information regarding genes and gene expression and highlights differences in gene expression patterns between individuals, tissues, or different environmental conditions (Wolf 2013; Cammen et al. 2016). For example, in a study on the southern California bottlenose dolphin, Trego et al. (2019) used RNA-sequencing together with the known genome of the species to identify contaminant-related alterations of genes involved in contaminant metabolism, hormone metabolism, and immune response. Even with species lacking a reference genome, de novo transcriptome assembly and differential expression analyses can be very powerful in identifying specific genes or pathways affected by contaminants. Using such techniques, Brown et al. (2017) were able to detect energy metabolism imbalances in ringed seals exposed to a local PCB source in Arctic Labrador, Canada. The RNA-seq data also allowed them to identify five new gene targets that may be used in future marine mammal molecular and toxicological studies.

Metabolomics

Metabolites result from gene expression and protein activity, and include amino acids, sugars, fatty acids, lipids, and nucleic acids. The metabolic profiles of blood serum/plasma, tissues, or cells may serve as biomarkers for physiological alterations that may occur following exposure to endocrine disrupting compounds (EDCs), such as PCBs, or other stressors. As the endocrine system is the major regulator of metabolism (Wallaschofski 2012), one may be able to detect whether exposure to EDCs has resulted in metabolism alterations. Field-based studies using metabolomics are limited. Using samples collected from harvested polar bears and multivariate statistical analyses, Morris et al. (2019) identified 29 metabolites that were altered in two populations of polar bears under different contaminant exposure regimes. Those included metabolites involved in fatty acid oxidation and inflammatory response. Additionally, Tartu et al. (2017) showed that PCBs, chlordanes, PBDEs, and fluorinated compounds were related to alterations of fatty acid metabolism, insulin utilization, and cholesterol homeostasis in polar bears. While contaminants appeared to play a role in the different metabolite profiles observed, the authors noted dietary variation, sex, and sea ice decline may have additional influence on their results (Tartu et al. 2017; Morris et al. 2018, 2019).

While metabolite profiling is in its infancy in marine mammal toxicology, the integration of metabolomics and other omics approaches (e.g., transcriptomics) offers promising tools to study the health of marine mammals and to provide information on the effects of contaminants and other environmental stressors.

Contaminants, Marine Mammals, and Conservation

Case of the Endangered St Lawrence Estuary Beluga Whales

The St. Lawrence Estuary (SLE) beluga population is the most southerly of the beluga populations and is a relict Arctic population genetically distinct from all other beluga populations (De March et al. 2002). Their current habitat range (2790 km^2, Lemieux Lefebvre et al. 2012) is 65% of what was historically described, and their annual core area distribution is at the lower end for any population of this species (DFO 2014). Despite a 40-year hunting ban, the SLE beluga is comprised of approximately 900 individuals and is listed as endangered under the Canadian Species at Risk Act (SARA). A Recovery Strategy (Canada, DFO 2012) has been identified and written for the SLE beluga population. The SLE beluga population has been studied since the mid-1970s, largely due to its low population size and loss of viable habitat (Hickie et al. 2000). Critical habitat has been identified and relates to the summer area occupied by females accompanied by calves and juveniles. Contaminants, anthropogenic disturbances, reduction in the abundance, quality, and availability of prey, and other degradation of habitat are currently considered the most serious threats to the population (DFO 2014).

The SLE belugas live downstream from a densely populated, highly industrialized region of Canada and the United States. High levels of contaminants including PCBs, PBDEs, mercury, legacy pesticides, polycyclic aromatic hydrocarbons (PAHs), short-chain chlorinated paraffins (SCCPs), and emerging flame retardants have been found in the SLE belugas (De Guise et al. 1995a, b; Lebeuf et al. 2001, 2004, 2007, 2010, 2014; Simond et al. 2020) and historically were associated with high rates of cancer and other diseases (Martineau et al. 1999, 2002; Lair 2007), and altered reproductive capacity (Martineau et al. 1988, 2002, 2003; Béland et al. 1992, 1993; De Guise et al. 1995a, b, 1996). More recently, altered mRNA transcript levels for genes involved in thyroid and steroid hormone regulation were associated with PCBs, organochlorine pesticides (OCs), and an emerging flame retardant (hexabromobenzene, HBB) (Simond et al. 2019), and altered lipid metabolism (Simond et al. 2020) was associated with short-chain chlorinated paraffins (SCCPs), despite reported decreases in some POPs over time (LeBeuf et al. 2014).

Case of the Endangered Southern Resident Killer Whales

The Southern Resident killer whale (SRKW) population primarily resides in the transboundary waters of British Columbia, Canada, and Washington State, USA (Haro Strait, Boundary Pass, the eastern portion of Juan de Fuca Strait, and southern portions of the Strait of Georgia), during the summer and fall. Some of the population remains year-round, while others travel great distances and have been sighted as far as central California and southeastern Alaska (Hauser et al. 2007). The population has fluctuated between 70 and 99 individuals since 1976 (DFO 2018), and as of December 2020 consisted of 74 individuals (Center for Whale Research 2020). The Southern Residents consist of three family groups or pods (J, K, and L pods) (Parsons et al. 2009). Critical habitat has been identified for the SRKWs and relates largely to summer and fall areas frequented by these cetaceans. Both Canada and the United States have listed SRKWs as endangered, and an Action Plan and Recovery Strategy (Canada, DFO 2017, 2018) and Recovery Plan (USA, NMFS 2008) have been identified and written. SRKWs forage primarily on salmonids between late spring and fall, with their dominant prey being Chinook salmon (*Oncorhynchus tshawytscha*) (Ford et al. 2016). Much of this information concerning the diet and habitat use has been gathered during the summer and fall months. Less is known about their distribution in winter months, when they are seen infrequently in the Canada/United States transboundary waters. The Northern Resident

Killer Whales (NRKW) and 'transient' or 'Bigg's' killer whales, which is another ecotype or group of killer whales, also frequent the coastal waters of British Columbia and adjacent areas. The two resident populations consume fish, while the Bigg's killer whale population consumes pinnipeds and cetaceans almost exclusively, but no fish (Ford et al. 1998).

Exposure to environmental contaminants, including PCBs, PBDEs, and legacy pesticides, as well as the reduction in the availability or quality of prey and acoustic disturbance are currently considered the three dominant threats to the SRKW population (DFO 2018). Similar to the SLE belugas, the SRKWs forage in waters that are surrounded by highly populated, industrialized urban areas of the United States and Canada, which exposes them to elevated levels of complex mixtures of contaminants. The SRKW and Bigg's killer whales are reported to be among the most contaminated cetaceans in the world (Ross et al. 2000). PCB burdens in this population are predicted to inhibit population growth or cause a decline of approximately 15% for reproductive or combined reproductive and immune effects, respectively (Desforges et al. 2018). Altered mRNA transcript levels in blubber of killer whales for a number of toxicological and metabolic related genes were associated with PCBs (Buckman et al. 2011), which provides evidence of physiological effects of POPs in the endangered SRKW population.

Comparison of the SLE Beluga and SRKW

While logistical, legal, and ethical constraints limit toxicological sampling in free-ranging SLE belugas and SRKWs, a number of biopsy-based samples have been collected over the years and analyzed for a select suite of contaminant classes, including PCBs and PBDEs. The most recent data (2008-2017) reveal that average ΣPCB and ΣPBDE concentrations in adult male SRKW are approximately 3.5–3.7-fold greater, respectively, than average concentrations in adult male SLE belugas (Fig. 2, Simond et al. 2019; Mongillo et al. 2016). Average ΣPCB concentrations for SRKW and SLE beluga exceed marine mammal immune and endocrine effects threshold concentrations (1300 ng/g lw, Mos et al. 2010) by 31- and 9-fold, respectively. Whereas, average ΣPBDE concentrations for SRKW exceed a marine mammal thyroid hormone disruption threshold (1500 ng/g lw, Hall et al. 2003) by 3-fold, and average PBDE concentrations in SLE beluga remains just below the threshold (Fig. 2). As mentioned above, changes in blubber mRNA levels for genes involved in immune and endocrine function, metabolism, and thyroid-related function were correlated with increasing blubber PCBs in resident and transient to Bigg's killer whales of central and northern British Columbia (Buckman et al. 2011). mRNA transcript levels for genes involved in growth, development, metabolism, and endocrine function in SLE beluga were correlated with increasing levels of PCBs, pesticides, and industrial by-products, HBB, and the halogenated flame retardant Dec-604 (Simond et al. 2019). Together, these findings suggest continued health risks associated with persistent and bioaccumulative contaminants to both endangered whale populations. While it is difficult to infer the population-level consequences from these findings, observations of a decrease in abundance in both populations continue to occur and are of concern (Mosnier et al. 2015; DFO 2018).

Conclusion

Marine mammals are exposed to complex mixtures of pollutants. As noted above, elevated levels of contaminants represent one of the top threats to marine mammal populations globally, and adverse effects have been observed in a number of species. Despite recent advances in sample collection and laboratory techniques to evaluate contaminant concentrations and effects, much remains to be learned on contaminant-mediated effects in marine mammals at both the individual and population levels. For example, toxicity reference values (TRVs) for marine mammals only exist for a few contaminants of concern, including

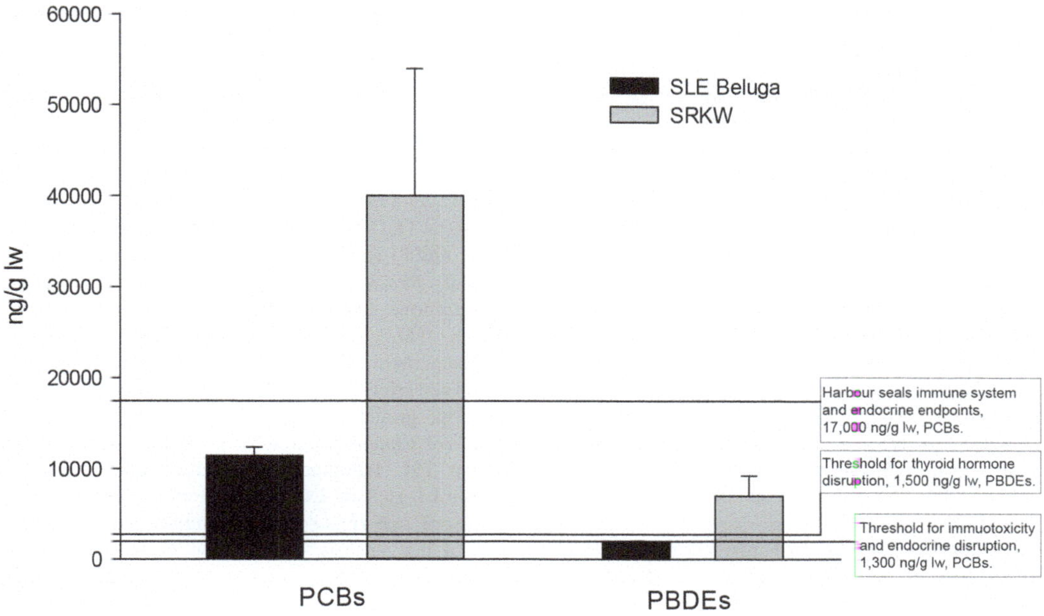

Fig. 2 Mean concentrations (+/– SE) for ΣPCBs and ΣPBDEs in the blubber of SLE beluga and SRKW. Effects values for PCBs are from free-ranging harbour seals from Ross et al. (1996) and Mos et al. (2010). Effects values for PBDEs are from grey seals from Hall et al. (2003). The PCB and PBDE concentrations for SLE beluga (2015–2017) and SRKW (2008–2013) are from Simond et al. (2020) and Mongillo et al. (2016), respectively

PCBs and PBDEs (Ross et al. 1996; Mos et al. 2010; Brown et al. 2014; Desforges et al. 2013; Hall et al. 2003), and little to no information exists on the cumulative effects of these contaminants on marine mammals. The lack of such thresholds impedes our ability to generate environmental quality guidelines that are protective of marine mammals. It is important to acknowledge and consider the use of alternate approaches for characterizing contaminant exposure and health effects in marine mammals given the inherent challenges in obtaining samples from certain populations. For example, the use of modelling species (e.g., alternate indicator species, reference populations, or prey) combined with modelling approaches (Alava et al. 2012, 2016; Hickie et al. 2000) may provide insight into contaminant concentrations and potential health effects for marine mammals. Further, many effects studies are often short-lived and do not include additional stressors (climate change, noise, and disturbance) in the analysis and interpretation of their data. There is a need for long-term monitoring programmes using multidisciplinary approaches to assess regional contaminant trends and effects in marine mammals in the context of other environmental stressors, including climate change.

References

Abbasi G, Buser AM, Soehl A, Murray MW, Diamond ML (2015) Stocks and flows of PBDEs in products from use to waste in the U.S. and Canada from 1970 to 2020. Environ Sci Tech 49:1521–1528

Ahrens L, Bundschuh M (2014) Fate and effects of poly- and perfluoroalkyl substances in the aquatic environment: a review. Environ Toxicol Chem 33 (9):1921–1929

Aksenov AA, Yeates L, Pasamontes A, Siebe C, Zrodnikov Y, Simmons J, McCartney MM, Deplanque P, Wells RS, Davis CE (2014) Metabolite content profiling of bottlenose dolphin exhaled breath. Anal Chem 86(21):10616–10624

Alava JJ, Ross PS, Lachmuth C, Ford JK, Hickie BE, Gobas FA (2012) Habitat-based PCB environmental quality criteria for the protection of endangered killer whales (*Orcinus orca*). Environ Sci Tech 46 (22):12655–12663

Alava JJ, Ross PS, Gobas FA (2016) Food web bioaccumulation model for resident killer whales from the Northeastern Pacific Ocean as a tool for the derivation of PBDE-sediment quality guidelines. Arch Environ Contam Toxicol 70(1):155–168

Avila IC, Kaschner K, Dormann CF (2018) Current global risks to marine mammals : taking stock of the threats. Biol Conserv 221:44–58

Barkay T, Miller SM, Summers AO (2003) Bacterial mercury resistance from atoms to ecosystems. FEMS Microbiol Rev 27(2–3):355–384

Bechshøft TØ, Sonne C, Dietz R, Born EW, Novak MA, Henchey E, Meyer JS (2011) Cortisol levels in hair of East Greenland polar bears. Sci Total Environ 409 (4):831–834

Bechshøft TØ, Riget FF, Sonne C, Letcher RJ, Muir DCG, Novak MA et al (2012) Measuring environmental stress in East Greenland polar bears, 1892–1927 and 1988–2009: what does hair cortisol tell us? Environ Int 45:15–21

Bechshoft T, Derocher AE, Richardson E, Mislan P, Lunn NJ, Sonne C et al (2015) Mercury and cortisol in Western Hudson Bay polar bear hair. Ecotoxicology 24(6):1315–1321

Beckmen KB, Blake JE, Ylitalo GM, Stott JL, O'Hara TM (2003) Organochlorine contaminant exposure and associations with hematological and humoral immune functional assays with dam age as a factor in free-ranging northern fur seal pups (*Callorhinus ursinus*). Mar Pollut Bull 46(5):594–606

Béland P, De Guise S, Plante R (1992) Toxicologie et pathologie des mammifères marins du Saint-Laurent. In: Fonds pour la toxicologie faunique du Fonds mondial pour la nature, 99p

Béland P, De Guise S, Girard C, Lagacé A, Martineau D, Michaud R, Muir DCG, Norstrom RJ, Pelletier E, Ray S, Shugart LR (1993) Toxic compounds and health and reproduction effects in St. Lawrence beluga whales. J Great Lakes Res 19:766–775

Beltran RS, Sadou MC, Condit R, Peterson SH, Reichmuth C, Costa DP (2015) Fine-scale whisker growth measurements can reveal temporal foraging patterns from stable isotope signatures. Mar Ecol Prog Ser 523:243–253

Berghe MV, Weijs L, Habran S, Das K, Bugli C, Pillet S et al (2013) Effects of polychlorobiphenyls, polybromodiphenylethers, organochlorine pesticides and their metabolites on vitamin A status in lactating grey seals. Environ Res 120:18–26

Boggs AS, Schock TB, Schwacke LH, Galligan TM, Morey JS, McFee WE, Kucklick JR (2017) Rapid and reliable steroid hormone profiling in Tursiops truncatus blubber using liquid chromatography tandem mass spectrometry (LC-MS/MS). Anal Bioanal Chem 409(21):5019–5029

Borras E, Aksenov AA, Baird M, Novick B, Schivo M, Zamuruyev KO et al (2017) Exhaled breath condensate methods adapted from human studies using longitudinal metabolomics for predicting early health alterations in dolphins. Anal Bioanal Chem 409 (28):6523–6536

Braathen M, Derocher AE, Wiig Ø, Sørmo EG, Lie E, Skaare JU, Jenssen BM (2004) Relationships between PCBs and thyroid hormones and retinol in female and male polar bears. Environ Health Perspect 112 (8):826–833

Brookens TJ, O'Hara TM, Taylor RJ, Bratton GR, Harvey JT (2008) Total mercury body burden in Pacific harbor seal, *Phoca vitulina richardii*, pups from central California. Mar Pollut Bull 56(1):27–41

Brown TM, Ross PS, Reimer KJ, Veldhoen N, Dangerfield NJ, Fisk AT, Helbing CC (2014) PCB related effects thresholds as derived through gene transcript profiles in locally contaminated ringed seals (*Pusa hispida*). Environ Sci Tech 48(21):12952–12961

Brown TM, Hammond SA, Behsaz B, Veldhoen N, Birol I, Helbing CC (2017) *De novo* assembly of the ringed seal (*Pusa hispida*) blubber transcriptome: a tool that enables identification of molecular health indicators associated with PCB exposure. Aquat Toxicol 185:48–57

Buckman AH, Veldhoen N, Ellis G, Ford JK, Helbing CC, Ross PS (2011) PCB-associated changes in mRNA expression in killer whales (Orcinus orca) from the NE Pacific Ocean. Environ Sci Tech 45 (23):10194–10202

Burgess EA, Hunt KE, Kraus SD, Rolland RM (2018) Quantifying hormones in exhaled breath for physiological assessment of large whales at sea. Sci Rep 8 (1):10031

Cammen KM, Andrews KR, Carroll EL, Foote AD, Humble E, Khudyakov JI et al (2016) Genomic methods take the plunge: recent advances in high-throughput sequencing of marine mammals. J Hered 107(6):481–495

Cavill R, Jennen D, Kleinjans J, Briedé JJ (2015) Transcriptomic and metabolomic data integration. Brief Bioinform 17(5):891–901

Center for Whale Research (2020) Southern Resident Killer Whale Population. Web. April 27, 2021. https://www.whaleresearch.com/orca-population

Champagne CD, Kellar NM, Crocker DE, Wasser SK, Booth RK, Trego ML, Houser DS (2017) Blubber cortisol qualitatively reflects circulating cortisol concentrations in bottlenose dolphins. Mar Mamm Sci 33(1):134–153

Cotham WE Jr, Bidleman TF (1991) Estimating the atmospheric deposition of organochlorine contaminants to the Arctic. Chemosphere 22(1–2):165–188

Debier C, Ylitalo GM, Weise M, Gulland F, Costa DP, Le Boeuf BJ et al (2005) PCBs and DDT in the serum of juvenile California sea lions: associations with vitamins A and E and thyroid hormones. Environ Pollut 134(2):323–332

De Guise S, Martineau D, Béland P, Fournier M (1995a) Possible mechanisms of action of environmental contaminants on St. Lawrence beluga whales (*Delphinapterus leucas*). Environ Health Perspect 103 (suppl 4):73–77

De Guise S, Flipo D, Boehm JR, Martineau D, Béland P, Fournier M (1995b) Immune functions in beluga whales (Delphinapterus leucas): evaluation of phagocytosis and respiratory burst with peripheral blood leukocytes using flow cytometry. Vet Immunol Immunopathol 47(3–4):351–362

De Guise S, Bernier J, Martineau D, Béland P, Fournier M (1996) Effects of in vitro exposure of beluga whale splenocytes and thymocytes to heavy metals. Environ Toxicol Chem 15(8):1357–1364

Del Vento S, Dachs J (2002) Prediction of uptake dynamics of persistent organic pollutants by bacteria and phytoplankton. Environ Toxicol Chem Int J 21 (10):2099–2107

De March BGE, Maiers LD, Friesen MK (2002) An overview of genetic relationships of Canadian and adjacent populations of beluga (Delphinapterus leucas) with emphasis on Baffin Bay and Canadian eastern Arctic populations. NAMMCO Sci Publ 4:17–38

Department of Fisheries and Oceans (DFO) (2014) Status of beluga (*Delphinapterus leucas*) in the St. Lawrence River Estuary. Canadian Science Advisory Secretariat (CSAS), 2013/076

Department of Fisheries and Oceans (DFO) (2018) Recovery strategy for the Northern and Southern resident killer whales (*Orcinus orca*) in Canada. Species at risk act recovery strategy series. Fisheries & Oceans Canada, Ottawa, x + 84 pp

Desforges J-PW, Ross PS, Dangerfield N, Palace VP, Whiticar M, Loseto LL (2013) Vitamin A and E profiles as biomarkers of PCB exposure in beluga whales (*Delphinapterus leucas*) from the western Canadian Arctic. Aquat Toxicol 142–143:317–328

Desforges JPW, Sonne C, Levin M, Siebert U, De Guise S, Dietz R (2016) Immunotoxic effects of environmental pollutants in marine mammals. Environ Int 86:126–139

Desforges J-PW, Hall A, Mcconnell B, Rosing-Asvid A, Barber JL, Brownlow A, De Guise S, Eulaers I, Jepson PD, Letcher RJ, Levin M, Ross PS, Samarra F, Vikingson G, Sonne C, Dietz R (2018) Predicting global killer whale population collapse from PCB pollution. Science 361:1373–1376

De Swart RL, Ross PS, Vos JG, Osterhaus AD (1996) Impaired immunity in harbour seals (*Phoca vitulina*) exposed to bioaccumulated environmental contaminants: review of a long-term feeding study. Environ Health Perspect 104(suppl 4):823–828

Dietz R, Heide-Jørgensen MP, Härkönen T (1989) Mass deaths of harbor seals (*Phoca vitulina*) in Europe. Ambio (Sweden) 18:258–264

Duce RA (1998) The input of atmospheric chemicals to the ocean. WMO Bull 47(1):51–60

Dupont A, Siebert U, Covaci A, Weijs L, Eppe G, Debier C et al (2013) Relationships between in vitro lymphoproliferative responses and levels of contaminants in blood of free-ranging adult harbour seals (*Phoca vitulina*) from the North Sea. Aquat Toxicol 142:210–220

Eisenreich SJ, Looney BB, Thornton JD (1981) Airborne organic contaminants in the Great Lakes ecosystem. Environ Sci Tech 15(1):30–38

Fair PA, Romano T, Schaefer AM, Reif JS, Bossart GD, Houde M et al (2013) Associations between perfluoroalkyl compounds and immune and clinical chemistry parameters in highly exposed bottlenose dolphins (*Tursiops truncatus*). Environ Toxicol Chem 32(4):736–746

Fauchald OK, Hunter D (2008) Part 2: The year in review, I: General development. In: Fauchald OK, Hunter D (eds) Yearbook of international environmental law, vol 17. Oxford, p 248

Ferreira EO, Loseto LL, Ferguson SH (2011) Assessment of claw growth-layer groups from ringed seals (Pusa hispida) as biomonitors of inter-and intra-annual Hg, δ15N, and δ13C variation. Can J Zool 89(9):774–784

Fisheries and Oceans Canada (DFO) (2012) Recovery strategy for the beluga whale (*Delphinapterus leucas*) St. Lawrence Estuary population in *Canada. Species at risk act recovery strategy series. Fisheries and Oceans Canada,* Ottawa, 88 pp + X pp

Fisheries and Oceans Canada (DFO) (2017) Action plan for the Northern and Southern Resident Killer Whale (*Orcinus orca*) in Canada. Species at risk act action plan series. Fisheries and Oceans Canada, Ottawa, v + 33 pp

Ford JKB, Ellis GM, Barrett-Lennard LG, Morton AB, Palm RS, Balcomb KC III (1998) Dietary specialization in two sympatric populations of killer whales (*Orcinus orca*) in coastal British Columbia and adjacent waters. Can J Zool 76:1456–1471

Ford MJ, Hempelmann J, Hanson MB, Ayres KL, Baird RW, Emmons CK, Lundin JI, Schorr GS, Wasser SK, Park LK (2016) Estimation of a killer whale (*Orcinus orca*) population's diet using sequencing analysis of DNA from feces. PLoS One. https://doi.org/10.1371/journal.pone.0144956

Fossi MC, Marsili L, Neri G, Casini S, Bearzi G, Politi E et al (2000) Skin biopsy of Mediterranean cetaceans for the investigation of interspecies susceptibility to xenobiotic contaminants. Mar Environ Res 50 (1–5):517–521

Fossi MC, Urban J, Casini S, Maltese S, Spinsanti G, Panti C et al (2010) A multi-trial diagnostic tool in fin whale (*Balaenoptera physalus*) skin biopsies of the Pelagos Sanctuary (Mediterranean Sea) and the Gulf of California (Mexico). Mar Environ Res 69:S17–S20

Frère CH, Krzyszczyk E, Patterson EM, Hunter S, Ginsburg A, Mann J (2010) Thar she blows! A novel method for DNA collection from cetacean blow. PLoS One 5(8):e12299

Galligan TM, Balmer BC, Schwacke L, Bolton JL, Quigley BM, Rosel P et al (2019) Examining the relationships between blubber steroid hormone and persistent organic pollutant measurements in common bottlenose dolphins. Environ Pollut 249

Gentry RL, Holt JR (1982) Equipment and techniques for handling northern fur seals

Government of Canada (2006) Polybrominated diphenyl ethers regulations. Regulatory impact analysis statement. Canadian Gazette (Part 1), 4294

Griesel S, Kakuschke A, Siebert U, Prange A (2008) Trace element concentrations in blood of harbor seals (Phoca vitulina) from the Wadden Sea. Sci Total Environ 392 (2–3):313–323

Haave M, Ropstad E, Derocher AE, Lie E, Dahl E, Wiig Ø et al (2003) Polychlorinated biphenyls and reproductive hormones in female polar bears at Svalbard. Environ Health Perspect 111(4):431–436

Hall AJ, Thomas GO (2007) Polychlorinated biphenyls, DDT, polybrominated diphenyl ethers, and organic pesticides in United Kingdom harbor seals (*Phoca vitulina*)—mixed exposures and thyroid homeostasis. Environ Toxicol Chem Int J 26(5):851–861

Hall AJ, Kalantzi OI, Thomas GO (2003) Polybrominated diphenyl ethers (PBDEs) in grey seals during their first year of life—are they thyroid hormone endocrine disrupters? Environ Pollut 126(1):29–37

Hall-Aspland SA, Rogers TL, Canfield RB (2005) Stable carbon and nitrogen isotope analysis reveals seasonal variation in the diet of leopard seals. Mar Ecol Prog Ser 305:249–259

Hasin Y, Seldin M, Lusis A (2017) Multi-omics approaches to disease. Genome Biol 18(1):83

Hauser DDW, Logsdon MG, Holmes EE, VanBlaricom GR, Osborne RW (2007) Summer distribution patterns of southern resident killer whales *Orcinus orca*: core areas and spatial segregation of summer groups. Mar Ecol Prog Ser 351:301–310

Hayden M, Bhawal R, Escobedo J, Harmon C, O'Hara TM, Klein D et al (2017) Nanospray liquid chromatography/tandem mass spectrometry analysis of steroids from gray whale blubber. Rapid Commun Mass Spectrom 31(13):1088–1094

Hickie BE, Kingsley MCS, Hodson PV, Muir DCG, Béland P, Mackay D (2000) A modelling-based perspective on the past, present, and future polychlorinated biphenyl contamination of the St. Lawrence beluga whale (*Delphinapterus leucas*) population. Can J Fish Aquat Sci 57(Suppl. 1):101–112

Hobbs KE, Muir DC, Michaud R, Béland P, Letcher RJ, Norstrom RJ (2003) PCBs and organochlorine pesticides in blubber biopsies from free-ranging St. Lawrence River Estuary beluga whales (*Delphinapterus leucas*), 1994–1998. Environ Pollut 122(2):291–302

Hopps HC (1977) The biologic bases for using hair and nail for analyses of trace elements. Sci Total Environ 7 (1):71–89

Hunt KE, Moore MJ, Rolland RM, Kellar NM, Hall AJ, Kershaw J et al (2013) Overcoming the challenges of studying conservation physiology in large whales: a review of available methods. Conservat Physiol 1(1)

Hunt KE, Rolland RM, Kraus S, Burgess E (2016) Development of novel noninvasive methods of stress assessment in baleen whales. New England Aquarium, Boston, MA

Jeffries SJ, Brown RF, Harvey JT (1993) Techniques for capturing, handling and marking harbor seals. Aquat Mamm 19:21–21

Kakuschke A, Valentine-Thon E, Griesel S, Fonfara S, Siebert U, Prange A (2005) Immunological impact of metals in harbor seals (Phoca vitulina) of the North Sea. Environ Sci Tech 39(19):7568–7575

Karpovich SA, Skinner JP, Kapronczai LA, Smith JA, Janz DM (2019) Examination of relationships between stable isotopes and cortisol concentrations along the length of phocid whiskers. Mar Mamm Sci 35 (2):395–415

Kehrig HA, Malm O, Akagi H (1997) Methylmercury in hair samples from different riverine groups, Amazon, Brazil. Water Air Soil Pollut 97(1–2):17–29

Kellar NM, Catelani KN, Robbins MN, Trego ML, Allen CD, Danil K, Chivers SJ (2015) Blubber cortisol: a potential tool for assessing stress response in free-ranging dolphins without effects due to sampling. PLoS One 10(2):e0115257

Kunisue T, Tanabe S (2009) Hydroxylated polychlorinated biphenyls (OH-PCBs) in the blood of mammals and birds from Japan: lower chlorinated OH-PCBs and profiles. Chemosphere 74(7):950–961

Lahvis GP, Wells RS, Kuehl DW, Stewart JL, Rhinehart HL, Via CS (1995) Decreased lymphocyte responses in free-ranging bottlenose dolphins (Tursiops truncatus) are associated with increased concentrations of PCBs and DDT in peripheral blood. Environ Health Perspect 103(suppl 4):67–72

Lair S (2007) Programme de nécropsie—Suivi de la santé de la population de béluga de l'estuaire du Saint-Laurent à l'aide de l'examen post-mortem des carcasses échouées. In: MPO (ed) Compterendu de l'atelier sur le béluga de l'estuaire du Saint-Laurent - Revue du programme des carcasses. Secrétariat canadien de consultation scientifique. Compte-rendu 2007/005, pp 11–14

Lebeuf M, Bernt KE, Trottier S, Noël M, Hammill MO, Measures L (2001) Tris (4-chlorophenyl) methane and tris (4-chlorophenyl) methanol in marine mammals from the Estuary and Gulf of St. Lawrence. Environm Pollut 111(1):29–43

Lebeuf M, Gouteux B, Measures L, Trottier S (2004) Levels and temporal trends (1988-1999) of polybrominated diphenyl ethers in beluga whales (*Delphinapterus leucas*) from the St. Lawrence Estuary, Canada. Environ Sci Technol 38:2971–2977

Lebeuf M, Noël M, Trottier S, Measures L (2007) Temporal trends (1987-2002) of persistent, bioaccumulative and toxic (PBT) chemicals in beluga whales (*Delphinapterus leucas*) from the St. Lawrence Estuary, Canada. Sci Total Environ 383:216–231

Lebeuf M, Trottier S, Noël M, Raach M, Measures L (2010) A twenty years (1987-2007) trend of PBDEs in beluga from the St. Lawrence Estuary, Canada. Organohalogen Comp 71:372–376

Lebeuf M, Measures L, Noel M, Raach M, Trottier S (2014) A twenty-one year temporal trend of persistent organic pollutants in St. Lawrence Estuary beluga, Canada. Sci Total Environ 485:377–386

Lemieux Lefebvre S, Michaud R, Lesage V, Berteaux D (2012) Identifying high residency areas of the threatened St. Lawrence beluga whale from fine-scale movements of individuals and coarse-scale movements of herds. Mar Ecol Prog Ser 450:243–257

Levin M, De Guise S, Ross PS (2005) Association between lymphocyte proliferation and polychlorinated biphenyls in free-ranging harbor seal (Phoca vitulina) pups from British Columbia, Canada. Environ Toxicol Chem Int J 24(5):1247–1252

Lie E, JØrgen S, Larsen H, Larsen S, Marie Johanse G, Derocher AE, Lunn NJ et al (2004) Does high organochlorine (OC) exposure impair the resistance to infection in polar bears (Ursus maritimus)? Part I: Effect of OCs on the humoral immunity. J Toxicol Environ Health Part A 67(7):555–582

Lie E, Larsen HJS, Larsen S, Johansen GM, Derocher AE, Lunn NJ et al (2005) Does high organochlorine (OC) exposure impair the resistance to infection in polar bears (Ursus maritimus)? Part II: Possible effect of OCs on mitogen-and antigen-induced lymphocyte proliferation. J Toxicol Environ Health A 68 (6):457–484

Loseto LL, Stern GA, Ferguson SH (2008) Size and biomagnification: how habitat selection explains beluga mercury levels. Environ Sci Tech 42 (11):3982–3988

Loseto LL, Pleskach K, Hoover C, Tomy GT, Desforges JP, Halldorson T, Ross PS (2017) Cortisol levels in beluga whales (Delphinapterus leucas): setting a benchmark for Marine Protected Area monitoring. Arctic Sci 4(3):358–372

Lunardi D, Abelli L, Panti C, Marsili L, Fossi MC, Mancia A (2016) Transcriptomic analysis of bottlenose dolphin (Tursiops truncatus) skin biopsies to assess the effects of emerging contaminants. Mar Environ Res 114:74–79

Lundin JI, Dills RL, Ylitalo GM, Hanson MB, Emmons CK, Schorr GS et al (2016) Persistent organic pollutant determination in killer whale scat samples: optimization of a gas chromatography/mass spectrometry method and application to field samples. Arch Environ Contam Toxicol 70(1):9–19

Macdonald R, Mackay D, Hickie B (2002) Peer reviewed: Contaminant amplification in the environment. Environ Sci Technol 36(23):456A–462A

Mack AG (2004) Flame retardants, halogenated. In: Kirk-Othmer encyclopedia of chemical technology, pp 454–483

Manchester-Neesvig JB, Andren AW (1989) Seasonal variation in the atmospheric concentration of polychlorinated biphenyl congeners. Environ Sci Tech 23(9):1138–1148

Mancia A, Ryan JC, Van Dolah FM, Kucklick JR, Rowles TK, Wells RS et al (2014) Machine learning approaches to investigate the impact of PCBs on the transcriptome of the common bottlenose dolphin (*Tursiops truncatus*). Mar Environ Res 100:57–67

Mancia A, Abelli L, Kucklick JR, Rowles TK, Wells RS, Balmer BC et al (2015) Microarray applications to understand the impact of exposure to environmental contaminants in wild dolphins (*Tursiops truncatus*). Mar Genomics 19:47–57

Marsili L, Jiménez B, Borrell A (2018) Persistent organic pollutants in cetaceans living in a hotspot area: the Mediterranean Sea. In: Marine mammal ecotoxicology. Academic, pp 185–212

Martineau D, Lagacé A, Béland P (1988) Pathology of stranded beluga whales (*Delphinapterus leucas*) from the St Lawrence Estuary, Quebec, Canada. J Comp Pathol 98(3):287–311

Martineau D, De Guise S, Fournier M, Shugart L, Girard C, Lagace A, Beland P (1994) Pathology and toxicology of beluga whales from the St. Lawrence Estuary, Quebec, Canada. Past, present and future. Sci Total Environ 154(2–3):201–215

Martineau D, Lair S, De Guise S, Lipscomb T, Béland P (1999) Cancer in beluga whales from the St. Lawrence Estuary, Quebec, Canada: a potential biomarker of environmental contamination. J Cetacean Res Manag 1(special issue):249–265

Martineau D, Lemberger K, Dallaire A, Labelle P, Lipscomb TP, Michel P, Mikaelian I (2002) Cancer in wildlife, a case study: Beluga from the St. Lawrence estuary, Québec, Canada. Environ Health Perspect 110 (3):285–292

Martineau D, Lemberger K, Dallaire A, Michel P, Béland P, Labelle P, Lipscomb TP (2003) Cancer in the Beluga: response. Environ Health Perspect 111: A78–A79

McHuron EA, Walcott SM, Zeligs J, Skrovan S, Costa DP, Reichmuth C (2016) Whisker growth dynamics in two North Pacific pinnipeds: implications for determining foraging ecology from stable isotope analysis. Mar Ecol Prog Ser 554:213–224

Mongillo TM, Ylitalo GM, Rhodes LD, O'Neill SM, Noren DP, Hanson MB (2016) Exposure to a mixture of toxic chemicals: implications for the health of endangered southern resident killer whales. NOAA Technical Memorandum, NMFS-NWFSC-135. https://doi.org/10.7289/V5/TM-NWFSC-135

Morel FM, Kraepiel AM, Amyot M (1998) The chemical cycle and bioaccumulation of mercury. Annu Rev Ecol Syst 29(1):543–566

Morris AD, Letcher RJ, Dyck M, Chandramouli B, Fisk AT, Cosgrove J (2018) Multivariate statistical analysis of metabolomics profiles in tissues of polar bears (Ursus maritimus) from the Southern and Western Hudson Bay subpopulations. Polar Biol 41 (3):433–449

Morris AD, Letcher RJ, Dyck M, Chandramouli B, Cosgrove J (2019) Concentrations of legacy and new contaminants are related to metabolite profiles in Hudson Bay polar bears. Environ Res 168:364–374

Mos L, Morsey B, Jeffries SJ, Yunker MB, Raverty S, Guise SD, Ross PS (2006) Chemical and biological pollution contribute to the immunological profiles of free-ranging harbor seals. Environ Toxicol Chem Int J 25(12):3110–3117

Mos L, Tabuchi M, Dangerfield N, Jeffries SJ, Koop BF, Ross PS (2007) Contaminant-associated disruption of vitamin A and its receptor (retinoic acid receptor α) in free-ranging harbour seals (Phoca vitulina). Aquat Toxicol 81(3):319–328

Mos L, Cameron M, Jeffries SJ, Koop BF, Ross PS (2010) Risk-based analysis of polychlorinated biphenyl toxicity in harbor seals. Integr Environ Assess Manag 6 (4):631–640

Mosnier A, Doniol-Valcroze T, Gosselin JF, Lesage V, Measures LN, Hammill MO (2015) Insights into processes of population decline using an integrated population model: the case of the St. Lawrence Estuary beluga (*Delphinapterus leucas*). Ecol Model 314:15–31

Muir DCG, Howard PH (2006) Are there other persistent organic pollutants? A challenge for environmental chemists. Environ Sci Tech 40:7157–7166

NMFS NMF (2008) Recovery plan for southern resident killer whales (*Orcinus orca*). National Marine Fisheries Service Northwest Region, Seattle, WA

Noël M, Loseto LL, Helbing CC, Veldhoen N, Dangerfield NJ, Ross PS (2014) PCBs are associated with altered gene transcript profiles in Arctic beluga whales (*Delphinapterus leucas*). Environ Sci Tech 48 (5):2942–2951

Noël M, Jeffries S, Lambourn DM, Telmer K, Macdonald R, Ross PS (2016) Mercury accumulation in harbour seals from the northeastern Pacific Ocean: the role of transplacental transfer, lactation, age and location. Arch Environ Contam Toxicol 70(1):56–66

Nyman M, Koistinen J, Fant ML, Vartiainen T, Helle E (2002) Current levels of DDT, PCB and trace elements in the Baltic ringed seals (Phoca hispida baltica) and grey seals (*Halichoerus grypus*). Environ Pollut 119 (3):399–412

Nyman M, Bergknut M, Fant ML, Raunio H, Jestoi M, Bengs C et al (2003) Contaminant exposure and effects in Baltic ringed and grey seals as assessed by biomarkers. Mar Environ Res 55(1):73–99

Oskam I, Ropstad E, Lie E, Derocher A, Wiig Ø, Dahl E et al (2004) Organochlorines affect the steroid hormone cortisol in free-ranging polar bears (*Ursus maritimus*) at Svalbard, Norway. J Toxicol Environ Health 67(12):959–977

Ostertag SK, Stern GA, Wang F, Lemes M, Chan HM (2013) Mercury distribution and speciation in different brain regions of beluga whales (Delphinapterus leucas). Sci Total Environ 456:278–286

Outridge PM, Macdonald RW, Wang F, Stern GA, Dastoor AP (2008) A mass balance inventory of mercury in the Arctic Ocean. Environ Chem 5(2):89–111

Pacyna EG, Pacyna JM, Sundseth K, Munthe J, Kindbom K, Wilson S et al (2010) Global emission of mercury to the atmosphere from anthropogenic sources in 2005 and projections to 2020. Atmos Environ 44(20):2487–2499

Parsons KM, Balcomb KC, Ford JKB, Durban JW (2009) The social dynamics of southern resident killer whales and conservation implications for this endangered population. Anim Behav 77(4):963–971

Pasamontes A, Aksenov AA, Schivo M, Rowles T, Smith CR, Schwacke LH et al (2017) Noninvasive respiratory metabolite analysis associated with clinical disease in cetaceans: a deepwater horizon oil spill study. Environ Sci Tech 51(10):5737–5746

Pedro S, Dietz R, Sonne C, Rosing-Asvid A, Hansen M, McKinney MA (2019) Are vitamins A and E associated with persistent organic pollutants and fatty acids in the blubber of highly contaminated killer whales (Orcinus orca) from Greenland? Environ Res 177:108602

Pirrone N, Cinnirella S, Feng X, Finkelman RB, Friedli HR, Leaner J et al (2010) Global mercury emissions to the atmosphere from anthropogenic and natural sources. Atmos Chem Phys 10(13):5951–5964

Prevedouros K, Cousins IT, Buck RC, Korzeniowski SH (2006) Sources, fate and transport of perfluorocarboxylates. Environ Sci Tech 40(1):32–44

Rea LD, Castellini JM, Correa L, Fadely BS, O'Hara TM (2013) Maternal Steller sea lion diets elevate fetal mercury concentrations in an area of population decline. Sci Total Environ 454:277–282

Renner R (2006) The long and the short of perfluorinated replacements

Richard JT, Schultz K, Goertz C, Hobbs R, Romano TA, Sartini BL (2017) Assessing the quantity and downstream performance of DNA isolated from beluga (*Delphinapterus leucas*) blow samples. Aquat Mamm 43(4):398

Rolland RM, McLellan WA, Moore MJ, Harms CA, Burgess EA, Hunt KE (2017) Fecal glucocorticoids and anthropogenic injury and mortality in North Atlantic right whales *Eubalaena glacialis*. Endangered Species Res 34:417–429

Ross PS (2002) The role of immunotoxic environmental contaminants in facilitating the emergence of infectious diseases in marine mammals. Hum Ecol Risk Assess Int J 8(2):277–292

Ross PS (2006) Fireproof killer whales (*Orcinus orca*): flame-retardant chemicals and the conservation imperative in the charismatic icon of British Columbia, Canada. Can J Fish Aquat Sci 63(1):224–234

Ross P, De Swart R, Addison R, Van Loveren H, Vos J, Osterhaus A (1996) Contaminant-induced immunotoxicity in harbour seals: wildlife at risk? Toxicology 112(2):157–169

Ross PS, Ellis GM, Ikonomou MG, Barrett-Lennard LG, Addison RF (2000) High PCB concentrations in free-ranging Pacific killer whales, *Orcinus orca*: effects of age, sex and dietary preference. Mar Pollut Bull 40 (6):504–515

Routti H, Arukwe A, Jenssen BM, Letcher RJ, Nyman M, Bäckman C, Gabrielsen GW (2010) Comparative endocrine disruptive effects of contaminants in ringed seals (Phoca hispida) from Svalbard and the Baltic Sea. Comp Biochem Physiol Part C Toxicol Pharmacol 152 (3):306–312

Sandala GM, Sonne-Hansen C, Dietz R, Muir DCG, Valters K, Bennett ER et al (2004) Hydroxylated and methyl sulfone PCB metabolites in adipose and whole blood of polar bear (Ursus maritimus) from East Greenland Sci Total Environ 331(1-3):125–141

Schroeder WH, Munthe J (1998) Atmospheric mercury—an overview. Atmos Environ 32(5):809–822

Schwacke LH, Zolman ES, Balmer BC, De Guise S, George RC, Hoguet J et al (2012) Anaemia, hypothyroidism and immune suppression associated with polychlorinated biphenyl exposure in bottlenose dolphins (*Tursiops truncatus*). Proc R Soc B Biol Sci 279(1726):48–57

Simms W, Ross PS (2000) Vitamin A physiology and its application as a biomarker of contaminant-related toxicity in marine mammals: a review. Toxicol Ind Health 16(7-8):291–302

Simond AE, Houde M, Lesage V, Michaud R, Zbinden D, Verreault J (2019) Associations between organohalogen exposure and thyroid-and steroid-related gene responses in St. Lawrence Estuary belugas and minke whales. Mar Pollut Bull 145:174–184

Simond AE, Houde M, Lesage V, Michaud R, Verreault J (2020) Metabolomic profiles in the endangered St. Lawrence Estuary beluga population and associations with organohalogen contaminants. Sci Total Environ 717:137204

Skaare JU, Bernhoft A, Wiig Ø, Norum KR, Haug E, Eide DM, Derocher AE (2001) Relationships between plasma levels of organochlorines, retinol and thyroid hormones from polar bears (Ursus maritimus) at Svalbard. J Toxicol Environ Health A 62(4):227–241

Sonne C, Jepson PD, Desforges J-P, Alstrup AKO, Olsen MT, Eulaers I, Hansen M, Letcher RJ, McKinney MA, Dietz R (2018) Pollution threatens toothed whales. Sci Lett 361(6408):1208

Sørmo EG, Jussi I, Jussi M, Braathen M, Skaare JU, Jenssen BM (2005) Thyroid hormone status in gray seal (Halichoerus grypus) pups from the Baltic Sea and the Atlantic Ocean in relation to organochlorine pollutants. Environ Toxicol Chem Int J 24(3):610–616

St. Aubin DJ, Deguise S, Richard PR, Smith TG, Geraci JR (2001) Hematology and plasma chemistry as indicators of health and ecological status in beluga whales, *Delphinapterus leucas.* Arctic 317–331

Stemmler I, Lammel G (2010) Pathways of PFOA to the Arctic: variabilities and contributions of oceanic currents and atmospheric transport and chemistry sources. Atmos Chem Phys Discuss 10:11577–11614

Stricker CA, Christ AM, Wunder MB, Doll AC, Farley SD, Rea LD et al (2015) Stable carbon and nitrogen isotope trophic enrichment factors for Steller sea lion vibrissae relative to milk and fish/invertebrate diets. Mar Ecol Prog Ser 523:255–266

Swackhamer DL, Skoglund RS (1993) Bioaccumulation of PCBs by algae: kinetics versus equilibrium. Environ Toxicol Chem Int J 12(5):831–838

Tabuchi M, Veldhoen N, Dangerfield N, Jeffries S, Helbing CC, Ross PS (2006) PCB-related alteration of thyroid hormones and thyroid hormone receptor gene expression in free-ranging harbor seals (Phoca vitulina). Environ Health Perspect 114(7):1024–1031

Tanabe S (1988) PCB problems in future: forsight from current knowledge. Environ Pollut 50:5–28

Tartu S, Lille-Langøy R, Størseth TR, Bourgeon S, Brunsvik A, Aars J et al (2017) Multiple-stressor effects in an apex predator: combined influence of pollutants and sea ice decline on lipid metabolism in polar bears. Sci Rep 7(1):16487

Thomas GO, Moss SE, Asplund L, Hall AJ (2005) Absorption of decabromodiphenyl ether and other organohalogen chemicals by grey seals (*Halichoerus grypus*). Environ Pollut 133(3):581–586

Trego ML, Whitehead A, Kellar NM, Lauf M, Lewison RL (2019) Tracking transcriptomic responses to endogenous and exogenous variation in cetaceans in the Southern California Bight. Conserv Physiol 7(1): coz01

U.S. Environmental Protection Agency (2010) An exposure assessment of polybrominated diphenyl ethers. U.S. EPA, National Center for Environmental Assessment, Washington, DC. EPA/600/R-08/086F. http://cfpub.epa.gov/ncea/cfm/recordisplay.cfm?deid=210404

Veldhoen N, Ikonomou MG, Helbing CC (2012) Molecular profiling of marine fauna: integration of omics with environmental assessment of the world's oceans. Ecotoxicol Environ Saf 76:23–38

Villanger GD, Lydersen C, Kovacs KM, Lie E, Skaare JU, Jenssen BM (2011a) Disruptive effects of persistent organohalogen contaminants on thyroid function in white whales (*Delphinapterus leucas*) from Svalbard. Sci Total Environ 409(13):2511–2524

Villanger GD, Jenssen BM, Fjeldberg RR, Letcher RJ, Muir DC, Kirkegaard M et al (2011b) Exposure to mixtures of organohalogen contaminants and associative interactions with thyroid hormones in East Greenland polar bears (*Ursus maritimus*). Environ Int 37(4):694–708

Voegborlo RB, Matsuyama A, Adimado AA, Akagi H (2010) Head hair total mercury and methylmercury levels in some Ghanaian individuals for the estimation of their exposure to mercury: preliminary studies. Bull Environ Contam Toxicol 84(1):34–38

Wallaschofski H (2012) What will metabolomics studies mean to endocrinology? J Endocrinol 215(1):1

Wania F (2003) Assessing the potential of persistent organic chemicals for long-range transport and accumulation in polar regions. Environ Sci Technol 37 (7):1344–1351

Wania F (2007) A global mass balance analysis of the source of perfluorocarboxylic acids in the Arctic Ocean. Environ Sci Tech 41:4529–4535

Wasser SK, Lundin JI, Ayres K, Seely E, Giles D, Balcomb K et al (2017) Population growth is limited by nutritional impacts on pregnancy success in endangered Southern Resident killer whales (*Orcinus orca*). PLoS One 12(6):e0179824

Weijs L, Zaccaroni A (2016) Toxicology of marine mammals: new developments and opportunities. Arch Environ Contam Toxicol 70(1):1–8

Weijs L, Das K, Siebert U, van Elk N, Jauniaux T, Neels H et al (2009) Concentrations of chlorinated and brominated contaminants and their metabolites in serum of harbour seals and harbour porpoises. Environ Int 35(6):842–850

Wolf JB (2013) Principles of transcriptome analysis and gene expression quantification: an RNA-seq tutorial. Mol Ecol Resour 13(4):559–572

Wu JP, Luo XJ, Zhang Y, Luo Y, Chen SJ, Mai BX, Yang ZY (2008) Bioaccumulation of polybrominated diphenyl ethers (PBDEs) and polychlorinated biphenyls (PCBs) in wild aquatic species from an electronic water (e-waste) recycling site in South China. Environ Int 34:1109–1113

The Intentional Use of Pesticides as Poison in Kenya: Conservation and Ecohealth Impacts

Martin Odino and Darcy Ogada

Abstract

In Kenya, deliberate poisoning of birds for human consumption, particularly the poisoning of flocking water birds and seedeaters, is a practice that takes place primarily on large-scale, irrigated cereal plantations. Mass poisonings of vultures are also of concern in Kenya when they feed on poisoned carcasses. This occurs when livestock farmers poison livestock carcasses to use as bait for wild predators suspected of killing livestock, with vultures falling victim to such retaliatory efforts. Pesticides and topical antiparasitics, readily available over the counter in Kenya, are commonly used, overused, and misused by farmers, with little to no regulatory controls and/or unknown impacts on biodiversity and ecosystem health. Social surveys were conducted across various parts of Kenya and a case study of quantifying intentional bird poisoning was conducted at Bunyala Rice Irrigation Scheme (Bunyala Rice Plantation), a known bird poisoning hot spot in Kenya, to determine the extent of poisoning by the carbamate pesticide carbofuran and other poisons. Results from these studies reveal the extensive use of carbofuran as an intentional poison agent for birds and scavenger-predator wildlife, with carbosulfan also used against the latter. Further, glyphosates and vermin poisons were found to result in unintentional poisonings in livestock, pets, and even humans. Despite the withdrawal of carbofuran from the Kenyan market in 2009 by the manufacturer, Farm Machinery and Chemicals (FMC), carbosulfan, which metabolizes to carbofuran, and other chemicals are readily available for purchase and use. A general lack of adequate legislation, control, and enforcement of pesticide use in Kenya has allowed for the widespread, indiscriminate use of pesticides beyond their intended means, with potentially devastating consequences from an ecological standpoint.

Keywords

Poison · Poaching · Carbofuran · Pesticides · Birds · Vultures · Carnovores · Predators

Introduction

The agricultural sector dominates Kenya's economy and is the largest contributor to Kenya's gross domestic product at 51% (26% directly and 25% indirectly). Agriculture accounts for 60% of employment and 65% of exports (World Bank 2018). As an agricultural economy that promotes mainly conventional agriculture,

M. Odino (✉) · D. Ogada
The Peregrine Fund, Boise, ID, USA

Ornithology Section, National Museums of Kenya, Nairobi, Kenya

S. C. Underkoffler, H. R. Adams (eds.), *Wildlife Biodiversity Conservation*,
https://doi.org/10.1007/978-3-030-64682-0_16

Kenya's demand for pesticides is relatively high and steadily increasing (Route to Food 2019). In 2018, Kenya imported 17,803 tons of pesticides valued at 128 million dollars (Route to Food 2019). These pesticides are an assortment of insecticides, fungicides, herbicides, fumigants, rodenticides, growth regulators, defoliators, proteins, surfactants, and wetting agents. It is remarkable that the volume of imported pesticides including insecticides, herbicides, and fungicides has more than doubled within 4 years, from 6400 tons in 2015 to 15,600 tons in 2018, at a growth rate of 144% (Route to Food 2019). Of the total pesticide imports, insecticides, fungicides, and herbicides account for approximately 87% in terms of volume, and 88% of the total cost of pesticide imports (Route to Food 2019).

Pesticides hold a dual power that allows them to both protect and potentially poison, and as such, they must be treated as hazardous materials and utilized with both caution and regard for the long-term environmental consequences (Carson 1962). In the 1960s, the organochloride pesticide known as Dichlorodiphenyltrichloroethane (DDT) was readily used, resulting in chronic bioaccumulating intoxication in the food chain. After DDT was banned in the United States in 1972, as well as in most developed countries in the 1970s and 1980s, additional toxic pesticides were introduced (Vail 2015). The focus on the detriments of DDT, a chlorinated hydrocarbon, permitted organophosphates and carbamates to slip through the legislation cracks (Oberemok et al. 2015). Today both carbamates and organophosphates are among the most widely used classes of pesticides (19% of the global market) and play a major role in the control of insect pests (Casida and Durkin 2013).

Carbamates have a similar mechanism of action to organophosphates and are often used as biocides in Kenya. Carbofuran, traded as Furadan 5G (referred to as Furadan) has allegedly played a significant role in the decimation of Kenya's lion population to near extermination, estimated at 2000 individuals by the early 2000s, down from an estimated 50,000 in the 1950s (R. Leakey *pers. comm.*). As vultures feed on carcasses laced with poisons, mostly targeting problem predators, they too are poisoned, often in large numbers, as dozens and, at times, upwards of one hundred vultures may feed on a single poisoned carcass. Thus, carbofuran is believed to have driven similar declines in vulture populations across Kenya over the past few decades (Virani et al. 2011; Kendall and Virani 2012). Historically, hundreds of vultures would feed on the carcass of a large herbivore; however, in recent years numbers have declined, with less than one hundred feeding on such a carcass (S. Thomsett *pers. comm.*). Poisoning is a primary (Ogada et al. 2012) and critical threat to vulture populations (Plaza et al. 2019).

In Bunyala, carbofuran was used in cases of intentional poisoning of birds (M. Odino *pers. obsvn.*), which was confirmed by analytical lab testing at the Government Chemist Laboratory. This type of poaching takes place either for bushmeat (hereafter referred to as poison-poaching) or in retaliation, as birds can be a nuisance for crops. Intentional or deliberate poisoning is the primary focus of this chapter.

Poison-poaching takes place when poison is used to kill wildlife for domestic bushmeat or for commercial trade. At Bunyala Rice Plantation, poachers have used carbofuran to kill a wide variety of birds, employing two different techniques. One technique involves lacing food with carbofuran granules and placing it where flocks naturally feed (Odino 2011). To target whistling ducks (*Dendrocygna spp.*) for example, rice gran bait laced with carbofuran is placed on plates, mixed with mud, and submerged in flooded rice paddies where they prefer to feed, whereas for the insect-eating Abdim's Stork, (*Ciconia abdimii),* termite baits laced with carbofuran are placed in open fields. Thus, carbofuran is used in a variety of ways that are species-specific to target birds based on their diet and feeding behavior. Baited birds die from intoxication or become disoriented and are pursued by the poachers, who kill them via strangulation or blunt trauma. Live decoys are also used to lure birds in to feed where bait has been placed. The following technique is specific for the African openbills stork (*Anastomus l. lamelligerus): Pila ovata* snails (the preferred food of openbills) are laced with carbofuran granules and placed in the preferred feeding habitat; one or two captive openbills are then placed in the site after having a few of their flight feathers removed, their legs restrained by string, and a rubber band secured around the bill, to prevent flight or feeding, respectively (Odino 2011) (Image 1).

Retaliatory poisoning occurs when farmers experience losses of farmed crops or livestock to wild or domestic animals (Image 2). For example, farmers in Mweiga, central Kenya, lace tomato fruits with carbofuran or other chemicals to kill frugivorous mousebirds (P. Muriithi, *pers. comm.*). In November 2007, near Lewa Wildlife Conservancy, Isiolo District, a camel killed by lions was subsequently laced with carbofuran by local pastoralists, with the aim of killing predators that returned to feed on the carcass. Two lions and 15 vultures died from ingesting poisoned meat from the carcass (I. Craig *pers. comm.*). Around this time, a pride of nine lions from the nearby Samburu reserve were poisoned, five of which

Image 1 African openbill live decoy

Image 2 Lioness and her cow kill

died, along with significant numbers of birds of prey and other scavengers (I. Craig *pers. comm.*).

Even among farmers growing crops for subsistence in Kenya, there is limited awareness of the usefulness, associated risks, safety requirements, and application-related issues of pesticides, thus intoxications are not uncommon. Measures should be implemented to ensure that pesticides are not misused, overused, or abused, especially in rural areas where indiscriminate use by pastoralists and farmers is common practice (USAID-KAVES PERSUAP 2013). In this chapter, we examine the findings from social surveys and a bird poisoning quantification study conducted from 2007 to 2017 illustrate Kenyan wildlife, pet, and human poisoning incidents.

The objectives of the study were to:

- wList and classify the agricultural chemicals responsible for poisoning and mortality of wild birds, other wildlife, and domestic animals, as reported on social surveys and observed during a bird poisoning quantification study.
- Document the poisoning of wild birds, other wildlife, and domestic animals at the survey sites in Kenya.
- Identify existing weaknesses in place that allow the problem to persist.

Methodology

Four studies were conducted by M. Odino across nine locations in Kenya (Fig. 1). Studies were conducted from 2007 to 2016. Three social surveys were carried out from November to December 2007, from April to June 2008, and from August 2016 to January 2017. The bird poisoning survey was conducted from February 2009 to January 2010. The initial study sites for the first survey were randomly selected, with subsequent study site selection based on the findings of the initial survey. Study sites first surveyed included: Nairobi town, including Kikuyu; Machakos town, including adjacent farmlands; Kisii town, including adjacent farmlands; and Naivasha town, including adjacent farmlands. The second social survey covered Kajiado, Isiolo-Marsabit, and Bunyala Rice Plantation. The third and last survey was conducted at Masai Mara conservation area, Laikipia West, and Voi bordering the Tsavo East and West National Parks. A bird poisoning quantification study was also conducted at the Bunyala Rice Irrigation Scheme, after the first and second surveys.

Information was derived from three social studies that employed two survey approaches as follows:

1. *Questionnaires*
   Three types of questionnaires were administered by M. Odino. Questionnaires were used to assess the availability and usage (or misuse) of carbofuran and other pesticides as poisons. Questions were tailored to different respondents as follows:
   (a) Distributor questionnaire—directed to agrovet store vendors.
   (b) Farmer/user/witness questionnaire—directed to farmers or other end-users to assess abuse of carbofuran and other chemicals being used, including purpose and reported effectiveness.
   (c) Legislation questionnaire (for the first social survey only)—directed to Pest Control Products Board (PCPB) with the aim of understanding the agency's role in the regulation of carbofuran. Specifically, to obtain information regarding carbofuran use, to determine the source (s) of carbofuran in Kenya, and to understand the agency's role in cases of misuse.
2. *Interviews*
   Interviews conducted by M. Odino were open-ended and based on responses initially provided from questionnaires to obtain information that was not captured by the more structured questionnaires.
   Both questionnaires and interview methods were used concurrently in obtaining results for the three social studies (Odino and Ogada 2008a; b; Odino 2017).

The following methods were used during the bird poisoning quantification study (Odino 2011):

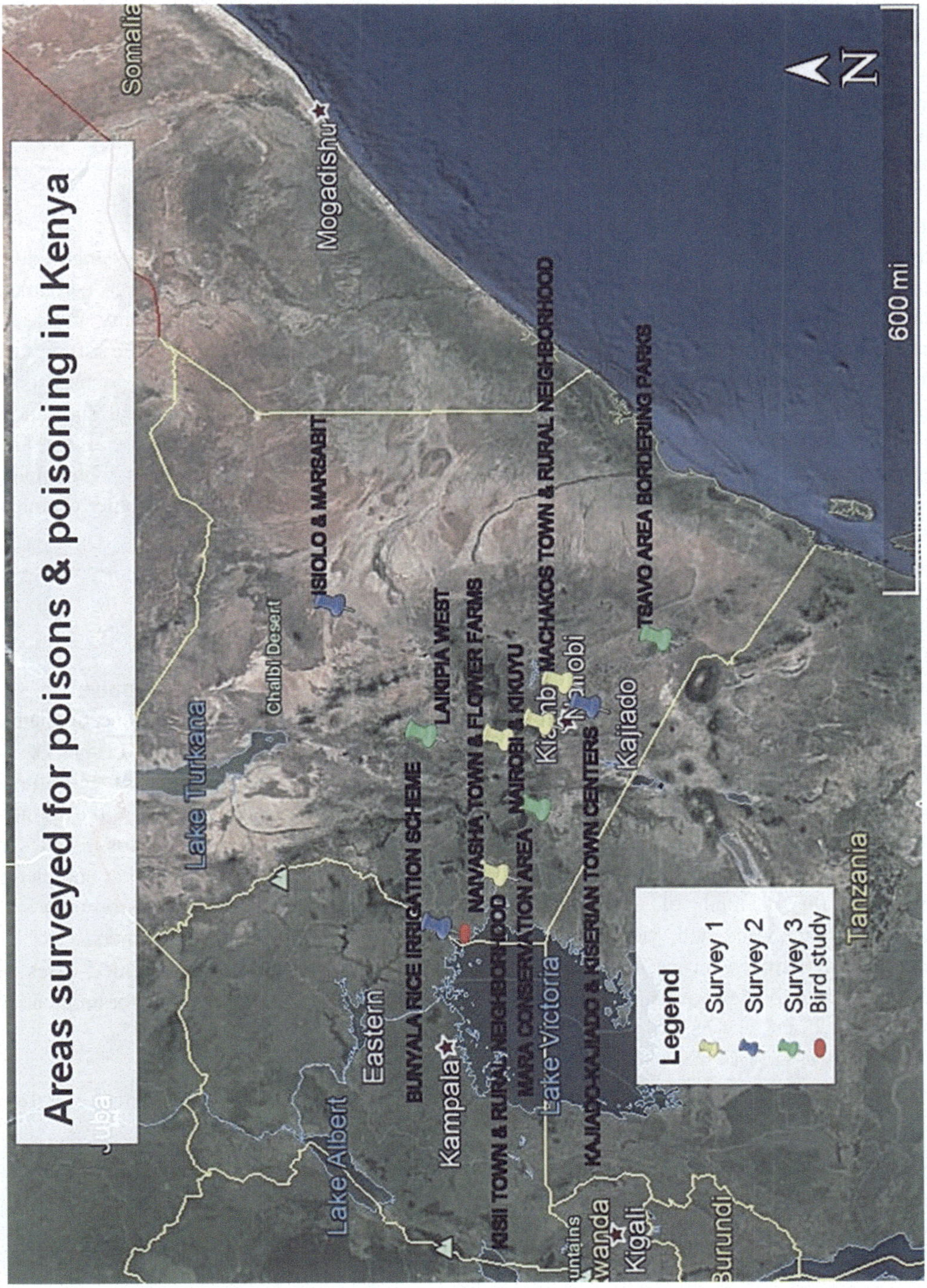

Fig. 1 Study Area

1. *Stratified sampling survey*
   This was employed at Bunyala rice plantation, a large plantation subdivided into small rectangular plots, which served as individual grids for sampling and observation during the study. Each plot was delineated by soil embankment and occupied an average area of 0.75 acres. Grids were opportunistically selected and observed for pesticide poisoning. Once bait had been confirmed as set or placed in a plot, number of birds was recorded from time of arrival (post-bait) to time of gathering carcasses. Birds that arrived during the "waiting period," i.e., the duration that a poacher kept away, and that passersby were directed to keep off the quadrant where bait had been laid, were included in the total. This comprised the total observed sample for each species. Individuals that left the site during this "waiting period" were still included in the number observed. Once a poacher had gathered his kill, surveyors recorded the number and species killed, with the poacher's permission. On the occasion that permission was not granted, species and numbers were estimated from a safe distance.
2. *Thin-Layer Chromatography (TLC) analysis*
   Nine bird carcasses were collected (four carcasses gathered from the study site and five additional carcasses purchased directly from poachers), along with one bait sample consisting of leftover granules from bait containers, for a total of ten samples. Carcasses were eviscerated, stored in glass tissue jars, kept frozen on ice in a cool box, and transported to the Government Chemist Laboratory, Nairobi. Five samples, including the bait sample, were analyzed for the presence of carbofuran (carbamate) using thin-layer chromatography (TLC) analysis. The remaining five samples were stored under refrigeration, in the event there was a need for further or repeat testing.

Results

The first and second social surveys focused on carbofuran or Furadan, a carbofuran carbamate pesticide, investigating its availability, uses, and especially its abuse as a poison at the survey sites in Kenya. The third social survey focused on a broad spectrum of all agrochemical substances used as poisons. All three social surveys investigated biodiversity mortalities, with emphasis on birds, due to carbofuran or other chemical ingestion.

Results of Social Surveys

Findings of Agrovet Questionnaire

All surveyed sites and completed questionnaires from the first survey are as follows (Table 1).

There was 100% availability of carbofuran packed in 200 g tins and sold across the counter in all 38 surveyed agrovet shops across all sites in the first social survey. All sellers that completed the questionnaire disclosed that carbofuran was the most preferred pesticide by farmers.

All vendors from Nairobi admitted cases of customers requesting the pesticide for unintended

Table 1 Summary of surveyed sites during first social survey

Site	Total questionnaires competed (total administered in parenthesis)		Total
	Agro-vet	Neighboring farmland	
Nairobi town and kikuyu	15 (25)	1 (7)	16
Machakos town and farmlands	5 (5)	2 (7)	7
Kisii town and farmlands	2 (3)	1 (24)	3
Naivasha town and farmlands	5 (5)	2 (2)	7
Legislative organization	**Total questionnaires filled**		
Pest Control Products Board	1		1
	Total		34

use and therefore abuse by farmers including: using the pesticide as a rodenticide; as a poison for problem animals namely, dogs, hyenas, domestic cats, and lions; as crocodile baits; or for killing bees and "nuisance" birds, namely vultures, eagles, and ravens. Most sellers suggested Mocap, an organophosphate whose active ingredient is enthroprophos, as a less toxic and ecologically safer alternative to Furadan.

The second survey found carbofuran in 15 of the 22 surveyed agrovet shops (Table 2).

In Isiolo, in addition to the 200 g package, the two surveyed agrovet shops also stocked 1 kg packages. In Kajiado, products with the active ingredient and trade name Thiram, a combination fungicide–ectoparasiticide, and Actara, an insecticide repellent whose active ingredient is Thiamethoxam, were reported as available alternatives to Furadan. Mocap was the only other available pesticide alternative to Furadan in supply around Isiolo. In Bunyala, the vendor of the certified agrovet shop at Nyadorera stocked only Furadan. The vendor reported that Furadan was used locally as a nematicide by rice farmers and as an avicide or poison for birds for subsistence.

The third survey found a variety of poisoning chemicals as follows:

Around Masai Mara,12 agrovet shops were surveyed with results showing that vermin poisons/pesticides were in high demand and were well-stocked in the shops. Several crop pesticides were also found to be in demand but were not used as poisons except for carbosulfan (trade name Marshal), and the glyphosate herbicide and 2-DCP derivative (trade name Round Up) (Table 3).

Carbofuran or Furadan was also mentioned to be of wildlife poisoning concern, although it was not found in any of the shops, and one vendor of the three where carbosulfan was found indicated that carbosulfan had taken over carbofuran's place as a pesticide and poison. However, according to one vendor, carbofuran could be found at Ewaso Township or Loita up until 2016, long after it was withdrawn from Kenya in 2009. The survey extended to Ewaso but did not find carbofuran.

One vermin poison/pesticide, Ectopor, a controlled pyrethroid insecticide/acaricide against ticks on cattle, camels, sheep, and goats, although not reportedly used in wildlife poisonings, is potentially a candidate for poisoning abuse because of its broad spectral biocide properties. The other crop pesticides reported to be in demand in Masai Mara but reportedly not used as agents of poisoning is Alpha Cymba, with the active ingredient alphacypermethrin; and Twigamectin, whose active ingredient is abamectin. Products with these active ingredients were rejected in a pesticides' safety assessment as Restricted Use Pesticides (USAID-KAVES PERSUAP 2013) posing intoxication risks if used freely without control.

In Laikipia, Round Up, a glyphosate herbicide, and Herbikill, a herbicide whose active ingredient is paraquat dichloride, were praised as effective in weed control; however, they were also reported to be responsible for herbivore and especially

Table 2 Summary of availability, knowledge, and use of carbofuran

	Distributors/Agrovets			Pastoralist/Farmer		
Site	Stock	Know/do not stock	Do not know	Use	Know/do not use	Do not know
Kajiado town and surrounding	3	–	–	6	–	1
Kiserian shopping center and surrounding	9	2	2	4	1	–
Isiolo town and surroundings	2	–	4	–	–	18
Maralal and surroundings	–	–	3	–	–	4
Nyadorera shopping center in Bunyala	1	–	–	16	–	–

Table 3 Pesticides of poisoning concern around Masai Mara

Chemical substance	Active ingredient	Classification/intended use	Agrovets that stocked	Agrovets that cited abuse	Reported poisoning abuse
Red cat	Zinc phosphide	Vermin poison/ rodenticide for rodents	7	2	Human suicide Poisoning hyenas
Almatix	Amitraz	Acaricide for ticks	3	4	Human suicide
Savina Dudu dust	Permethrin	Vermin poison (pyrethroid) for cockroaches and other crawling insects	2	1	Dusted on sheep for flea and mite control; could lead to poisoning of predators/ scavengers
Doom powder	Permethrin, Imiprothrin, D-trans Allethrin, Cypermethrin	Vermin poison (pyrethroid) for mosquitoes, cockroaches, and ants indoors.	2	1	Dusted on sheep for flea and mite control that could lead to poisoning of predators/ scavengers
Diazinon (Duduzol)	Diazinon	Vermin poison/ insecticide for control of fleas and lice on walls, floors, and furniture.	2	1	Human suicide
Promax (Baygon)	Propoxur	Vermin poison/household insecticide for cockroaches, fleas, and ants.	1	1	Human suicide
Carbosulfan (Marshal 350 STD)	Carbosulfan	Carbamate Nematicide—insecticide	3	5	Vulture poisoning
Queletox	Fenthion	Vermin poison against locust birds and insects	–	1	Indiscriminate poisoning of birds and grazers Human intoxication
Round Up	Glyphosate	Post-emergence systemic herbicide for weed control in plantation crops	–	1	Miscarriage in sheep and goats Wasting away of cattle

domestic livestock deaths. There were also single accounts of human suicide using Red Cat, a zinc phosphide rodenticide, and Diazinon, a bed-bug control poison. Cypermethrin was also reported to be in demand in Laikipia but not used for wildlife poisoning.

Around Tsavo, including the area between Voi and Kibwezi, one agrovet shop vendor reported that she stocked Mocap, an organophosphate insecticide. This was reported to have been purchased to kill baboons and hyenas. The same vendor also reported that an unknown herbicide was used as a quick means of poisoning elephants, allowing for the removal of tusks before rangers would be alerted to the carcass. Another agrovet vendor reported that Furadan used to be in high demand but had been missing in the market since 2011.

Findings of the Farmer/User/Witness/ Project Officer Questionnaire

During the first social survey, two respondents around the Nairobi–Machakos study area reported having used Furadan to euthanize a sick dog and to poison squirrels in retaliation for destroying tuber crops. Farmhands working on a flower farm in Naivasha reported heavy use of Furadan as a pesticide. All respondents reported the pesticide as very effective, and no respondents recommended alternatives to Furadan, as they believed the pesticide was very reliable.

In the second social survey, three of the 10 respondents in Kajiado confessed to abusing Furadan as a poison for problem animals, including hyenas, leopards, lions, wild dogs, rabid dogs/ cats, warthogs, wildebeests, and rodents. At Nyadorera in Bunyala, carbofuran was well known to all the farmers interviewed. All 16 farmers reported that, in addition to its use as a nematicide in the rice farms, it was used to poison a variety of birds for subsistence (Table 4). Dogs, cats, snakes, and monitor lizards were also unintentionally poisoned through the consumption of poison bait or the consumption of poisoned bird carcasses.

During the third survey, responses were consistent with those provided by vendors, with a few exceptions:

- A poisoning case in Masai Mara, Kenya, reported by a project officer, with multiple herbivore species killed, including impala, zebra, and giraffe. The poison used was a mixture of unknown plant toxins mixed with agrochemicals and laced on spears and arrows. Animals were targeted due to crop-raiding of maize and wheat farms. Some wildlife were speared with poison-tipped arrows to check the potency of the mixture, as the weapons would also be used to deter or kill cattle rustlers attempting to raid the pastoralists' livestock. Additionally, poachers deliberately contaminated water holes with poison, in an effort to kill vultures, whose mass presence at poaching sites serve as an alert system for potential illegal activity. They also laced leftover carcasses with poison so that other poachers could not retrieve them, resulting in additional vulture deaths.
- A pastoralist recounted that his family survived accidental carbofuran poisoning in 2012 after one of his children used an empty Furadan container to draw drinking water; his chickens also died from drinking the water.
- In Laikipia West, between 2014 and 2016, Round Up was used to poison starlings, helmeted guineafowls, vultures, and cows.
- In Rumuruti Town, locals reported that farmers continued to import carbofuran from Tanzania to poison livestock-raiding carnivores and crop-raiding herbivores, which resulted in vulture poisoning at ADC Mutara Ranch in 2016. Additionally, a respondent recounted an incident in 1996 where a carbofuran-laced Zebra carcass resulted in the death of approximately 500 vultures at Olmaisor Ranch.
- Dog poisoning using strychnine was prevalent. Strychnine was reported to be used for dog poisoning by uncertified personnel, despite its classification as a restricted product.
- The use of acaricides for human poisoning was also reported.
- In Voi, bordering both Tsavo East and West, a KWS officer had witnessed poisoning incidents involving an herbicide, informally referred to as "DCK" by poachers. This was laced on spears and arrows and used to kill elephants within minutes, after which the tusks would be retrieved and sold on the black market. The officer speculated that the "DCK" was a derivative of 2, 4-D, basing on speedy putrefaction of the carcass, and the chemical was an herbicide, insinuating DCP. He also observed that the population of hooded vultures in Kibwezi had plummeted.

Findings from the Questionnaire on Legislation Regarding Furadan

As part of the first survey, the questionnaire was completed by an employee at the Pest Control Products Board. However, the agency declined our request for an interview.

The agency's role regarding the regulation of Furadan or carbofuran was reported to be ascribed in the Pest Control Products Acts. The respondent stated it was according to PCPB's vision and mission statement and included regulating pest control products, including their manufacture, storage, distribution, importation, disposal, and monitoring to ensure safety and quality of products. They also investigate and prosecute those found in contravention of the Pest Control Products Act.

The officer disclosed that carbofuran, Furadan 5G, was being manufactured abroad at the time by FMC, although he referred to the fourth edition of the Pest Control Products List at the time for all queries regarding carbofuran, including country of origin of the supplier, while referring us to the Pest Control Products Act Cap 346 of 1984 and 2006 regarding usage of carbofuran and

Table 4 Birds poisoned using Furadan at Bunyala Rice Plantation

Species observed	Scientific Name	No. Alive	No. Dead	% mortality
Black-tailed Godwit	*Limosa l. limosa*	500	215	43
Ringed Plover	*Charadrius hiaticula tundra*	5	2	40
Common Greenshank	*Tringa nebularia*		2	
Wood Sandpiper	*Tringa glareola*	467	126	26.98
Marsh Sandpiper	*Tringa stagnatilis*	3	2	66.67
∗Green Sandpiper	*Tringa Ochropus*		1	
Common Snipe	*Gallinago g. gallinago*	16	10	62.5
Yellow Wagtail	*Motacilla flava*	14	4	28.57
Abdim's Stork	*Ciconia abdimii*	29	26	89.69
Glossy Ibis	*Plegadis f. falcinellus*	39	6	15.38
African Openbill	*Anastomous l. lamelligerus*	5848	2261	38.66
Marabou Stork	*Leptoptilos crumeniferus*	10	3	30
Hadada Ibis	*Brastrychia hagedash*	6	1	16.67
Cattle Egret	*Bubulcus i. ibis*	21	9	42.86
Little Egret	*Egretta g. garzetta*	8	3	37.5
∗Senegal Plover	*Vanellus tectus*		4	
Greater Painted-Snipe	*Rostratula b. benghalensis*	4	1	25
∗Long-crested Eagle	*Lophaetus occipitalis*		2	
∗Black Kite	*Milvus m. migrans*		1	
Dusky Turtledove	*Streptopelia l. lugens*			
Ring-necked Dove	*Streptopelia capicola*			
African Mourning Dove	*Streptopelia d. perspicillata*			
Red-eyed Dove	*Streptopelia semitorquata*			
Speckled Pigeon	*Columba g. guinea*			

(Doves/pigeons)	Refer above	1570	476	30.32
*Yellow-throated Longclaw	*Macronyx croceus*		1	
Grassland Pipit	*Anthus cinnamomeus*	2	1	50
*Wattled Starling	*Creatophora cinerea*		2	
Red-billed Quelea	*Quelea quelea aethiopica*	50	12	24
Fan-tailed Widowbird	*Euplectes axillaris*	11	8	72.72
*Southern Red bishop	*Euplectes orix nigrifrons*		3	
Parasitic Weaver	*Anomalospiza imberbis*	6	4	66.67
Village Weaver	*Ploceus cucullatus*	50	2	4
TOTAL		**8,659**	**3,186**	

Key: Palearctics seen on incoming migration
Palearctics seen on return migration
Resident African species
Palearctics seen both journeys of migration
Intra sites/African migrants
Collection of doves and pigeons
[a]Species whose live individuals were not counted/observed

any other agrochemical product, as well as terms of revocation of license. Finally, the respondent reported no knowledge of misuse of pest control products (outside of registered uses), including carbofuran, and cited their prohibition in the Pest Control Products Act. He noted this information was summarized on the product label under "notice to user"

Results from the Bird Poisoning Quantification Study

This study counted 3186 dead birds belonging to 32 different species intentionally poisoned using Furadan poisoning out of 8659 birds recorded visiting the field plots during the observed poisoning sessions at the irrigation scheme (Table 4). This represented a 37% mortality rate across the 32 species. Four hundred and fifty-two of 1005 Palearctic migrants were killed, representing an annual mortality rate of 45% during the study. The remainder, 2734 of the observed poisoned birds, out of a total 7654, were resident African and intra-African migrants, which represented a mortality rate of 38%.

Results of Bird Poisoning Questionnaires and Interviews

Poacher Responses

Poachers primarily used Furadan to poison birds at the irrigation plantation. However, Mocap, whose active ingredient is the organophosphorus compound ethoprophos, was used as a replacement for Furadan during July 2009, at the peak of Furadan's buy-back program, following the withdrawal of the product by FMC.

Most poachers reported a preference for the African openbill because they were abundant, therefore relatively cheap for their size with a ready market, followed by seedeaters and smaller waders. Poachers specialized in one of these categories, but opportunistically poisoned birds in the other categories.

Consumer Responses

The study (Odino 2011) revealed that poaching families consumed poison-killed birds on an almost daily basis (Images 3 and 4).

Consumers purchased bird meat opportunistically when provided by poachers and vendors at the local market. All bird meat for sale was obtained by poison-poaching, a fact that consumers were aware of. However, they maintained that draining fluids from the carcasses by slow roasting, or hanging the carcasses over hot coals while cooking "detoxified" the flesh. Paradoxically, they acknowledged that the pesticide being used, Furadan, was a deadly toxin.

Interviewees also reported eating many of the 32 poisoned species. However, the African openbill (Image 5) and doves/pigeons (Image 6) were clearly favored, having been named by 219 of the 242 consumers (90%). Wild ducks were a delicacy, as they were a seasonal find, and thus more expensive to purchase.

Image 3 Poacher's child holding a fulvous whistling duck carcass poisoned that day

Results of Thin-Layer Chromatography Analysis

Five gut samples and one bait sample were submitted for carbofuran testing. Of the samples tested, four gut samples and the bait sample tested positive for carbofuran (Image 7).

Discussion

Four organophosphates, two carbamates, an inorganic rodenticide, and an alkaloid are the primary environmental pesticide threats to Kenya's biodiversity (Tables 5 and 6).

However, cases of misuse of most of these products as poisons are partly due to legislative inefficiencies and deficiencies, as was established by the first survey, and based upon the author's experiences during this research.

Despite some products identified in the surveys being classified as restricted, they could be easily purchased over the counter. At the time of the first survey, the PCPB respondent indicated that carbofuran was categorized as a highly restricted pesticide. This would accord the chemical the same handling regulations as strychnine, which was also allegedly abused by uncertified persons in our surveys. Strychnine was suspended for use in Kenya in August 2019, and thus can only be accessed and administered through District Veterinary Officers. Carbofuran, having been readily available and easily acquired over the counter, was not treated as a restricted product. From observation at agrovet stores, carbofuran was easily accessible. M. Odino purchased a 200 g package of the pesticide without any special authorizing commendation. Likewise, Red Cat, whose toxic phosphine gas is effused when zinc phosphide encounters moisture, is sold over the counter, yet is categorized under the restricted class WHO 1b, and labeled as highly hazardous with a red color band around the packaging (PCPB 2018). PCPB has classified it under Conventional Pest Control Products for Use in Public Health (PCPB 2018) and is readily procured over the counter in Kenya. Unfortunately,

Image 4 Poacher's child with a roasted, poisoned white-faced whistling duck

Image 5 Poisoned African openbilled storks

Image 6 Different species of doves

the rules of acquisition and use of these restricted products are blatantly disregarded in Kenya.

There is also a lack of restricting legislation and enforcement of products with evidence of abuse as poisons, even though they may not have been initially classified as restricted. Chemicals like Marshal, and glyphosates such as Round Up, are readily predisposed to abuse by ill-motivated persons.

There is also no clear way to inform PCPB regarding ongoing abuse or emanating dire outcomes of a product. It is thus no surprise that the respondent from PCPB was confident that Furadan was not being used to poison wildlife. However, Furadan poisoning at rice schemes has been documented as early as 10 years prior to the current research (Simpson 1996). If the respondent genuinely believed that Furadan was not being abused as a poison, it serves to illustrate the disconnect between policy and reality, given the scope and magnitude of restricted products freely used for off-label and detrimental purposes.

Transparency and accountability measures are also lacking, such as the recording of purchaser's name and purpose. M. Odino's Furadan purchase did not warrant any justification of purchase and was not recorded. It is this lack of accountability that makes it easy for some pastoralists to use

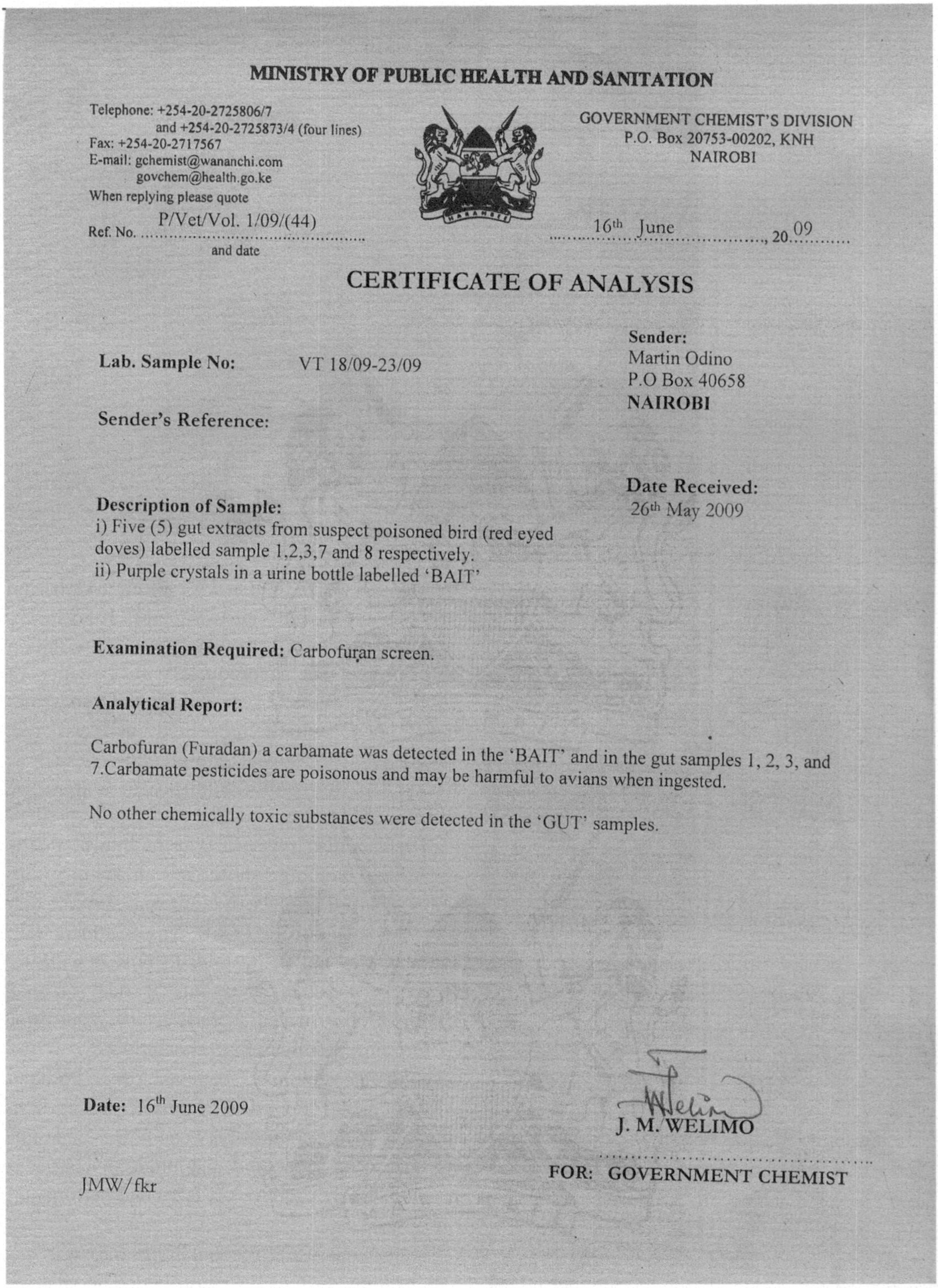

MINISTRY OF PUBLIC HEALTH AND SANITATION

Telephone: +254-20-2725806/7
and +254-20-2725873/4 (four lines)
Fax: +254-20-2717567
E-mail: gchemist@wananchi.com
govchem@health.go.ke
When replying please quote
Ref. No. P/Vet/Vol. 1/09/(44)
and date

GOVERNMENT CHEMIST'S DIVISION
P.O. Box 20753-00202, KNH
NAIROBI

16th June, 20 09

CERTIFICATE OF ANALYSIS

Lab. Sample No: VT 18/09-23/09

Sender's Reference:

Sender:
Martin Odino
P.O Box 40658
NAIROBI

Date Received:
26th May 2009

Description of Sample:
i) Five (5) gut extracts from suspect poisoned bird (red eyed doves) labelled sample 1,2,3,7 and 8 respectively.
ii) Purple crystals in a urine bottle labelled 'BAIT'

Examination Required: Carbofuran screen.

Analytical Report:

Carbofuran (Furadan) a carbamate was detected in the 'BAIT' and in the gut samples 1, 2, 3, and 7.Carbamate pesticides are poisonous and may be harmful to avians when ingested.

No other chemically toxic substances were detected in the 'GUT' samples.

Date: 16th June 2009

J. M. WELIMO

FOR: GOVERNMENT CHEMIST

JMW/fkr

Image 7 Certificate of analysis

Table 5 Classification of poisonous chemicals identified during studies

Agrochemical	Active ingredient	Class	Information and threats
1. Furadan (5G)	Carbofuran	Carbamate	Neurotoxin through oral and dermal pathways. Believed to be primary biocide in Kenya. Currently "voluntarily withdrawn" by PCPB-certified manufacturer, FMC. Not explicitly listed as "banned" or "deregistered" (higher concentration 10G is listed as banned by PCPB). – Used in intentional bird poisoning at Bunyala. – Intentional predator killing and accidental vulture poisoning. – Poisoning in human beings.
2. Marshal 350ST/ Carbosulfan	Carbofuran	Carbamate	Neurotoxin. Used in Masai Mara, a critical ecosystem for Kenya's wildlife. – Intentional predator killing and accidental vulture poisoning.
3. Round Up	Glyphosate *N*-(phosphonomethyl)glycine)	Organophosphate	Especially used in vegetable plantation farms such as in Laikipia in northern Kenya and Ewaso Nyiro. – Kills mostly herbivores when they graze on treated crops/overspray to pastures including when washed into watering holes. There are anecdotal claims by pastoralists that it is used to induce miscarriage in livestock.
4. Mocap GR 10.	Enthroprophos	Organophosphate	Proposed as a safer pesticide than carbofuran. – Effective in baboon control that are otherwise tough to bait with many other poisons.
5. DCP	2, 4-D	Organophosphate	Described as "DCK" possibly to conceal identity and now corrupted identity is widely used; should have been DCP (2, 4-D); alleged to result in quick putrefaction of carcasses.
6. Queletox	Fenthion	Organophosphate	Restricted and administered only by the desert locust control department. – Poisons by dermal pathway but also pungent and irritating to respiratory system to people, birds, and other animals.
7. Red cat	Zinc phosphide	Inorganic rodenticide/ pesticide	Kills by releasing toxic phosphine gas when it reacts with moisture. – Used to poison hyenas. – Known in human suicide.
8. Strychnine sulfate	Strychnine	Alkaloid	Potent convulsant historically used for dog control in Kenya but starting 2019 its use in Kenya was suspended; entry into an organism via inhalation, ingestion, and dermal pathways

pesticides in off-label ways without fully understanding the toxicological and ecological consequences. Adding measures such as purchase and use logs, as well as the enforcement of consequences for unrestricted usage, could help to quell the overuse and abuse of such chemicals.

Although Furadan 5G has been banned in Kenya, it is still supplied into neighboring Uganda and Tanzania and is readily smuggled into Kenya for continued use as a poison for nuisance animals.

Furthermore, there is also weakness in Kenya's ability to ascertain the identity of poisons, due to a lack of suitable laboratory capacity for toxin analysis. For example, samples collected during the bird poisoning quantification study were analyzed by thin-layer chromatography, which can indicate the presence of a carbamate; however, it cannot determine which carbamate was present. To determine which carbamate was present, high-performance liquid chromatography is needed, which to the author's

Table 6 Animals reported to have been killed, and poisons used during social surveys

Poisoned animals		Social survey and poison		
		Survey 1	Survey 2	Survey 3
Wild birds	Scientific name	Poison	Poison	Poison
1. Yellow-necked Spurfowl	*Pternistis leucoscepus*			Seed dresser likely to be Marshal 250 EC
2. African Openbill (stork)	*Anastomus l. lamelligerus*		Furadan	Queletox
3. White stork	*Ciconia ciconia*			Queletox
4. Unspecified vultures	Various			Furadan, Marshal, Round Up
5. Hooded vulture	*Necrosyrtes monachus*			2,4-D product
6. Wattled Starling	*Creatophora cinerea*		Furadan	
7. Unspecified Starling	Various			Round Up
8. Helmeted Guineafowl	*Numida meleagris*	Furadan	Furadan	Round Up
9. Unspecified eagles	Various	Furadan		
10. Unspecified ravens	Various	Furadan		
11. Specked pigeon	*Columba g. guinea*		Furadan	
12. African mourning dove	*Streptopelia d. perspicillata*		Furadan	
13. Laughing dove	*Spilopelia senegalensis*		Furadan	
14. Fan-tailed widowbird	*Euplectes axillaris*		Furadan	
15. Parasitic weaver	*Anomalospiza imberbis*		Furadan	
16. White-faced whistling duck	*Dendrocygna viduata*		Furadan	
Wild animals and insects		**Survey 1**	**Survey 2**	**Survey 3**
		Poison	**Poison**	**Poison**
1. Lion	*Panthera leo*		Furadan, strychnine	Marshal, Furadan
2. Elephant	*Laxodonta africanus*			2,4-D product
3. Leopard	*Panthera pardus*	Furadan		
4. Impala	*Aepyceros melampus*			Unknown concoction
5.Zebra	*Equus burcheli*			Unknown concoction
6. Masai giraffe	*Giraffa camelopardalis tippelskirchi*			Unknown concoction
7. Spotted hyena	*Crocuta crocuta*	Furadan	Furadan	Marshal Mocap
8. African wild dog	*Lycaon pictus*		Furadan	
9. Wildebeest	*Connochaetes taurinus*	Furadan		
10.Warthog	*Phacochoerus africanus*	Furadan		
11. Yellow baboon	*Papio cyanocephalus*			Mocap
12. Nile crocodile	*Crocodylus niloticus*	Furadan		
13. Nile monitor (lizard)	*Varanus niloticus*		Furadan	

(continued)

Table 6 (continued)

Poisoned animals		Social survey and poison		
		Survey 1	Survey 2	Survey 3
Wild birds	Scientific name	Poison	Poison	Poison
12. Unspecified snakes	Unspecified species		Furadan	
13. Bees	*Apis mellifera scutellata*	Furadan		
14. Rat	Unspecified species	Furadan		
12. Striped ground squirrel	*Euxerus erythrops*			
Domestic animals and man		**Survey 1**	**Survey 2**	**Survey 3**
		Poison	**Poison**	**Poison**
1. Human being	*Homo s. sapiens*		Furadan	Queletox, Red cat, Diazinon, Unknown concoction
2. Chicken	*Gallus domesticus*			Furadan
3. Domestic dog	*Canis lupus familiaris*	Furadan	Furadan	Strychnine
4. Domestic cat	*Felis catus*	Furadan	Furadan	
5. Sheep	*Ovis spp.*			Round Up
6. Cow	*Bos indicus*			Round Up

knowledge is not used by The Government Chemist Laboratory in Kenya.

In Kenya, there is an overall lack of stewardship regarding the use of agrochemicals as poisons. Individuals are required to use products as per instructions on the label, yet public education, supervision, and guidance is lacking, especially in the handling and usage of these dangerous chemicals. Results from our surveys revealed an apparent lack of supervision and guidance from agricultural officers regarding the proper use of agricultural chemicals, leaving citizens to use, overuse, and abuse chemicals as they wish.

Conclusion

Eight agricultural chemicals were identified as a health threat to animals, humans, and the environment in Kenya, as demonstrated by this research. Furadan, a carbamate pesticide poison, was primarily responsible for the intoxications observed during this research.

Birds were the most affected fauna according to respondent reports and surveyor (M. Odino) observation, as compared to other classifications of wildlife.

Carbosulfan and Round Up are emerging pesticide poisons of concern. Carbosulfan was a poison of concern around Masai Mara and may be used in place of carbofuran to poison hyenas and lions. Round Up is of concern in both the Masai Mara and Laikipia, where a heavy human–livestock–wildlife interface exists, thus its overuse may result in toxic effects for free-ranging wildlife.

The Pest Control Products Board is too distanced from the public and does not appear to be aware of what is happening regarding the use of products under their legislation. It is recommended that a means of communication be established whereby the public can readily communicate to the Board, including the anonymous reporting of misuse or abuse of toxins.

Future Recommendations

- PCPB should carefully reassess and consider revising legislation for chemicals used as poisons, while also revising the over-the-counter procurement procedure for all products, whereby the purchaser should clearly state why they require the product they are purchasing, where they will use it, as well as provide their address and contact details.

- Legislation for Furadan 5G, the carbofuran withdrawn from Kenya, needs to be clearly redefined, including its unlawful importation from neighboring countries into Kenya.
- Additional research should focus on chemicals that pose a higher poisoning risk than previously known, including Marshal, Round Up, and unidentified poisons made from combined substances.
- There is a critical need for toxicological studies in order to establish the long-term health effects of these toxins from a one health standpoint.
- The agricultural sector in Kenya is encouraged to consider a shift to organic farming that will minimize the use of inorganic chemicals, many of which are harmful to the environment.

References

Carson R (1962) Silent Spring. Houghton Mifflin Company, Boston

Casida J, Durkin K (2013) Neuroactive insecticides: targets, selectivity, resistance, and secondary effects. Annual Rev Entomol 58:99–117. https://doi.org/10.1146/annurev-ento-120811-153645

Kendall CJ, Virani MZ (2012) Assessing mortality of African vultures using wing tags and GSM-GPS transmitters. J Raptor Res 46:135–140

Oberemok V, Laikova K, Gninenko Y, Aleksei Z, Nyadar P, Adeyemi T (2015) A short history of insecticides. J Plant Protect Res 55:221–226. https://doi.org/10.1515/jppr-2015-0033

Odino M (2011) A chronicling of long-standing carbofuran use and its menace to wildlife in Kenya: measuring the conservation threat that deliberate poisoning poses to birds in Kenya: the case of pesticide hunting with Furadan in the Bunyala Rice irrigation scheme in Carbofuran and wildlife poisoning. In: Richards N (ed) Global perspectives and forensic approaches. John Wiley & Sons, Chichester, pp 53–70

Odino M (2017) Poisons and poisoning affecting vultures in Masai Mara, Laikipia and Tsavo in Kenya. Report to Alterra Wageningen University and the Peregrine Fund

Odino M, Ogada DL(2008a Jan) Furadan use in Kenya and its impacts on birds and other wildlife: a survey of the regulatory agency, distributors, and end-users of this highly toxic pesticide. Report to the Bird Committee of Nature Kenya, 17 p

Odino M, Ogada DL (2008b July) Furadan use in Kenya: a survey of the distributors and end-users of toxic Carbofuran (Furadan) in pastoralist and rice growing areas. Report to Kenya Wildlife Trust, 19 p

Ogada DL, Keesing F, Virani MZ (2012) Dropping dead: causes and consequences of vulture population declines worldwide. Ann N Y Acad Sci 1249:57–71. https://doi.org/10.1111/j.1749-6632.2011.06293.x

PCPB (2018) Pest control products registered for use in Kenya

Plaza P, Martínez-López E, Lambertucci S (2019) The perfect threat: Pesticides and vultures. Sci Total Environ 687:1207–1218. https://doi.org/10.1016/j.scitotenv.2019.06.160

Route to Food (2019) Pesticides in Kenya: Why our health, environment and food security are at stake. https://routetofood.org/wp-content/uploads/2019/08/RTFI-White-Paper-Pesticides-in-Kenya.pdf

Simpson D (1996, Oct) Life in the wild: talk is cheap. SWARA, Sept/Oct issue

USAID Kenya Agricultural Value Chain Enterprises (2013) Pesticide evaluation report and safer use action plan (PERSUAP). https://www.rti.org/sites/default/files/kaves_persuap.pdf

Vail DD (2015) Toxicity abounds: new histories on pesticides, environmentalism, and silent spring. Stud Hist Philos Sci Part C: Stud Hist Philos Biol Biomed Sci 53:118–121

Virani MZ, Kendall C, Njoroge P, Thomsett S (2011) Major declines in the abundance of vultures and other scavenging raptors in and around the Masai Mara ecosystem, Kenya. Biol Conserv 144:746–752. https://doi.org/10.1016/j.biocon.2010.10.024

World Bank Group (2018) Kenya economic update, April 2018, No. 17: policy options to advance the big 4. World Bank, Nairobi. https://openknowledge.worldbank.org/handle/10986/29676

A Primer to the Global Trade of Reptiles: Magnitude, Key Challenges, and Implications for Conservation

Jordi Janssen

Abstract

Reptiles are among the most intensively harvested and traded species groups globally. The global trade of reptiles includes the trade of live reptiles as pets, as well as the trade of reptiles or their parts for use in traditional medicine, for reptile skins, or for human consumption. Reptiles have been widely used to treat a large variety of ailments, and medicinal use has been documented for at least 284 species. Reptiles are also an important food source, with the heaviest exploitation for this purpose in tropical and subtropical regions. In addition to the consumption of reptile meat for nutritional value, it is often intertwined with cultural beliefs, or consumed for medicinal purposes. The international trade of reptile skins is the largest trade in volume for all uses of reptile species and included 75 species from 2000 to 2017. However, the trade in live reptiles impacts significantly more species, with international trade documented for at least 642 taxa. Available data on the reptile trade may be unreliable due to frequently occurring discrepancies and a general lack of data on the volume and included species. Additionally, the reptile trade increases the risk of disease transmission globally, as well as the introduction of invasive species. All these factors have resulted in population declines across the globe. However, approximately 80% of the world's population also relies on the use of natural resources for traditional medicine, food, or additional income. In order to effectively reduce the negative impacts of trade on reptiles, it is vital to address the underlying drivers of the trade.

Keywords

Wildlife trade · CITES · Traditional medicine · Bushmeat · Reptile skins · Pets

Introduction

The global trade of reptiles, as for other species, is primarily regulated through the Convention on International Trade in Endangered Species of Wild Fauna and Flora (CITES). The Convention subjects international trade of selected species to a licensing system with varying levels of trade controls or restrictions. CITES Appendix I includes species threatened with extinction and generally prohibits commercial trade of wild-caught specimens. Appendix II includes species not necessarily threatened with extinction, but for which trade needs to be regulated in order to ensure that the species does not become threatened by trade. Export permits are required for all Appendix II species, and additional import

J. Janssen (✉)
Monitor Conservation Research Society, Big Lake Ranch, BC, Canada
e-mail: jordi.janssen@mcrsociety.org

S. C. Underkoffler, H. R. Adams (eds.), *Wildlife Biodiversity Conservation*,
https://doi.org/10.1007/978-3-030-64682-0_17

permits may be required, depending on local laws (e.g., in the European Union). Appendix III includes species that are nationally protected in their range states, and for which the range states request the assistance of other Parties to regulate trade and prevent illegal trade. The trade of such species can only be conducted with a certificate of origin and/or export permit from the range state. As of May 2020, 11,243 reptile species are recognized (Uetz and Etzold 1996), of which 97 species (including 5 subspecies) are listed on CITES Appendix I, 771 species are listed on CITES Appendix II, and 85 (6 subspecies) species are listed on Appendix III (June 2020; http://www.speciesplus.net). The international trade of approximately 8.5% of all reptile species is thus covered under the CITES Convention, while the remaining, and majority, of species are not protected when traded internationally (Auliya et al. 2016). This chapter serves as a primer to the global reptile trade and will cover the magnitude of the trade, challenges of regulating the trade, as well as the implications for the conservation of reptiles globally. The global trade of reptiles is complex, with many stakeholders and challenges, and as such, a limited number of examples are presented in this chapter. The commercial trade has been documented as one of the sources of population decline for many reptile species (O'Brien et al. 2003; Lyons and Natusch 2011, 2013; Lyons et al. 2013; D'Cruze et al. 2015; Parusnath et al. 2017). This chapter covers the trade of live reptiles as pets, as well as the trade of reptiles or their parts for use in traditional medicine, for their skins, or for human consumption.

Magnitude

Reptiles are sourced in large volumes for both the legal and illegal market. They are sourced for a variety of purposes, including for food (Klemens and Thorbjarnarson 1995), skins (Caldwell 2012; Kasterine 2012), as pets (Auliya 2003), and for use in traditional medicines (da Nóbrega Alves et al. 2008; Ferreira et al. 2009; Alves et al. 2013). Many species are used for multiple purposes, e.g., tokay geckos, *Gekko gecko,* are widely used in traditional medicine, but are also exported to be kept as pets (Nijman and Shepherd 2015). Similarly, reticulated pythons, *Malayopython reticulatus,* are popular pets around the world (Shiau et al. 2006; Luiselli et al. 2012) but are primarily traded for the trade in reptile skins (Kasterine 2012; Natusch et al. 2016, 2017).

Reptiles in Traditional Medicine

The use of wild plant and animal species for medicinal purposes is widespread and has been in practice for thousands of years (Bauer 2009; Chen et al. 2009). Approximately 80% of the world's population still relies on the use of these natural resources for traditional medicine (Alves and Rosa 2005). This is particularly evident in developing countries, as it is often more affordable in comparison to modern medicine. In remote areas, it may be the only form of medicine available (Sofowora 1996; da Nóbrega Alves and Pereira Filho 2007; Alves and Rosa 2010). Traditional medicine is strongly influenced by social, economic, and cultural practices, which in some cultures has led to a higher preference for traditional medicine over modern medicine (Ngokwey 1995; Nazarea et al. 1998).

Of all animal species used in traditional medicine, reptiles are one of the most frequently and ubiquitously encountered on a global scale (da Nóbrega Alves et al. 2008; Zhou and Jiang 2004; Enríquez Vázquez et al. 2006; Mahawar and Jaroli 2008; Bauer 2009; Alves and Rosa 2010; Whiting et al. 2013; Nijman and Bergin 2017). Alves et al. (2013) reported the use of at least 284 reptile species representing 46 families in traditional medicine around the world. However, as the medicinal use of many species is undocumented, the actual number of species used for medicinal purposes is likely to be higher.

Reptiles have been widely used to treat a large variety of ailments such as coughs (Read 1934), asthma (Sheu 1977; Chuang et al. 1999), diabetes (Read 1934), cancer (Chen and Huang 2001), hyperglycemia (You and Wang 2000), skin ailments, bruises, sprains, arthrosis, rheumatism

(Alves et al. 2009), and many others. Whereas certain parts are used from some species (e.g., fat or skin), multiple parts are used from others, thus an individual species may have up to two dozen indications (Alves et al. 2009). The choice of species is often based on cultural relevance and perception (Bauer 2009). Reptiles have historically been regarded as mysterious, and have long been associated with special abilities (Bauer 2009). A reptile may bring good fortune (Frazer 1951) or can be the harbinger of bad luck (Mahendra 1936). They are also blamed for spreading diseases such as skin disorders (Bauer 2009). In some cases, their use in traditional medicine is to cure the ailment that the particular species is accused of causing in the first place (Bauer 2009).

It is currently unclear how many reptiles are traded for medicinal purposes, as a significant portion of the trade takes place domestically, that is to say, the trade remains within the boundaries of the range country. One exception lies in the trade of tokay geckos (Fig. 1). Nijman et al. (2012) reported the export of up to 1.2 million dried tokay geckos to China from three traders in Indonesia, despite a ban on the export of dried animals (Caillabet 2013). Indonesia has a quota for the captive breeding of this species, but doubts have arisen on the biological and economic feasibility of such operations (Nijman and Shepherd 2015). Approximately seven metric tons of illegally harvested tokay geckos were seized en route from Indonesia to Hong Kong (Caillabet 2013), suggesting the actual trade is magnitudes higher than reported. Large volumes of tokay geckos are also exported from other range states, but these numbers are not known.

Fig. 1 Live tokay geckos, *Gekko gecko*, offered for sale for medicinal purposes in Indonesia © Monitor

International Trade for Traditional Medicine

The United Nations Environment Programme World Conservation Monitoring Centre (UNEP-WCMC) CITES Trade Database provides some additional indications for the international trade of reptiles for traditional medicine. Between the years 2000 and 2017, international trade for medicinal purposes was reported for 28 different taxa, some of which are on the genus level. This trade was reported in different units of measurement, e.g., gram, kilogram, milliliter, liter, or no attached unit at all. Discrepancies between importer-reported numbers (I) and exporter-reported numbers (E) differ significantly. For example, importers report 24 taxa with a total of **2,342,697 units** (unit of measurement not reported), while exporter-reported quantities only account for four taxa of **405,123 units** (unit of measurement not reported). Importers also reported nine taxa traded in grams (g), with **6445 grams** in total. Exporter-reported quantities report **42,424 grams** of merely one taxon. As this taxon (Chinese pond turtles, *Mauremys reevesii*) is listed on CITES Appendix II, no import permits are required unless by local legislation (e.g., in European Union). If it can be assumed that all quantities traded as liquids had a similar density to water, and all units (milligram, gram, milliliters, liters) are converted to kilogram, this results in a total amount of **5067.63 kilogram** (I) – **31,642.91 kilogram** (E), in addition to the reported trade with no units. Four of the reported taxa, Cheloniidae spp., saltwater crocodile (*Crocodylus porosus*), Siamese crocodiles (*Crocodylus siamensis*), and *Testudinidae* spp., are currently listed on CITES Appendix I. Sources were from registered breeding facilities (source code "D"), seized specimens (source code "I"), or bred in captivity (source code "C").

Reptile Skin Trade

International Trade of Reptile Skins

Reptile skins are primarily traded for the international market, with very limited local or domestic trade. The magnitude of this trade could represent the largest trade volume for all uses of reptile species, yet a general lack of transparency, combined with gaps in data collection, make it difficult to obtain a representative picture of the total trade. Trade in reptile skins of CITES-listed species is documented under multiple terms: skins (SKI), skin pieces (SKP), scraps (SKS), or leather items (SKO), as well as large and small leather products (LPL and LPS). When looking at items documented as skins (SKI), these items are traded in multiple units, i.e., per piece, per m^2, cm, ft^2, pairs, or even in kilogram. These discrepancies significantly hinder the analysis of the global trade of reptile skins.

When the number of skins (SKI) in units other than per surface area (m^2, ft^2 or cm) or weight (e.g., kilogram) are examined, the UNEP-WCMC CITES trade database shows that importing Parties (I) have reported the importation of **86,146,397** reptile skins for commercial purposes (purpose code "T") from 2000 to 2017 (Fig. 2). Exporting Parties (E) reported the export of **77,857,026** reptile skins over the same time period. Additionally, the database reports a further **3,078,215** (E) – **23,045,188** (I) m^2, **5** (E) – **46,648** ft^2, **3234** (I) – **19,412** (E) cm, **800** (E) – **113,105** (I) g and **122,580–154,062** kg of reptile skins. Measurements in length or weight are excluded from the numbers mentioned in the paragraph below. The differences between importer-reported and exporter-reported quantities are often substantial (e.g., **5** [E] – **46,648** ft^2[I]) (Robinson and Sinovas 2018) (see *Issues*).

From 2000 to 2017, the trade of reptile skins of **75** different species was documented, and for 13 genera, trade was documented to the genus (e.g., *Tupinambis* spp) or family level (e.g., Aligatoridae spp.); the latter two accounting for between **339,127** (importer-reported) and **331,468** (exporter-reported) reptile skins (Table 1).

Alligatoridae (alligators and caimans) are the most heavily traded for reptile skins with **26,780,476** (E) – **27,274,401** (I) skins traded from 2000 to 2017, followed by Pythonidae (**17,590,246** [E] – **23,469,656** [I]), Varanidae (**15,894,116** [I] – **16,301,700** [E]), Crocodylidae (**7,021,938** [E] – **7,777,690** [I]) and Teiidae (**5,418,044** [E] – **6,195,285** [I]).

For both the importer-reported and exporter-reported quantities, the top five species traded are the same, though the order differs slightly (Table 2).

For both importer-reported and exporter-reported quantities, wild-caught reptiles (source code "W") are the most important source (**62% [I] and 59% [E]**) for skins. Reptile skins from captive-bred reptiles (source code "C") accounted for **32% (I) – 34% (E)** of all reptile skins traded. Skins from ranched reptiles (source code "R") totaled **3% (E) – 4% (I),** with other sources accounting for approximately 3% of all skins traded. Reptile skins of species listed on CITES Appendix I almost exclusively belong to the Crocodylidae, with the exception of importing Parties reporting the import of 73 skins of *Brachylophus* spp., a group of severely threatened Iguanidae from the Fiji Islands. This may be due to a reporting error, as this species is not known to be used for skins. While a total of 155 countries have been documented to import reptile skins, the main importers of reptile skins are Singapore (**18,660,993** [E] – **19,049,020** [I]) and Mexico (**10,062,471** [E] – **11,658,905** [I]). The combined total from these two countries accounts for 37% – 39% of all reptile skins imported. The top five importers further consist of Japan (**10,351,113** skins), Italy (**8,766,107** skins), and France (**6,073,027** skins), according to importer-reported statistics.

The main exporter of reptile skins is Singapore (**18,639,123** [I] – **20,793,427** [E]), suggesting that nearly all skins imported by Singapore are also exported, with Singapore acting mainly as a transit region (Kasterine et al. 2012). However,

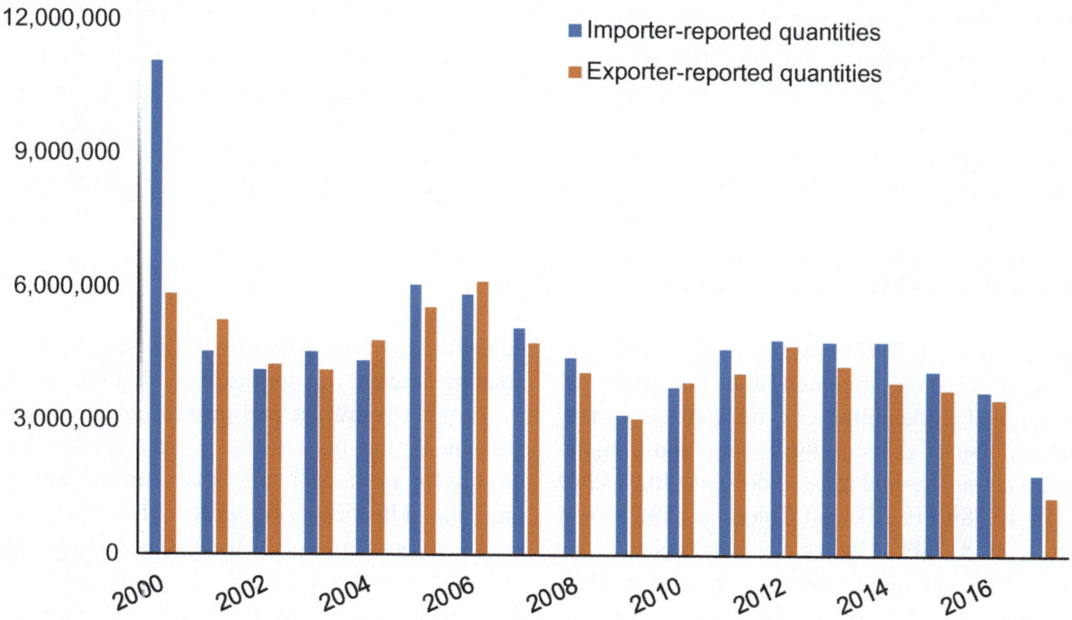

Fig. 2 Global trade of reptile skins displayed in importer- and exporter-reported quantities, from 2000 to 2017. Source: UNEP-WCMC CITES Trade Database

Table 1 Trade of CITES-listed reptiles, reported on the family level, traded as skins from 2000 to 2017

Family	Importer-reported quantity	Exporter-reported quantity
Acrochordidae	20,986	1206
Agamidae	1608	
Alligatoridae	27,274,401	26,780,487
Boidae	76,655	79,172
Colubridae	3,629,485	2,515,743
Cordylidae	2	
Crocodylidae	7,805,146	7,082,277
Elapidae	1,700,157	2,001,640
Geoemydidae	54	
Helodermatidae	1	1
Hydrophiidae	3377	19
Iguanidae	9905	12,849
Pythonidae	23,469,656	17,590,246
Teiidae	6,195,285	5,418,044
Testudinidae	7	45
Varanidae	15,894,116	16,301,700
Viperidae	36,398	73,560
Xenopeltidae	29,100	
(blank)	58	37
Grand Total	**86,146,397**	**77,857,026**

Both importer-reported quantities and exporter-reported quantities are displayed
Source: UNEP-WCMC CITES Trade Database

Table 2 Five most frequently traded CITES-listed reptile species for the trade of skins from 2000 to 2017

Top 5 Importer-reported	Quantity	Top 5 Exporter-reported	Quantity
Malayopython reticulatus	17,077,674	*Caiman crocodilus fuscus*	16,118,216
Caiman crocodilus fuscus	15,422,371	*Varanus salvator*	13,629,168
Varanus salvator	13,141.717	*Malayopython reticulatus*	11,504,670
Alligator mississippiensis	10,631,364	*Alligator mississippiensis*	9,685,728
Crocodylus niloticus	5,435,049	*Crocodylus niloticus*	4,781,708

Both importer-reported quantities and exporter-reported quantities are displayed
Source: UNEP-WCMC CITES Trade Database

Kasterine et al. (2012) suggests that a large proportion of skins imported into Singapore are stockpiled. Stockpiling can be done to control the number of skins on the market, and to maintain a certain desired price. Indonesia (**10,073,541** [I] – **10,784,119** [E]) and Colombia (**10,255,641** [E] – **10,817,497** [I]) are the numbers two and three main exporters, alternating spots depending on the use of importer-reported or exporter-reported statistics, followed by Italy (**2,437,612** [E] – **8,364,952** [I]) and the United States (**7,351,101** [I] – **7,395,299** [E]) or Malaysia (**5,755,357** [E] – **6,444,281** [I]).

Reptiles for Human Consumption

Reptiles are an important food source for many humans around the world, with the heaviest exploitation for this purpose in tropical and subtropical regions (Klemens and Thorbjarnarson 1995). Consumption of reptiles varies considerably among different reptile groups, including turtles (freshwater and sea turtles), tortoises, larger lizards (Klemens and Thorbjarnarson 1995), and crocodilians. The availability of meat is often a by-product of the skin industry, e.g., pythons or crocodilians (Klemens and Thorbjarnarson 1995). The consumption of reptile meat is not merely for nutritional purposes, as it is often intertwined with cultural beliefs or consumed for medicinal purposes. Yet consuming reptile meat is not without risks (Magnino et al. 2009), as both farmed and wild-caught reptiles are reservoirs for zoonotic bacteria (e.g., Salmonella spp.) zoonotic parasites, and biotoxins (Magnino et al. 2009). Magnino et al. (2009) found that crocodile meat is often contaminated with pathogenic Salmonella spp., creating a health risk for the general public.

Turtles (freshwater and marine) and tortoises are among the most heavily exploited species groups for meat, and scientists estimate that at one time in history, every turtle and tortoise species have been used as a food source (Klemens and Thorbjarnarson 1995). For many cultures, turtle meat is an important source of protein, although in some regions it may only be available on a seasonal basis (Klemens and Thorbjarnarson 1995). Live reptiles are often traded for human consumption, as fresh meat is preferred.

Whereas the trade of turtles for subsistence used to be localized or regional in scope (Klemens and Thorbjarnarson 1995), globalization facilitated the worldwide movement of large quantities of turtles, with China as the largest importer (van Dijk et al. 2000; Liou 2006; Schneider et al. 2011; Wu 2015). In addition to importing enormous numbers of turtles and tortoises (Schneider et al. 2011), China has mass-scale commercial breeding facilities (Haitao et al. 2007, 2008). Haitao et al. (2008) estimated, based on a survey of 684 Chinese turtle farms, that a total of >300 million turtles are sold each year, worth approximately 750 million USD. The majority of these comprise Chinese softshell turtles, *Pelodiscus sinensis*, but also include species native to North America, and species designated as critically endangered on the International Union for Conservation of Nature (IUCN) Red List of Threatened Species (Haitao et al. 2008). Globalization of the trade of turtle meat provides additional pressure to species as demonstrated from the large-scale imports of *Podocnemis* spp. by Hong Kong and China. Hong Kong reportedly imported 2,052,858 live yellow-spotted river turtles, *Podocnemis unifilis,* from 2005 to 2017, while China imported 902,193 live yellow-spotted river turtles over the same period. This genus is considered threatened throughout its range as a result of local subsistence hunting. Bates (1863) estimated up to 48 million eggs of this species were harvested every year in the nineteenth century, which equates to the total production of 400,000 nesting females. Since the onset of significant population declines, *Podocnemis* spp. are now "ranched" (Cantarelli et al. 2014) and are part of community-based management programs that include nest protection details (Miorando et al. 2013) after suffering significant population declines (Kemenes and Pezzuti 2007). Klemens and Thorbjarnarson (1995) reported that a single shop in Tefé (Northern Brazil) selling roasted turtle reportedly sold approximately 12,000 turtles a year. As a significant proportion of trade of reptile meat and eggs for human consumption is local or regional, figures obtained for international trade are likely a small proportion of the actual volume.

Sea turtles are heavily exploited for their meat and eggs (Lagueux 1991) and are considered important economic resources for humans living in coastal regions. The collection of eggs and consumption of sea turtle meat happens both on a local subsistence level as well as through larger-scale commercial activities. Species most frequently targeted for meat and eggs are flatback sea turtles (*Natator depressa)*, green sea turtles (*Chelonia mydas)*, loggerhead sea turtles (*Caretta caretta*), and olive ridley sea turtle (*Lepidochelys olivacea)*. For other species, consumption is only local, e.g., leatherback sea turtles (*Dermochelys coriacea)*, as the meat is considered to be oily, and thus undesirable (Klemens and Thorbjarnarson 1995). This species is targeted for its eggs instead (Klemens and Thorbjarnarson 1995).

Many tortoises are consumed for meat and for oil (Klemens and Thorbjarnarson 1995), and this has been a significant cause of decline for many species (van Dijk et al. 2000, 2000; Morgan 2018). Much of this trade is local or regional subsistence hunting, and as it is largely illegal, it is clandestine in nature, making it difficult to grasp the true scale of hunting for meat. Numerous reports of consumption being the major cause of decline provides a measure of the true magnitude. Consumption of turtle meat is also considered to be the primary cause of decline for Mexican gopher tortoise, *Gopherus flavomarginatus* (Morafka et al. 1989), Florida gopher tortoise, *Gopherus polyphemus,* and many giant tortoises (Swingland and Klemens 1989). In Africa, the largest mainland tortoise, African spurred tortoise, *Centrochelys sulcata,* has suffered relatively little from overconsumption, as the consumption of turtle flesh was prohibited according to local religious beliefs (Broadley 1989). However, the population of critically endangered Madagascan radiated tortoise, *Astrochelys radiata,* has been decimated by poaching, partially because its flesh is considered a local delicacy (Goodman et al. 1994; O'Brien et al. 2003). Almost all Asian tortoise species are consumed by humans, with consumption listed as the major cause of decline, e.g., Asian forest tortoise (*Manouria emys)*, Burmese star tortoise (*Geochelone platynota* (Moll 1989a)), and Elongated tortoises (*Indotestudo elongata* (Moll 1989b)).

Snakes are hunted for subsistence in many parts of the world, and especially in Asia (Klemens and Thorbjarnarson 1995). The Chinese city of Guangzhou is home to one of the largest snake markets in the world (Pope 1961). The market for snake meat includes many species such as cobras (*Naja* spp.), kraits (*Bungarus* spp.), rat snakes (e.g., *Ptyas* and *Elaphe* spp.), and pythons (*Python* and *Malayopython* spp.). The consumption of snake meat is widely associated with Cantonese culture in Asia, but other cultures also eat snakes. Large-bodied snakes are preferable for their flesh. In Africa, large venomous snake species like rhinoceros viper, *Bitis nasicornis,* are popular for consumption as they are readily available in palm oil plantations (Klemens and Thorbjarnarson 1995). In North America, *Crotalus* spp. are consumed, often after so-called "roundups," where large numbers are captured and killed during public festivals (McCormick 1996; Franke 2000).

The human consumption of lizards is significantly related to cultural beliefs and perceived medicinal value (as is the case with snakes and some crocodilians). Trade for these purposes overlaps. The perceived medicinal value of lizards can be traced back to Java during the Pleistocene era (Auffenberg 1988). Today, live lizards and their parts may be purchased in markets throughout Southeast Asia, where they are boiled, consumed in the form of satay, or barbequed. Throughout South America, Iguanidae (Green iguanas (*Iguana iguana*) or *Ctenosaura* spp.) are the primary lizard species consumed (Cotí and Ariano-Sánchez 2008; Stephen et al. 2011; Pasachnik et al. 2014). *Varanus* spp. are another group of lizards frequently consumed by humans (Asibey 1974; Magnino et al. 2009; Hidelaratchi et al. 2010); however, these are primarily consumed locally. Overexploitation of lizards for human consumption is considered to have contributed to the extinction of several species (Wiewandt 1977).

Crocodilian meat and eggs are widely consumed by mostly rural communities in some countries, and within some cultures (Klemens and Thorbjarnarson 1995). The majority of literature on the commercial trade of crocodilian products has placed a focus on the commercial skin trade (Caldwell 2012), and international trade of meat appears to be a by-product of the skin trade (Klemens and Thorbjarnarson 1995). The consumption of crocodilian meat is influenced significantly by local cultural beliefs and varies from region to region (Klemens and Thorbjarnarson 1995). In some areas, cultural and religious beliefs prohibit the consumption and killing of crocodiles, while in others the killing of these reptiles is fueled by their perceived medicinal value (Thorbjarnarson 1988). Aboriginal people in Australia have used the meat and eggs of crocodiles for up to 40,000 years (Webb et al. 1987), and many South American caimans are a popular source of protein (Klemens and Thorbjarnarson 1995; Alves et al. 2009; Fernandes et al. 2014). In Venezuela, caiman meat and eggs are consumed and harvested on a seasonal basis (Thorbjarnarson 1991), reducing the population of crocodilians around human settlements (Ouboter 1989). For the Yaruro Indians in Venezuela, both Caiman meat, eggs and river turtles are the most important sources of protein (Klemens and Thorbjarnarson 1995). On an interesting note, Klemens and Thorbjarnarson (1995) report that the Orinoco crocodile, *Crocodylus intermedius,* is avoided by these communities, as the meat allegedly tastes bad. In West Africa, the dwarf crocodile, *Osteolaemus tetraspis,* is hunted for food (Kofron and Steiner 1994; Van Vliet and Nasi 2008; Zoer 2012), yet in areas where only the larger Nile crocodile, *Crocodylus niloticus,* occurs, the consumption of crocodilian meat is almost absent (Child 1987). In the United States, the American alligator, *Alligator mississipiensis*, is harvested and ranched for meat (Nichols 1976; Klemens and Thorbjarnarson 1995; Rice et al. 1999). Consumption of crocodilian meat and eggs is thus widespread across the globe, but varies by region, is primarily confined to rural areas, and based on available species and cultural beliefs.

International Trade for Human Consumption

While most reptile consumption appears to be local or regional, the UNEP-WCMC CITES trade database contains records for the international sale of reptile meat. This includes reptile meat traded in boxes and cans and measured in kilograms or grams. Between **9,408,334** (E) and **254,616,403** (I) kilograms of reptile meat from 55 taxa were traded internationally from 2000 to 2017 (Table 3).

The most-traded reptile meat commodity is crocodilians, with exporting countries reporting up to **7,458,148** kilograms, and importing countries **253,668,179** kilograms. The least reported internationally traded are turtles and tortoises, with **11,490 (I) – 43,057 (E)** kilograms of meat. Note that massive volumes of tortoise and freshwater turtle meat have been, and continue to be (though in lower volumes) trafficked from Southeast Asia to East Asia, and between Southeast Asian countries, in violation of CITES, and therefore without permits. As a result, few records exist for this trade. In addition, many of the live turtles and tortoises traded are destined

Table 3 Reported trade of reptile meat in kilograms from 2000 to 2017

Family	Importer-reported quantity	Exporter-reported quantity
Crocodilia	253,668,179	7,458,148
Sauria	489,604	252,207
Serpentes	447,128	1,654,922
Testudines	11,490	43,057
Grand Total	**254,616,403**	**9,408,334**

Source: UNEP-WCMC CITES Trade Database

for turtle farms or human consumption (Haitao et al. 2008).

Importing parties recorded the highest quantities for the Siamese crocodile, *Crocodylus siamensis,* while it is only the second most traded species according to exporting countries (Table 4). Interestingly, the number of kilograms traded for this species by importing countries is a staggering 153 times higher than the number reported by exporting countries (there are several potential explanations for this difference; see *Issues*). According to data provided by importing countries, **97.55%** of all reptile meat traded internationally originates from species listed on CITES Appendix I, although this is mainly the Siamese crocodile, which is farmed in large numbers. CITES Appendix II species make up **2.44%,** and species on CITES Appendix III, **0.1%** of the total amount of reptile meat. These percentages are different for records provided by exporting countries: CITES Appendix I only accounts for **17.44%,** CITES Appendix II for **79.88%,** CITES Appendix III for **2.65%,** and non-CITES-listed species for **0.65%.**

This difference is also noticed when it comes to the reported source of trade. According to importing countries, "*Appendix-I animals bred in captivity for commercial purposes in operations included in the Secretariat's Register, in accordance with Resolution Conf. 12.10 (Rev. CoP15)*" or Source Code "D" make up for **97.55%** of all reported trade of reptile meat, followed by animals bred in captivity at **2.44%**. According to importing countries, reptile meat from reptiles caught in the wild only comprise **0.62%** of all reptile meat traded. In contrast, exporting countries report that **39.45%** of meat originated from animals bred in captivity (source code "C"), followed by animals caught in the wild (source code "W") with **29%**. Reptile meat from reptiles either ranched (source code "R") or from registered breeding facilities (source code "D") only make up to **13.55%** and **17.45%**, respectively.

Live Reptile Trade

The global trade of live reptiles is enormous, and reptiles are considered among the most heavily harvested terrestrial fauna in the world (Harfoot et al. 2018). Data on the global trade of live reptiles is significantly hindered by several factors; for instance, only approximately 8.5% of all reptile taxa are currently listed on the CITES Appendices (Auliya et al. 2016), meaning that trade of the majority of species goes relatively undocumented, and therefore unnoticed (Janssen and Shepherd 2018; Jensen et al. 2018) unless countries have adopted stricter legislation (e.g., Annexes of the European Union include some non-CITES-listed species, and US Fish and Wildlife Law Enforcement LEMIS contains details of non-CITES-listed species). The European Union is a good example of how much this influences trade analysis: from 2002 to 2017, the CITES Trade Database contains records of **4,658,562** live CITES-listed reptiles imported into the European Union, while the EUROSTAT database (https://ec.europa.eu/eurostat; commodity group number 0106 20 00) reports the import of **26,472,570** live reptiles over the same time period. Thus, most of the trade consists of species not listed on the CITES Appendices, thus actual trade levels may be significantly higher than reported. However, reptiles

Table 4 Top 5 of species traded as reptile meat in kilograms from 2000 to 2017

Top 5 Importer-reported	Quantity	Top 5 Exporter-reported	Quantity
Crocodylus siamensis	248,367,494	*Crocodylus niloticus*	3,858,863
Crocodylus niloticus	3,827,001	*Crocodylus siamensis*	1,623,298
Alligator mississippiensis	632,514	*Crocodylus porosus*	746,502
Crocodylus porosus	531,508	*Alligator mississippiensis*	739,114
Malayopython reticulatus	369,762	*Naja sputatrix*	525,801

Source: UNEP-WCMC CITES Trade Database

can be traded alive for different purposes such as traditional medicine, food, or to supply the global pet trade, and these numbers cannot be retrieved from trade records.

International Trade of Live Reptiles

Harfoot et al. (2018) reported that from 1975 and 2014, 152 million CITES-listed reptiles were traded globally. When looking at the UNEP-WCMC CITES Trade Database in more recent times, and excluding all records with units that do not allow identification of the number of animals traded, e.g., units such as shipments or kilograms, between **23,095,425** (I) and **29,979,113** (E) CITES-listed reptiles were traded from 2000 to 2017 (Fig. 3).

The reptile order with the highest numbers traded differs slightly between importer and exporter-reported quantities. Lizards (reported as Sauria) are the most intensively traded, at **10,465,514**, according to importer countries. This is followed by Testudines (turtles and tortoises) at **7,685,292** specimens, snakes at **4,077,015** specimens, and crocodilians at **900,506** specimens. According to exporter-reported quantities, the highest quantity was reported for Testudines at **12,574,087**, versus **11,899,705** reported lizards. This was followed by snakes (**4,823,004**) and crocodilians (**1,356,605** specimens).

The UNEP-WCMC CITES Trade database contains data for 642 taxa, of which 60 are only reported to the genus level (spp.). The five species traded in the highest quantities were the same for both importer- and exporter-reported quantities, although the order differed (Tables 5 and 6). However, the Green iguana, *Iguana iguana*, was the species traded in the highest quantities for both (**7,331,919 [I]** – **8,475,693 [E]**). For all five species, exporter-reported quantities were higher compared to quantities reported by importing countries.

For both importer-reported and exporter-reported quantities, captive-bred reptiles (source code "C") are the most important source (**44.43% [I]** and **46.99% [E]**) for live reptiles. Live reptiles from a wild-caught origin (source code "W") made up **23.67% (E)** – **26.55% (I)** of all reported trade, followed by reptiles with a ranched origin (source code "R") with **22.56% (E)** – **22.58% (I)**. The remaining source codes comprised less than 3% of the reported trade per source code.

Most live reptiles traded between 2000 and 2017 are listed on CITES Appendix II (**81.21% [E]** – **88.92% [I]**), followed by CITES Appendix III (**8.35% [I]** – **14.26% [E]**). Between **2.3% (I),** and **2.79% (E)** of all reptiles were listed on the highest protection category, Appendix I. Reptiles not listed on the CITES Appendices (e.g., listed on EU Annexes) comprised **0.4% (I)** – **1.74% (E)** of the trade records. Commercial trade of live reptile species listed in Appendix I was almost exclusively Siamese crocodiles *Crocodylus siamensis* (**97.17% [E]** – **97.81% [I]**); the remaining taxa (n = 43) comprise the additional percentages (Fig. 4).

While a total of 167 countries have been documented to import live reptiles, the main importers of live reptiles are United States (**8,634,235** [I] – **9,180,233** [E]), and Hong Kong (**3,349,470** [I] – **4,515,588** [E]), combined accounting for 52% – 59% of all live reptiles imported. The top five importers differ slightly between importer-reported and exporter-reported statistics, with Mexico (**1,541,683** reptiles), China (**1,345,498** reptiles), and Germany (**1,260,235**

Fig. 3 Global trade of CITES-listed live reptiles between 2000 and 2017. Source: UNEP-WCMC CITES Trade Database

reptiles) making up the remaining countries for importer-reported statistics. For exporter-reported statistics the remaining countries are China (**4,265,726** reptiles), Germany (**1,398,058** reptiles), and Mexico (**1,138,058** reptiles).

The UNEP-WCMC CITES Trade Database contains data on the export of live reptiles from 165 different countries. The main exporter of live reptiles is El Salvador (**5,591,889** [I] – **6,693,389** [E]). The top five, according to importer-reported statistics, is completed by the United States (**1,998,752** reptiles), Peru (**1,664,093** reptiles), Togo (**1,579,852** reptiles), and Ghana (**1,492,328** reptiles). For exporter-reported statistics the top five is completed by the United States (**3,246,676** reptiles), Peru (**2,614,366** reptiles), Vietnam (**1,961,256** reptiles), and Ghana (**1,678,608** reptiles) (Table 6).

Issues

Attempting to understand the global trade of reptiles is not without its challenges, many of which are well documented (Schlaepfer et al. 2005; Janssen and Shepherd 2018; Robinson and Sinovas 2018). Some challenges have been covered above in sections regarding the scale of the trade, but many more issues exist, often with a significant impact on what is known about the

Table 5 Top five of CITES-listed species with the highest quantity traded based on importer-reported quantities (left) and exporter-reported quantities (right) between 2000 and 2017

Top 5 Importer-reported	Quantity	Top 5 Exporter-reported	Quantity
Iguana iguana	7,331,919	*Iguana iguana*	8,475,693
Python regius	2,983,078	*Podocnemis unifilis*	2,601,623
Podocnemis unifilis	1,653,301	*Python regius*	2,504,285
Testudo horsfieldii	1,134,360	*Graptemys pseudogeographica*	1,576,273
Graptemys pseudogeographica	866,126	*Testudo horsfieldii*	1,309,334

Source: UNEP-WCMC CITES Trade Database

Fig. 4 Whipsnakes (Ahaetulla spp.) offered for sale in a plastic bottle at a market in Indonesia © Monitor

global trade of reptiles. Besides widely known problems of corruption (Smith and Walpole 2005), weak governance, weak legislation (Carpenter et al. 2005; Rasheed 2013), or lack of enforcement (Lee et al. 2005; Bennett 2011; Challender and MacMillan 2014), the issues to be discussed in the following sections should be given equal consideration.

Availability of Data and the Challenges in Analysis

As mentioned previously, approximately 8.5% of all reptiles are listed on the CITES Appendices, and therefore trade of such species, are found in the UNEP-WCMC CITES Trade Database (https://trade.cites.org/). The trade of CITES-listed species has been the primary focus in the published literature (Auliya et al. 2016; Foster et al. 2016; Harfoot et al. 2018; Poole and Shepherd 2017; Robinson and Sinovas 2018), yet trade of non-CITES-listed species is typically only detected during physical or online market surveys, unless regions have adopted stricter regulations. Thus, the majority of the trade in reptiles remains poorly documented.

The United States of America (USA) and the European Union (EU), both key players in the global reptile trade (Robinson et al. 2015), have gone beyond the implementation of CITES regulations by adopting their own, stricter measures. The USA documents international trade of wildlife in the Law Enforcement Information Management System (LEMIS). Within LEMIS, each import or export shipment receives a species code, a genus code, or a more general code (e.g., NONR = Non-CITES reptile), with the latter more common in larger shipments (Schlaepfer et al. 2005). The EU implements wildlife trade legislation through Council Regulation (EC) No. 338/97, listing species on EU Annex A, B, C, and D. Annex A includes CITES Appendix I species, some Appendix II and III species, for which the EU has adopted stricter measures, and some non-CITES listed species. Annex B includes all other CITES Appendix II species, and some Appendix III and non-CITES species. Annex C includes the remaining CITES Appendix III species. EU Annex D is a stricter measurement from the EU and includes some species listed on CITES Appendix III and non-CITES listed species for which the EU requires additional monitoring (Janssen and Shepherd 2018). Data for non-CITES listed species is often confusing, irregular, and far from complete, and therefore only able to provide an indication of the quantities traded (Janssen and Shepherd 2018). This is particularly relevant for non-CITES listed species, as the sheer lack of information in the trade of these species significantly hinders the timely detection of any negative effects trade might have. It is imperative that trade of non-CITES species be documented, in order to preemptively detect any negative impact trade might have on such species (Janssen and Shepherd 2018).

The analysis of CITES trade data is not without its hurdles (Robinson et al. 2015). Many studies have highlighted the challenge with

Table 6 Global trade of live reptiles from 2000 to 2017

Family	Importer-reported quantity	Exporter-reported quantity
Acrochordidae	30	
Agamidae	443,657	506,924
Alligatoridae	63,228	139,185
Anguidae	29	29
Boidae	459,288	527,293
Carettochelyidae	2651	3573
Chamaeleonidae	1,072,905	1,074,003
Chelidae	183	341
Cheloniidae	103	216
Chelydridae	161,863	499,925
Colubridae	188,201	939,701
Cordylidae	104,448	96,550
Crocodylidae	837,277	1,217,420
Dermatemydidae	142	315
Dermochelyidae		22
Elapidae	103,342	307,178
Emydidae	1,102,796	1,865,798
Gavialidae	1	
Gekkonidae	279,879	274,860
Geoemydidae	1,568,331	2,370,670
Gerrhosauridae	855	944
Helodermatidae	921	1585
Hydrophiidae	15	
Iguanidae	7,339,932	8,476,600
Lacertidae	72	69
Loxocemidae	200	181
Pelomedusidae	65,501	148,927
Platysternidae	1679	181
Podocnemididae	1,655,583	2,605,798
Pythonidae	3,316,927	3,047,532
Scincidae	15,141	6213
Teiidae	104,938	123,083
Testudinidae	2,713,260	3,245,794
Trionychidae	407,199	1,158,239
Tropidophiidae	541	52
Varanidae	1,080,962	1,338,314
Viperidae	2197	1047
Xenopeltidae	197	
Xenosauridae	885	531
N/A	66	20
Grand Total	**23,095,425**	**29,979,113**

relying on CITES trade data, as it is full of discrepancies that compromise the accuracy of the data (Robinson et al. 2015; Janssen and Blanken 2016). The reliability of CITES trade data is dependent upon the accuracy and compliance of the CITES Parties responsible for reporting the trade. Discrepancies between reported numbers may be caused by a difference in reported source, purpose, terms, or units (UNEP-WCMC 2013), possibly due to reporting errors (Fialho et al. 2016). Moreover, nonstandard units are often used to document trade, such as in boxes or cans of reptile meat. Many of the units cannot be easily converted, e.g., leather products to number of animals, or kilogram to number of specimens, which impedes the reliability of the data (Mieres and Fitzgerald 2006; Jiang et al. 2013; Arroyo-Quiroz et al. 2007). Trade reports may be filed or not, and animals may be exported in one year but not received by the importing Party until the following year (UNEP-WCMC 2013). In addition, some countries will report trade by number of permits issued or the actual number of specimens traded, and this is often not clear from the annual reports (UNEP-WCMC 2013), although CITES requires Parties to report the actual numbers of animals traded (Annex to CITES Notification 2017/007). The actual number of animals traded is often lower than recorded on permits, resulting in lower imported-reported quantities (Robinson et al. 2015). In addition, species listed on CITES Appendix II do not require an import permit unless required by the destination country (e.g., the EU requires import permits). This leads to underreporting of importer-reported quantities for CITES Appendix II species (Robinson et al. 2015), unless imported by Parties with stricter regulations. The quantities traded, as reported in the section on the scale of the reptile trade, provide both the importer-reported and exporter-reported quantities in order to illustrate the impact of this issue on the reported data.

While the CITES Trade Database is a useful tool for understanding and monitoring the trade of wildlife and their products, caution is required when analyzing and interpreting the data. Moreover, it should be acknowledged that this comprises only international trade, despite the knowledge that many species are traded and used in the confines of the domestic market. Nonetheless, it is currently the most comprehensive open-source tool available, and thus essential in understanding the global trade of wildlife.

Fraudulent Source Declarations

Commercial captive breeding has been considered by many as a useful conservation tool (Challender and MacMillan 2014; Jepson and Ladle 2009), with the aim of reducing pressure on the wild population. The number of CITES-listed reptiles reported as bred or born in captivity increased during the 1990s, and currently exceeds those declared as wild-caught. From 1996 to 2012 a reported 9.6 million live CITES Appendix II-listed reptiles were captive-bred (source code "C"). Yet, captive breeding facilities require a substantive investment in equipment and supplies, employment, and operation costs (Snyder et al. 1996; Nijman and Shepherd 2015), compared to relatively low costs for wild-sourced specimens. Captive-bred animals should be at least equally or more profitable in order for traders to favor them over wild-sourced animals (Bulte and Damania 2005). In addition, not all species are biologically able to meet the demand and set quotas (Janssen and Chng 2018), and are thus not suitable for commercial captive breeding.

Captive-bred animals are less scrutinized, and thus provide incentives for traders to declare wild-sourced animals as captive-bred. This so-called laundering of wildlife, where wild-sourced animals are fraudulently declared as captive-bred, reportedly occurs on a large scale (Nijman and Shepherd 2009; Lyons and Natusch 2011; Nijman and Shepherd 2015; Janssen and Chng 2018). Laundering of wildlife can potentially have serious impacts on wild populations (Natusch and Lyons 2012). Fraudulently declaring wild-sourced animals as captive-bred not only undermines the CITES regulations, it also undermines national legislation, and can lead to overexploitation and loss of community benefits (Janssen and Chng 2018). Additionally, it creates a false sense of sustainability, erroneously claiming little to no impact on the wild population. Any potential declines due to overharvesting may therefore go unnoticed.

In 2016, Resolution Conf. 17.7 was passed at the CITES Conference of the Parties 17 (CoP17), for the review of trade of animal species reported as produced in captivity. Resolution Conf. 17.7 highlights the concern of a growing body of evidence that wild-sourced specimens are fraudulently declared as captive-bred, and brings attention to doubts about the legal origin of parental stocks of both native and non-native captive-bred specimens (CITES 2016).

It is important that wild-caught animals can reliably be differentiated from animals bred in captivity. Stable isotope ratios of nitrogen ($\delta^{15}N$) and carbon ($\delta^{13}C$) have proved to be a useful tool to estimate the trophic position of a species within the food chain (van Schingen et al. 2016; Janssen et al. 2016). In recent years, this technique has been applied to see whether it can differentiate between wild and captive animals (van Schingen et al. 2016; Ziegler 2016; Natusch et al. 2017). van Schingen et al. (2016) found that the isotope ratio of skin samples of captive specimens of the CITES Appendix I-listed Vietnamese crocodile lizard *Shinisaurus crocodilurus vietnamensis* were significantly enriched in $\delta^{13}C$ and $\delta^{15}N$ as compared to the wild population. This allowed for a high degree of accuracy in distinguishing wild specimens from captive-bred animals. Similar results were obtained by Natusch et al. (2017), who found that skins from pythons allowed for the differentiation between wild and captive-bred pythons, in particular when animals were fed a special diet. Ziegler (2016) also found that feeding regimes in captivity (for *Varanus* spp.) can result in smaller isotopic ranges compared to wild populations due to a more homogenous diet, allowing for differentiation between captive-bred and wild-caught animals. These studies point to the importance of diet specificity as a necessary component allowing for the differentiation of wild-caught versus captive-bred animals of the same species. Diet specificity may be tied to ecological niche, and thus, for species with a broad range of ecological habitats, and thus a broader range of dietary components, the technique may not be suitable (van Schingen et al. 2016; Ziegler 2016; Natusch et al. 2017). Moreover, studies have shown that isotope ratios may differ between age classes, as juveniles and adults have access to different dietary items, and thus may differ in dietary makeup (Janssen et al. 2015; Ziegler 2016). The development of a database of reference ratios may allow for the use of this technique on species with larger ecological ranges. A substantive investment will be required to sample species across their range and within different age classes.

Other techniques such as isotopic marking (e.g., ^{15}N marker glycine by van Schingen et al. 2016; Natusch et al. 2017) showed promising results, with isotopic markers detectable in shed epidermal fragments of lizards after more than 3 months (Ziegler 2016). However, the authors argue this technique may only be useful for adult animals (juveniles shed much faster) or species that do not shed their epidermis on a regular basis, such as tortoises (Ziegler 2016). In addition to isotopic techniques, several studies researched the feasibility of elemental analysis to differentiate between captive-bred and wild-caught animals. Natusch et al. (2017) found that elemental markers (e.g., heavy metals) were useful in differentiating between both source and origin of specimens. They highlight the large difference observed between captive-bred and wild-caught reptiles, which is influenced by the high levels of groundwater pollution in Vietnam's Mekong Delta. Captive animals fed wild prey also showed elevated elemental levels, in contrast to samples from animals in a less polluted environment. In addition, Brandis et al. (2018) found that elemental analysis outperformed stable isotope ($\delta^{13}C$, $\delta^{15}N$) analyses, resulting in a 100% classification accuracy for a mammal, short-beaked echidna, *Tachyglossus aculeatus*. While these results are promising, more work is required in order to reliably and cost-effectively differentiate between captive-bred and wild-caught reptiles.

Lack of International Protection for Nationally Protected Species

As previously mentioned, only a small percentage of reptile species (~8.5%) are currently listed on the CITES Appendices, and thus the trade in these species is at least somewhat monitored and regulated (Auliya et al. 2016). The remaining species can be traded internationally with virtually no restrictions. Many of these are protected in their range states, and export is technically prohibited. Australian reptiles provide an example of this challenge. While Australia does not allow the export of live native reptiles, Australian reptiles are some of the most common reptiles in overseas markets, with the central bearded dragon, *Pogona vitticeps,* as one of the most popular reptile pet species in general. While many of the Australian reptiles are now bred in captivity, parental stock for almost all species have been smuggled out of Australia. Yet, the absence of a CITES listing for many of these species means they are freely traded once smuggled out of Australia. Due to the lack of international protection for nationally protected species, large numbers of endemic and protected reptiles are found on the reptile markets. Previous studies have documented nationally protected reptiles from at least 19 non-EU countries within the European Union (Auliya 2003; Altherr 2014, 2016; Janssen and Indenbaum 2018; Shepherd et al. 2019).

The United States utilizes the *Lacey Act*, which prevents the direct import of species in violation of national (foreign) legislation. Therefore, the direct import of nationally protected species into the United States is prohibited. In order to circumvent this, nationally protected species are first trafficked to other countries, where these species are not considered protected. From there, they are legally shipped to the United States and other markets, effectively legalizing smuggled animals (Vinke and Vinke 2015). Despite their protected status in their range states, and an export ban, international trade of such animals is not prohibited.

Conservation Implications

As has been shown above, the scale of the global trade of reptiles is enormous and encompasses many different, and frequently overlapping, purposes (Nekaris et al. 2010; Natusch and Lyons 2012; Robinson et al. 2018a, b). Population declines as a result of overharvesting have been widely documented (Gibbons et al. 2000; O'Brien et al. 2003; Shepherd and Ibarrondo 2005; Todd et al. 2010; Clark et al. 2011; Lyons and Natusch 2011; Parusnath et al. 2017). This is particularly true for turtles and tortoises, for which the impact of unsustainable use has been highlighted by many (Van Dijk and Palasuwan 2000; Haitao et al. 2007; Nijman and Shepherd 2014; D'Cruze et al. 2015; Morgan 2018). Many of these species are used as a local food source (Klemens and Thorbjarnarson 1995; Blasco et al. 2011), sold as pets (O'Brien et al. 2003; Morgan 2018), or used in traditional medicine (Chen et al. 2009). For the most part, the trade is extensive and unregulated (Sharma 1999). Overexploitation and the unsustainable harvesting of reptiles for trade have serious conservation implications. Additionally, deforestation and habitat loss place further stressors on many species of reptiles globally (Gibbons et al. 2000; Iskandar and Erdelen 2006; Todd et al. 2010). For example, the construction of roads increases poaching and trade of bushmeat and other wildlife products (Laurance et al. 2009; Suarez et al. 2009). Combined impacts between habitat loss and trade are likely severely underestimated (Symes et al. 2018). As previously shown, trade is recorded for only a small percentage of reptile species. Therefore, a large proportion of trade remains undetected for many species, resulting in the underestimation of population losses and ecosystem impacts.

Introduction of Diseases

The potential conservation implications of the global trade of reptiles are not only limited to impacts on the species being traded. The anthropogenic spread of a pathogen into an area in which it has not previously been found may have significant consequences for local populations. The most well-known examples are found within the amphibians, with the introduction of *Batrachochytrium dendrobatidis* (Garner et al. 2006), *B. salamandrivorans* (Martel et al. 2013) and *ranavirus* (Hyatt et al. 2000, 2002). The global trade for pets is often considered an important source for the anthropogenic spread of pathogens (Kolby et al. 2014; Nguyen et al. 2017). The transfer of pathogens from captive populations to wild animals has been reported for several reptile species, e.g., *Gopherus* sp. and *Terrapene* sp. (Jacobson et al. 1991; Brown et al. 2001; Feldman et al. 2006) and Iguanids (Hellebuyck et al. 2017). Inclusion Body Disease (IBD) has been found in captive Boidae and Pythonidae around the world, and wild populations (Chang and Jacobson 2010). Snake Fungal Disease (SFD) is a fungal disease resulting in severe skin infections in snakes and is often fatal (Lorch et al. 2016). This disease was first discovered in a population of timber rattlesnakes *Crotalus horridus* in the north-eastern part of the United States (Clark et al. 2011) and has subsequently been documented in over 30 species (Franklinos et al. 2017). It has been documented in wild populations in Canada (Lorch et al. 2016) and Europe (Franklinos et al. 2017), and in captive snakes in North America, Europe, and Australia (Lorch et al. 2016).

Invasive Species Risk

The introduction of non-native reptiles has the potential to significantly disrupt native reptile communities (Cole et al. 2005; Reed 2005; Platenberg 2007). The Global Invasive Species Database (http://www.iucngisd.org) currently lists 32 reptile species considered invasive around the world. Not all non-native species introduced to an ecosystem are considered invasive; the term "invasive species" specifically implies that this species causes harm to that specific new environment. One of the most well-known examples is that of the brown tree snake, *Boiga irregularis*, introduced to the island of Guam. In its introduced habitat, it managed to thrive and decimate the local bird and lizard communities, and also caused the extirpation of the gecko Pacific slender-toed gecko, *Nactus pelagicus* (Rodda and Fritts 1992). Burmese pythons, *Python bivittatus*, have been introduced through the pet trade into southern Florida (United States) and are currently flourishing (Holbrook and Chesnes 2011; Willson et al. 2011; Reed et al. 2012). Florida now hosts a staggering list of 31 non-native lizard species, one turtle (red-eared slider, *Trachemys scripta elegans*), one crocodilian (spectacled caiman, *Caiman crocodilus*) and three snake species, for a large part considered to be released pets (Hardin 2007). The Green iguana, *Iguana iguana*, introduced to the Lesser Antillean islands is now threatening the native and threatened Lesser Antillean iguana, *Iguana delicatessima*, facilitated by faster reproduction and the ability to hybridize with the native species (Knapp et al. 2014; Vuillaume et al. 2015). Yet, the impacts of invasive species can also be indirect, as shown with the impact of Brown Anoles *Anolis sagrei* on Grand Cayman anoles *Anolis conspersus*, of Grand Cayman. Here the Brown Anoles caused behavioral changes and shifts in habitat use of the native species (Losos et al. 1993).

Conclusion

The global trade of reptiles is enormous and faced with numerous issues and challenges. Overharvesting fuels the decline of species around the world, yet, approximately 80% of the world's population relies on the use of these natural resources for traditional medicine, food, or additional income.

Reptiles are an important food source for many people around the world, with the heaviest exploitation for this purpose in tropical and subtropical regions. The consumption of reptiles varies considerably among different reptile groups, and its meat is not merely for nutritional purposes but is often intertwined with cultural beliefs or used as traditional medicine. The use of traditional medicine is strongly influenced by social, economic, and cultural practices, which in some cultures has led to a higher preference for traditional medicine over modern medicine. Literature shows that at least 284 reptile species from 46 families are known to be used in traditional medicine around the world. However, as the medicinal use of many species is undocumented, the actual number of species used for medicinal purposes is likely to be higher.

The international trade of reptile skins is the largest by volume and involves over 75 different species and up to 13 genera. More than half of all the reptile skins traded are from reptiles caught in the wild. In contrast, the majority of live reptiles traded were animals bred in captivity. For the live reptile trade, data showed that 642 reptile taxa have been traded, of which 60 are only reported at the genus level. So, while the skin trade is the largest by volume, the live reptile trade affects significantly more species.

Many countries are involved in the global trade of reptiles (e.g., 155 countries are documented to trade reptile skins), yet the countries involved differ strongly depending on the item traded. Singapore and Mexico are the largest importers of reptile skins, and Singapore is also the largest exporter of reptile skins. The primary importers of live reptiles are the United States and Hong Kong, and the primary exporter is El Salvador.

The analysis of trade data and attempts to understand the global trade in reptiles are frequently hindered by both poor availability and quality of data. Data is only collected for species listed on the CITES Appendices, unless stricter domestic measures require so (e.g., for the United States and European Union). The remaining species can therefore be traded internationally with virtually no restrictions. Yet, many of these are protected in their range states, and export is technically prohibited. In addition, trade of many of these protected species is only permitted, or less scrutinized, when it comprises captive-bred animals, thus providing an incentive for traders to declare wild-sourced animals as captive-bred. This so-called laundering of wildlife appears to occur frequently and on a large scale. To date, there is no reliable method that can distinguish between wild and captive animals that can be used on a large scale. While techniques like stable isotope analysis and elemental markers have shown promising results in differentiating between captive-bred and wild-caught animals, more research is necessary before this is method can be used widely.

The unsustainable global trade of reptiles has many conservation implications and is responsible for widespread population declines of several free-ranging reptile species. The risk of zoonotic disease spread from reptiles from the trade poses not just significant risks to human health but also to other reptiles, and in particular, threatened or endangered species in captivity and in the wild. The trade has allowed for the establishment of a number of invasive reptile species that threaten the ecological integrity and biodiversity of native habitats. However, although community dependence on the usage of reptiles may clash with conservation goals (Gareau 2007), it cannot be overlooked. Many people depend on these species for either medicinal purposes, food sources, or simply as an additional source of income. In addition, the motivations of those involved in the trade of reptiles may differ strongly between communities, households, and individuals. In order for any conservation method to be successful, this must be taken into account. In order for conservation solutions to be effective, the underlying drivers of the trade must be identified and addressed.

References

Altherr S (2014) Stolen Wildlife – Why the EU needs to tackle smuggling of nationally protected species Report by Pro Wildlife, Munich, p 32

Altherr S (2016) Stolen wildlife II – Why the EU still needs to tackle smuggling of national protected species. Report by Pro Wildlife. Report by Pro Wildlife, Munich, Germany

Alves RRN, Rosa IL (2005) Why study the use of animal products in traditional medicines? J Ethnobiol Ethnomed 1:5

Alves RRN, Rosa IL (2010) Trade of animals used in Brazilian traditional medicine: trends and implications for conservation. Hum Ecol 38:691–704

Alves RRN, Santana G, Almeida W, Neto NL, Vieira W (2009) Reptiles used for medicinal and magic religious purposes in Brazil. Appl Herpetol 6:257–274

Alves RRN, Vieira WLS, Santana GG, Vieira KS, Montenegro PFGP (2013) Herpetofauna used in traditional folk medicine: conservation implications. In: Animals in traditional folk medicine. Springer, Berlin, pp 109–133

Arroyo-Quiroz I, PÈrez-Gil R, Leader-Williams L (2007b) Mexico in the international reptile skin trade: a case study. Biodivers Conserv 16:931–952

Asibey EO (1974) Wildlife as a source of protein in Africa south of the Sahara. Biol Conserv 6:32–39

Auffenberg W (1988) Gray's monitor lizard. University of Florida Press, Gainesville, pp 419

Auliya M (2003) Hot trade in cool creatures: a review of the live reptile trade in the European Union in the 1990s with a focus on Germany. TRAFFIC Europe, Brussels

Auliya M, Altherr S, Ariano-Sanchez D, Baard EH, Brown C, Brown RM, Cantu J-C, Gentile G, Gildenhuys P, Henningheim E (2016) Trade in live reptiles, its impact on wild populations, and the role of the European market. Biol Conserv 204:103–119

Bates HW (1863) The naturalist on the river amazons: a record of adventures, habits of animals, sketches of Brazilian and Indian life, and aspects of nature under the equator, during eleven years of travel. J. Murray, London

Bauer A (2009) Geckos in traditional medicine: forensic implications. Appl Herpetol 6:81–96

Bennett EL (2011) Another inconvenient truth: the failure of enforcement systems to save charismatic species. Oryx 45:476–479

Blasco R, Blain H-A, Rosell J, Díez Fernández-Lomana JC, Huguet Pàmies R, Rodríguez J, Arsuaga JL, Bermúdez de Castro JM, Carbonell E (2011) Earliest evidence for human consumption of tortoises in the European early Pleistocene from Sima del Elefante. Sierra de Atapuerca, Spain

Brandis KJ, Meagher PJ, Tong LJ, Shaw M, Mazumder D, Gadd P, Ramp D (2018) Novel detection of provenance in the illegal wildlife trade using elemental data. Sci Rep 8:15380

Broadley DG (1989) *Geochelone sulcata*. The conservation biology of tortoises. In: Swingland IR, Klemmens NW (eds) Occas, vol 5. Papers IUCNSSC, Gland, pp 47–48

Brown MB, Brown DR, Klein PA, McLaughlin GS, Schumacher IM, Jacobson ER, Adams HP, Tully JG (2001) *Mycoplasma agassizii* sp. nov., isolated from the upper respiratory tract of the desert tortoise (*Gopherus agassizii*) and the gopher tortoise (*Gopherus polyphemus*). Int J Syst Evol Microbiol 51:413–418

Bulte EH, Damania R (2005) An economic assessment of wildlife farming and conservation. Conserv Biol 19 (4):1222–1233

Caillabet OS (2013) The trade in tokay geckos *Gekko Gecko* in South-East Asia: with a case study on novel medicinal claims in peninsular Malaysia. TRAFFIC, Petaling Jaya, Selangor, Malaysia

Caldwell J (2012) World trade in crocodilian skins 2008–2010. UNEP World Conservation Monitoring Centre, Cambridge

Cantarelli VH, Malvasio A, Verdade LM (2014) Brazil's *Podocnemis expansa* conservation program: retrospective and future directions. Chelonian Conserv Biol 13:124–128

Carpenter AI, Robson O, Rowcliffe JM, Watkinson AR (2005) The impacts of international and national governance changes on a traded resource: a case study of Madagascar and its chameleon trade. Biol Conserv 123:279–287

Challender DW, MacMillan DC (2014) Poaching is more than an enforcement problem. Conserv Lett 7:484–494

Chang L-W, Jacobson ER (2010) Inclusion body disease, a worldwide infectious disease of boid snakes: a review. J Exotic Pet Med 19:216–225

Chen M, Huang JH (2001) Present status of study on gecko used as traditional Chinese medicine. Shijie Kexue Jishu-Zhongyiyao Xiandaihua 3:53–56

Chen T-H, Chang H-C, Lue K-Y (2009) Unregulated trade in turtle shells for Chinese traditional medicine in east and Southeast Asia: the case of Taiwan. Chelonian Conserv Biol 8:11–18

Child G (1987) The management of crocodiles in Zimbabwe. In: Wildlife management: crocodiles and alligators. Surrey Beatty and Sons, Chipping Norton, pp 49–62

Chuang I-C, Huang Y-L, Lin T-H (1999) Determination of lead and cadmium in Chinese crude drugs by graphite-furnace atomic absorption spectrometry. Anal Sci 15:1133–1136

Clark RW, Marchand MN, Clifford BJ, Stechert R, Stephens S (2011) Decline of an isolated timber rattlesnake (*Crotalus horridus*) population: interactions between climate change, disease, and loss of genetic diversity. Biol Conserv 144:886–891

Cole NC, Jones CG, Harris S (2005) The need for enemy-free space: the impact of an invasive gecko on island endemics. Biol Conserv 125:467–474

Cotí P, Ariano-Sánchez D (2008) Ecology and traditional use of the Guatemalan black iguana (Ctenosaura palearis) in the dry forests of the Motagua Valley, Guatemala. Iguana 15:142–149

D'Cruze N, Singh B, Morrison T, Schmidt-Burbach J, Macdonald DW, Mookerjee A (2015) A star attraction: the illegal trade in Indian star tortoises. Nat Conserv 13:1–19

da Nóbrega Alves RR, Pereira Filho GA (2007) Commercialization and use of snakes in north and northeastern Brazil: implications for conservation and management. Biodivers Conserv 16:969–985

da Nóbrega Alves RR, da Silva Vieira WL, Santana GG (2008) Reptiles used in traditional folk medicine: conservation implications. Biodivers Conserv 17:2037–2049

Enríquez Vázquez P, Mariaca Méndez R, Guiascón R, Gustavo Ó, Piñera N, Jorge E (2006) Uso medicinal de la fauna silvestre en los Altos de Chiapas. México Interciencia 31:491–499

Feldman SH, Wimsatt J, Marchang RE, Johnson AJ, Brown W, Mitchell JC, Sleeman JM (2006) A novel mycoplasma detected in association with upper respiratory disease syndrome in free-ranging eastern box turtles (*Terrapene carolina carolina*) in Virginia. J Wildl Dis 42:279–289

Fernandes VRT, Souza Franco MLR, Mikcha JMG, de Souza VLF, Gasparino E, Coutinho ME, Tanamati A, Del Vesco AP (2014) Yacare caiman (*Caiman yacare*) trim hamburger and sausage subjected to different smoking techniques. J Sci Food Agric 94:468–472

Ferreira FS, Brito SV, Ribeiro SC, Saraiva AA, Almeida WO, Alves RR (2009) Animal-based folk remedies sold in public markets in Crato and Juazeiro do Norte, Ceará, Brazil. BMC Complement Altern Med 9:17

Fialho SM, Ludwig G, ValenÁa-Montenegro MM (2016) Legal international trade in live neotropical primates originating from South America. Primate Conserv 30:1–6

Foster S, Wiswedel S, Vincent A (2016) Opportunities and challenges for analysis of wildlife trade using CITES data–seahorses as a case study. Aquat Conserv Mar Freshwat Ecosyst 26(1):154–172

Franke J (2000) Rattlesnake roundups: uncontrolled wildlife exploitation and the rites of spring. J Appl Anim Welf Sci 3:151–160

Franklinos LHV et al (2017) Emerging fungal pathogen *Ophidiomyces ophiodiicola* in wild European snakes. Sci Rep 7:3844

Frazer JG (1951) The golden bough. A study in magic and religion. Part 6. In: The scapegoat, vol 9. Macmillan, London

Gareau BJ (2007) Ecological values amid local interests: natural resource conservation, social differentiation, and human survival in Honduras. Rural Soc 72:244–268

Garner TW, Perkins MW, Govindarajulu P, Seglie D, Walker S, Cunningham AA, Fisher MC (2006) The emerging amphibian pathogen *Batrachochytrium dendrobatidis* globally infects introduced populations of the north American bullfrog, *Rana catesbeiana*. Biol Lett 2:455–459

Gibbons JW, Scott DE, Ryan TJ, Buhlmann KA, Tuberville TD, Metts BS, Greene JL, Mills T, Leiden Y, Poppy S (2000) The global decline of reptiles, Déjà vu amphibians reptile species are declining on a global scale. Six significant threats to reptile populations are habitat loss and degradation, introduced invasive species, environmental pollution, disease, unsustainable use, and global climate change. Bioscience 50:653–666

Goodman SM, Pidgeon M, O'Connor S (1994) Mass mortality of Madagascar radiated tortoise caused by road construction. Oryx 28:115–118

Haitao S, Parham JF, Lau M, Tien-Hsi C (2007) Farming endangered turtles to extinction in China. Conserv Biol 21:5–6

Haitao S, Parham JF, Zhiyong F, Meiling H, Feng Y (2008) Evidence for the massive scale of turtle farming in China. Oryx 42:147–150

Hardin S (2007) Managing non-native wildlife in Florida: state perspective, policy and practice. Managing Vertebrate Invasive Species, p 14

Harfoot M, Glaser SA, Tittensor DP, Britten GL, McLardy C, Malsch K, Burgess ND (2018) Unveiling the patterns and trends in 40 years of global trade in CITES-listed wildlife. Biol Conserv 223:47–57

Hellebuyck T, Questel K, Pasmans F, Van Brantegem L, Philip P, Martel A (2017) A virulent clone of *Devriesea agamarum* affects endangered lesser Antillean iguanas (*Iguana delicatissima*). Sci Rep 7:12491

Hidelaratchi MDP, Riffsy MTM, Wijesekera JC (2010) A case of eosinophilic meningitis following monitor lizard meat consumption, exacerbated by anthelminthics. Ceylon Med J 50:84–86

Holbrook J, Chesnes T (2011) An effect of Burmese pythons (*Python molurus bivittatus*) on mammal populations in southern Florida. Florida Scientist 74:17–24

Hyatt AD, Gould AR, Zupanovic Z, Cunningham AA, Hengstberger S, Whittington RJ, Kattenbelt J, Coupar BEH (2000) Comparative studies of piscine and amphibian iridoviruses. Arch Virol 145:301–331

Hyatt AD, Williamson M, Coupar BEH, Middleton D, Hengstberger SG, Gould AR, Selleck P, Wise TG, Kattenbelt J, Cunningham AA (2002) First identification of a ranavirus from green pythons (*Chondropython viridis*). J Wildl Dis 38:239–252

Iskandar DT, Erdelen WR (2006) Conservation of amphibians and reptiles in Indonesia: issues and problems. Amphibian Reptile Conserv 4:60–87

Jacobson ER, Gaskin JM, Brown MB, Harris RK, Gardiner CH, LaPointe JL, Reggiardo C (1991) Chronic upper respiratory tract disease of free-ranging desert tortoises (Xerobates agassizii). J Wildl Dis 27 (2):296–316

Janssen J, Blanken LJ (2016) Going Dutch: an analysis of the import of live animals from Indonesia by the Netherlands. Traffic

Janssen J, Chng SC (2018) Biological parameters used in setting captive-breeding quotas for Indonesia's breeding facilities. Conserv Biol 32:18–25

Janssen J, Indenbaum RA. 2018. Endemic Vietnamese reptiles in commercial trade. Journal of Asia-Pacific Biodiversity

Janssen J, Shepherd CR (2018) Challenges in documenting trade in non CITES-listed species: a case study on crocodile skinks (Tribolonotus spp.). J Asia-Pacific Biodiversity 11:476–481

Janssen J, Towns DR, Duxbury M, Heitkönig IM (2015) Surviving in a semi-marine habitat: dietary salt exposure and salt secretion of a New Zealand intertidal skink. Comp Biochem Physiol A Mol Integr Physiol 189:21–29

Jensen TJ, Auliya M, Burgess ND, Aust PW, Pertoldi C, Strand J (2018) Exploring the international trade in African snakes not listed on CITES: highlighting the role of the internet and social media. Biodivers Conserv 28:1–19

Jepson P, Ladle RJ (2009) Governing bird-keeping in Java and Bali: evidence from a household survey. Oryx 43:364–374

Jiang Z, Zhou Z, Meng Z, Meng X, Li L, Ping X, Zeng Y, Mallon DP (2013) Domestic and CITES regulations controlling the international snake trade in China. Oryx 47:532–534

Kasterine A (2012) The trade in southeast Asian Python skins. ITC, Geneva

Kasterine A, Caillabet O,Natusch D (2012) The Trade in South-East Asian python skins. International trade centre. Switzerland

Kemenes A, Pezzuti JCB (2007) Estimate of trade traffic of Podocnemis (Testudines, pedocnemididae) from the middle Purus River, Amazonas, Brazil. Chelonian Conserv Biol 6:259–262

Klemens MW, Thorbjarnarson JB (1995) Reptiles as a food resource. Biodivers Conserv 4:281–298

Knapp C, Breuil C, Rodriguez C, Iverson J, Debrot AO (2014) Lesser Antillean Iguana: Iguana delicatissima: conservation action plan, 2014–2016. IUCN/SSC Primate Specialist Group, Gland

Kofron CP, Steiner C (1994) Observations on the African dwarf crocodile, *Osteolaemus tetraspis*. Copeia 1994:533–535

Kolby JE, Smith KM, Berger L, Karesh WB, Preston A, Pessier AP, Skerratt LF (2014) First evidence of amphibian chytrid fungus (*Batrachochytrium dendrobatidis*) and ranavirus in Hong Kong amphibian trade. PLoS One 9:e90750

Lagueux CJ (1991) Economic analysis of sea turtle eggs in a coastal community on the Pacific coast of Honduras. University of Chicago Press, Chicago

Laurance WF, Goosem M, Laurance SG (2009) Impacts of roads and linear clearings on tropical forests. Trends Ecol Evol 24:659–669

Lee RJ, Gorog AJ, Dwiyahreni A, Siwu S, Riley J, Alexander H, Paoli GD, Ramono W (2005) Wildlife trade and implications for law enforcement in Indonesia: a case study from North Sulawesi. Biol Conserv 123:477–488

Liou C (2006) The state of wildlife trade in China: information on the trade in wild animals and plants in China 2006. TRAFFIC East Asia, China

Lorch JM, Knowles S, Lankton JS, Michell K, Edwards JL, Kapfer JM, Staffen RA, Wild ER, Schmidt KZ, Ballmann AE (2016) Snake fungal disease: an emerging threat to wild snakes. Philos Trans R Soc B: Biol Sci 371:20150457

Losos JB, Marks JC, Schoener TW (1993) Habitat use and ecological interactions of an introduced and a native species of Anolis lizard on grand Cayman, with a review of the outcomes of anole introductions. Oecologia 95:525–532

Luiselli L, Bonnet X, Rocco M, Amori G (2012) Conservation implications of rapid shifts in the trade of wild African and Asian pythons. Biotropica 44:569–573

Lyons JA, Natusch DJ (2011) Wildlife laundering through breeding farms: illegal harvest, population declines and a means of regulating the trade of green pythons (*Morelia viridis*) from Indonesia. Biol Conserv 144:3073–3081

Lyons JA, Natusch DJ (2013) Effects of consumer preferences for rarity on the harvest of wild populations within a species. Ecol Econ 93:278–283

Lyons JA, Natusch DJD, Shepherd CR (2013) The harvest of freshwater turtles (Chelidae) from Papua, Indonesia, for the international pet trade. Oryx 47(2):298–302

Magnino S, Colin P, Dei-Cas E, Madsen M, McLauchlin J, Nöckler K, Maradona MP, Tsigarida E, Vanopdenbosch E, Van Peteghem C (2009) Biological risks associated with consumption of reptile products. Int J Food Microbiol 134:163–175

Mahawar MM, Jaroli DP (2008) Traditional zootherapeutic studies in India: a review. J Ethnobiol Ethnomed 4:17

Mahendra BC (1936) Geckos and superstition. J Bombay Nat Hist Soc Mumbai 88:631–633

Martel A, Spitzen-van der Sluijs A, Blooi M, Bert W, Ducatelle R, Fisher MC, Woeltjes A, Bosman W, Chiers K, Bossuyt F (2013) *Batrachochytrium salamandrivorans* sp. nov. causes lethal chytridiomycosis in amphibians. Proc Natl Acad Sci 110:15325–15329

McCormick C (1996) Eating fried rattler: the symbolic significance of the rattlesnake roundup. South Folk 53:41

Mieres MM, Fitzgerald LA (2006b) Monitoring and managing the harvest of tegu lizards in Paraguay. J Wildl Manag 70:1723–1734

Miorando PS, Rebêlo GH, Pignati MT, Brito Pezzuti JC (2013) Effects of community-based management on Amazon river turtles: a case study of *Podocnemis sextuberculata* in the lower Amazon floodplain, Pará, Brazil. Chelonian Conserv Biol 12:143–150

Moll EO (1989a) *Geochelone platynota* Burmese star tortoise. The Conservation Biology of Tortoises. Occasional Papers IUCN Species Survival Commission

Moll EO (1989b) *Indotestudo elongata,* Elongated tortoise. Page The Conservation Biology of Tortoises

Morafka DJ, Aguirre G, Adest GA (1989) *Gopherus flavomarginatus* Bolson tortoise. The Conservation Biology of Tortoises Switzerland: Occasional Paper of the IUCN Species Survival Commission 5:10–13

Morgan J (2018) Jakarta's flourishing trade in threatened turtles and tortoises under the spotlight again. TRAFFIC, Kuala Lumpur, Malaysia. Available from https://www.traffic.org/publications/reports/slow-and-steady-the-global-footprint-of-jakartas-tortoise-and-freshwater-turtle-trade/. Accessed 22 Jan 2019

Natusch DJ, Lyons JA (2012) Exploited for pets: the harvest and trade of amphibians and reptiles from Indonesian New Guinea. Biodivers Conserv 21:2899–2911

Natusch DJD, Lyons JA, Mumpuni RA, Shine R (2016) Jungle giants: assessing sustainable harvesting in a difficult-to-survey species (Python reticulatus). PLoS One 11:e0158397

Natusch DJ, Carter JF, Aust PW, Van Tri N, Tinggi U, Riyanto A, Lyons JA (2017) Serpent's source: determining the source and geographic origin of traded python skins using isotopic and elemental markers. Biol Conserv 209:406–414

Nazarea V, Rhoades R, Bontoyan E, Flora G (1998) Defining indicators which make sense to local people: intra-cultural variation in perceptions of natural resources. Hum Organ 57:159–170

Nekaris KAI, Shepherd CR, Starr CR, Nijman V (2010) Exploring cultural drivers for wildlife trade via an ethnoprimatological approach: a case study of slender and slow lorises (Loris and Nycticebus) in south and Southeast Asia. Am J Primatol 72:877–886

Ngokwey N (1995) Home remedies and doctors' remedies in Feira (Brazil). Soc Sci Med 40:1141–1153

Nguyen TT, Van Nguyen T, Ziegler T, Pasmans F, Martel A (2017) Trade in wild anurans vectors the urodelan pathogen *Batrachochytrium salamandrivorans* into Europe. Amphibia-Reptilia 38:554–556

Nichols JD. 1976. Simulation of a commercially harvested alligator population in Louisiana

Nijman V, Bergin D (2017) Reptiles traded in markets for medicinal purposes in contemporary Morocco. Contrib Zool 86:39–50

Nijman V, Shepherd CR (2009) Wildlife trade from ASEAN to the EU: issues with the trade in captive-bred reptiles from Indonesia. In: TRAFFIC Europe Report for the European Commission. Brussels, Belgium

Nijman V, Shepherd CR (2014) Analysis of a decade of trade of tortoises and freshwater turtles in Bangkok, Thailand. Biodivers Conserv 24:309–318

Nijman V, Shepherd CR (2015) Adding up the numbers: an investigation into commercial breeding of tokay geckos in Indonesia. TRAFFIC Southeast Asia, Petaling Jaya, Selangor

Nijman V, Shepherd CR, Mumpuni SKL (2012) Over-exploitation and illegal trade of reptiles in Indonesia. Herpetol J 22:83–89

O'Brien S, Emahalala ER, Beard V, Rakotondrainy RM, Reid A, Raharisoa V, Coulson T (2003) Decline of the Madagascar radiated tortoise *Geochelone radiata* due to overexploitation. Oryx 37:338–343

Ouboter PE (1989) The impact of an Indian Village on caimans. Crocodile Spec Group Newsl 8:28

Parusnath S, Little IT, Cunningham MJ, Jansen R, Alexander GJ (2017) The desolation of Smaug: the human-driven decline of the Sungazer lizard *(Smaug giganteus)*. J Nat Conserv 36:48–57

Pasachnik SA, Dannof-Burg JA, Antunez EE, Corneil JP (2014) Local knowledge and use of the Valle de Aguán spiny-tailed iguana, *Ctenosaura melanosterna*, in Honduras. Herpetol Conserv Biol 9:436–447

Platenberg RJ (2007) Impacts of introduced species on an island ecosystem: non-native reptiles and amphibians in the US Virgin Islands

Pope CH (1961) The giant snakes: the natural history of the *boa constrictor*, the anaconda, and the largest pythons, including comparative facts about other snakes and basic information on reptiles in general. Random House, New York

Poole CM, Shepherd CR (2017) Shades of grey: the legal trade in CITES-listed birds in Singapore, notably the globally threatened African grey parrot Psittacus erithacus. Oryx 51(3):411–417

Rasheed T (2013) Illegal reptilian trade in Chagai Desert, Pakistan: a narrative of bad governance and weakening of traditional institutions. Traffic Bull 25:8

Read BE (1934) Chinese materia medica: dragons and snakes. Peking Nat Hist Bull 4:297–357

Reed RN (2005) An ecological risk assessment of nonnative boas and pythons as potentially invasive species in the United States. Risk Anal 25:753–766

Reed RN, Willson JD, Rodda GH, Dorcas ME (2012) Ecological correlates of invasion impact for Burmese pythons in Florida. Integr Zool 7:254–270

Rice KG, Percival HF, Woodward AR, Jennings ML (1999) Effects of egg and hatchling harvest on American alligators in Florida. J Wildl Manag 63:1193–1200

Robinson JE, Sinovas P (2018) Challenges of analyzing the global trade in CITES-listed wildlife. Conserv Biol 32(5):1203–1206

Robinson JE, Griffiths RA, John FAS, Roberts DL (2015) Dynamics of the global trade in live reptiles: shifting trends in production and consequences for sustainability. Biol Conserv 184:42–50

Robinson JE, Fraser IM, John FAS, Randrianantoandro JC, Andriantsimanarilafy RR, Razafimanahaka JH, Griffiths RA, Roberts DL (2018a) Wildlife supply chains in Madagascar from local collection to global export. Biol Conserv 226:144–152

Robinson JE, Griffiths RA, Fraser IM, Raharimalala J, Roberts DL, St John FA (2018b) Supplying the wildlife trade as a livelihood strategy in a biodiversity hotspot. Ecol Soc 23:13

Rodda GH, Fritts TH (1992) The impact of the introduction of the colubrid snake *Boiga irregularis* on Guam's lizards. J Herpetol 26:166–174

Schlaepfer MA, Hoover C, Dodd CK (2005) Challenges in evaluating the impact of the trade in amphibians and reptiles on wild populations. Bioscience 55:256–264

Schneider L, Ferrara CR, Vogt RC, Burger J (2011) History of turtle exploitation and management techniques to conserve turtles in the Rio Negro Basin of the Brazilian Amazon. Chelonian Conserv Biol 10:149–157

Sharma DS (1999) Tortoise and freshwater turtle trade and utilisation in peninsular Malaysia, vol 39. TRAFFIC Southeast Asia, Petaling Jaya

Shepherd CR, Ibarrondo B (2005) The trade of the roti island snake-necked turtle *Chelodina mccordi*, Indonesia. TRAFFIC Southeast Asia, Petaling Jaya

Shepherd CR, Janssen J, Noseworthy J (2019) A case for listing the Union Island gecko *Gonatodes daudini* in the appendices of CITES. Global Ecol Conserv 17: e00549

Sheu HY (1977) Research on Chinese animal crude drugs. New Medical Drug, Taipei

Shiau T-W, Hou P-C, Wu S-H, Tu M-C (2006) A survey on alien pet reptiles in Taiwan. Taiwania 51:71–80

Smith RJ, Walpole MJ (2005) Should conservationists pay more attention to corruption? Oryx 39:251–256

Sofowora A (1996) Research on medicinal plants and traditional medicine in Africa. J Altern Complement Med 2:365–372

Stephen C, Pasachnik S, Reuter A, Mosig P, Ruyle L, Fitzgerald L (2011) Survey of status, trade, and exploitation of Central American iguanas. Available from https://www.researchgate.net/publication/271443622_Survey_of_Status_Trade_and_Exploitation_of_Central_American_Iguanas

Suarez E, Morales M, Cueva R, Bucheli VU, Zapata-Ríos G, Toral E, Torres J, Prado W, Olalla JV (2009) Oil industry, wild meat trade and roads: indirect effects of oil extraction activities in a protected area in North-Eastern Ecuador. Anim Conserv 12:364–373

Swingland IR, Klemens MW (1989) The conservation biology of tortoises. IUCN, Gland

Symes WS, Edwards DP, Miettinen J, Rheindt FE, Carrasco LR (2018) Combined impacts of deforestation and wildlife trade on tropical biodiversity are severely underestimated. Nat Commun 9:4052

Snyder NF, Derrickson SR, Beissinger SR, Wiley JW, Smith TB, Toone WD, Miller B (1996) Limitations of captive breeding in endangered species recovery. Conserv Biol 10:338–348

Thorbjarnarson JB (1988) The status and ecology of the American crocodile in Haiti. Bulletin of the Florida State Museum. Biol Sci (USA) 33:1–86

Thorbjarnarson JB (1991) An analysis of the spectacled caiman (*Caiman crocodilus*) harvest program in Venezuela. Neotropical wildlife use and conservation. University of Chicago Press, Chicago, pp 217–235

Todd BD, Willson JD, Gibbons JW (2010) The global status of reptiles and causes of their decline. Ecotoxicology of amphibians and reptiles, 2nd edn. CRC Press, Boca Raton

Uetz P, Etzold T (1996) The EMBL/EBI reptile database. Herpetol Rev 27:174–175

UNEP-WCMC (2013) A guide to using the CITES trade database. Version 8. Cambridge, United Kingdom

Van Dijk PP, Palasuwan T (2000) Conservation status, trade, and management of tortoises and freshwater turtles in Thailand. Chelonian Res Monogr 2:137–144

Van Dijk PP, Stuart BL, Rhodin AG (2000) Asian turtle trade: proceedings of a workshop on conservation and trade of freshwater turtles and tortoises in Asia–Phnom Penh, Cambodia, 1–4 December 1999. Chelonian Research Foundation, Lunenburg, MA

van Schingen M, Ziegler T, Boner M, Streit B, Nguyen TQ, Crook V, Ziegler S (2016) Can isotope markers differentiate between wild and captive reptile populations? A case study based on crocodile lizards (Shinisaurus crocodilurus) from Vietnam. Glob Ecol Conserv 6:232–241

Van Vliet N, Nasi R (2008) Hunting for livelihood in Northeast Gabon: patterns, evolution, and sustainability. Ecol Soc 13:33

Vinke T, Vinke S (2015) Can illegal be legal within the European Union? Schildkröten im Fokus 1:1–6

Vuillaume B, Valette V, Lepais O, Grandjean F, Breuil M (2015) Genetic evidence of hybridization between the endangered native species *Iguana delicatissima* and the invasive *Iguana iguana* (Reptilia, Iguanidae) in the Lesser Antilles: management implications. PLoS One 10:e0127575

Webb GJW, Manolis SC, Whitehead PJ (1987) The management of crocodiles in the northern territory of Australia. Wildlife management: crocodiles and alligators'. In: Webb GJW, Manolis SC, Whitehead PJ (eds) . Surrey Beatty & Sons, Sydney, pp 107–124

Whiting MJ, Williams VL, Hibbitts TJ (2013) Animals traded for traditional medicine at the faraday market in South Africa: species diversity and conservation implications. In: Animals in traditional folk medicine. Springer, Berlin, pp 421–473

Wiewandt TA (1977) Ecology, behavior, and management of the Mona Island ground iguana, *Cyclura stejnegeri*. PhD thesis. Cornell University USA

Willson JD, Dorcas ME, Snow RW (2011) Identifying plausible scenarios for the establishment of invasive Burmese pythons (*Python molurus*) in southern Florida. Biol Invasions 13:1493–1504

Wu Y-S (2015) Not beating around the bush: understanding China and South Africa's illegal wildlife trade

You L, Wang G (2000) 65 Chinese traditional and herbal medicines with effect to blood glucose. Chinese journal of information on traditional. Chin Med 7:32–34

Zhou Z, Jiang Z (2004) International trade status and crisis for snake species in China. Conserv Biol 18:1386–1394

Ziegler S (2016) Monitoring lizards part 2 – pilot study testing the applicability of isotopic markers in scales to discriminate between captive and wild individuals. TRAFFIC, United Kingdom

Zoer PR (2012) The bush meat and conservation status of the African dwarf crocodile *Osteolaemus tetraspis*. PhD thesis. University of Pretoria

RhODIS® (The Rhinoceros DNA Index System): The Application of Simple Forensic and Genetic Tools Help Conserve African Rhinoceros

Cindy Kim Harper

Abstract

All rhinoceros populations are under severe threat due to habitat encroachment and illegal hunting for their horns. Rhinoceros horn has been used for centuries in Traditional Chinese Medicine (TCM) and more recently to produce high-value items such as jewelry. Rhinoceros are killed for their horns which are trafficked through the various levels of crime syndicates to their final destination in primarily the Asian consumer countries. The demand for rhinoceros horn has made poaching of rhinoceros a highly profitable and nearly risk-free criminal activity due to difficulties in connecting confiscated horns to a specific crime scene. Rhinoceros poaching has increased in all African rhinoceros range states including South Africa, which saw a rise in poaching from 2008, reaching a peak in 2014. Rhinoceros poaching has become an increasingly sophisticated activity involving organized crime syndicates that operate across international borders. The development of an extraction method to obtain nuclear DNA from rhinoceros horn was a first step in producing the DNA evidence that links the horn to a specific crime scene. RhODIS® (The Rhinoceros DNA Index System) was launched in 2010 and the system includes an STR (Short Tandem Repeat)-based reference database for African rhinoceros, representing multiple populations across almost all the range of African black and white rhinoceros. Using data from the RhODIS® database in rhinoceros poaching and rhinoceros horn seizure cases has affirmed the value of the RhODIS® approach in criminal prosecutions, through the successful prosecution, conviction, and sentencing of suspects in South Africa and other countries. The same DNA data may additionally be used to support the management of rhinoceros populations, by evaluating the genetic viability and selection of individuals to ensure that genetic diversity of the remaining, increasingly isolated and fragmented populations, is maintained. Genetic information captured within a comprehensive database, therefore, directly supports the conservation of the species by enhancing both law enforcement and reproductive efficiency to offset losses due to poaching.

Keywords

Rhinoceros · DNA profiling · RhODIS® · Poaching · DNA database · Rhinoceros horn · DNA matches · GENETIC variability · Population management

C. K. Harper (✉)
Veterinary Genetics Laboratory, Faculty of Veterinary Science, University of Pretoria, Pretoria, South Africa
e-mail: cindy.harper@up.ac.za

S. C. Underkoffler, H. R. Adams (eds.), *Wildlife Biodiversity Conservation*,
https://doi.org/10.1007/978-3-030-64682-0_18

Introduction: The Beginning

The story of RhODIS® (The Rhinoceros DNA Index System) began with a chance meeting and a poster. In early 2010, an investigator from the Environmental Crime Investigation section of the Kruger National Park in South Africa visited the primary author at the Veterinary Genetics Laboratory at the Faculty of Veterinary Science in Pretoria. The purpose of his visit was to discuss the use of DNA testing as evidence in an increasing number of rhinoceros poaching cases in the park. The author had just finalized a project that included extracting DNA from rhinoceros horn and obtaining an STR-based DNA profile from the extract. The preliminary data was summarized in a poster which was presented in 2009. Upon reading the poster, the investigator promptly removed it from the wall at reception, rolling it up and putting it under his arm, stating that it was exactly what he was looking for. The method described on the poster provided a simple and practical tool to obtain powerful data linking horn, or a piece of horn (whether removed from a carcass or a live rhinoceros) back to its origin, thereby providing a link between the person in possession of the horn and the original crime scene. Such information could assist investigators in linking poachers and poaching syndicates back to a specific or multiple rhinoceros poaching scenes, tracing of movements, the connecting of criminals, and the addition of illegal hunting to their criminal charges. Because rhinoceros horn was traditionally believed to be a clump of hair, no attempt had been made prior to this study to obtain a nuclear DNA-based individual profile from this material in order to link the horn to the animal that it originated from.

Inter-laboratory and between platform STR profile standardization has been very limited in terms of animals, and particularly wild animals. Even when specific species have been extensively DNA profiled, these databases may not represent all isolated populations (Johnson et al. 2014). STR reference databases have been developed for various animal species mainly in domesticated animals such as cats (Menotti-Raymond et al. 2005), dogs (Kun et al. 2013; Wictum et al. 2013), cattle (Van De Goor et al. 2009), horses (Chen et al. 2010), and pigs (Lin et al. 2014), but is limited in wild species. Newer technologies, including next-generation sequencing methods (Jäger et al. 2017) are evolving and becoming more cost-effective and useful across a broader spectrum of users, but STR-based technology will remain a highly effective, widely used technology for the purpose of forensic testing for the foreseeable future with large databases of human and animal STR profiles available to various research and investigative authorities (Butler 2015; Johnson et al. 2014).

The use of dinucleotide repeats in forensic testing is generally not recommended due to increased stutter in these loci and heterozygote imbalance that can make allele calling difficult. The Veterinary Genetics Laboratory had been using high-throughput, STR-based DNA profiling systems for various animal species for many years and had been involved in standardizing these tests for mainly domestic species. The International Society of Animal Genetics (ISAG), of which the laboratory has been a member since its inception, provides a mechanism for standardization for domestic animal STR data sharing through a biennial comparison testing system (http://www.isag.us/). The application of this technology to rhinoceros and rhinoceros horn, and the setting up of a robust and efficient system to DNA profile large numbers of samples from these animals, was therefore merely an expansion of existing services and expertise.

Individual identification and assignment in human and animal forensics currently use mainly STR loci and more recently SNP (Single Nucleotide Polymorphism) loci. RhODIS® provides such an STR-based reference database for African rhinoceros and is distinctive in that it represents multiple populations across almost all the range of African rhinoceros. The use of a single reference database of STR loci requires extensive sharing of data, control samples to provide reference genotypes for standardization, and regular inter-laboratory proficiency testing.

In 2010, SAPS (The South African Police Service) submitted the first case of horns seized from OR Tambo International Airport, in Johannesburg, South Africa to the laboratory to identify a possible link to rhinoceros illegally killed on a private reserve in South Africa. A total of seven horn samples were submitted, and two of these horns were linked to two rhinoceros from a farm in the Vaalwater district in South Africa. The DNA evidence supported the conviction and sentencing of the suspect to 10 years imprisonment. Following this case, the number of samples from rhinoceros crime cases submitted to the laboratory increased, and with the support of SANParks (South African National Parks) and the provincial wildlife authorities in South Africa, a decision was made to support the development of a single rhinoceros database. This database was named RhODIS®, an acronym for the Rhinoceros DNA Index System, based on the example of the CODIS program of the FBI in the USA. A decision was also made to increase the size of the database as rapidly as possible by sampling all live rhinoceros, rhinoceros horns, and rhinoceros carcasses from both legal hunts and illegal killings, in order to ensure the database could provide a traceability system for wildlife and enforcement authorities, as numbers of poached rhinoceros were increasing. The profiling data would also provide statistical support to matches found.

In 2012, the South African Department of Environmental Affairs (now the Department of Environment Forestry and Fisheries or DEFF) published a new set of regulations for the marking of rhinoceros and rhinoceros horns. These regulations required that samples from rhinoceros being translocated, dehorned, notched, or hunted, as well as those that died from natural mortalities and illegal hunting, must be collected in sampling kits supplied by the laboratory, and submitted for DNA profiling, followed by addition into the RhODIS® database. As a result, the number of samples submitted increased significantly, reaching 50,000 by August 2018 and exceeding 60,000 in 2019.

The application of DNA profiling as a forensic tool depends upon the statistical power of the composite genotypes to identify and differentiate individuals in the population. Rhinoceros population structure, historic contractions or expansions, migration, translocation, and population fragmentation caused by poaching and habitat reduction influence the genetic composition of the population, and thus the results of statistical analyses. Population data of African rhinoceros has been reported but not on a comprehensive scale (Anderson-Lederer et al. 2012; Guerier et al. 2012; Karsten et al. 2011; Kotzé et al. 2014; Moodley et al. 2017; Van Coeverden de Groot et al. 2011). The partitioning of population genetic variation within and between free-ranging white and black rhinoceros populations influences the forensic match probability of DNA profiles, and therefore the effectiveness of DNA evidence in rhinoceros criminal prosecutions, as well as decisions related to the genetic management of the extant populations. A brief summary of the history of the black and white rhinoceros is therefore provided in the following section.

Rhinoceros: The Victim

The African rhinoceros is a charismatic megafauna that has in recent centuries endured extreme pressure to survive, beginning with the historical tradition of hunting during colonial times, and continuing as uncontrolled hunting even into modern times (Walker and Walker 2012). The illegal killing of rhinoceros is referred to as poaching. Rhinoceros poachers are financed through organized criminal syndicates (Milliken and Shaw 2012). Rhinoceros are killed for their horns which are removed from the carcass and are trafficked through the various levels of crime syndicates to their final destination, mainly the Asian consumer countries.

There are five remaining rhinoceros species. Three Asian rhinoceros species including the *Rhinoceros unicornis* (Greater One-horned rhinoceros) that occurs in India and Nepal (approximately 3000 remain), the *Rhinoceros sondaicus* (Javan rhinoceros) that is found in a single population in the Ujung Kulon National

Park in Java, Indonesia, and is critically endangered (approximately 60 remain) and the *Dicerorhinus sumatrensis* (Sumatran rhinoceros) that occurs on the islands of Borneo and Sumatra and is critically endangered (approximately 100 remain).

Africa has two species of rhinoceros, the black rhinoceros (*Diceros bicornis*) and the white rhinoceros (*Ceratotherium simum*) that show clear morphological distinction. The African Rhinoceros Specialist Group (AfRSG) of the IUCN (International Union for Conservation of Nature) recognizes four subspecies of black rhinoceros of which *Diceros bicornis longipes* was declared extinct in 2011 (Emslie and Brooks 1999). Analyses of 883 black rhinoceros in Harper et al. (2018) support this division of the extant *Diceros bicornis* species.

These are listed with the approximate numbers remaining in 2015 given in brackets (Emslie et al. 2016);

Diceros bicornis bicornis; Also named the South-western subspecies that occurs in Namibia, the Northern Cape and Eastern Cape provinces of South Africa (2200).

Diceros bicornis minor; The South-central subspecies that occurs in Zimbabwe, Botswana, KwaZulu-Natal province of South Africa, the Kruger National Park and remaining provinces of South Africa, Malawi, Southern Tanzania, Swaziland, and Zambia (2164).

Diceros bicornis michaeli; The Eastern subspecies that occurs in Kenya, Northern Tanzania, and an extralimital population in South Africa (886).

Diceros bicornis longipes; The Western subspecies that occurred in central and western Africa, Cameroon, Chad, the Central African Republic, and Sudan and is now extinct.

By the late 1960s about 70,000 black rhinoceros remained on the African continent. By 1981 only 10,000 to 15,000 remained and by 1995 this number had decreased to 2410, the population having been decreased by over 90% in 30 years. The continental black rhinoceros population has subsequently shown a steady increase to 5055 in 2013, with 98% of black rhinoceros present in just four countries: Namibia; Kenya; Zimbabwe; and South Africa.

In 1935 the black rhinoceros population in South Africa was less than 110 animals, all of the *D.b. minor* subspecies. These animals represented 0.1% of the African continental black rhinoceros population. By 2007 the number of black rhinoceros had increased to 1488 in South Africa representing 35% of the continental numbers and included all 3 of the extant subspecies, *D.b. minor, D.b. bicornis*, and *D.b. michaeli* (Knight and Kerley 2009). South Africa is the only country in which all three remaining black rhinoceros subspecies, *D.b. minor*, *D.b. bicornis,* and *D.b. michaeli* occur. All three have shown an increase in numbers since 1992. Recently black rhinoceros have been translocated from South Africa to Rwanda (18 *D.b. michaeli* animals) and Chad (6 *D.b. minor* animals), increasing the number of African rhinoceros range states from 11 (Botswana, Kenya, Malawi, Mozambique, Namibia, South Africa, Swaziland, Uganda, Tanzania, Zambia, and Zimbabwe) to 13. Figure 1 is a map indicating the 13 African rhinoceros range states.

Black rhinoceros populations have remained in small pockets of isolated animals following their widespread decimation throughout their previous range on the African continent. The rapid decline in numbers of both the white and black rhinoceros resulted in a significant loss of genetic diversity (Moodley et al. 2017). The survival of the black rhinoceros in South Africa, similar to the white rhinoceros, is due essentially to the pioneering work of the Natal Parks Board. The last black rhinoceros in the Cape Province was shot near Addo in 1853, the last in the Free State in 1842, and the last living black rhinoceros was seen in the area of the now Kruger Park in 1936. Less than 100 black rhinoceros survived in the area of the Hluhluwe, Umfolozi, and Mkuzi Game Reserves in 1930. In 1971, 20 black rhinoceros (10 males and 10 females) were moved from Hluhluwe to the Kruger National Park, followed by 12 animals in 1972 from the Zambesi Valley in Zimbabwe (then Rhodesia) in order to reintroduce these animals into the area. These

Fig. 1 Map of Africa with the 13 rhinoceros range states colored

animals were all classified as the *D.b. minor* subspecies.

Black rhinoceros were returned to the Addo Elephant National Park in 1961 and 1962 from the Kiboko region in Kenya. These animals were of the Eastern black rhinoceros (*D.b. michaeli*) subspecies, and this reintroduction represented an extralimital movement of animals into a region in which the specific subspecies had not previously occurred. In 1979 rhinoceros were introduced from KwaZulu-Natal (Zululand) into the Great Fish River reserve in the Eastern Cape, an area that fell between the former range of the Cape rhinoceros (*D.b. bicornis*) to the west and the Zululand rhinoceros (*D.b. minor*) to the east and could thus have formerly been an area in which intergradation of the subspecies was expected (Hall-Martin 2009).

The black rhinoceros in Namibia were under similar threat by 1966 with only about 90 remaining in the Northwest corner of the country. From 1970 to 1972, an effort was made to protect these animals and most were moved to the Etosha National Park, with about 20 remaining in the Kaokoveld region and 30 in Damaraland (Hall-Martin 2009).

Historically the range of the black rhinoceros thus extended from the Cape, through southern Africa, Namibia, Angola, Botswana, Zimbabwe, Mozambique, Zambia, Malawi, Uganda, Tanzania, Kenya, Somalia, Ethiopia, Sudan, Chad, Central African Republic, and North-Eastern Nigeria along the southern edge of the Sahara (Rookmaaker and Antoine 2012). The subspecies ranges most likely overlapped with individuals representing a gradient of phenotypic and adaptive characteristics occurring at the interface of geographically defined areas. The severe and rapid contraction of the black rhinoceros ranges created pockets of isolated populations where nominal to no natural exchange of genetic material occurred (Karsten et al. 2011; Kotzé et al. 2014; Muya et al. 2011; Van Coeverden de Groot et al. 2011).

The southern white rhinoceros (*Ceratotherium simum simum*) is a grazer and its numbers are believed to have dropped to below 50 in the early 1900s with only a few animals remaining in the Hluhluwe-Imfolozi game reserve, where these animals were protected from being hunted to extinction due to the presence of the tsetse fly in the area. Extensive efforts to increase the numbers of the species in the 1960s by the Natal Parks Board were extremely successful, with numbers of southern white rhinoceros at around 20,000 in 2012 (Player 2013). The northern white rhinoceros (*C.s. cottoni*) numbers have dwindled to only two, remaining in a small population under intense protection in the Ol Pejeta Conservancy in Kenya (Howard 2015). Advanced reproductive technologies are being attempted to obtain embryos from the remaining two females, which are now unable to reproduce naturally, and sperm frozen from the last male that died in 2018 (Hildebrandt et al. 2018).

Rhinoceros Horn: The Product

In traditional Chinese Medicine (TCM), rhinoceros horn (Xi Jiao/Cornu Rhinoceri) is described as a cold drug, used to cool the blood and thus reduce fevers, stop bleeding and counteract toxins. Rhinoceros horn has been valued in Chinese tradition as an ingredient to increase longevity as well as for its esthetic value as drinking cups and other ornamental items since the seventh century AD. The use of rhinoceros horn in TCM was banned by the Chinese government in 1993 and rhinoceros horn was removed from the Chinese Pharmacopeia. A more recent and nontraditional claim for the use of rhinoceros horn was its use as a cure for cancer. This rumor, started by an unnamed Vietnamese politician, is thought to have been one of the contributing factors in the spike in rhinoceros poaching. Rhinoceros horn was never used as an aphrodisiac in TCM. This belief appears to have originated in the West and may have its origins in the influence of Indian culture on the myths surrounding the rhinoceros and its horn (Patton 2011).

Yemen used to be a significant importer of rhinoceros horn, as the horn was used for the handles of daggers known as jambiyas. These daggers were given to Yemeni boys at the age of 12 as a sign of manhood. The imports of rhinoceros horn were banned in Yemen in 1982 and the economic decline in this country also reduced its influence on the decline of the rhinoceros population significantly.

As the number of rhinoceros decreases, the rarity value of the horn increases. In the early 1990s, the price of rhinoceros horn was about \$250–\$500 per kilogram but this has increased to over \$60,000 per kilogram on the black market, making it the most valuable illegally traded commodity in the world (Caulderwood 2014). Vietnam and China are the dominant end-user markets of rhinoceros horn (Crosta et al. 2017, Emslie et al. 2016, Moneron et al. 2017). The insatiable demand is driven by increasing wealth and disposable income in consumer countries. Along with the increase in wealth, there has been an increasing demand for rarities, including carved items such as cups, bangles, beads, and ornaments. Due to the high value placed on horn, even the powder generated as a by-product of these items is sold at lower prices for medicinal and social use (Crosta et al. 2017; Emslie et al. 2016; Moneron et al. 2017). The cancer rate continues to rise in Vietnam, possibly due to increasingly unhealthy lifestyles and the effects of pollution. This is believed to be one of the reasons for the increasing use of traditional and non-western medicine in Vietnam and other countries in the region.

Some consider TCM to be of decreasing importance in the use of rhinoceros horn, and believe the use of horn as a lifestyle enhancing drug, which includes adding powdered horn to alcoholic drinks and serving it to special guests, stimulates continued demand. It is believed by many to cure and prevent hangovers. It is also powdered in special ceramic bowls and served by adding it to water (Karl Ammann, personal communication, 2014). Rhinoceros horn is also considered a status symbol and a rare item in a country where luxury goods are sought after and provided as gifts to secure political favor (Guilford 2013).

Traceability of the carved products and powder is more complicated since horns are smuggled in pieces, and thus cannot reliably be identified physically. Additionally, microchips placed in the horn for tracing purposes are often lost in the process of cutting and shaping horn products. DNA profiling remains the only method to trace the origin of these items to a specific animal or animals, providing evidence of source and trade routes. Emslie et al. (2016) state that the origin of horn seizures would improve knowledge of source and trade routes if these could be sampled and identified more extensively and this information was made publicly available.

The carving of horn into various trinkets does not only occur in the destination countries but is also done in the supply countries such as South Africa. These re-worked items are extremely difficult to identify as originating from rhinoceros horn if a person is not familiar with the products. In many countries, charges of illegal possession of rhinoceros horn cannot be made without DNA species verification of seized horn items.

The cutting of horns into sections provide opportunities to export the horn in innovative ways that make it more difficult for authorities to identify the material. Horn has been illegally shipped in wax ornaments, in chocolate wrappers, and inside books with the pages cut out. Blenders lined with sanding paper, chisels, drills, and other tools used to cut, shape, and rework horn into various items, including beads and bangles, have been confiscated from smugglers. Beads must be identified individually, as they may come from different horns, thus several animals may be represented by a single rhinoceros horn bracelet. When presenting this evidence, we always report the number of animals involved based on the number of observed DNA profiles.

Hieronymus et al. (2006) used computed tomography (CT scanning), gross observation of sectioned horn, and light microscopy to examine rhinoceros horn structure. Rhinoceros horns lack a bony core and are anchored to the dermis through keratinized cells that form tubular cellular structures and grow from a generative layer of epidermis covering dermal papillae. All horn growth takes place from the base of the horn, and damage to the growth plate causes abnormalities in the structure of the horn. The structure of the horn has been compared to ungulate hoof walls, baleen plates, and cockatoo bills. Rhinoceros horn continues to grow at a rate of approximately 5–6 cm per year and can be harvested without harm to the animal as long as the cut is made above the conically shaped growth plate at the base. The horn matrix consists of calcium salts and melanin. Keratins are damaged by prolonged exposure to UV light, and the melanin's role appears to be to protect the horn from such damage. The center of the horn has an area of increased melanization and calcification, resulting in a greater resistance to damage and wear, and also contributing to the conical shape of the horn (Hieronymus et al. 2006).

A study by Boy et al. (2015) shows the horn originating from elongated dermal papillae and consisting of tightly packed corneocytes from the base to the tip of the horn. The corneocytes show vertical and horizontal orientation and no hollow spaces or intercellular matrix are present. This study also shows the epidermal base membrane dipping into the underlying connective tissue at the edges of the horn base, resulting in an inward slant to the outer dermal papillae which contributes to the conical shape of the horn (Boy et al. 2015). The presence of corneocytes throughout the horn structure supports the extraction of DNA and the ability to obtain a nuclear DNA profile from any part of the horn.

Poaching: The Problem

The demand for rhinoceros horn has made exploitation of rhinoceros for their horns highly profitable and nearly risk-free due to difficulties in connecting confiscated horns to a specific crime scene and routine imposition of very light sentencing upon conviction (Milliken and Shaw 2012). Rhinoceros horn is considered one of the most valuable illegally trafficked wildlife items, with a per weight value more than gold.

Poaching in Zimbabwe increased following independence in 1980. Following concerted efforts to consolidate rhinos into Intensive Protection Zones and widespread dehorning, poaching decreased and remained relatively low in Zimbabwe and South Africa between 1994 and 2002. Between 1990 and 2005 an average of 14 rhinoceros were poached in South Africa per year. Poaching in Zimbabwe started to increase again in 2003, driven by illegal trade dynamics in South Africa that included legal internal trade of horns, pseudo-hunting, and illegal hunting in Zimbabwe due to political destabilizing factors that undermined the security of wildlife in the area. At this time private game farmers in South Africa set up safari hunting operations in Zimbabwe that were linked to rhinoceros poaching. These factors drove up horn prices until the poaching spilled over into neighboring South Africa and eventually Namibia as well (Raoul du Toit, personal communication, 2017). In February 2009 a moratorium prohibiting the legal sale of rhinoceros horn was implemented in South Africa and this, in addition to the factors already mentioned, appeared to trigger the dramatic increase in poaching levels in South Africa after 2008 (Milliken and Shaw 2012).

The most extensive losses have occurred in the Kruger National Park with the greatest threat coming from its eastern border with Mozambique (Milliken and Shaw 2012). An increase in poaching across the African continent has caused concern for the survival of the species, with more than 7000 animals killed in the last 10 years in the African rhinoceros range states despite significant security measures to protect the animals (Moneron et al. 2017).

In 2008, South Africa experienced a steady increase in the numbers of rhinoceros poached, when 83 animals were poached, increasing to over 1000 animals poached each year from 2013 to 2017. In 2014 rhinoceros poaching in South Africa peaked at 1215 animals and has shown a steady decline since, to only 594 in 2019. The decrease is believed to be primarily a result of anti-poaching and law enforcement efforts. This decrease has also been attributed to a relative decrease in the total number of animals in the population due to poaching and droughts in the country. However, as the overall value of horn has not diminished, poaching remains a widespread and serious threat to the African species across the continent, including other rhinoceros range states, with decreases in one region inevitably accompanied by increases in another.

Rhinoceros poaching has become an increasingly sophisticated activity involving organized crime syndicates that operate across international borders. Rhinoceros horn trade syndicates are believed to be divided into five levels of operation, with the poacher at level 1. Level 2 represents the local buyer, courier, or organizer who supplies the horn to level 3, the national buyer, courier, or exporter. The local exporter connects with level 4, the international courier or buyer who sells to the international buyer or end consumer at level 5. The national buyer, importer, and international buyer are mainly of Asian nationality; however, horns have also been exported to Europe and the USA. Poaching operations have even included using local sex workers in pseudo-hunting operations and recruiting foreign hunters to obtain horns for the illegal market (Milliken and Shaw 2012).

Smuggling of horn by syndicates is highly organized, adaptive, and utilizes various methods of travel and routes, including hand luggage, shipping containers, courier freight, and postal services. The relative ease of cross-border movement is facilitated by high levels of corruption, lack of cooperation between enforcement and border control agencies, and delayed response times in transit countries (Moneron et al. 2017).

RhODIS®: One Part of the Solution

The development of an extraction method to obtain DNA from rhinoceros horn was a first step in producing the DNA evidence that links the horn to a specific crime scene. Underpinning the application of this information in the prosecution of rhinoceros crime was the need for: (1) an extensive and representative rhinoceros DNA profile database; (2) a sampling methodology that follows the chain of custody principles, and (3) statistical support of DNA matches in cases.

The same DNA data can be used to support the management of rhinoceros populations by evaluating the genetic viability and selection of individuals to ensure the diversity of the remaining, increasingly isolated and fragmented populations, is maintained. The genetic information captured within a comprehensive database, therefore, supports the conservation of the species by enhancing both law enforcement and reproductive efficiency to offset losses due to poaching.

The Department of Environmental Affairs Forestry and Fisheries (DEFF) in South Africa implemented a ban on internal trade in rhinoceros horn through a moratorium in 2009. The moratorium was lifted by the Constitutional Court in April 2017, after lengthy litigation by private rhinoceros owners, and internal legal trade was initiated shortly after (Reuters 2017). The changes in legislation allowed for the legal trade of rhinoceros horn in South Africa under strictly controlled conditions. DNA-based identification of the existing rhinoceros population in the country and the extensive rhinoceros horn stockpile forms an integral part of managing the legal trade. Currently, legislation requires that samples be collected from each animal when sedated for treatment, during translocation, or dehorning. Additionally, all horns must be sampled, photographed, and microchipped. RhODIS® has produced sampling kits that also follow forensic principles to collect routine samples, to ensure the integrity of the system and database derived from such samples. Data is currently stored on RhODIS® and forms part of the national rhinoceros database managed by DEFF in South Africa. The number of individual samples, including horn, blood, tissue and hair samples, on RhODIS® exceeded 60,000 in 2019.

All Rhinocerotidae are listed as Appendix I (threatened by extinction), except *Ceratotherium simum simum* in South Africa and Swaziland, which are listed in Appendix II (controlled trade allowed) by CITES (The Convention on International Trade in Endangered Species of Wild Fauna and Flora). The legal shipment of samples, rhinoceros, and rhinoceros horns between CITES signatory countries is governed by a permitting system issued by the CITES authority in the destination country and the country of origin. This has made the comparison of samples from seizures of horns from Asian transit and consumer countries with the RhODIS® data quite difficult. A few international seizures have been sampled, and in most cases, some or all the horns could be traced to specific crime scenes in one of the rhinoceros range states. While previously the extraction of mitochondrial DNA from rhinoceros horn could be used to identify the species of origin of the horn (Hsing-Mei et al. 2003), the RhODIS® system, through the application of simple and existing techniques, could identify a sample as originating from a specific rhinoceros, whether the sample was obtained from a white or black rhinoceros, and could provide the gender of the animal from which the sample originated.

RhODIS®: Sampling System

RhODIS® sampling kits were developed in collaboration with the Environmental Crime Investigation Unit (ECI) of South African National Parks (SANParks) and the South African Police Services (SAPS), and the kits were launched in 2011 when 1000 were officially handed to SANParks. The kits were developed in order to ensure that chain of custody (CoC) requirements would be fulfilled during both the routine collection of samples from live animals and the collection of samples during the investigation of a crime scene from a rhinoceros carcass. By October 2017 more than 60,000 kits had been distributed throughout Southern Africa and other African rhinoceros range states.

Several training courses followed on the use of the RhODIS® DNA sampling kits, particularly in rhinoceros crime scene investigation. These courses included wildlife and law enforcement authorities from Namibia, Swaziland, Kenya, India, and South Africa. Three kits were developed: a horn kit to sample rhinoceros horns; a routine kit to sample live and hunted rhinoceros; and a forensic kit to sample from rhinoceros that had been illegally killed. An instructional video showing the use of the kits and DNA sampling as part of the crime scene investigation was developed and launched. The kits were reduced to two types in 2017—the horn sampling kit, and an animal sampling kit that could be used for routine and forensic sampling. The kits are adapted to the needs of individual clients and packed and distributed by the laboratory. Figure 2 shows a routine RhODIS® sampling kit being used during a training course.

The sampling kits contain all the sample collection material and containers required for collecting a full set of DNA samples from an animal. Needles, syringes, EDTA vacutainer blood collection tubes, gloves, and disposable scalpels are commercially obtained and presterilized. The sample bottles, drill bits, plastic containers, and bleeding shoulders are decontaminated, and all packing is done in a clean and separate area. Each kit includes a form in a sealable plastic pouch that must be completed so that the data can be entered into the RhODIS® database (if not collected using the eRhODIS® app [see below for description]). The form also includes the instructions for use of the kit and shipment information to the laboratory. The kits are sent in a sealed forensic evidence bag with a unique identification number, which is also applied as a barcoded label to all containers in the kit, to link them to a specific kit. Each kit contains a second unsealed forensic evidence bag with related number that must be used to return the samples and sampling items. Samples must be sealed at the side of the animal or at the crime scene. All steps in sampling at the crime scene and receipt of the kit in the laboratory are photographed. A drill bit that is sharp and decontaminated is included in the kits to collect horn material from the rhinoceros when providing a hole for the insertion of a microchip into the horn.

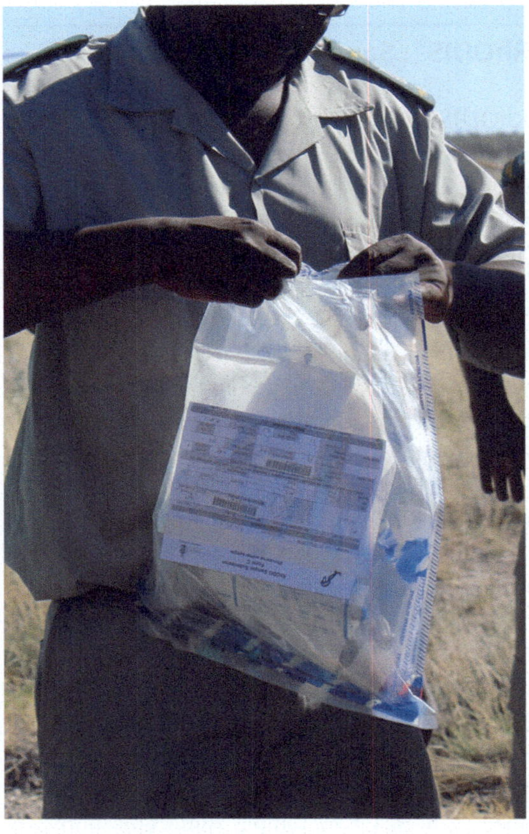

Fig. 2 A RhODIS® animal sampling kit being used during a training course

In order to improve the accuracy and efficiency of the system, a data collection app was developed for android devices and named eRhODIS™. Donor support assisted in the distribution of initial tablets (installed with the eRhODIS app) to specific authorities and officials to assist in collecting the biodata. By August 2020 more than 31,000 eRhODIS™ submissions had been received.

The majority of samples submitted to RhODIS® are collected in RhODIS® sampling kits during the forensic investigation of poaching scenes, during routine translocation, notching, dehorning for identification, hunting, or from rhinoceros horn stockpile identification operations, according to the RhODIS® guidelines. Samples received in the laboratory are assigned individual barcoded sample numbers. Sample quality can vary from highly degraded, particularly in the case of samples from old carcasses, to highest quality blood samples from live animals.

RhODIS® also receives tools used in the execution of rhinoceros crimes, including knives, axes, pangas, and blood-stained clothing from poachers. Several vehicle carpet samples and swabs from vehicles, as well as bags and cloths used to cover the horns, have also been received, and the DNA profiles obtained from the blood and tissue recovered from these items are matched to horns and poached animals.

The tool items, horns, and carcass material are rarely received together, and usually received in separate cases, often months or years apart. Continuous matches are run against the whole database of stored profiles, and links much like CODIS are made with samples previously received. This assists investigators in linking a trafficking syndicate to one or several unrelated poaching incidents, as well as to poachers, or poachers with other poachers. Several cases have been received where horns have been cut, or the back and front horn divided into separate shipments, and DNA evidence has linked the shipments through matches between recovered items.

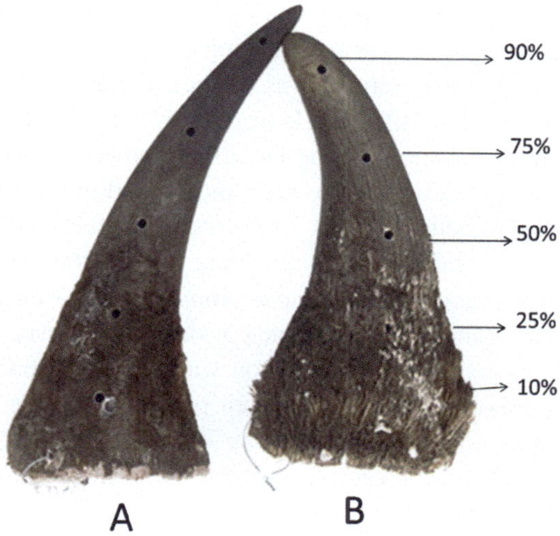

Fig. 3 Two horns, one from a black rhinoceros (**a**) and one from a southern white rhinoceros (**b**) used to validate the optimal drill position for sampling

RhODIS®: Test Methodology

DNA Extraction

When validating the extraction method from rhinoceros horn, we obtained two horns, one from a black rhinoceros (*Diceros bicornis minor*) and one from a southern white rhinoceros (*Ceratotherium simum simum*), which were donated by the Ezemvelo KZN Wildlife from two rhinoceros of approximately the same age and size. Each horn was mounted in a drill press so that the median plane of the horn was horizontal. Drillings were performed at distances of approximately 10%, 25%, 50%, 75%, and 90% from the base to the tip of the horn. The horns with drill sites are shown in Fig. 3. Up to three separate samples representing drillings at different depths were collected into separate sample tubes and extracted individually. The horn sampling method is now used as a standard method to collect samples from recovered horns, horn stockpiles, and ornaments made from horn, and is included in the RhODIS® training guidelines. Samplers are urged to keep the drill speed slow and pressure low, so that the horn is not burned and compacted, as this invariably results in a failed profiling test.

Samples collected from the center of the two horns anywhere from the base to the tip of the horn consistently produced DNA extracts with the highest concentration. Extractions further from the center of the horn were less efficient, and extractions closest to the outside surface of the horn sometimes resulted in incomplete DNA profiles. This supported the theory that horn was cellular throughout, and the melanin-rich center protected cells from environmental and specifically UV damage. Outer layers are more exposed, and thus more damaged, affecting the efficiency of the amplification. It was recommended that when collecting horn samples from live rhinoceros by drilling into the horn from the outside, only the drillings from deeper in the horn should be collected. Drillings collected from the center of the horn are easy to identify, as they have a darker brown to black color as compared to the white material from the periphery of the horn.

The amount of horn required to provide a complete DNA profile was determined by using various amounts of powdered horn in the extraction assay from 0.1 to 35 mg and evaluating the

DNA profile for amplification success and allelic dropout. Approximately 200 mg of the rhinoceros horn drilling sample was collected. The horn, being much softer than bone or ivory, was shaved off in thin slices during the drilling process. The importance of using a sharp and clean drill bit is emphasized, as the test is extremely sensitive, and even a small particle remaining on the drill bit may contaminate a sample. Horn shavings were then homogenized to a fine powder using a tissue homogenizer. The extraction was performed on approximately 20 mg of horn powder using the Prepfiler extraction kit of Life Technologies. The DNA extraction was completed on a Kingfisher 96 Magnetic Particle Processor.

This technique has consistently provided excellent results, allowing for complete DNA profiles to be obtained from pieces of rhinoceros horn about 1 mm in length, recovered from dust left in bags in which the horns were transported, or attached to tape used to lift suspected horn from a table surface. Other samples including tissue and blood are also routinely extracted using the Prepfiler kit and magnetic bead method in the laboratory, while hair is extracted using a sodium hydroxide and heat method described previously (Rudbeck and Dissing 1998). DNA has also been obtained from bone samples collected from old carcasses of animals suspected to have been poached. We use a freezer mill to powder the bone in preparation for the lysis step during the extraction process and have obtained between 30 and 50% successful amplification from these samples.

The effect of horn age on DNA amplification success was evaluated by testing horns of various ages received during routine trophy sampling. A walking stick dated 1888 which appeared to be made of a single rhinoceros horn was tested by drilling and provided a complete DNA profile. Another horn in a trophy stockpile was tested and provided a complete DNA profile. The horn was a hunting trophy from an animal shot in the Sudan in 1938 and was from a northern white rhinoceros. This indicates the potential value of horn stockpiles in providing a genetic reconstruction of rhinoceros populations that are now extinct. Authorities are therefore encouraged to ensure that samples are collected before confiscated products are destroyed, so that the genetic information can be retrieved and utilized to study genetic changes in populations that may inform future management decisions.

In order to demonstrate that DNA profiles obtained from horn could be compared to other tissues from the same animal, samples of blood and horn collected from white rhinoceros during routine capture operations in the Kruger National Park (as part of a SANParks approved project) were obtained.

The sensitivity of the DNA extraction method was tested using variable amounts of horn powder in the extraction protocol. An amount of 0.1 mg, 1 mg, 2 mg, 5 mg, 10 mg, 15 mg, 20 mg, 25 mg, and 35 mg of powdered horn from a single animal was processed. All extracts provided a complete DNA profile except the 0.1 mg of horn powder, which gave a profile with 21 of the possible 23 loci.

The recommended volume of substrate in the Prepfiler™ kit is 20 mg, which is also the recommended volume of powder for DNA extraction from rhinoceros horn, and which has provided the most consistent results. However, pieces of material resembling rhinoceros horn may be substantially smaller than this. These results demonstrate that one can generate complete DNA profiles from extremely small amounts of rhinoceros horn that may be of value in matching a horn, or part of a horn, back to the animal from which it originated.

DNA Profiling and Data Analysis

The markers used in the RhODIS® test panel were selected based on previously published research, mostly population studies. The selected markers were amplified successfully in both black and white rhinoceros species (Brown and Houlden 1999; Cunningham et al. 1999; Florescu et al. 2003; Nielsen et al. 2008; Scott et al. 2004; Scott 2008). The Zinc Finger locus found on the X and Y chromosomes was included in the same multiplex test to identify gender while obtaining the DNA profile (Peppin et al. 2010).

The RhODIS® DNA profiling test includes 23 short tandem repeats (STR), or microsatellite, markers, and a gender marker (ZF1). The markers are multiplexed in four sets, as determined to be the optimal setup. The laboratory has attempted to combine the markers into a single multiplex by redesigning the primers. However, since the test is meant to be used simultaneously on white and black rhinoceros in samples often of unknown origin, the single test often includes overlap when analyzing the electropherogram data. Genetic variability of the extant species is limited, but new alleles are often identified when an old horn is profiled, which may overlap with other markers, as their size ranges could not be anticipated. The principle of the RhODIS® test is to ensure rapid, high-throughput results that support the generation of forensic results in multiple, simultaneous cases, as well as to ensure data from live rhinoceros or rhinoceros horns is provided efficiently in a high-throughput, high-volume system. Using a set of markers that identify both the species and gender in a single test allows for such efficiency.

The STR markers used include loci originally identified in white, black, Indian (*Rhinoceros unicornis*), and Sumatran (*Dicerorhinus sumatrensis*) rhinoceros (Brown and Houlden 1999; Cunningham et al. 1999; Florescu et al. 2003; Scott et al. 2004; Scott 2008). Eighteen of the 23 markers are polymorphic, with between two and four alleles observed in white rhinoceros. Four markers (BlRh37D, DB23, IR22, and SR74) are monomorphic in the white rhinoceros but are polymorphic in the black rhinoceros. These four markers were all originally isolated from the black (BlRh37D, DB23), Indian (IR22), and Sumatran (SR74) rhinoceros. The marker 32F originally isolated from the white rhinoceros was polymorphic, with four alleles in the white rhinoceros, but was monomorphic in the black rhinoceros. In the case of SR74, the monomorphic allele in the white rhinoceros was of similar size to one of the three alleles observed in the black rhinoceros, whereas for all other monomorphic loci the size of the monomorphic allele was unique in the species in which it was monomorphic, providing a mechanism for confirming the species of origin for the sample investigated. The marker DB66 was highly polymorphic and informative in the black rhinoceros. In the white rhinoceros, however, this marker amplified four alleles but appears to have complex sequence variation that warrants further investigation through sequencing.

Published reference sequence data of the markers selected for DNA profiling in the RhODIS® panels were aligned to the genome scaffolds of the southern white rhinoceros (*C.s. simum*) (http://www.ncbi.nlm.nih.gov/genome/24631), available on GenBank. Using a reference-assisted chromosome alignment tool, Chromosomer (Tamazian et al. 2016), the rhinoceros scaffolds were aligned to the domestic horse (*Equus ferus caballus*) genome as reference (Wade et al. 2009). The horse was selected as the reference species, as it is the closest relative of the rhinoceros with a complete and annotated genome. The two X-linked markers (IR12 and SR74) were consistently hemizygous in males, and either homo- or heterozygous in females, in both black and white rhinoceros genotyped, supporting the referenced alignment of these two markers to the X-chromosome using this technique. The chromosome assignment, albeit indirectly involving two distantly related Perissodactyla species, allowed for an indication of likely chromosome linkage in detecting linkage disequilibria between STR loci.

Details of the markers, multiplexes, scaffold number, and imputed chromosome position are presented in Table 1. Loci DB66, IR22, and SR262 could not be placed on chromosomes, and locus IR10 could not be placed on a scaffold. Table 1 also shows the citations for each STR marker isolation (Brown and Houlden 1999; Cunningham et al. 1999; Florescu et al. 2003; Harper et al. 2013; Nielsen et al. 2008; Scott et al. 2004; Scott 2008) and the sequence motif of each locus in the rhinoceros species of origin. The monomorphic STR loci, used for species identification, were consistently identified in all white (BlRh37D, DB23, SR74, and IR22) and black (32F) rhinoceros samples.

The PCR master mix used in the RhODIS® system consists of 5 μl of KAPA2G Fast

Table 1 Summary of the forward and reverse primers, repeat motifs, GenBank accession numbers, reference, dye label, size range, multiplex, imputed chromosome assignment, and scaffold position of STR loci used in the RhODIS® test for genotyping of white and black rhinoceros

Locus	Forward primer (5′-3′)	Reverse primer (5′-3′)	Repeat motif	Accession	References	Dye	Size (Bp)	Multiplex	Chromosome	Scaffold
BlRh1B	GATCAGTAACACCAAAGTCC	AGTGAAGACAGAAGGATCAC	$(GT)_{13}GCA(TG)_3$	AY606078	Nielsen et al. (2008)	NED	230–250	3	Chr 4	JH767724:8,299,703-8,299,802
BlRh1C	AGATTCTTGGAAAGGTCACT	AACATTGGGTTTCACCTC	$(AC)_{17}G(CA)_4$	AY606079	Nielsen et al. (2008)	NED	120–160	2	Chr 5	JH767759:3,308,328-3,308,518
BlRh37D	ACATGTGTAAACTTGGGAAC	TGGTTCATTGATCTCTTCTC	$(TG)_6(AG)_{11}GA(AG)_5$	AY606083	Nielsen et al. (2008)	NED	200–250	1	Chr 3	JH767760:5,803,569-5,803,831
BR6	TCATTTCTTTGTTCCCCATAGCAC	AGCAATATCCACGATATGTGAAGG	$(CA)_{15}$		Cunningham et al. (1999)	PET	150–165	3	Chr 13	JH767793:958,941-959,116
DB1	TAAGTCACAGGGACTAATCTG	GAGGGTTTATTGTGAATGAG	$(CA)_{14}$	AF129724	Brown and Houlden (1999), Nielsen et al. (2008)	VIC	230–250	3	Chr 14	JH767726:55,212,707-55,212,874
DB23	ATCTTCCTCAGCAATAAGG	ATCATCAGAGTTTCCAGTTC	$(CA)_{12}$	AF129734	Brown and Houlden (1999), Nielsen et al. (2008)	FAM	180–214	4	Chr 15	JH767779:4,018,091-4,018,257
DB44	AGGGTGGAATGTCAAGTAG	CTTCTAGAGGGAGACTAGGAG	$(TG)_4C\ (GT)_{16}$	AF129730	Brown and Houlden, (1999), Nielsen et al. (2008)	VIC	200–230	3	Chr 11	JH767765:2,229,186-2,229,302
DB52	CATGTGAAATGGACCGTCAGG	ATTTCTGGGAAGGGGCAGG	$(CA)_{21}$	AF129732	Brown and Houlden (1999), Nielsen et al. (2008)	PET	110–140	1	Chr 1	JH767728:49,254,161-49,254,310
DB66	CCAGGTGAAGGGTCTTATTATTAGC	GGATTGGCATGGATGTTACC	$(CA)_7TA(CA)_{16}$	AF129733	Brown and Houlden (1999)	PET	210–230	3	Not found	JH767780:644,242-644,385
IR10	CAGTGAGGAAGATTGGTTGC	CCTGACTCACACATCACCAG	$(CA)_{22}$		Scott (2008)	NED	120–140	4	Unknown	Unplaced scaffold00031
IR12	GAATGCTGATCATTTAGTGAC	GGGTCCAGTTGAGATATCAC	$(CA)_{18}$		Scott (2008)	PET	170–200	4	Chr X	JH767774:6,674,158-6,674,305
IR22	ATGGTGGAAGAAGTGCAGCC	ACTTCTGTGTCTCTAGCGCC	$(CA)_{22}$		Scott (2008)	VIC	200–230	2	Not found	JH767747:9,369,432-9,369,535
SR63	CTTGAGCAGAGTAGAATTTGG	CTCTGTATCCACCTCATTCC	$(AC)_{19}$	AY427965	Scott et al. (2004)	FAM	180–210	4	Chr 14	JH767726:49,561,075-49,561,304

SR74	CAGCACAATGTTTGGCACTTG	TTGGAGTCTTATGTCACCACC	$(CA)_{19}$	AY427967	Scott et al. (2004)	NED	160–180	2	Chr X	JH767831:1,676,337-1,676,424
SR262	CTGCCTTAACAACTGAACTGC	TGGAGGTTATCTCATGCCAC	$(TG)_{28}$	AY606077	Scott (2008)	FAM	80–110	3	Not found	JH767723:55,939,772-55,939,942
SR268	GTTTATACTATGCCCTGCAC	GGATGCTACCGAATAGATTG	$(CA)_{25}$	AY427972	Scott (2008)	VIC	170–200	4	Chr 9	JH767757:15,973,899-15,974,168
SR281	AGGTGATTAGGGAATTGCTGG	TTCTTCTGTCCTGGCATTGC	$(GT)_{23}$	AY427974	Scott et al. (2004)	FAM	220–250	2	Chr 13	JH767802:6,346,286-6,346,528
7B	AACCAACTTGTAATGAGAGG	AATGAACAGGAAGGAAGAC	$(TG)_{16}A(GT)_5$	AY138544	Florescu et al. (2003), Nielsen et al. (2008)	PET	220–230	3	Chr 7	JH767853:1,478,532-1,478,698
7 C	GTCAGTTCAAGTTTTTGCTC	CTCATCCATGCTTCTTCTAC	$(CT)_{14}(AT)_{11}$	AY138543	Florescu et al. (2003), Nielsen et al. (2008)	FAM	130–170	3	Chr 2	JH767727:10,444,144-10,444,335
12F	ACAGCTAGAATCACCAAAAC	TCCTGCTGCATAAATCTC	$(TA)_8(AA)_4$	AY138545	Florescu et al. (2003), Nielsen et al. (2008)	VIC	220–240	1	Chr 15	JH767782:5,765,753-5,766,031
32A	CTAGCAAAATCTCAAAGAGG	TTACTAAGGGAATCACCAAG	$(AC)_6$.. $.(AC)_{15}$	AY138541	Florescu et al. (2003), Nielsen et al. (2008)	FAM	190–210	1	Chr 6	JH767776:13,563,190-13,563,319
32F	GGCAAAACTAAGAGAACTTG	GATACCAAACTGGAAATGG	$(AC)_{18}$	AY138542	Florescu et al. (2003), Nielsen et al. (2008)	VIC	170–240	1	Chr 17	JH767732:27,885,389-27,885,530
RH12	CTGGTGCATTCATCAGGGCT	AGAAGAGGTAGGAGAGGAAGTCA			(https://www.ncbi.nlm.nih.gov/nuccore/37496513)	VIC	108-140	2	Chr 20	JH767735:34,492,757-34,492,857
ZF1	GATTTGGAASCTAGGCATTTCC	GCCATGATACTCATGAATGACA			Peppin et al. (2010)	FAM	95–105	4		

Multiplex PCR Kit (Roche). Having tested several polymerases, this kit consistently gave the best results, amplifying all loci even in degraded samples and bones. PCR cycling conditions were standardized as follows for all samples: 3 min at 95 °C; 30 cycles of 95 °C for 15 s; 60 °C for 30 s; and 72 °C for 30 s; followed by an extension step at 72 °C for 10 min. Amplification success is determined by the number of markers that have successfully amplified, the absence of allelic dropout, a peak height of more than 200 relative fluorescence units (RFU), and a balanced profile.

DNA profiles in forensic analysis may be defined as a full profile (contains all alleles in the genotype), a partial profile (a genotype with at least 1 allele missing), and a mixed profile (DNA profiles of more than 1 donor are present) (Puch-Solis et al. 2012).

The primary author used STRand software (University of California, Davis) (Toonen and Hughes 2001) to analyze the data, including the rhinoceros data. This software has allowed storage of the electropherogram data across multiple platforms over the years while still allowing for analysis and comparison of profiles. Known control samples for both black and white rhinoceros are included as a standard in each rhinoceros run, including routine runs to ensure the accuracy of allele calls between runs, and identification of any sequencer problems that may affect the accuracy of the reads. These control samples have shown almost 100% consistency over hundreds of runs and on different instruments used in the laboratory.

When a matching DNA profile is found between a carcass sample and a horn, or tool sample from a white rhinoceros, the random match probability is calculated using 18 of the 23 markers (the four monomorphic markers and the marker DB66 are excluded from the calculation). Using these data, the random match probability calculated for five white rhinoceros profiles randomly selected from the RhODIS® database ranged from 1:1.56 $\times$ 10^8 to 1:2.1 $\times$ 10^{11} without any correction for inbreeding and from 1:1.7 $\times$ 10^5 to 6.0 $\times$ 10^6 using a Theta (θ) of 0.3 to correct for significant inbreeding (Linacre et al. 2011). With Theta (θ) set at 0.1 in the five animals investigated, the estimated random match probability ranged from 1:7.3 $\times$ 10^6 to 1:3.0 $\times$ 10^8. Given that the total population of white rhinoceros in the world is between 18,000 and 20,000 (Emslie et al. 2016), such random match probabilities indicate that the genotyping system described provides data that can be used for evidentiary purposes.

The between-population allelic variation was quantified from RhODIS® data by $F_{ST.}$ Based on these data a Theta (θ) value of 0.1 was selected for use in the calculation of the match probability of African rhinoceros species. The use of a sufficiently conservative Theta (θ) also compensates for departures from Hardy–Weinberg equilibrium at specific loci due to allelic variation in populations sampled (Buckleton et al. 2016).

Matching of specific DNA profiles provide evidentiary support that two samples are derived from the same individual if underlying data are available to permit an estimate of the rarity of the profile (Weir 1996). Single locus match probability in RhODIS® is calculated using the formulae of Balding and Nichols (Balding and Nichols 1995).

The current southern white rhinoceros population in Africa is between 18,000 and 20,000 (Emslie et al. 2016) and all are descended from a single founder population. This historic data is supported by the overall genetic uniformity of the southern white rhinoceros (*C.s. simum*) population. Southern white rhinoceros comprise over 90% of the criminal cases received by the laboratory, supporting the use of the product rule for this species without consideration of possible sub-structuring. The large dataset in this study also confirmed the utility of monomorphic loci for species identification. DNA profile matches are made using all amplified loci, and by comparing the DNA profile to all genotype data on the RhODIS® database using the Cervus identity match feature, confirmed both manually and with electropherogram data. Match probabilities for white and black rhinoceros matches are done using the subspecies specific allele frequencies and a conservative Theta (θ) of 0.1 following The Second National Research Council report on forensic DNA evidence recommendation 4.2

for estimating random match probabilities in human populations (Weir 1996). This provides the most conservative value for individual genotype matches.

Additional analysis of the population data revealed the mean number of alleles per locus (2.478) was lowest in the white rhinoceros, as compared to the three black rhinoceros subspecies, as might be expected given the early twentieth century population bottleneck of white rhinoceros. The levels of observed heterozygosity (Ho) in the black rhinoceros subspecies were lowest in the *D.b. minor* in KwaZulu-Natal (KZN) (0.360), that, like the white rhinoceros (0.349), also went through a population bottleneck in the early twentieth century. *D.b. michaeli* had the highest mean number of alleles per locus (5.522) and highest Ho of the black rhinoceros subspecies (0.558). *D.b. bicornis* had a low mean number of alleles per locus (3.348) and a low Ho (0.406) which may be due to a smaller number of founders compared to the Zimbabwean *D.b. minor* and *D.b. michaeli* populations (Emslie and Brooks 1999).

This data indicates that the genetic variability of the African rhinoceros species is generally low when examining an extensive dataset of STR profiles of these animals, thus remaining populations must be carefully managed to avoid further population contractions, particularly through continued poaching pressure, to a level from which they may not be able to recover. The effect of such low variability on a megafauna with a slow reproductive rate is also not well understood.

RhODIS®: Forensics, the Primary Application

The role of the forensic scientist as an expert in criminal proceedings is to present evidence from an unbiased and balanced perspective and provide an opinion related to the evidence within their scope of expertise (see chapter "Processing the Wildlife Crime Scene and Evidence of Forensic Importance", this volume), including the strength of calculated match values. The presentation of the DNA match evidence should be carefully worded and should indicate the relative rarity, or lack thereof, of a profile in a population.

Between 2010 and 2019 the laboratory had received over 60,000 individual rhinoceros animal and rhinoceros horn samples, and had submitted over a thousand case reports to law enforcement authorities in South Africa, Malawi, Namibia, Swaziland, Uganda, Zambia, Botswana, and Kenya. Successful prosecutions have confirmed the value of the RhODIS® data and approach in criminal prosecutions of rhinoceros cases. Some of these cases have been selected and summarized here to illustrate the use of this system.

Two suspected poachers were apprehended in the Kruger National Park with three rhinoceros horns in their possession. These horns were identified as being from male and female southern white rhinoceros animals using DNA profiles obtained from the horns. Two of the horns were matched to one rhinoceros carcass and the other horn to a second rhinoceros carcass at a single poaching incident in the park. The evidence assisted in the conviction and sentencing of the accused to 29 years imprisonment each in 2012. The match probabilities, in this case, were 4.20×10^{-9} and 2.03×10^{-10}, respectively.

In another similar case involving suspects apprehended in the Kruger National park with horns linked to two black rhinoceros carcasses, the accused was sentenced to 14 years imprisonment in 2013 with match probabilities in this case 4.18×10^{-12} and 1.03×10^{-12}, respectively. A poaching case in Limpopo province in South Africa was examined and the DNA evidence recovered from blood on clothing matched the carcass of a white rhinoceros, resulting in a sentence of 8 years in 2015.

In another case, DNA profiling of three rhinoceros horns recovered from a Chinese citizen apprehended at OR Tambo International airport, Johannesburg, South Africa, in 2016 showed that the horns came from three different female southern white rhinoceros animals poached in the Hluhluwe-imFolozi Game Reserve in the KwaZulu-Natal province in South Africa. These horns were a single horn of the pair that had been removed from the rhinoceros. In a separate case received several months after this incident, horn drilling samples from three horns sampled in Cambodia matched the same animals, indicating that the horn pairs were split and shipped separately, with the three back horns having successfully been illegally trafficked to Asia while the front horn shipment was intercepted before export.

Using the RhODIS® system in yet another case, DNA obtained from a sample of horn recovered from suspects in Kenya matched blood on a piece of carpet, resulting in the successful conviction of poaching suspects and sentencing to 11 years in prison by a Kenyan court. The DNA profile indicated the samples were from a black rhinoceros, and specifically from *D.b. michaeli,* with a match probability of 8.98×10^{-22}.

In 2016, Chinese traffickers were sentenced to 14 years in prison by a Namibian court following the matching of rhinoceros horns to a carcass of a *D.b. bicornis* animal in the Etosha National Park in Namibia. The samples from 14 horns recovered from the traffickers at the Hosea Kutako International Airport, Windhoek, Namibia were submitted to RhODIS® and could not initially be linked to a specific rhinoceros individual. However, a STRUCTURE analysis of the data showed that the horns were of the *D.b.bicornis* subspecies that occurred mainly in Namibia and the Northern Cape Province of South Africa. At that time no poaching incidents had been reported from the Northern Cape Province. Namibian authorities subsequently found carcasses of poached rhinoceros in the Etosha National Park. These samples were also submitted to the laboratory and the matching carcass was subsequently found.

RhODIS®: Population Management, a Useful Bonus

RhODIS® genetic data also allows the researcher to investigate populations of rhinoceros to assist reserves in managing their populations at a metapopulation scale. Legislation requires private populations to include F2 (second generation) animals in order to be registered as a captive population breeding operation and DNA confirmation of parentage is essential in these cases. RhODIS® data may be utilized directly to confirm parentage in rhinoceros, negating the need to resample and profile animals for this purpose.

The genotypes for 3085 White rhinoceros (*Ceratotherium simum*) and 883 Black rhinoceros (*Diceros bicornis*), sampled since 2010, were selected from the RhODIS® database to provide a background dataset from which to determine the random match probability statistics for forensic cases and to investigate the population sub-structuring of the extant white and black African rhinoceros populations for forensic and population management purposes. Samples were submitted to the laboratory to be added to the RhODIS® database from all the South African reserves and private game farms and various African rhinoceros range states. Recognized living subspecies of black rhinoceros were included: (1) 51 samples of the Eastern subspecies (*D.b. michaeli*); (2) 357 samples of the Southwestern subspecies (*D.b. bicornis*); and (3) 475 samples of the South-Central subspecies (*D.b. minor*). Because only two animals remain of the northern white subspecies they were not included in this dataset.

The study population was examined using different approaches to identify sub-structuring: (1) an individual-based tree was constructed; (2) Principal Component Analysis was performed; and (3) the STRUCTURE algorithm was applied. Allele frequencies were obtained from this data for the southern white rhinoceros and the black rhinoceros population as well as the individual black rhinoceros subpopulations identified using the population partitioning software approaches. Black rhinoceros are generally subdivided into three modern subspecies also recognized by the Species Specialist Group of the IUCN, *D.b. bicornis, D.b. michaeli and D.b. minor* (Moodley et al. 2017). Population structure of white and black rhinoceros based upon three different analyses (Composite STR phylogeny, Principal Component Analysis, and STRUCTURE) affirmed the partition of white versus black rhinoceros species and further the separation of the three black rhinoceros subspecies.

Figure 4 shows the output of the STRUCTURE algorithm that also revealed a fine grain distinctiveness between black rhinoceros, *D.b. minor* populations from Zimbabwe (green in Fig. 4), and Kwazulu-Natal (KZN), South Africa (red in Fig. 4). STRUCTURE analysis further indicated that black rhinoceros in the Kruger National Park (KNP) are comprised of a mix of KZN and Zimbabwe rhinoceros as expected, since KNP black rhinoceros founders originated from these two areas (Hall-Martin 1988). *D.b. bicornis* (blue in Fig. 4) occur in Namibia and in South Africa in the Northern Cape and Eastern Cape provinces. All *D.b. bicornis* in South Africa were introduced from Namibia (Hall-Martin 1988). The population differentiation (F_{ST}) between white and black rhinoceros subspecies calculated from the data support the recognition of one southern white rhinoceros subspecies (*Ceratotherium simum simum*), and three black rhinoceros subspecies, *D.b. bicornis*, *D.b. michaeli*, and *D.b. minor*, with significant partitioning of the Zimbabwe versus Kwazulu-Natal (KZN) *D.b. minor* populations in the present African rhinoceros populations. Similar partitions were observed previously with fewer (8–11) STR loci (Moodley et al. 2017).

RhODIS® Provides Evidence of Subspecies Admixture and Genetic Rescue of a Population of Black Rhinoceros in South Africa

How a species or subspecies is defined can have a major impact on how that species is managed and even protected. The problem with keeping small vulnerable populations isolated for long periods without the addition of genes that would likely have occurred in natural circumstances though migrants, leads to a loss of diversity and evolutionary potential over time. Keller and Waller (2002) note that populations kept small and isolated for many generations suffer the deleterious effects of inbreeding that include reduced survival rate, reduced reproduction, reduced resistance to disease, reduced adaptability to environmental stressors, and increased predation. The question, therefore, is whether anthropogenic admixture in small vulnerable populations of threatened species is more detrimental to species survival than anthropogenic species separation leading to inbreeding, genetically depauperate animals, and eventually local extinction. The inadvertent genetic rescue of the black rhinoceros in the Addo Elephant National Park (AENP) is an example of the beneficial effects of subspecies admixture and, more uniquely, in a large, slow reproducing mammal like the rhinoceros. RhODIS® data was used to identify this admixture and confirm that not all suspected admixed animals had been removed from the population before they were translocated to a new breeding site.

The black rhinoceros (*Diceros bicornis bicornis*) occurred in the Addo area of the Eastern Cape Province of South Africa until 1863, when the last animal was apparently shot (Hall-Martin and Penzhorn 1977). In 1961 and 1962, seven black rhinoceros animals from Kenya, of the *D. b. michaeli* subspecies, were introduced to the Addo Elephant National Park (AENP) in the Eastern Cape Province of South Africa. When the last adult bull in this population died in 1977, the African Rhino Specialist Group (AfRSG) of the IUCN made the decision to introduce three bulls of the *D.b. minor* subspecies from KwaZulu-Natal to the population. This decision was reversed in 1981 and the *D.b. minor* bulls and their presumed offspring were removed so that the population would remain pure *D.b. michaeli*. Between 1998 and 2003, the majority of these animals were translocated to a private game ranch in South Africa to be replaced by *Diceros bicornis bicornis* animals that were translocated from Namibia, via the Augrabies and Vaalbos National Parks, in an attempt to return the subspecies to its historic range.

The population of *D.b. michaeli* translocated to a private game ranch represents an individual extralimital population of this black rhinoceros subspecies in South Africa. A STRUCTURE analysis of the black rhinoceros data collected as part of the RhODIS® program (Harper et al.

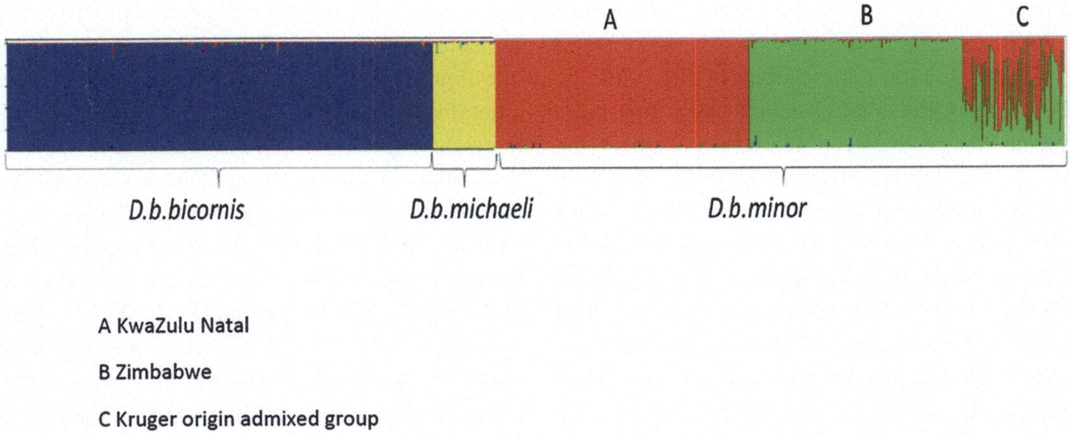

Fig. 4 The Structure diagram supports the black rhinoceros subdivision with an additional partition between the *D.b. minor* subspecies originating in Kwazulu-Natal (South Africa) (**a**), Zimbabwe (**b**), and a third group that is an admixture of these two groups (**c**) originating from the Kruger National Park in South Africa

2018) provided evidence of a *D.b. minor* signature in the translocated *D.b. michaeli* population that indicated a historic admixture between the *D. b. minor* and *D.b. michaeli* populations that had not been completely cleared, as had been reported (Hall-Martin 1984). Historic and anecdotal information was used to determine the origin and relationships of specific animals that were the founders of this population, and a near-complete pedigree could be constructed of the population at the private game ranch, where the level of admixture indicated by the STRUTURE analysis was confirmed by pedigree assignment.

The private game ranch population showed greater genetic diversity than the three *D.b. minor* populations in Zimbabwe, KwaZulu-Natal, and the Kruger National Park, and the *D. b. bicornis* in Namibia. The number of alleles identified and observed heterozygosity of the population in the private game ranch (5.722 and 0.705) is similar to that of the *D.b. michaeli* population in Kenya (6.444 and 0.644), and markedly greater than the number of alleles identified and observed heterozygosity of the black rhinoceros population in KwaZulu-Natal (2.500 and 0.377), indicating that the inadvertent admixture had improved the genetic variability of the population. The benefits were also observed in the general health and reproductive success of the animals.

In May 2017, 18 black rhinoceros were translocated from the private game ranch to Akagera National Park in Rwanda, returning the species to the area following an absence of 10 years. Taking into account the critically endangered status of the species and subspecies, the IUCN African Rhino Specialist Group recommended selecting animals with a Q_iStr of 0.85 and higher *D.b. michaeli* ancestry to provide a selection of founders for a newly translocated population with sufficient variability (Richard Emslie, personal communication, 2018). The translocation has proven to be very successful with the first calf born within 4 months of arrival, in August 2017. The success of the translocation would support the view that the admixture did not negatively affect the adaptability and reproductive potential of these animals.

This analysis affirms the value of the RhODIS® program to verify a pedigree and identify subspecies admixture in a black rhinoceros population, and to provide the ancestry coefficients from a Bayesian clustering analysis, and directly from the pedigree. The verified pedigree information enabled an assessment of the reproductive fitness of the population, and these data have informed management decisions relating to the successful establishment of a new black rhinoceros population in a former range state. The RhODIS® data, therefore, provides an effective management tool to support rhinoceros breeding programs, as the focus on the species shifts toward increasing the reproductive success and population growth rates to counteract the progressive population decline caused by an escalation in rhinoceros poaching across the African continent.

Conclusion

The successful prosecution, conviction, and sentencing of suspects in South Africa and in other countries in rhinoceros poaching and rhinoceros horn seizure cases affirm the utility of the RhODIS® approach in criminal prosecutions.

In addition, the RhODIS® data has not only been used as a forensic tool but has become a management tool for individual rhinoceros population management. The RhODIS® sample bank, DNA profile database, and test methodologies provide the opportunity to support a number of future research projects, including the development of additional marker systems for forensic and population management purposes, genome sequencing, re-sequencing, and mitochondrial DNA studies to investigate black rhinoceros subspecies and individuals carrying unique and valuable genetic types. These data can be used as a meta-population management tool to inform translocation decisions and the establishment of new rhinoceros populations, and to support a comparative study to investigate the relationship between different rhinoceros populations in Africa, to provide a review of the current understanding of the rhinoceros species and subspecies classification. Additional opportunities exist to investigate the use and value of the RhODIS® data as forensic evidence in rhinoceros cases, by following cases from sampling through reporting to conclusion in court. The project has grown into a self-sustaining program with an inventory of African rhinoceros and rhinoceros horns throughout the continent.

References

Anderson-Lederer RM, Linklater WL, Ritchie PA (2012) Limited mitochondrial DNA variation within South Africa's black rhino (*Diceros bicornis minor*) population and implications for management. Afr J Ecol 50:404–413

Balding DJ, Nichols RA (1995) A method for quantifying differentiation between populations at multi-allelic loci and its implications for investigating identity and paternity. Genetica 96:3–12

Boy SC, Raubenheimer EJ, Marais J, Steenkamp G (2015) White rhinoceros *Ceratotherium simum* horn development and structure: a deceptive optical illusion. J Zool 296:161–166

Brown SM, Houlden BA (1999) Isolation and characterization of microsatellite markers in the black rhinoceros (*Diceros bicornis*). Mol Ecol 8:1559–1561

Buckleton J, Curran J, Goudet J, Taylor D, Thiery A, Weir BS (2016) Population-specific FST values for forensic STR markers: a worldwide survey. Forensic Sci Int Genet 23:91–100

Butler JM (2015) Advanced topics in forensic DNA typing: interpretation. Academic Press, San Diego

Caulderwood K (2014) Like cocaine minus the risk, rhino horn trade explodes in Africa; http://www.ibtimes.com/cocaine-minus-risk-rhino-horn-trade-explodes-africa-1569192 [Online]. Accessed 15 June 2018

Chen J-W, Uboh CE, Soma LR, Li X, Guan F, You Y, Liu Y (2010) Identification of racehorse and sample contamination by novel 24-plex STR system. Forensic Sci Int Genet 4:158–167

Crosta, A., Sutherland, K. & Talerico, C. 2017. Grinding-rhino: an undercover investigation on rhino horn trafficking in China and Vietnam

Cunningham J, Harley EH, O'Ryan C (1999) Isolation and characterization of microsatellite loci in black rhinoceros (*Diceros bicornis*). Electrophoresis 20:1778–1780

Emslie RH, Brooks M (1999) African Rhino. Status survey and conservation action plan. IUCN/SSC African Rhino Specialist Group, IUCN, Gland and Cambridge

Emslie RH, Milliken T, Talukdar B, Ellis S, Adcock K, Knight MH (2016) African and Asian rhinoceroses – status, conservation and trade. A report from the IUCN Species Survical Commission (IUCN/SSC) African and Asian Rhino Specialist Groups and TRAFFIC to the CITES Secretariat pursuant to Resolution Conf. 9.14 (Rev. CoP15). Report to CITES

Florescu A, Davila JA, Scott C, Fernando P, Kellner K, Morales JC, Melnick D, Boag PT, Van Coeverden De Groot P (2003) Polymorphic microsatellites in white rhinoceros. Mol Ecol Notes 3:344–345

Guerier AS, Bishop JM, Crawford SJ, Schmidt-Küntzel A, Stratford KJ (2012) Parentage analysis in a managed free ranging population of southern white rhinoceros: genetic diversity, pedigrees and management. Conserv Genet 13:811–822

Guilford G (2013). http://www.theatlantic.com/business/archive/2013/05/why-does-a-rhino-horn-cost-300-

000-because-vietnam-thinks-it-cures-cancer-and-hangovers/275881/ [Online]. Accessed 20 March 2014
Hall-Martin A (1984) Kenya's black rhino in Addo, S. Africa. Newsl Afr Elephant Rhino Group 3:11
Hall-Martin A (1988) Conservation of the black rhino: the strategy of the National Parks Board of South Africa. Quagga 1:12–17
Hall-Martin A (2009) Black Rhinoceros in southern Africa. Oryx 15:27–32
Hall-Martin A, Penzhorn BL (1977) Behaviour and recruitment of translocated black rhinoceros *Diceros bicornis*. Koedoe 20:147–162
Harper CK, Vermeulen GJ, Clarke AB, De Wet JI, Guthrie AJ (2013) Extraction of nuclear DNA from rhinoceros horn and characterization of DNA profiling systems for white (*Ceratotherium simum*) and black (*Diceros bicornis*) rhinoceros. Forensic Sci Int Genet 7:428–433
Harper C, Ludwig A, Clarke A, Makgopela K, Yurchenko A, Guthrie A, Dobrynin P, Tamazian G, Emslie R, van Heerden M, Hofmeyr M, Potter R, Roets J, Beytell P, Otiende M, Kariuki L, du Toit R, Anderson N, Okori J, Antonik A, Koepfli K-P, Thompson P, O'Brien SJ (2018) Robust forensic matching of confiscated horns to individual poached African rhinoceros. Curr Biol 28:R13–R14
Hieronymus TL, Witmer LM, Ridgely RC (2006) Structure of white rhinoceros (*Ceratotherium simum*) horn investigated by X-ray computed tomography and histology with implications for growth and external form. J Morphol 267:1172–1176
Hildebrandt TB, Hermes R, Colleoni S, Diecke S, Holtze S, Renfree MB, Stejskal J, Hayashi K, Drukker M, Loi P, Göritz F, Lazzari G, Galli C (2018) Embryos and embryonic stem cells from the white rhinoceros. Nat Commun 9:2589
Howard BC (2015) Only three northern White rhinos remain. National Geographic [Online]
Hsing-Mei H, Huang L-H, Tsai L-C, Kuo Y-C, Meng H-H, Linacre A, Lee JC-I (2003) Species identification of rhinoceros horns using the cytochrome b gene. Forensic Sci Int 136:1–11
Jäger AC, Alvarez ML, Davis CP, Guzmán E, Han Y, Way L, Walichiewicz P, Silva D, Pham N, Caves G, Bruand J, Schlesinger F, Pond SJK, Varlaro J, Stephens KM, Holt CL (2017) Developmental validation of the MiSeq FGx forensic genomics system for targeted next generation sequencing in forensic DNA casework and database laboratories. Forensic Sci Int Genet 28:52–70
Johnson RN, Wilson-Wilde L, Linacre A (2014) Current and future directions of DNA in wildlife forensic science. Forensic Sci Int Genet 10:1–11
Karsten M, Jansen van Vuuren B, Goodman P, Barnaud A (2011) The history and management of black rhino in KwaZulu-Natal: a population genetic approach to assess the past and guide the future. Anim Conserv 14:363–370
Keller LF, Waller DM (2002) Inbreeding effects in wild populations. Trends Ecol Evol 17:230–241
Knight MH, Kerley GIH (2009) Black rhino translocations within Africa. Africa Insight 39:70–83
Kotzé A, Dalton DL, Du Toit R, Anderson N, Moodley Y (2014) Genetic structure of the black rhinoceros (*Diceros bicornis*) in South-Eastern Africa. Conserv Genet 15:1479–1489
Kun T, Lyons LA, Sacks BN, Ballard RE, Lindquist C, Wictum EJ (2013) Developmental validation of mini-DogFiler for degraded canine DNA. Forensic Sci Int Genet 7:151–158
Lin Y-C, Hsieh H-M, Lee JC-I, Hsiao C-T, Lin D-Y, Linacre A, Tsai L-C (2014) Establishing a DNA identification system for pigs (Sus scrofa) using a multiplex STR amplification. Forensic Sci Int Genet 9:12–19
Linacre A, Gusmão L, Hecht W, Hellmann AP, Mayr WR, Parson W, Prinz M, Schneider PM, Morling N (2011) ISFG: recommendations regarding the use of non-human (animal) DNA in forensic genetic investigations. Forensic Sci Int Genet 5:501–505
Menotti-Raymond MA, David VA, Wachter LL, Butler JM, O'Brien SJ (2005) An STR forensic typing system for genetic individualization of domestic cat (*Felis catus*) samples. J Forensic Sci 50:1061–1070
Milliken T, Shaw J (2012) The South Africa – Viet Nam rhino horn trade Nexus: a Traffic report. Traffic
Moneron, S., Okes, N. & Rademeyer, J. 2017. Pendants, powder and pathways
Moodley Y, Russo IRM, Dalton DL, Kotzé A, Muya S, Haubensak P, Bálint B, Munimanda GK, Deimel C, Setzer A, Dicks K, Herzig-Straschil B, Kalthoff DC, Siegismund HR, Robovský J, O'Donoghue P, Bruford MW (2017) Extinctions, genetic erosion and conservation options for the black rhinoceros (*Diceros bicornis*). Sci Rep 7:41417
Muya SM, Bruford MW, Muigai AW-T, Osiemo ZB, Mwachiro E, Okita-Ouma B, Goossens B (2011) Substantial molecular variation and low genetic structure in Kenya's black rhinoceros: implications for conservation. Conserv Genet 12:1575–1588
Nielsen L, Meehan-Meola D, Kilbourn A, Alcivar-Warren A (2008) Characterization of microsatellite loci in the black rhinoceros (*Diceros bicornis*) and white rhinoceros (*Ceratotherium simum*): their use for cross-species amplification and differentiation between the two species. Conserv Genet 9:239–242
Patton, F. 2011. The medicinal value of rhino horn – a quest for the truth
Peppin L, McEwing R, Ogden R, Hermes R, Harper C, Guthrie A, Carvalho GR (2010) Molecular sexing of African rhinoceros. Conserv Genet 11:1181–1184
Player I (2013) The White rhino Saga. Jonathan Ball Publishers, Johannesburg
Puch-Solis R, Roberts P, Pope S, Aitken C (2012) Communicating and interpreting statistical evidence in the Administration of Criminal Justice 2. Assessing the Probative value of DNA Evidence, Royal Statistical Society, London
Reuters (2017) Constitutional Court lifts ban on domestic sales in rhino horn: http://www.enca.com/south-africa/

constitutional-court-lifts-ban-on-domestic-sales-in-rhino-horn. *eNCA*
Rookmaaker K, Antoine P (2012) New maps representing the historical and recent distribution of the African species of rhinoceros: *Diceros bicornis*, Ceratotherium simum and *Ceratotherium cottoni*. Pachyderm 52:91–96
Rudbeck L, Dissing J (1998) Rapid, simple alkaline extraction of human genomic DNA from whole blood, buccal epithelial cells, semen and forensic stains for PCR. BioTechniques 25:588–592
Scott CA (2008) Microsatellite variability in four contemporary rhinoceros species: implications for conservation. MSc Dissertation MSc Dissertation, Queen's University
Scott C, Foose T, Morales JC, Fernando P, Melnick DJ, Boag PT, Davila JA, Van Coeverden De Groot PJ (2004) Optimization of novel polymorphic microsatellites in the endangered Sumatran rhinoceros (*Dicerorhinus sumatrensis*). Mol Ecol Notes 4:194–196
Tamazian G, Dobrynin P, Krasheninnikova K, Komissarov A, Koepfli K-P, O'Brien SJ (2016) Chromosomer: a reference-based genome arrangement tool for producing draft chromosome sequences. GigaScience 5:38
Toonen RJ, Hughes S (2001) Increased throughput for fragment analysis on an ABI prism® 377 automated sequencer using a membrane comb and STRand software. BioTechniques 31:1320–1324
Van Coeverden de Groot PJ, Putnam AS, Erb P, Scott C, Melnick D, O'Ryan C, Boag PT (2011) Conservation genetics of the black rhinoceros, *Diceros bicornis bicornis*, in Namibia. Conserv Genet 12:783–792
Van De Goor LHP, Panneman H, Van Haeringen WA (2009) A proposal for standardization in forensic bovine DNA typing: allele nomenclature of 16 cattle-specific short tandem repeat loci. Anim Genet 40:630–636
Wade CM, Giulotto E, Sigurdsson S, Zoli M, Gnerre S, Imsland F, Lear TL, Adelson DL, Bailey E, Bellone RR, Blöcker H, Distl O, Edgar RC, Garber M, Leeb T, Mauceli E, MacLeod JN, Penedo MCT, Raison JM, Sharpe T, Vogel J, Andersson L, Antczak DF, Biagi T, Binns MM, Chowdhary BP, Coleman SJ, Della Valle G, Fryc S, Guérin G, Hasegawa T, Hill EW, Jurka J, Kiialainen A, Lindgren G, Liu J, Magnani E, Mickelson JR, Murray J, Nergadze SG, Onofrio R, Pedroni S, Piras MF, Raudsepp T, Rocchi M, Røed KH, Ryder OA, Searle S, Skow L, Swinburne JE, Syvänen AC, Tozaki T, Valberg SJ, Vaudin M, White JR, Zody MC, Lander ES, Lindblad-Toh K (2009) Genome sequence, comparative analysis, and population genetics of the domestic horse. Science 326:865–867
Walker C, Walker A (2012) The rhino keepers. Johannesburg, Jacana Media
Weir BS (1996) The second National Research Council report on forensic DNA evidence. Am J Hum Genet 59 (3):497–500
Wictum E, Kun T, Lindquist C, Malvick J, Vankan D, Sacks B (2013) Developmental validation of DogFiler, a novel multiplex for canine DNA profiling in forensic casework. Forensic Sci Int Genet 7:82–91

The manufacturer's authorised representative in the EU is Springer Nature Customer Service Centre GmbH, Europaplatz 3, 69115 Heidelberg, Germany. If you have any concerns regarding our products, please contact ProductSafety@springernature.com

Printed and bound by CPI Group (UK) Ltd, Croydon, CR0 4YY
30/09/2026
02225691-0020